# Handbook of Operational Amplifier Circuit Design

# OTHER McGRAW-HILL HANDBOOKS OF INTEREST

*American Institute of Physics* · American Institute of Physics Handbook
*Baumeister and Marks* · Standard Handbook for Mechanical Engineers
*Beeman* · Industrial Power Systems Handbook
*Blatz* · Radiation Hygiene Handbook
*Brady* · Materials Handbook
*Burington and May* · Handbook of Probability and Statistics with Tables
*Condon and Odishaw* · Handbook of Physics
*Coombs* · Basic Electronic Instrument Handbook
*Coombs* · Printed Circuits Handbook
*Croft, Carr, and Watt* · American Electricians' Handbook
*Dean* · Lange's Handbook of Chemistry
*Etherington* · Nuclear Engineering Handbook
*Fink* · Electronics Engineers' Handbook
*Fink and Carroll* · Standard Handbook for Electrical Engineers
*Gruenberg* · Handbook of Telemetry and Remote Control
*Hamsher* · Communication System Engineering Handbook
*Harper* · Handbook of Electronic Packaging
*Harper* · Handbook of Materials and Processes for Electronics
*Harper* · Handbook of Thick Film Hybrid Microelectronics
*Harper* · Handbook of Wiring, Cabling, and Interconnecting for Electronics
*Henney* · Radio Engineering Handbook
*Hicks* · Standard Handbook of Engineering Calculations
*Hunter* · Handbook of Semiconductor Electronics
*Huskey and Korn* · Computer Handbook
*Ireson* · Reliability Handbook
*Jasik* · Antenna Engineering Handbook
*Juran* · Quality Control Handbook
*Klerer and Korn* · Digital Computer User's Handbook
*Koelle* · Handbook of Astronautical Engineering
*Korn and Korn* · Mathematical Handbook for Scientists and Engineers
*Kurtz* · The Lineman's and Cableman's Handbook
*Landee, Davis, and Albrecht* · Electronic Designer's Handbook
*Machol* · System Engineering Handbook
*Maissel and Glang* · Handbook of Thin Film Technology
*Markus* · Electronics and Nucleonics Dictionary
*Markus* · Handbook of Electronic Control Circuits
*Markus and Zeluff* · Handbook of Industrial and Electronic Circuits
*Perry* · Engineering Manual
*Skolnik* · Radar Handbook
*Smeaton* · Motor Application and Maintenance Handbook
*Terman* · Radio Engineers' Handbook
*Truxal* · Control Engineers' Handbook
*Tuma* · Engineering Mathematics Handbook
*Tuma* · Handbook of Physical Calculations
*Tuma* · Technology Mathematics Handbook
*Watt and Summers* · NFPA Handbook of the National Electrical Code

# Handbook of Operational Amplifier Circuit Design

**DAVID F. STOUT**
Senior Engineering Specialist
Aeronutronic Ford Corporation

Edited by
**MILTON KAUFMAN**
President, Electronic Writers and Editors, Inc.

**McGRAW-HILL BOOK COMPANY**
New York   St. Louis   San Francisco   Auckland   Düsseldorf
Johannesburg   Kuala Lumpur   London   Mexico   Montreal
New Delhi   Panama   Paris   São Paulo   Singapore
Sydney   Tokyo   Toronto

Library of Congress Cataloging in Publication Data

Stout, David F.
   Handbook of operational amplifier circuit design.

    1. Operational amplifiers. 2. Electronic circuit design. I. Kaufman, Milton. II. Title.
TK7871.58.06S76      621.3815'35      76-3491
ISBN 0-07-061797-X

Copyright © 1976 by McGraw-Hill, Inc. All rights reserved. Printed in the United States of America. No part of this publication may be reproduced, stored in a retrieval system, or transmitted, in any form or by any means, electronic, mechanical, photocopying, recording, or otherwise, without the prior written permission of the publisher.

1234567890 KPKP 785432109876

*The editors for this book were Harold B. Crawford, Ross J. Kepler, and Betty Gatewood, the designer was Naomi Auerbach, and the production supervisor was George E. Oechsner. It was set in Caledonia by The Kingsport Press.*

*Printed and bound by The Kingsport Press.*

*To Mildred Stout and David M. Kaufman*

# Contents

Preface .................................................... xiii

## 1. INTRODUCTION TO OPERATIONAL AMPLIFIERS .......... 1-1

1.1 Overview of Operational Amplifiers ............................................. 1-1
    The Op Amp Model ................................................................. 1-3
    Parameters of the Ideal Op Amp ................................................ 1-4
1.2 Applications of Op Amps ......................................................... 1-5
    Linear Amplifiers .................................................................... 1-5
    Nonlinear Amplifiers ............................................................... 1-5
    Comparators ........................................................................ 1-5
    Filters .................................................................................. 1-6
    Logarithmic Applications ......................................................... 1-6
    Multivibrators ....................................................................... 1-6
    Oscillators – Generators .......................................................... 1-7
    Regulators ........................................................................... 1-7
    Sampling Circuits .................................................................. 1-8
1.3 The Real Op Amp ................................................................. 1-8
    Op Amp Packaging ................................................................ 1-8
    Parameters of Real Op Amps ................................................... 1-9
        Voltage Gain .................................................................... 1-9
        Bandwidth ....................................................................... 1-10
        Slew Rate ....................................................................... 1-11
        Input Resistance .............................................................. 1-11
        Input Bias Current ............................................................ 1-11
        Input Offset Current .......................................................... 1-11
        Input Offset Voltage ......................................................... 1-11
1.4 Op Amps Compared with Transistors ......................................... 1-12
    Gain .................................................................................. 1-12
    Frequency Response ............................................................. 1-12
    Input Impedance .................................................................. 1-12
    Output Resistance ................................................................ 1-12
1.5 Basic Op Amp Circuits ........................................................... 1-12
    Basic Inverting Op Amp Circuit ............................................... 1-12
    Basic Noninverting Op Amp Circuit .......................................... 1-14
1.6 Real and Ideal Op Amps Compared ........................................... 1-14
    Finite Open-loop Gain ........................................................... 1-14
    Finite Bandwidth .................................................................. 1-15
    Finite Input Resistance .......................................................... 1-16
    Output Resistance $> 0$ ......................................................... 1-17
1.7 The Feedback Equation .......................................................... 1-17
1.8 Large-Signal Behavior of Op Amps ............................................ 1-19
1.9 Open-Loop Characteristics of Op Amps ...................................... 1-21
    Calculations Using Decibels .................................................... 1-21

## viii CONTENTS

|  |  |  |
|---|---|---|
|  | Op Amp Open-loop Phase Shift | 1-21 |
|  | Gain and Phase of a Zero | 1-22 |
|  | The Bode Approximation | 1-22 |
| 1.10 | Closed-Loop Characteristics of Op Amps | 1-23 |
| 1.11 | Loop Gain and Phase | 1-24 |
| 1.12 | Circuits Inside an Op Amp | 1-25 |
|  | Input Differential Amplifier | 1-25 |
|  | Second Differential Amplifier | 1-27 |
|  | Level-Shifting Amplifier | 1-27 |
|  | Output-Power Amplifier | 1-27 |

## 2. FUNDAMENTALS OF CIRCUIT DESIGN USING OP AMPS . . . . . . . 2-1

|  |  |  |
|---|---|---|
| 2.1 | Basic Rules Which Simplify Design | 2-1 |
|  | Inverting Amplifier | 2-1 |
|  | Noninverting Amplifier | 2-2 |
| 2.2 | How to Minimize Op Amp Errors | 2-3 |
|  | Input Offset Voltage $V_{io}$ | 2-3 |
|  | Input Bias Current $I_b$ | 2-5 |
|  | Input Offset Current $I_{io}$ | 2-8 |
|  | Equivalent Input Noise $V_n$ and $I_n$ | 2-9 |
|  | Input Resistance $R_{id}$ and $R_{ic}$ | 2-12 |
|  | Input Capacitance $C_{id}$ and $C_{ic}$ | 2-12 |
|  | Output Resistance $R_o$ | 2-13 |
|  | Open-loop Gain $A_v$ and Open-loop DC Gain $A_{vo}$ | 2-14 |
|  | Bandwidth $f_u, f_{cp}, f_f$ | 2-16 |
|  | Slew Rate $S$ | 2-17 |
|  | Common-Mode Rejection Ratio (CMRR) | 2-18 |
|  | Power Supply Rejection Ratio (PSRR) | 2-21 |
| 2.3 | General Method to Compute $A_{vc}$ | 2-21 |
|  | Y Parameters | 2-22 |
|  | Computing $A_{vc}$ with Y Parameters | 2-24 |

## 3. FEEDBACK STABILITY . . . . . . . . . . . . . . . . . . . . . 3-1

|  |  |  |
|---|---|---|
| 3.1 | Review of Feedback Theory | 3-1 |
|  | Results of Positive and Negative Feedback | 3-2 |
|  | First-Cut Stability Analysis | 3-3 |
|  | Development of the Loop-gain Equation | 3-3 |
|  | Gain Margin and Phase Margin | 3-10 |
| 3.2 | Compensation Circuits | 3-11 |
|  | Lag Compensation | 3-11 |
|  | Lead Compensation | 3-13 |
|  | Lead-lag Compensation | 3-15 |
| 3.3 | The Seven Major Causes of Op Amp Instability | 3-18 |
|  | First—Compensation Recommended by Data Sheet Not Used | 3-19 |
|  | Second—Closed-Loop Gain Too Low for Type (and Amount) of Compensation Used | 3-20 |
|  | Third—Excessive Capacitive Load on Op Amp | 3-22 |
|  | Fourth—Incorrect Phase Lead/Lag in Feedback Network | 3-24 |
|  | Fifth—Excessive Resistance Between Ground and Op Amp Positive Input | 3-25 |
|  | Sixth—Excessive Stray Capacitance Between Op Amp Output and Balance Terminals | 3-26 |
|  | Seventh—Inadequate Power-Supply Bypassing | 3-26 |
| 3.4 | Feedback Stability-Design Examples | 3-27 |
|  | First Example—Inverting Amplifier with Gain of 10 | 3-27 |
|  | Second Example—Wide Bandwidth Amplifier | 3-31 |
| 3.5 | Feedback-Stability Measurements | 3-35 |
|  | Loop-gain Measurement Method | 3-35 |
|  | Closed-loop AC Method | 3-37 |
|  | Transient-Response Method | 3-38 |

## 4. AMPLIFIERS . . . . . . . . . . . . . . . . . . . . . . . . . 4-1

|  |  |  |
|---|---|---|
| 4.1 | Basic Inverting Amplifier | 4-1 |
| 4.2 | Basic Noninverting Amplifier | 4-10 |

|     |                                      |      |
| --- | ------------------------------------ | ---- |
| 4.3 | Current Amplifier                    | 4-12 |
| 4.4 | Transresistance Amplifier            | 4-14 |
| 4.5 | Transconductance Amplifier           | 4-16 |
| 4.6 | AC Coupled Inverting Amplifier       | 4-17 |
| 4.7 | Charge-Sensitive Amplifier           | 4-18 |
| 4.8 | Summing Amplifier                    | 4-20 |

## 5. COMPARATORS — 5-1

| 5.1 | Zero-Crossing Detector                 | 5-1  |
| --- | -------------------------------------- | ---- |
| 5.2 | Zero-Crossing Detector with Hysteresis | 5-5  |
| 5.3 | Level Detector                         | 5-9  |
| 5.4 | Level Detector with Hysteresis         | 5-11 |

## 6. CONVERTERS — 6-1

| 6.1 | Dual-Slope A/D Converter    | 6-1 |
| --- | --------------------------- | --- |
| 6.2 | Digital-to-Analog Converter | 6-8 |

## 7. DEMODULATORS AND DISCRIMINATORS — 7-1

| 7.1 | Synchronous AM Demodulator  | 7-1 |
| --- | --------------------------- | --- |
| 7.2 | FM Demodulator              | 7-4 |
| 7.3 | Pulse-Width Discriminator   | 7-8 |

## 8. DETECTORS — 8-1

| 8.1 | Positive-Peak Detector | 8-1 |
| --- | ---------------------- | --- |
| 8.2 | Phase Detector         | 8-4 |

## 9. DIFFERENTIAL AMPLIFIERS — 9-1

| 9.1 | Basic Differential Amplifier | 9-1 |
| --- | ---------------------------- | --- |
| 9.2 | Instrumentation Amplifier    | 9-8 |

## 10. LOW-PASS FILTERS — 10-1

| 10.1 | Second-Order Low-Pass Filter | 10-1 |
| ---- | ---------------------------- | ---- |
| 10.2 | Third-Order Low-Pass Filter  | 10-8 |

## 11. HIGH-PASS FILTERS — 11-1

| 11.1 | Second-Order High-Pass Filter | 11-1 |
| ---- | ----------------------------- | ---- |
| 11.2 | Third-Order High-Pass Filter  | 11-7 |

## 12. BANDPASS FILTERS — 12-1

| 12.1 | Multiple-Feedback Bandpass Filter | 12-1 |
| ---- | --------------------------------- | ---- |
| 12.2 | State-Variable Bandpass Filter    | 12-7 |

## 13. BANDSTOP FILTERS — 13-1

| 13.1 | Active Inductor Bandstop Filter | 13-1 |
| ---- | ------------------------------- | ---- |
| 13.2 | Twin-Tee Bandstop Filter        | 13-4 |

## 14. FREQUENCY CONTROL — 14-1

| 14.1 | Frequency Doubler             | 14-1 |
| ---- | ----------------------------- | ---- |
| 14.2 | Frequency-Difference Detector | 14-6 |

x CONTENTS

## 15. INTEGRATORS AND DIFFERENTIATORS ... 15-1
  15.1 Differentiator ... 15-1
  15.2 Integrator ... 15-9

## 16. LIMITERS AND RECTIFIERS ... 16-1
  16.1 Amplitude Limiter ... 16-1
  16.2 Precision Half-Wave Rectifier ... 16-5

## 17. LOGARITHMIC CIRCUITS ... 17-1
  17.1 Differential Logarithmic Amplifier ... 17-1
  17.2 Antilogarithmic Amplifier ... 17-8

## 18. MODULATORS ... 18-1
  18.1 Amplitude Modulator ... 18-1
  18.2 Pulse-Amplitude Modulator ... 18-7
  18.3 Pulse-Width Modulator ... 18-9

## 19. MULTIPLIERS AND DIVIDERS ... 19-1
  19.1 FET-Controlled Multiplier ... 19-1
  19.2 Log-Antilog Multiplier/Divider ... 19-6

## 20. MULTIVIBRATORS ... 20-1
  20.1 Astable Multivibrator ... 20-1
  20.2 Bistable Multivibrator ... 20-4
  20.3 Monostable Multivibrator ... 20-7

## 21. OSCILLATORS ... 21-1
  21.1 Wien-Bridge Sine-Wave Oscillator ... 21-1
  21.2 Voltage-Controlled Oscillator ... 21-5

## 22. PARAMETER ENHANCEMENT AND SIMULATION ... 22-1
  22.1 Capacitance Multiplier ... 22-1
  22.2 Inductance Simulator ... 22-3

## 23. POWER CIRCUITS ... 23-1
  23.1 Op Amp Bandwidth/Power Booster ... 23-1
  23.2 Op Amp Output-Voltage Booster ... 23-7

## 24. REGULATORS ... 24-1
  24.1 Current-Limited Voltage Regulator ... 24-1
  24.2 High-Voltage Regulator ... 24-9
  24.3 Shunt Voltage Regulator ... 24-11
  24.4 Precision Voltage Reference ... 24-13
  24.5 Dual Voltage Regulator ... 24-13
  24.6 Switching Voltage Regulator ... 24-15
  24.7 Floating-Load Current Regulator ... 24-17
  24.8 Grounded-Load Current Regulator ... 24-19

## 25. SAMPLING CIRCUITS ............................. 25-1

    25.1  FET Multiplexer ................................................................... 25-1
    25.2  Precision Gate Multiplexer ................................................... 25-4
    25.3  Sample-and-Hold Circuit ..................................................... 25-6

## 26. TIME AND PHASE CIRCUITS ................... 26-1

    26.1  Phase Lead/Lag Circuit ....................................................... 26-1
    26.2  Adjustable Lead/Lag Circuits ............................................... 26-3
    26.3  Analog Timer ..................................................................... 26-5

## 27. WAVEFORM GENERATORS ...................... 27-1

    27.1  Voltage Ramp Generator ..................................................... 27-1
    27.2  Voltage Triangle Generator ................................................. 27-4
    27.3  Voltage Staircase Generator ............................................... 27-8

## Appendix I. OPERATIONAL AMPLIFIER PARAMETERS ........... I-1

## Appendix II. OPERATIONAL AMPLIFIER MAXIMUM RATINGS ....... II-1

## Appendix III. CIRCUIT FABRICATION TECHNIQUES ............. III-1

    III.1  Protection Circuits ................................................. III-1
            Input Terminals ..................................................... III-1
            Output Terminal .................................................... III-2
            Power Supply Terminals ......................................... III-3
    III.2  Noise Prevention Techniques .................................. III-3
            Grounding/Bypassing Problems ............................... III-4
            Shielding/Guarding ................................................ III-5
    III.3  Passive Devices in Op Amp Circuits ........................ III-6
            Resistors .............................................................. III-6
            Capacitors ........................................................... III-6

## Appendix IV. NOTATION USED IN HANDBOOK ................. IV-1

## Appendix V. DECIBEL CALCULATIONS ....................... V-1

## Appendix VI. RC CIRCUIT CHARACTERISTICS ................ VI-1

**Index follows Appendix VI.**

# Preface

Operational amplifiers have become one of the most popular electronic devices in use today. They are versatile, easy to use, and have many applications. This Handbook takes advantage of the easily systematized nature of op amp design. In a clear, step-by-step way, it provides engineers, technicians, and scientists with time- and work-saving procedures for designing or analyzing a circuit using op amps. The Handbook thus eliminates the need for time-consuming searches through the available literature to learn which equations are applicable to a particular circuit.

To help designers quickly determine which equations are applicable to their particular needs, the Handbook presents all such equations in quick-reference tabular form. This kind of simplified format makes the Handbook easily used by technicians as well as engineers and scientists.

The *Handbook of Operational Amplifier Circuit Design* provides concise, easy-to-follow information and procedures that include such labor-saving design aids as:

1. A basic introduction to op amps for those new to the field (chapter 1).
2. The two fundamental rules needed to design op amp circuits (chapter 2).
3. A comprehensive list of op amp error sources and methods by which they can be minimized (chapter 2).
4. A method of designing an amplifier with any required frequency response (chapter 2).
5. A detailed description of commonly used frequency compensation circuits and their effect on loop gain, phase margin, and gain margin (chapter 3).
6. A listing of the seven major causes of op amp instability, and remedies for this problem (chapter 3).
7. Design information on 68 op amp circuits (chapters 4 through 27). A list of design equations is provided for each circuit. The exact meaning and applicability of each equation is also given.
8. A parameter list accompanying each set of design equations, which carefully explains what each parameter or component is and what it does to circuit performance.
9. A comprehensive design procedure for one or two circuits in each of the design chapters (chapters 4 through 27). Each design procedure shows a recommended set of design steps that can be used to quickly construct a working model of the circuit.

10. A detailed numerical example for each circuit for which a design procedure is given. This example clearly shows how the design steps are carried out.

11. Six appendixes covering op amp parameters, maximum ratings, circuit fabrication techniques, notation used in the handbook, decibel calculation hints, and a comprehensive table of RC transfer functions.

The author wishes to acknowledge the ever-helpful support of his wife, Mildred, who painstakingly typed the entire manuscript. The author and editor also wish to thank Analog Devices, Inc., Fairchild Semiconductor Corp., National Semiconductor Corp., and Teledyne Philbric for allowing us to use some of their material.

*David F. Stout*
*Milton Kaufman*

# Handbook of Operational Amplifier Circuit Design

# Chapter 1

# Introduction to Operational Amplifiers

## 1.1 OVERVIEW OF OPERATIONAL AMPLIFIERS

The operational amplifier is a direct-coupled high-gain amplifier which uses feedback to control its performance characteristics. Internally, it consists of several series-connected transistor amplifiers. Externally, it is represented by the symbol shown in Fig. 1.1. The operational amplifier (op amp) is widely popular with analog-circuit designers because of its nearly ideal characteristics.

The op amp is capable of amplifying, controlling, or generating any sinusoidal or nonsinusoidal waveform over frequencies from dc to many megahertz. All classical computational functions are possible such as addition, subtraction, multiplication, division, integration, and differentiation. It is useful for innumerable applications in control systems, regulating systems, signal processing, instrumentation, and analog computation.

Functionally, as shown in Fig. 1.1, the op amp contains one output terminal which is controlled by two input terminals. If a positive voltage is applied to the positive (+) input, the op amp output will go positive. Likewise, a positive voltage on the negative (−) input will cause the output to go negative. A simplified model of the op amp is shown in Fig. 1.2. It indicates that an op amp can be represented by a voltage source which is controlled by two "floating" terminals. The op amp has high voltage gain for differential

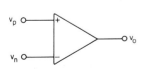

**Fig. 1.1** Standard op amp symbol.

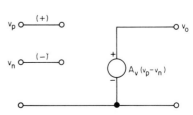

**Fig. 1.2** Alternate symbol for an op amp.

signals effective between the two inputs. The op amp also has very low gain for signals applied to both inputs simultaneously. These are called common-mode input signals.

The two inputs are labeled positive and negative or noninverting and inverting. The positive input is in phase with the output, and the negative input is 180° out of phase with the output. If we connect two resistors to the op amp as shown in Fig. 1.3, we have the basic noninverting amplifier circuit. The circuit gain from $v_i$ to $v_o$ is determined only by the resistors $R_f$ and $R_1$. If we replace $R_f$ by a short circuit and remove $R_1$, we have a very popular circuit known as the voltage follower. It has a gain of exactly 1. The basic inverting amplifier is shown in Fig. 1.4. Again, its gain is determined only by $R_f$ and $R_1$.

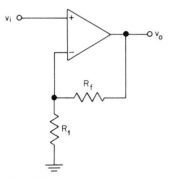

**Fig. 1.3** Noninverting amplifier.

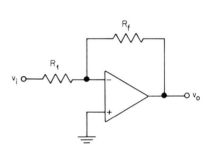

**Fig. 1.4** Inverting amplifier.

Both inputs to the op amp can be used simultaneously for differential-amplifier circuits, as shown in Fig. 1.5. The output voltage is proportional to the difference in the voltages applied to the two inputs. The constant of proportionality depends only on the size of the $R_f$ and $R_1$ resistors.

The operational amplifier is most versatile in simulating mathematical functions. Figure 1.6 shows a circuit which produces the sum of the inputs. Figure 1.7 produces an output voltage which is proportional to the integral of the input voltage. (This is an integrator.) Likewise, Fig. 1.8 produces an output voltage which is proportional to the derivative of the input voltage. (This is a differentiator.)

Op amps can be used to convert voltage to voltage, as in Figs. 1.3 and 1.4, or voltage to current, as in Fig. 1.9, or current to voltage, as in Fig. 1.10, or current to current, as in Fig. 1.11.

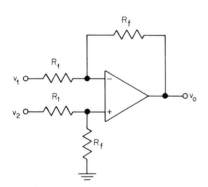

**Fig. 1.5** Differential amplifier.

This chapter is intended primarily for readers who are only slightly familiar with the op amp. Topics covered will be a comparison of ideal and real op amps, applications of op amps, calculations with basic circuits, and gain/phase plots over frequency.

**THE OP AMP MODEL** The operational amplifier model shown in Fig. 1.1 is the normal symbol seen in simplified schematics. More detailed schematics show more terminals on the op amp, as shown in Fig. 1.12. Some op amps require most of these extra terminals, while some require only the addition of power-supply terminals.

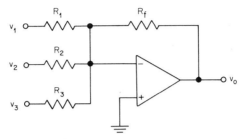

**Fig. 1.6** Summing circuit.

The op amp triangle symbol points in the direction of increasing signal level. The main output terminal is at one corner of the triangle. Sometimes a lower-power-level output is available and is shown next to the main output terminal. The inputs are always on the side opposite the output terminal. The other two sides of the triangle are for all the remaining terminals.

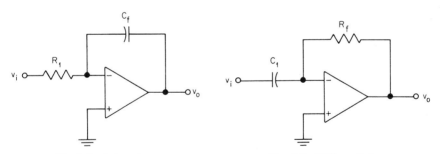

**Fig. 1.7** Integrator.  **Fig. 1.8** Differentiator.

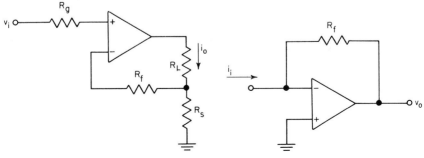

**Fig. 1.9** Voltage-to-current converter.  **Fig. 1.10** Current-to-voltage converter.

**1-4  INTRODUCTION TO OPERATIONAL AMPLIFIERS**

**PARAMETERS OF THE IDEAL OP AMP**  An op amp will perform most satisfactorily in the circuits described in this book if it has the following ideal properties (see Sec. 1.3 and Chap. 2 for a discussion of each parameter):
1. Differential voltage gain = ∞.
2. Common-mode voltage gain = 0.
3. Bandwidth = ∞.
4. Input impedance = ∞.
5. Output impedance = 0.
6. Output voltage = 0 when input voltage = 0.
7. Parameter drift with temperature = 0.
8. Equivalent input noise = 0.

None of these ideal parameters are achieved by any operational amplifier, nor will they ever be achieved. Manufacturers of op amps are continually improving their products, and some of these parameters are now so close to the ideal that the difference is hardly discernible. It would not be worthwhile to tabulate the state of the art, since improvements in op amp technology are reported in the literature every day.

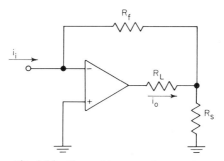

**Fig. 1.11**  Current-to-current converter.

When selecting an op amp for a specific application, the designer usually will be looking for optimum performance of two or three parameters. After an op amp with the best possible values for these parameters is found, it is often discovered that many of the other parameters are much less than optimum. This is the trade-off that every designer must go through during

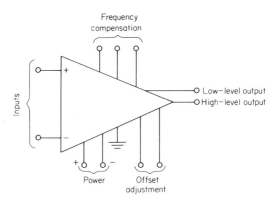

**Fig. 1.12**  Symbol showing some of the possible terminals found on different types of op amps.

the design phase of a circuit. It is not possible to find an op amp which has state-of-the-art performance of every parameter. For example, low-noise op amps are not usually wide-bandwidth devices. Likewise, wide-bandwidth op amps may not have a very high input impedance.

## 1.2 APPLICATIONS OF OP AMPS

Op amps can be used in nearly every area of linear and nonlinear electronics and also in a few digital applications.

**LINEAR AMPLIFIERS** The primary use of op amps is for linear amplifiers where highly stable gain is the main requirement. The high gain of the op amp combined with heavy feedback results in amplification which is almost independent of temperature, time, and op amp gain changes. For example, assume an op amp has a gain of $10^5$ and feedback is used to reduce the circuit gain to 10. The ratio of op amp gain to circuit gain is $10^5/10 = 10^4$. Factors which affect the op amp gain will have $10^4$ times less effect on the circuit gain. If the op amp gain changes from $10^5$ to $2 \times 10^5$ (a 100 percent change), the circuit gain will change $100\%/10^4 = 0.01\%$. This is the major benefit of high-gain op amps. The circuit can also be designed so that gain is traded for gain stability. If higher gains are desired, one automatically gets less stability.

In addition to stability, the heavy feedback mentioned above has several other benefits. Nonlinearity is reduced, bandwidth is increased, input and output impedances are changed, and the op amp can be replaced without affecting circuit gain.

Linear amplifiers using op amps can be constructed with either positive or negative gain. The magnitude of gain can range from less than one to several million, the upper limit depending on the particular op amp used. As mentioned above, however, the very high gain circuit will have poor gain stability and its bandwidth will be very narrow.

**NONLINEAR AMPLIFIERS** Many types of nonlinear amplifiers are possible using op amps. One of the most common is an op amp circuit which precisely amplifies signals of one polarity but not the other. This is known as a precision rectifier. Several variations of this basic circuit are possible, such as rectification in either direction. Adding a capacitor in one arrangement makes the circuit a filtered precision rectifier. Slight rearrangements of the diode and capacitor change the circuit to a peak detector, peak-to-peak detector, or average-value detector. Another slight modification turns this circuit into an absolute-value amplifier (full-wave rectifier).

**COMPARATORS** The op amp is a natural choice for a comparator owing to its high gain. The output terminal changes from plus saturation to minus saturation, or vice versa, with a millivolt or less change across the input terminals. By definition, this is a zero cross detector: the output changes polarity whenever the input passes through zero voltage. With a small bias on one input or the other, the circuit becomes a level detector. In this case the output changes state only when the signal input passes through the value of the bias on the other input.

Comparators have other more complex capabilities such as double-ended level detectors, level detectors with prescribed hysteresis, window detectors, and pulse-height analyzers. Most types of analog-to-digital converters require a comparator. This is such a widely required device that many specialized op amps are simply called comparators on the data sheet.

## 1-6 INTRODUCTION TO OPERATIONAL AMPLIFIERS

**FILTERS** Filter design has been revolutionized (and revitalized) because of the operational amplifier. One of the disadvantages of conventional filter design was its reliance on inductors. A simple op amp circuit is able to behave as a very stable, highly linear inductor. Modern filters, accordingly, use resistors, capacitors, and op amps as long as the application is within the frequency limitations of the op amps. These simulated inductors do not have all the disadvantages of real inductors: nonlinearity, hysteresis, core loss, radiation, unwanted coupling, large size, and difficult fabrication.

All the popular filter types are realizable with op amp filters (usually called active filters). A few of these types include:
  Low-pass filters
  High-pass filters
  Bandpass filters
  Bandstop filters

The ripple, phase, and rolloff of these filters can be tightly controlled with Chebyshev, Butterworth, or Bessel characteristics. Any combination of these classical characteristics is also possible with active filters.

Several other important advantages are offered by active filters. First, their output impedance (for most configurations) is less than 1 $\Omega$. This relieves the designer of the task of impedance matching. The input impedance of active filters is usually fairly high; so impedance matching is not required there either. Second, active filters can be designed to provide gain and/or supply large amounts of power.

Passive *RLC* filters require impedance matching at both the input and output and have less than unity gain at all frequencies. They can supply large amounts of power by appropriate choice of components. However, they cannot provide power gain as is possible with active filters.

**LOGARITHMIC APPLICATIONS** If nonlinear elements (diodes or transistors) are used in the feedback circuit, a host of logarithmic circuits are possible. The basic circuit provides an output voltage proportional to the logarithm of the input voltage. Either dc or ac signals can be converted in this manner. A simple part rearrangement turns the circuit into an antilog circuit.

Log and antilog circuits are sometimes known as compressors and expanders, respectively, when utilized for audio and video signals. By use of appropriate combinations of log and antilog circuits, the following types of functions are possible:
  Multiplier
  Divider
  Squaring circuit
  $Y = X^n$ circuit
  Square-root circuit
  Square root of the sum of squares

All these functions can be performed on either ac or dc signals.

**MULTIVIBRATORS** All three basic types of multivibrators are possible with the op amp. These basic types are the bistable (flip-flop), monostable (one-shot or single-shot), and the astable (rectangular-waveform generator) multivibrators. It is true that these are really digital functions and are much easier to build using digital microcircuits. However, more flexibility is offered with the op amp approach. For example, most digital microcircuits operate between specific voltages such as +5 and ground or −10 and ground. The maximum power level of these devices is also restricted. A multivibrator built with an op amp can operate between a wide range of minimum voltages (such as −20 to zero) and a wide range of maximum voltages (such as zero to +20 V). Thus, a multivibrator with a 40-V (±20-V) pulse output or a 1-V output could be designed. The power level could be increased to any desired level by inserting buffer transistors after the op amp but inside the feedback loop.

**OSCILLATORS-GENERATORS** Waveforms of many shapes, sizes, and frequencies are realizable using op amps. Sine-wave oscillators are implemented by using phase-shift feedback, inductor-capacitor feedback, or twin-tee feedback. The oscillator frequency and/or amplitude can be voltage-controlled. Negative feedback can be simultaneously incorporated to provide a highly stable output amplitude. Two- or three-phase circuits are commonly seen. Frequency stability can be assured by utilizing a quartz crystal or tuning fork.

Waveform generators have been designed which provide all the commonly required waveshapes. Rectangular shapes with either fixed or independently adjustable width and period adjustments are possible. Sawtooth generators with either fixed or independently adjustable rise and fall times are also commonly seen. The same comments can be made regarding triangle generators, which are merely a special class of sawtooth generators. Staircase generators can also be implemented with an op amp and several other parts. The step size, step period, and reset time can all be made independently adjustable.

Any of the adjustable parameters discussed above can be performed with voltage control. The op amp makes voltage control of parameters much easier than is possible using discrete parts.

**REGULATORS** Tight control of some parameter such as voltage, current, or temperature is easily performed using the op amp as a comparator. The op amp is such a useful device for voltage regulation that a whole class of specialized op amps has been developed for this application. These are commonly called monolithic voltage regulators. Since it is beyond the scope of this book to discuss special offshoots of the op amp device, we will discuss only how regulators are implemented using standard op amps. The monolithic voltage regulators utilize the same theory and electronic parts as regulators using op amps and discrete parts. However, monolithic regulators place most of the other electronic parts on the same monolithic chip as the op amp.

Voltage regulators of the following types can be designed around the ordinary op amp:

    Series pass regulation
    Shunt regulation
    Positive output
    Negative output
    Switching
    Foldback
    Current-limited
    Floating
    High voltage
    Precision

Other specialized voltage regulators are also possible.

Current regulators can be designed to handle many specialized applications. They are often seen supplying current to floating loads, grounded loads, or even active complex loads.

**1-8    INTRODUCTION TO OPERATIONAL AMPLIFIERS**

**SAMPLING CIRCUITS**   In this age of computerized control of analog processes, sampling circuits are indispensable. We will present design information on a number of sampling circuits such as precision analog gates, sample-and-hold circuits, and analog-to-digital converters. The sampling portion of these circuits requires one or two field-effect transistors in conjunction with an op amp. Analog-to-digital converters often require two or three op amps.

## 1.3   THE REAL OP AMP

**OP AMP PACKAGING**   The operational amplifier is available in many shapes and sizes. The designer's choice of shape and size depends on packaging and performance requirements. In general, the larger op amps have the best performance. The small units, such as those shown in Figs. 1.13 to 1.15, usually contain a single monolithic chip. The larger packages, such as that shown in Fig. 1.16, contain a printed circuit with discrete parts. Each part can be optimally selected to produce op amps in which some of the parameters are nearly ideal.

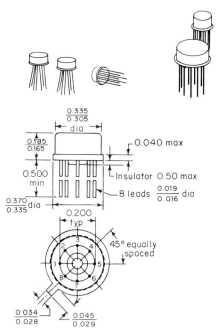

**Fig. 1.13**   The TO-99 op amp package. (*National Semiconductor Corp.*)

Many hybrid-microcircuit op amps are also available in relatively small packages. The package sometimes used is similar to Fig. 1.13. Hybrid op amps have many of the advantages of discrete op amps, but their package size is close to that of the monolithic devices. It is not nearly as easy to select resistors and transistors in a hybrid op amp as it is in a discrete op amp. However, if one is willing to pay the cost, hybrid op amps with discrete op amp performance can be obtained.

## THE REAL OP AMP 1-9

**PARAMETERS OF REAL OP AMPS** Now that we have discussed both op amp packaging and the ideal op amp parameters (Sec. 1.1), we will discuss the relationship between packaging and real parameters. It should become obvious that larger op amps usually have better parameters. We will also briefly explain the meaning of each parameter.

**Voltage gain** This parameter is often called open-loop gain and is specified at dc. It is defined as the ratio of an output-voltage change to an input-

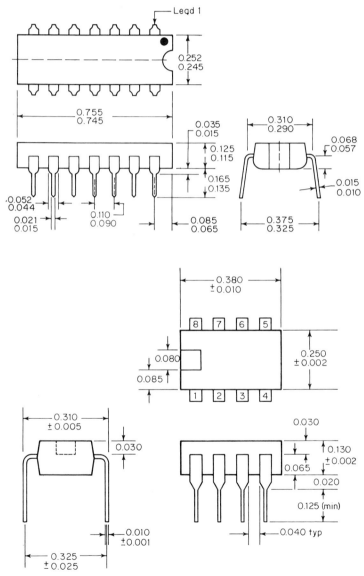

**Fig. 1.14** Several types of dual in-line packages commonly used for op amps. (*National Semiconductor Corp.*)

voltage change. One should not judge the quality of an op amp with this parameter, since it is an easy matter to cascade transistors to obtain large amounts of voltage gain. Throughout the text we will call it $A_{vo}$. The value of $A_{vo}$ may be only 1,000 for some wide-bandwidth devices, or $A_{vo}$ might have a maximum over one hundred million for chopper-stabilized op amps.

$A_{vo}$ is the differential voltage gain at dc, that is, the ratio of output voltage to the voltage between the two input terminals. The gain from both inputs (tied together) to the output is called common-mode voltage gain $A_{cmo}$. $A_{cmo}$ is usually thousands of times smaller than $A_{vo}$.

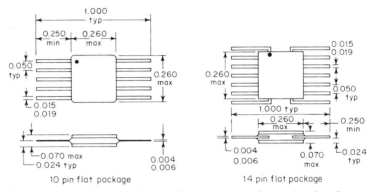

**Fig. 1.15** For high-density packaging the op amp may be put inside a flat pack.

**Bandwidth** The bandwidth of a device can be defined in several ways. The three methods commonly seen on op amp data sheets are:

1. *Unity-Gain Crossover Frequency.* This is obviously the frequency at which the voltage gain passes through a gain of 1 (0 db). We will call this frequency $f_u$ throughout the text. Values of $f_u$ from 1 kHz to 100 MHz are

**Fig. 1.16** The highest-quality op amps utilize discrete parts and are placed in packages as shown above. (*Analog Devices.*)

available. The typical value for monolithic op amps is in the range of 0.5 to 5 MHz. Special-purpose discrete-component op amps such as electrometers sometimes have very low bandwidths, or special wide-bandwidth devices may have an $f_u$ of 100 MHz.

2. *Unity-Gain Risetime.* Often the op amp is connected as a unity-gain noninverting circuit and its small-signal risetime $t_r$ is measured. The bandwidth is then computed from BW = $0.35/t_r$. This bandwidth will be quite close to the $f_u$ defined above.

3. *Full-Power Bandwidth.* This bandwidth is usually 10 to 100 times lower in frequency than $f_u$. It is defined as the maximum frequency at which a full-sized undistorted sine wave can be observed at the op amp output. If the op amp is using ±15-V supplies, a full output is approximately ±10 V. Monolithic op amps do not have full-power bandwidth beyond 0.5 MHz, while discrete-part op amps may extend up to 10 MHz.

**Slew rate** This parameter is related to full-power bandwidth. When an op amp is overdriven with a high frequency sine wave, the output appears to be a triangular waveform. The slope of this triangular-output waveform is the slew rate. It is expressed in volts per microsecond. The best monolithic op amps have slew rates of 100 V/$\mu$s, while the typical is 1 V/$\mu$s. Discrete-part op amps can be optimized for this parameter and achieve over 1,000 V/$\mu$s slew rates.

**Input resistance** This parameter is defined as either differential input resistance $R_{id}$ (between the two input terminals) or common-mode input resistance $R_{ic}$ (from either input terminal to the negative power-supply terminal). In monolithic op amps $R_{id}$ ranges from several hundred kilohms to several megohms. $R_{ic}$ is not specified very often, but it is often $10^8$ $\Omega$ or more. Some monolithic op amps have FET input transistors diffused on the same chip. The input resistance may approach $10^{12}$ $\Omega$ in these devices. Discrete-part op amps with FET input transistors may possess an $R_{id}$ and/or $R_{ic}$ of $10^{13}$ $\Omega$.

**Input bias current** This parameter, called $I_b$, is the average value of the two op amp input currents. Ideally, this current should be zero so that the input and feedback circuitry will not be disturbed. In monolithic op amps this current ranges from 1 nA to 1 $\mu$A. FET input monolithic op amps require slightly less than 1 nA. Good discrete-part op amps require less than 1 pA input bias current.

**Input offset current** The currents going into the two op amp inputs are always slightly different. The difference between these two currents $I_{io}$ is defined as the input offset current. This current should also be zero in the ideal case. Monolithic op amps have input offset currents ranging from less than 1 to several hundred nanoamperes. Discrete-part op amps are better, being much less than 1 pA.

**Input offset voltage** The voltage required across the op amp inputs to drive the output to zero is called the input offset voltage $V_{io}$. In monolithic op amps this parameter is often quite high if the input bias current is low. It usually ranges from 1 to 100 mV. Discrete-part op amps can be designed so that both types of offsets are somewhat optimized. Quite often, however, producers of op amps do not specify input offset voltage, since it can be nulled out using offset-adjustment terminals. Rather, they specify the temperature coefficient of the offset voltage. Discrete-part op amps may possess offset-voltage temperature coefficients of only 0.1 $\mu$V/°C. The best monolithic op amp is around 1 $\mu$V/°C, and most of these devices are up around 5 to 10 $\mu$V/°C.

## 1.4 OP AMPS COMPARED WITH TRANSISTORS

In this section we will take a look at the advantages and disadvantages of op amps compared with transistors. This comparison may be used as a foundation for expanding one's design capabilities from transistors into the area of op amps. The same type of transition was required by many circuit designers when the electronic industry started changing from vacuum tubes to transistors. In that case, the transition was difficult, since vacuum tubes are easier to understand than transistors. The transition from transistors to op amps is different. Here we are changing from a device which has many design difficulties to a device which is relatively easy to use. Even though the op amp contains from 10 to 100 transistors inside its envelope, it is probably the easiest electronic device with which one may perform a "cookbook" circuit design.

**GAIN** Transistors provide voltage gain from zero to 100, depending on the circuit and the input and output terminals used. Op amps can provide voltage gain from zero to over one million, depending on the circuit arrangement and type of op amp used. These are the voltage gains at dc.

**FREQUENCY RESPONSE** Individual transistors can be obtained which have little degradation of gain (−3-dB) at frequencies up to 5 or 10 GHz. The −3-dB frequencies for op amps are in the 1-Hz to 10-kHz range. However, op amps usually operate with such large amounts of feedback that the effective −3-dB frequency may be over 10 Mhz.

**INPUT IMPEDANCE** The input resistance of a FET input op amp may be as high as $10^{13}$ Ω. The typical value is over 1 MΩ for bipolar monolithic op amps. The input capacitance is typically 3 pF. Bipolar transistors have input resistances of only a few thousand ohms. FET transistors, however, have input resistances comparable with FET input op amps. The input capacitances of transistors may be only 1 pF or less for small-geometry devices.

**OUTPUT RESISTANCE** The output resistance of most op amps is 100 Ω or less. When feedback is incorporated, this output resistance may be reduced to below 1 Ω. This is an almost ideal situation. Transistors, on the other hand, have output resistances of thousands or tens of thousands of ohms. Since their gain is not too high, the output resistance cannot be lowered more than a few octaves when feedback is incorporated. A high output resistance is one undesirable characteristic of transistors.

## 1.5 BASIC OP AMP CIRCUITS

Two simple op amp circuits which have wide usage will be discussed in this section. These are called the basic inverting and noninverting op amp circuits. Many of the more complex circuits in this handbook are merely extensions of these two basic circuits.

**BASIC INVERTING OP AMP CIRCUIT** Figure 1.17 shows the basic inverting op amp circuit. The voltage gain at dc and low frequencies is called $A_{vco}$. This is equal to

$$A_{vco} = \frac{v_o}{v_i} = -\frac{R_f}{R_1} \qquad (1.1)$$

The circuit voltage gain as a function of frequency is called $A_{vc}$. Since $R_f$ and $R_1$ represent resistances, the amplifier gain is somewhat independent of frequency, as shown in Fig. 1.18. At higher frequencies, where the op amp gain has fallen off to the point where it equals $A_{vco}$, the circuit gain falls off at the same rate.

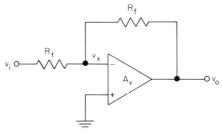

**Fig. 1.17** Basic inverting op amp circuit. The voltage gain is $v_o/v_i = -R_f/R_1$.

If $R_f$ and $R_1$ become two impedances $Z_f$ and $Z_1$, the ratio $Z_f/Z_1$ may be a function of frequency. Simple filters may be constructed by using appropriate reactive components for $Z_f$ and $Z_1$. For instance, if $Z_1$ is a capacitor ($Z_1 = 1/j2\pi fC$) and $Z_f$ is a resistor ($Z_f = R$),

$$A_{vc} = -\frac{Z_f}{Z_1} = \frac{-R}{1/j2\pi fC} = -j2\pi fRC$$

The circuit gain therefore increases with frequency until the op amp gain is reached. Then the circuit gain will fall with frequency at the same rate the op amp gain falls with frequency.

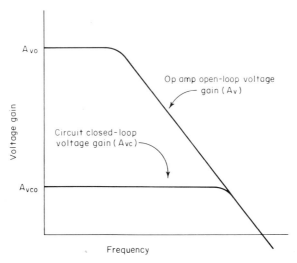

**Fig. 1.18** Open-loop gain of op amp shown in Fig. 1.17 and the closed-loop gain after the resistors $R_f$ and $R_1$ are added.

**1-14 INTRODUCTION TO OPERATIONAL AMPLIFIERS**

The circuit of Fig. 1.17 has an input impedance of $Z_1$ (or $R_1$). The output impedance is very small, usually less than 1 Ω.

**BASIC NONINVERTING OP AMP CIRCUIT** If we apply the input voltage $v_i$ to the positive op amp input as shown in Fig. 1.19, we have a noninverting amplifier. Negative feedback is still required, however, to stabilize the circuit and to set the gain. For this circuit the closed-loop voltage gain at dc and low frequencies is

$$A_{vco} = \frac{v_o}{v_i} = 1 + \frac{R_f}{R_1} \quad (1.2)$$

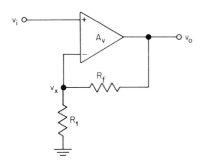

**Fig. 1.19** Basic noninverting op amp circuit.

Note that no minus sign is used. If impedances are used instead of resistances, the closed-loop gain as a function of frequency is

$$A_{vc} = \frac{v_o}{v_i} = 1 + \frac{Z_f}{Z_1} \quad (1.3)$$

$Z_f$ and $Z_1$ may be either linear or nonlinear. The above equation holds true no matter how nonlinear they become.

The input impedance of a noninverting op amp circuit is very high—approximately equal to the op amp input impedance times the ratio $A_v/A_{vc}$. The output impedance is very low, usually below 1 Ω.

## 1.6 REAL AND IDEAL OP AMPS COMPARED

We will now explore what happens to the closed-loop gain [Eqs. (1.1) to (1.3)] when real op amp parameters are incorporated.

**FINITE OPEN-LOOP GAIN** The relationship between $A_v$ (open-loop gain) and $A_{vc}$ (closed-loop gain) is

$$A_{vc}(\text{inverting}) = \frac{-Z_f/Z_1}{1 + 1/A_v + Z_f/A_v Z_1} \quad (1.4)$$

and

$$A_{vc}(\text{noninverting}) = \frac{1 + Z_f/Z_1}{1 + 1/A_v + Z_f/A_v Z_1} \quad (1.5)$$

The numerators of both above equations are simply the ideal closed-loop gains. The degradation to the ideal closed-loop gains comes about from the terms in the denominator. Ideally, the term $A_v = \infty$. This means that $1/A_v = 0$ and $Z_f/A_v Z_1 = 0$. The denominator then becomes $1 + 0 + 0$, and the effect of the open-loop gain vanishes.

If $A_v \neq \infty$, the effect of $A_v$ on closed-loop gain $A_{vc}$ becomes larger as $A_v$ becomes smaller. We will first examine the significance of this statement for the dc case (using $A_{vo}$ and $A_{vco}$). Later we will examine the ac case (using $A_v$ and $A_{vc}$). As a specific example, assume $Z_f/Z_1 = 100$ and $A_{vo} = 1,000$. Equation (1.4) becomes

$$A_{vco}(\text{inverting}) = \frac{-100}{1 + 1/1,000 + 100/1,000}$$

$$= \frac{-100}{1.101} = -90.83$$

This indicates a 9.2 percent reduction of dc gain from the 100 desired. It also means (and this is probably more important) that a 100 percent change in $A_{vo}$ will result in a change in $A_{vco}$ of approximately 9 percent. The exact change is determined by using Eq. (1.4) twice, once for each value of $A_{vo}$. If $A_{vo}$ is increased to 10,000, $A_{vco}$ is only reduced to

$$A_{vco} = \frac{-100}{1 + 1/10{,}000 + 100/10{,}000} = -99.00$$

Thus the gain error is only 1 percent if the op amp has a gain of 10,000. By similar calculations we find that if the required $A_{vco}$ is only 10 and $A_{vo}$ is still 10,000, the error is reduced to 0.1 percent. We conclude that the gain error depends on the ratio of open- to closed-loop gains. Table 1.1 summarizes this conclusion.

**TABLE 1.1  Gain Error at DC Caused by Finite Open-Loop DC Gain**

| $A_{vo}/A_{vco}$ | % gain error (dc) |
|---|---|
| 1 | −50 |
| 10 | −9 |
| $10^2$ | −1 |
| $10^3$ | −0.1 |
| $10^4$ | $-10^{-2}$ |
| $10^5$ | $-10^{-3}$ |
| $10^6$ | $-10^{-4}$ |

We can also make a table of gain errors for the ac case. As shown in Figs. 1.18 and 1.21, the op amp open-loop gain has a −90° phase shift over much of the usable range of frequencies. This causes a phase difference of 90° between $A_v$ and $A_{vc}$ for these frequencies. As a result, the degradation of $A_{vc}$ by $A_v$ is much less than shown for the dc case. Table 1.2 shows $A_{vc}$ reduction as a function of the ratio $A_v/A_{vc}$. We will explore the equations behind this table in Chap. 2. It applies only for the region between the first pole of $A_v$ and the first pole of $A_{vc}$ (i.e., from 10 to $10^5$ Hz in Fig. 1.21).

**TABLE 1.2  AC Gain Error Caused by Finite Open-Loop AC Gain**

| $A_v/A_{vc}$ | % gain error (ac) |
|---|---|
| 1 | −33 |
| 10 | −0.6 |
| $10^2$ | −0.006 |
| $10^3$ | $-5 \times 10^{-5}$ |
| $10^4$ | $-5 \times 10^{-7}$ |
| $10^5$ | $-5 \times 10^{-9}$ |
| $10^6$ | $-5 \times 10^{-11}$ |

**FINITE BANDWIDTH**  If the amplifier is required to handle ac signals also, the degradation of gain at various frequencies must be considered. Referring back to Fig. 1.18, we note that the ratio $A_v/A_{vc}$ is the largest at low frequencies. The ratio gets progressively smaller until it equals 1 at the point where the $A_v$ and $A_{vc}$ curves intersect. The error when $A_v/A_{vc} = 1$ is −33 percent. This explains why the intersection of the two curves is not abrupt. Since the error

## 1-16 INTRODUCTION TO OPERATIONAL AMPLIFIERS

gets large very fast in the $A_v/A_{vc} = 1$ region, the $A_{vc}$ curve falls off and gradually blends in with the $A_v$ curve.

**FINITE INPUT RESISTANCE** This parameter lowers the closed-loop gain slightly and also limits the high input impedance expected with the noninverting amplifier. The equation for reduction in closed-loop gain is similar to the equation for finite $A_v$. For the following equations $R_{id}$ is the op amp differential input resistance (i.e., between the + and − input terminals) and $R_{ic}$ is the op amp common-mode input resistance (i.e., from either input terminal to the negative power-supply terminal).

$$A_{vc}(\text{inverting}) = \frac{-R_f/R_1}{1 + 1/A_v + R_f/A_vR_1 + R_f/A_vR_{id} + R_f/A_vR_{ic}} \quad (1.6)$$

and

$$A_{vc}(\text{noninverting}) = \frac{1 + R_f/R_1}{1 + 1/A_v + R_f/A_vR_1 + R_f/A_vR_{id} + R_f/A_vR_{ic}} \quad (1.7)$$

Equations (1.6) and (1.7) are identical to Eqs. (1.4) and (1.5) except that the factor $R_f/A_vR_{id} + R_f/A_vR_{ic}$ has been added to the denominators. If the three terms $A_v$, $R_{id}$, and $R_{ic}$ are all very large, this entire factor is very small and does not affect gain. This is one case where one op amp parameter may help another. That is, if for some reason $R_{id}$ is low but $A_v$ is very large, the net result is a large product $A_vR_{id}$. The above factor would then be negligible.

It is a little more difficult to show a table indicating the percent effect on gain due to finite $R_{id}$ and $R_{ic}$, since $A_v$ is also included. For purposes of illustration, assume $R_{ic} \gg R_{id}$ (as is usually the case). We can therefore drop $R_f/A_vR_{ic}$ for this example. We may now make a table of percent error of $A_{vc}$ as a function of $A_v$ and $R_{id}$. If $A_v$ is high and $R_{id}$ is low, $A_{vc}$ will have little degradation. If both $A_v$ and $R_{id}$ are high, $A_{vc}$ will be essentially undisturbed. However, if both $A_v$ and $R_{id}$ are low, severe degradation of $A_{vc}$ will occur. For simplicity, the accompanying table is computed only at dc, so $A_{vo}$ is used

| $A_{vo}$ | $R_{id}$, Ω | % gain error |
|---|---|---|
| 1 | $10^4$ | −92.3 |
| 1 | $10^6$ | −91.7 |
| 1 | $10^8$ | −91.7 |
| $10^2$ | $10^4$ | −10.7 |
| $10^2$ | $10^6$ | − 9.92 |
| $10^2$ | $10^8$ | − 9.91 |
| $10^4$ | $10^4$ | − 0.12 |
| $10^4$ | $10^6$ | − 0.11 |
| $10^4$ | $10^8$ | − 0.11 |
| $10^6$ | $10^4$ | − 0.0012 |
| $10^6$ | $10^6$ | − 0.0011 |
| $10^6$ | $10^8$ | − 0.0011 |

NOTE: This table assumes $R_f = 10\text{k}\Omega$, $R_1 = 1\text{k}\Omega$, and $A_{vco} = 10$.

in place of $A_v$. This table provides the interesting result that gain error is much more sensitive to $A_{vo}$ than it is to $R_{id}$. Changes to $R_{id}$ are only 0.01 to 0.1 percent as influential on gain error as $A_{vo}$. The error at frequencies above the first op amp pole must be computed with the 90° lag of $A_v$ considered.

**OUTPUT RESISTANCE > 0** The output resistance of an op amp affects the circuit output resistance, gain, and stability. We will save the stability discussion for Chap. 3.

The heavy feedback usually incorporated in an op amp circuit makes the circuit output resistance effectively very low. If the op amp output resistance is $R_o$, the circuit output resistance is

$$R_{out} = \frac{R_o}{\beta A_v} \qquad (1.8)$$

where

$$\beta = \frac{R_1}{R_1 + R_f}$$

$R_o$ is commonly 100 Ω or less. $A_{vo}$ is quite large in most op amps—typically $5 \times 10^4$ to $10^6$. Therefore the circuit output resistance at dc may be less than

$$R_{out} = \frac{100}{5 \times 10^4} = 2 \times 10^{-3} \; \Omega$$

(This result assumes $\beta = 0.1$ and $A_{vo} = 5 \times 10^5$.) At high frequencies the situation is not so good. $R_o$ often goes up to several hundred ohms near unity open-loop gain. Thus $A_v$ is small and $R_o$ is large at the same time. At these frequencies $R_{out}$ may be nearly as large as $R_o$.

Many other nonideal op amp parameters could be listed here and their effects on circuit performance summarized. We will save these details for Chap. 2. The point to be emphasized here is: All parameters tabulated in an op amp data sheet should be carefully studied and their effects on circuit performance calculated. Compromises must often be resorted to in the design of many circuits because of the nonideal behavior of the op amp. The designer must continually refresh his approach so that the important parameters are optimized and time is not wasted on attempts to optimize less important parameters.

## 1.7 THE FEEDBACK EQUATION

The feedback equation, which is fundamental to all op amp circuit analysis and design, will be derived and explained in this section. This equation is of the form

$$A_{vc} = \frac{A_v}{1 + \beta A_v} \qquad (1.9)$$

for the noninverting circuit, and

$$A_{vc} = \frac{A_v(\beta - 1)}{1 + \beta A_v} \qquad (1.10)$$

for the inverting circuit, where

$$\beta = \frac{R_1}{R_1 + R_f}$$

as before. Some textbooks manipulate the algebra slightly and state these equations as

$$A_{vc} = \frac{1/\beta}{1 + 1/A_v \beta} \qquad (1.9A)$$

for the noninverting circuit, and

$$A_{vc} = \frac{1 - 1/\beta}{1 + 1/A_v\beta} \quad (1.10A)$$

for the inverting circuit, but the outcome is identical. [Note that Eqs. (1.9A) and (1.10A) are exactly the same as Eqs. (1.4) and (1.5), except $\beta = R_1/(R_1 + R_f)$ is substituted.]

Equation (1.9) [or (1.9A)] is derived as follows:
By inspection of Fig. 1.19, we get

$$v_x = \beta v_o = \frac{R_1 v_o}{R_1 + R_f} \quad (1.11)$$

by using simple voltage-divider theory. Since the op amp has a voltage gain of $A_v$,

$$v_o = A_v(v_i - v_x) \quad (1.12)$$

Substituting $v_x$ from Eq. (1.11) into Eq. (1.12), we get

$$v_o = A_v(v_i - \beta v_o)$$

Solving this for $A_{vc} = v_o/v_i$, the final result becomes

$$A_{vc} = \frac{v_o}{v_i} = \frac{A_v}{1 + \beta A_v}$$

which is identical to Eq. (1.9).

Equation (1.10) [or (1.10A)] is derived using Fig. 1.17. In this case, however, $v_x$ is found by superimposing the voltage at $v_x$ caused by $v_i$ with the voltage at $v_x$ caused by $v_o$. The voltage-divider action is opposite in these two cases.

$$v_x = \frac{R_f v_i}{R_1 + R_f} + \frac{R_1 v_o}{R_1 + R_f} \quad (1.13)$$

Since $v_x$ is the only signal driving the op amp,

$$v_o = -A_v v_x$$

Rearranging this, we get

$$v_x = \frac{-v_o}{A_v}$$

Substituting this result into Eq. (1.13), the composite formula is

$$-\frac{v_o}{A_v} = \frac{R_f v_i}{R_1 + R_f} + \frac{R_1 v_o}{R_1 + R_f}$$

Rearranging terms and using the facts that $1 - 1/\beta = -R_f/R_1$ and $\beta = R_1/(R_1 + R_f)$, we get

$$A_{vc} = \frac{v_o}{v_i} = \frac{A_v(\beta - 1)}{1 + \beta A_v}$$

This is identical to Eq. (1.10).

The meaning and usage of the feedback equation can now be explained. We will begin with Eq. (1.9), since it is simpler than the others. It is repeated here for convenience.

$$A_{vc} = \frac{A_v}{1 + \beta A_v} \quad (1.9)$$

The term $A_v$ is usually much larger than $10^4$. The term $\beta$ is usually in the range of 0.001 to 1. If the product $\beta A_v$ is very large, say 100 or more, the 1 in the denominator can be discarded and we get

$$A_{vc} \approx \frac{A_v}{\beta A_v} = \frac{1}{\beta}$$

In the noninverting case the gain $A_{vc}$ is therefore

$$A_{vc} = \frac{1}{\beta} = \frac{R_1 + R_f}{R_1} = 1 + \frac{R_f}{R_1}$$

The same comments apply to the inverting amplifier [eq. (1.10)]. If we drop the 1 in the denominator,

$$A_{vc} \approx \frac{A_v(\beta - 1)}{\beta A_v} = \frac{\beta - 1}{\beta} = -\frac{R_f}{R_1}$$

What do the above results mean? In Sec. 1.6 we showed that the difference between real and ideal closed-loop gain becomes very small as $A_v$ or $A_{vo}$ becomes very large. Values of $A_{vo} > 1,000\, A_{vco}$ are required to keep the real dc gain within 0.1 percent of the ideal dc gain. A large $A_v$ or $A_{vo}$ has many other benefits. For example, if $A_{vo} = 1,000\, A_{vco}$, a 100 percent change in $A_{vo}$ will affect $A_{vco}$ by only $100\%/1,000 = 0.1\%$. Examination of any typical op amp data sheet shows that a 100 percent change in $A_{vo}$ is very likely. One must therefore be sure that $A_{vo} \gg A_{vco}$ (and $A_v \gg A_{vc}$) for all temperatures within which the op amp must perform.

## 1.8 LARGE-SIGNAL BEHAVIOR OF OP AMPS

Many applications of op amps require saturation in the positive or negative direction. A clear understanding of the basic nonlinear behavior of the op amp is a necessary prerequisite to any large-signal design using the saturation characteristics of the device.

A typical input-output voltage-transfer function of an op amp is shown in Fig. 1.20. This particular op amp was operated with $\pm 15$-V power supplies. The maximum $\pm$op amp output is usually several volts less than the power-supply voltages. Thus this device saturates at $\pm 13$ V. These limits cannot be exceeded no matter how hard the inputs are driven (within the maximum input capability of the op amp, naturally). The output limits can be raised only by increasing the $\pm$power-supply voltages. These voltages, however, must not exceed the maximums stated on the data sheet or damage will result.

The op amp of Fig. 1.20 is linear only over the $v_{out} = \pm 10$-V range. The open-loop gain $A_{vo}$ in this linear region can be computed from the transfer function. Since a $+10$-V output requires only a $+0.1$-mV input, the gain is

$$A_{vo} = \frac{\Delta v_{out}}{\Delta v_{in}} = \frac{10\text{V}}{10^{-4}\text{V}} = 100,000$$

If a nonlinear application required the output to rise all the way up to $+13$ V, $+0.3$-mV input is required. The voltage gain in this case is

$$A_{vo} = \frac{\Delta v_{out}}{\Delta v_{in}} = \frac{13\text{V}}{3 \times 10^{-4}\text{V}} = 43,000$$

Whenever operation is required in the saturation region, this reduction in gain must be recognized.

**1-20 INTRODUCTION TO OPERATIONAL AMPLIFIERS**

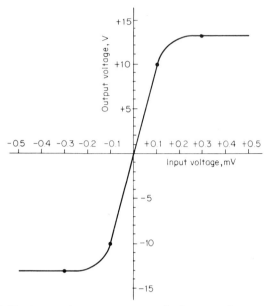

**Fig. 1.20** A typical input-output transfer function of an op amp.

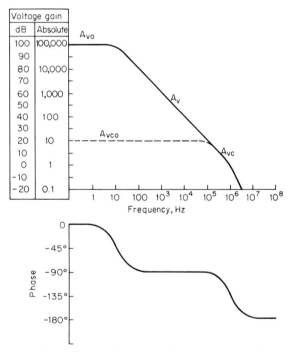

**Fig. 1.21** Gain and phase of a typical op amp. The output $v_o$ is relative to the positive input $v_1$. The negative input is assumed to be grounded. A typical closed-loop gain curve ($A_{vco} = 10$) is also shown.

## 1.9 OPEN-LOOP CHARACTERISTICS OF OP AMPS

Whenever the op amp is used to amplify ac signals, the plot of open-loop frequency response as shown in Fig. 1.21 must be consulted. This plot provides a wealth of information useful in dc, ac, and feedback analysis. The horizontal line on top of the plot is the dc gain of the op amp $A_{vo}$. The $-3$-dB point (open-loop bandwidth) is that point where the gain is 3 dB below the horizontal line. Unity open-loop gain frequency $f_u$ is where the gain passes through zero dB($A_v = 1$).

**CALCULATIONS USING DECIBELS** A review of operations using decibels (dB) will be useful at this point. We will be working with voltage (or sometimes current) ratios in this book, so the relationship to remember is

$$dB = 20 \log\left(\frac{v_2}{v_1}\right)$$

or

$$dB = 20 \log\left(\frac{i_2}{i_1}\right)$$

A table of dB vs. $v_2/v_1$ (or $i_2/i_1$) is given in Appendix V. The table is much easier to use than the equations above. For example, suppose an op amp has an open-loop gain at dc of 94.7 dB. What is the corresponding voltage ratio ($v_2/v_1 = v_{\text{out}}/v_{\text{in}}$)? We merely find a set of dB values which add up to 94.7 dB and record the corresponding voltage ratio for each as follows:

| dB | $v_2/v_1$ |
|---|---|
| 90 | 31,623 |
| 4.0 | 1.5849 |
| 0.7 | 1.0839 |
| 94.7 | Product = 31,623 × 1.5849 × 1.0839 = 54,324.3 |

The voltage ratio corresponding to 94.7 dB is the product of the 3 $v_2/v_1$ ratios above. Thus

$$94.7 \text{ dB} = 54{,}324.3 \text{ V/V}$$

The table is used in the other direction as per the following example: Suppose we are told the gain ($v_2/v_1$) of a device is 39,450. How many dB is this? We begin by finding the largest $v_2/v_1$ which is less than 39,450, i.e., 31,623, which is equal to 90 dB. Then we divide 39,450 by 31,623 to get 1.2475 and look for the next ratio under this on the table. We find that 1.1220, corresponding to 1.0 dB, is the next lower number. Again we divide 1.2475 by 1.1220 to get 1.1119, which corresponds to approximately 0.9 dB. Now we add all the dB's found above to get our answer:

$$90 \text{ dB} + 1.0 \text{ dB} + 0.9 \text{ dB} = 91.9 \text{ dB}$$

This is equivalent to 39,450.

**OP AMP OPEN-LOOP PHASE SHIFT** A plot of the op amp open-loop phase is very valuable for feedback-stability calculations. We will explore this in detail in Chap. 3. Many op amp data sheets do not show phase plots, so we will describe how to make one using the open-loop gain plot. First, however, we will describe the meaning of the phase plot.

## 1-22  INTRODUCTION TO OPERATIONAL AMPLIFIERS

Referring back to Fig. 1.21, it is noted that the gain plot has two points where the slope of the plot becomes steeper (i.e., at 10 and $10^6$ Hz). These points are called poles. At frequencies between the first pole (we will always count from the left to the right) and the second pole the slope is $-20$ dB/decade. This is equal to $-6$ dB/octave. At frequencies above the second pole the slope is $-40$ dB/decade ($-12$ dB/octave). One should learn to use both dB/decade and dB/octave, since literature on op amps use both methods.

As shown in Fig. 1.22, each pole causes a 90° phase lag in the op amp gain. The 90° lag does not occur at the pole, but at frequencies above the pole. The phase lag due to a pole is 45° right at the frequency of the pole. The full 90° lag occurs at all frequencies greater than 10 times the pole frequency.

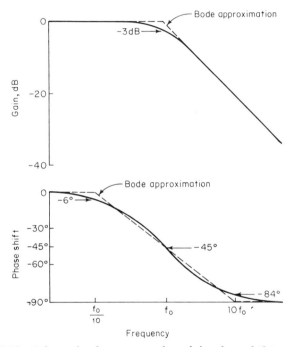

**Fig. 1.22** Relationship between a pole and the phase shift it causes.

A second pole causes another 90° phase lag. Figure 1.21 shows the shape of a phase plot for an op amp having two poles.

**GAIN AND PHASE OF A ZERO** The function opposite to that of a pole is called a zero. A zero causes the gain to have a $+20$ dB/decade slope for all frequencies above the zero. A zero causes a $+45°$ phase shift at the frequency of the zero and a $+90°$ phase shift for all frequencies greater than 10 times the zero frequency. Figure 1.23 shows the relationship between the gain and phase of a zero.

**THE BODE APPROXIMATION** When constructing gain and phase plots for some complex circuits, it becomes quite cumbersome to plot exact gain and phase. Many designers use the Bode-approximation method for constructing gain and phase for their "first-cut" design. After the design has progressed somewhat, effort is (sometimes) put into constructing exact gain and phase plots.

The Bode-approximation method uses the dashed lines of Figs. 1.22 and 1.23. Note that all plots can be made with straight lines. The straight lines intersect at the pole or zero frequency in the gain plots. The straight lines intersect at 1/10 and 10 times the pole or zero frequency in the phase plots.

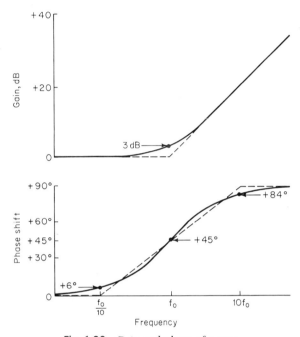

**Fig. 1.23** Gain and phase of a zero.

The straight lines on the gain plots always have a slope of 0, ±20 dB/decade, ±40 dB/decade, etc. The straight lines on the phase plot always have slopes of 0, ±45°/decade, ±90°/decade, etc.

## 1.10 CLOSED-LOOP CHARACTERISTICS OF OP AMPS

When feedback circuitry is connected from the op amp output to the input, a closed-loop circuit results. The closed-loop frequency characteristics are dramatically different from the open-loop characteristics. Figure 1.24 shows some of the possible ways the closed-loop frequency characteristics can be shaped. The first thing one notices is that all closed-loop curves fall within the open-loop curve. This is true of all op amp circuits which utilize passive feedback networks.

The phase response of closed-loop op amp circuits is also much different from that of open-loop op amps. In all the closed-loop curves shown in Fig. 1.24, each pole will produce a −90° phase shift and each zero will cause a +90° phase shift. Often there might be two poles or two zeros at one frequency. In these cases a −180° phase shift occurs for the double pole and

1-24   INTRODUCTION TO OPERATIONAL AMPLIFIERS

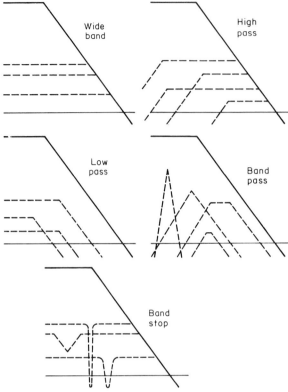

**Fig. 1.24** Five possible classes of closed-loop frequency characteristics. Others are possible.

+180° for the double zero. The full 180° phase shift takes effect over a 2-decade frequency range, as did single poles and zeros.

## 1.11  LOOP GAIN AND PHASE

Loop gain is another parameter which is extensively used in op amp circuits. Its principal application is for feedback analysis, which we will explore in Chap. 3. Loop gain is merely the product of the op amp gain and the feedback-network transfer function. This requires one passage around the loop—hence the name "loop gain."

It would be beneficial at this point to digress slightly and show in Table 1.3 the types of gain discussed so far. This will help avoid confusing one type of gain with another.

Briefly, the way loop gain is used in feedback analysis is as follows: (1) The loop gain is calculated (or measured) and plotted as a function of frequency. (2) The phase of loop gain is also plotted as a function of frequency. The resultant plots may look something like Fig. 1.25, although the exact shape depends on the circuit, op amp, load, and other factors. (3) The frequency at which the loop gain passes through 0 dB (loop gain of 1) is noted. (4) The loop phase is also noted at this same frequency. If the phase is more than 45° above the $-180°$ line, the amplifier will be stable. If the phase is within $\pm 5°$ of the $-180°$ line, the circuit will probably oscillate.

**TABLE 1.3  Types of Gain Used in Op Amp Circuits**

| Name | Description | Typical schematic |
|---|---|---|
| Open-loop gain | The gain of the op amp from $v_1$ to $v_o$. This is called $$\frac{v_o}{v_1} = A_v$$ | |
| Closed-loop gain | The gain of the entire op amp circuit after feedback and input networks are added. This is called $$\frac{v_o}{v_1} = A_{vc}$$ | |
| Loop gain | The gain through the feedback network and the op amp, i.e., from $v_1$ to $v_o$. This is called $$\frac{v_o}{v_1} = A_f A_v$$ ($A_f$ defined below) | |
| Feedback-network gain (usually an attenuation) | The voltage-transfer function of the feedback network. This is called $$\frac{v_o}{v_1} = A_f = \frac{Z_1}{Z_1 + Z_f}$$ | |

The number of degrees that the phase shift is above the $-180°$ line is called the phase margin $\phi_m$. In Fig. 1.25 the phase margin is approximately 90°. A design which results in a nominal phase margin of 45°, or a worst-case minimum of 30°, is considered an adequate design.

## 1.12 CIRCUITS INSIDE AN OP AMP

To those who are already familiar with transistor circuits, a brief description of the transistor circuits inside an op amp will be enlightening. Those who are not interested in the internal workings of an op amp may skip this section without loss of continuity.

A large number of circuits are presently in use inside op amps. We could not possibly discuss all these circuits here. Many monolithic op amps, however, use the general scheme as shown in Fig. 1.26.

**INPUT DIFFERENTIAL AMPLIFIER** This stage determines the ultimate gain stability, common-mode rejection, bias drift, input impedance, slewing rate,

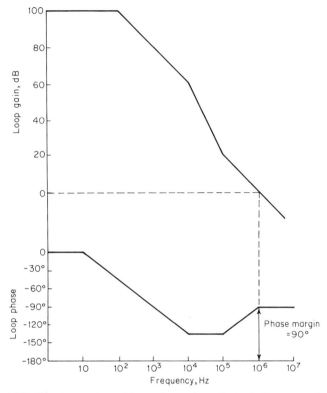

**Fig. 1.25** The loop-gain and loop-phase plots of a typical op amp circuit.

bandwidth, and noise of the op amp. Subsequent stages have little effect on these parameters. That is, if the first stage has a voltage gain of 10, the errors in the second stage will only be 1/10 as noticeable as errors of equal size in the first stage. It is therefore mandatory that the input differential amplifier be carefully designed and produced with repeatable quality.

The input-differential-amplifier stage usually has a circuit similar to Fig. 1.27A. This is greatly simplified when compared with a real op amp. The current source is required so that the circuit will have a large common-mode rejection (this will be explained in Chap. 2).

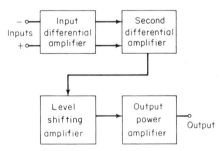

**Fig. 1.26** Block diagram of a popular method for constructing an op amp.

**SECOND DIFFERENTIAL AMPLIFIER** This stage (Fig. 1.27B) is almost identical to the input differential amplifier except that a resistor takes the place of the current source. The common-mode rejection requirement is not as great in the second stage; so a simple resistor current source is sufficient. The seven critical parameters determined by the input stage are 10 to 100 times less critical in this second stage. Only one output is used on this second stage.

**LEVEL-SHIFTING AMPLIFIER** The quiescent output voltage of an op amp should be zero if the input differential voltage is zero. The output of the two differential amplifiers discussed previously does not usually provide this zero output voltage. Often, another stage is required to provide a dc level shift and to provide some more gain. A standard PNP common-emitter amplifier as shown in Fig. 1.27C is often used. The load resistor is made up of two diodes and a resistor. This provides a temperature-compensated voltage divider for driving the output-power amplifier. A small resistor is placed in the emitter to provide some negative feedback. This stage also transforms the high impedance of the second differential amplifier to a low impedance capable of driving the output-power amplifier.

**OUTPUT-POWER AMPLIFIER** This stage is usually an emitter follower of the complementary type; that is, the NPN transistor in Fig. 1.27D handles the

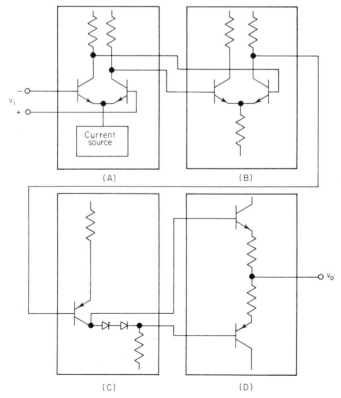

**Fig. 1.27** The four main stages of a typical op amp. (A) The input differential amplifier. (B) The second differential amplifier. (C) The level-shifting amplifier. (D) The output-power amplifier.

positive output signal and the PNP transistor handles the negative output signal. An emitter follower provides high current gain, wide bandwidth, high input impedance, and a low output impedance. Since this stage must drive devices outside the op amp, it has substantial current-driving capability. Current limiting beyond a fixed value is usually provided to protect the op amp.

## REFERENCES

1. Giles, J. N.: "Fairchild Semiconductor Linear Integrated Circuits Applications Handbook," Fairchild Semiconductor, 1967.
2. Millman, J., and C. C. Halkias: "Integrated Electronics: Analog Digital Circuits and Systems," McGraw-Hill Book Company, New York, 1972.

Chapter **2**

# Fundamentals of Circuit Design Using Op Amps

## 2.1 BASIC RULES WHICH SIMPLIFY DESIGN

By using the ideal properties of op amps listed in Sec. 1.1, we arrive at two basic rules which greatly simplify op amp circuit design. These rules are:
 1. *Op amp input terminals draw no current.*
 2. *Voltage across input terminals is zero.*

These rules are adequate for most design work. They are also adequate to use for an initial design in those cases where op amp parameter drifts must later be considered. Circuits of any complexity can be handled with these rules. Circuits which require a more careful design afterward are usually the following types:

 1. Low-level dc and high-precision dc circuits must consider input offset voltage, input bias current, input offset current, equivalent input noise, and input resistance.

 2. Low-level ac, high-precision ac, and wide-bandwidth ac circuits must consider equivalent input noise, finite bandwidth, finite slew rate, and input capacitance.

 3. All precision circuits must consider output resistance, common-mode rejection ratio, and power-supply rejection ratio.

These additional considerations will be discussed at length in the next section. In the meantime, several examples using the two basic rules are in order.

**INVERTING AMPLIFIER** The most fundamental version of an inverting op amp circuit is shown in Fig. 2.1. Let us consider each of the two basic rules individually to see how easily they are applied.

BASIC RULE 1: No current goes into positive or negative input terminals: This means that the current passing through $R_1$ must be identical with the current through $R_f$. Two equations can therefore be developed by just using Ohm's law:

$$v_i - v_x = iR_1 \tag{2.1}$$

$$v_x - v_o = iR_f \tag{2.2}$$

## 2-2  FUNDAMENTALS OF CIRCUIT DESIGN USING OP AMPS

BASIC RULE 2: Voltage across input terminals is zero: Since the positive input terminal is grounded (at zero voltage), the negative input terminal must be at zero voltage also. Thus $v_x = 0$ and Eqs. (2.1) and (2.2) become

$$v_i = iR_1$$
$$-v_o = iR_f$$

or, by manipulating the algebra on these two equations, we get

$$i = \frac{v_i}{R_1}$$

$$i = -\frac{v_o}{R_f}$$

Setting these two equations equal to each other, the final results are

$$i = \frac{v_i}{R_1} = -\frac{v_o}{R_f}$$

or

$$\frac{v_o}{v_i} = -\frac{R_f}{R_1} \tag{2.3}$$

Equation (2.3) is the fundamental gain equation for inverting amplifiers. It should be committed to memory.

**NONINVERTING AMPLIFIER**  The fundamental noninverting-amplifier circuit is shown in Fig. 2.2. We can use the same approach used in the first example.

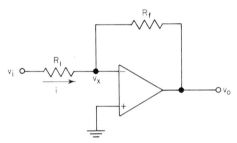

Fig. 2.1  Fundamental inverting-amplifier circuit.

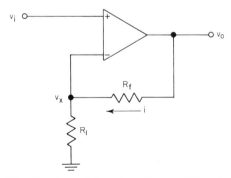

Fig. 2.2  Fundamental noninverting-amplifier circuit.

BASIC RULE 1: No current goes into positive or negative input terminals: the same current $I$ therefore flows in $R_1$ and $R_f$. We can develop two equations just by using Ohm's law:

$$v_o - v_x = iR_f \tag{2.4}$$

$$v_x - 0 = iR_1 \tag{2.5}$$

BASIC RULE 2: Voltage across input terminals is zero: This means that $v_i = v_x$. Incorporating this into Eqs. (2.4) and (2.5), we get

$$v_o - v_i = iR_f$$

$$v_i - 0 = iR_1$$

Solving for $i$ in each case,

$$i = \frac{v_o - v_i}{R_f} = \frac{v_o}{R_f} - \frac{v_i}{R_f}$$

$$i = \frac{v_i}{R_1}$$

Setting these equal to each other,

$$i = \frac{v_o}{R_f} - \frac{v_i}{R_f} = \frac{v_i}{R_1}$$

Solving for $v_o/v_i$, we get the final result

$$\frac{v_o}{v_i} = 1 + \frac{R_f}{R_1} \tag{2.6}$$

This is the fundamental equation of noninverting amplifiers. It should also be committed to memory.

In the discussions to follow we will be using both upper- and lower-case voltages and currents, i.e., $V_i$, $v_i$, $I_b$, $i_o$, etc. The upper-case notation represents either complex ac signals or dc signals. The lower-case notation is reserved for instantaneous incremental signals, which may include nonlinear, ac, and/or dc signals.

## 2.2 HOW TO MINIMIZE OP AMP ERRORS

We now consider the effects of each op amp parameter on Eqs. (2.3) and (2.6). A method (or methods) to minimize these adverse effects will be outlined in each case. Methods for measuring each parameter are described in reference 2 of Chap. 3.

**INPUT OFFSET VOLTAGE** $V_{io}$  All op amps have a slight mismatch of the emitter-base forward bias voltages of the two input transistors. This results in a voltage offset at the op amp output. The input offset voltage $V_{io}$ (between the bases of the two input transistors) is related to the output offset $V_o$ by

$$V_o = \pm V_{io}\left(1 + \frac{R_f}{R_1}\right)$$

This equation is true for both the inverting and noninverting configurations. If, for example, the circuit has a voltage gain of $-1{,}000$ (an inverting amplifier), we divide the output offset voltage by 1,001 to obtain the input offset voltage. Op amp data sheets always specify the offset voltage at the input

## 2-4 FUNDAMENTALS OF CIRCUIT DESIGN USING OP AMPS

terminals, since the magnitude of output offset depends on the circuit gain. The input offset is independent of the circuit. By definition, the input offset voltage is that voltage required across the input terminals which nulls the output. The input offset voltage $V_{io}$ is typically in the range of a fraction of a millivolt to several millivolts. In high-gain circuits the output offset voltage may therefore be several volts. This offset will vary with temperature and could cause problems in a dc-coupled system. It is often neglected in ac circuits unless the sum of output offset voltage and peak ac voltage is a voltage approaching either power-supply voltage. Clipping of the output signal would then begin to occur.

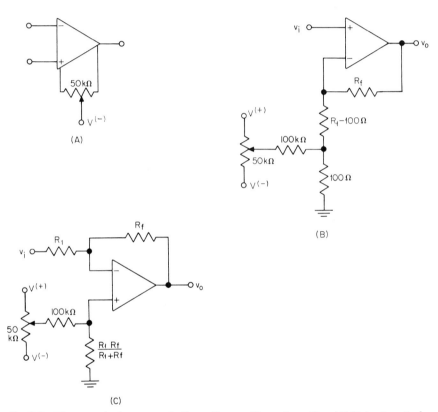

**Fig. 2.3** Three methods to cancel effect of input offset voltage $V_{io}$. (A) Using terminals provided on op amp. (B) Typical method for noninverting amplifiers. (C) Typical method for inverting amplifiers.

Many op amps have special offset-adjustment terminals such as that shown in Fig. 2.3A. A potentiometer is placed between these two terminals with the wiper connected to the minus power-supply terminal. Adjustments of ±15 mV equivalent input offset voltage are possible with this method. This adjustment merely places the output offset at the desired value. The temperature effect on offset voltage, such as that shown in Fig. 2.4, is still present. The temperature coefficient of input offset voltage is typically 5 to 10 $\mu$V/°C for bipolar monolithic op amps. For chopper-stabilized op amps this coefficient may be only 0.1 to 1.0 $\mu$V/°C.

If the op amp does not provide offset terminals, the circuits of Fig. 2.3B and C are recommended. Other schemes are possible, such as temperature-sensitive circuits which cancel out most of the offset-voltage drift.

**INPUT BIAS CURRENT** $I_b$  The input transistors in the first-differential-amplifier stage of the op amp must be forward-biased. This requires a small current into each of their bases. The input bias current $I_b$ is defined as one-half the sum of these two currents, in other words, the average of the two currents. This definition applies only if the output terminal is balanced or nulled to zero volts.

The typical input bias current for bipolar monolithic op amps ranges from 10 to several thousand nanoamperes. High-quality chopper-stabilized or parametric op amps may have input bias currents under 10 pA.

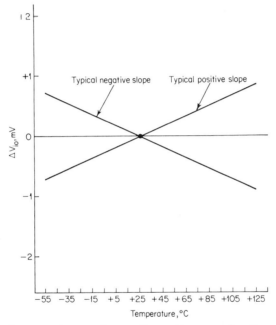

**Fig. 2.4** Typical curves showing input offset voltage as a function of temperature. Note: Either positive or negative slope is possible. The op amp is assumed to be nulled at +25°C.

As with input offset voltage, the input bias current is a dc parameter and may not affect the design of an ac amplifier. If the circuit must amplify ac and dc, the bias current must be considered. If the amplifier is for ac only, one must determine whether the resulting output offset plus the peak ac signal gets near the saturation region of the op amp. If this happens, the peaks of the ac signal will be clipped.

By reference to Fig. 2.5A we will show how the input bias current is a potential error source. All currents and voltages are assumed to be dc to simplify the discussion. Suppose we require an inverting amplifier with a gain of $-1,000$ and an input resistance of $1,000$ Ω. The input resistor $R_1$ must be $1,000$ Ω to satisfy the input-resistance requirement. $R_f$ must be $1,000$ times greater to satisfy the gain requirement:

## 2-6 FUNDAMENTALS OF CIRCUIT DESIGN USING OP AMPS

$$A_{vco} = -\frac{R_f}{R_1} = -1{,}000$$

Thus $R_f$ must equal 1 MΩ. Assume the input bias current is 100 nA. This current will pass through $R_1$ and $R_f$. We then have $I_b = I_1 + I_f$:

$$I_b = \frac{V_i - V_b}{R_1} + \frac{V_o - V_b}{R_f}$$

We assume $V_i = 0$ so that the bias current can be isolated from the signal. The term $V_b$ will be less than $V_o$ by a factor $A_v$ (10,000 or more). We can therefore drop $V_b$ and solve for $V_o$:

$$V_o = I_b R_f$$
$$= 10^{-7} \times 10^6 = 0.1 \text{ V}$$

The error caused by input bias current can be almost totally canceled if an extra resistor, having a resistance $R_p$, is added to the circuit as shown in Fig. 2.5B. This resistor is often made adjustable with a maximum value two or three times the computed $R_p$. The $I_b$ into the op amp positive input will develop a voltage across $R_p$ which will cancel the effects of $I_b$ into the negative input. Perfect cancellation does not take place, however, because the two

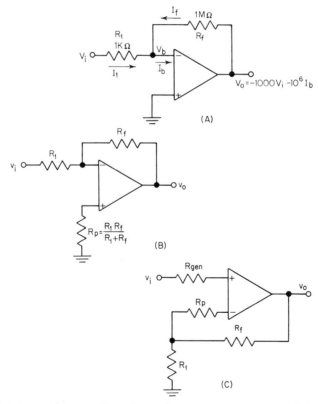

**Fig. 2.5** (A) Circuit showing effect of input bias current (inverting amplifier). (B) The most popular method for reducing the error caused by input bias current. (C) Solution to problem for noninverting circuit.

bias currents are not exactly equal. We will look into this problem in the next section.

The output error caused by $I_b$ is often larger than that caused by the input offset voltage. The size of the error caused by bias current depends upon the size of $R_p$. If $R_p$ is larger than $V_{io}/I_b$, the output-voltage error caused by $I_b$ will be larger than that caused by $V_{io}$. If, in the above example, we had a $V_{io}$ of 5 mV,

$$\frac{V_{io}}{I_b} = \frac{5 \times 10^{-3}}{10^{-7}} = 50 \text{ k}\Omega$$

Any $R_p$ in this op amp circuit greater than 50 k$\Omega$ will create an error due to input bias current which is larger than the error caused by input offset voltage.

The solution to input-bias-current error described above and shown in Fig. 2.5B is for an inverting-amplifier circuit. If a noninverting amplifier needs error correction for the same reason, the circuit of Fig. 2.5C is recommended. As before, the source resistance seen by both op amp input terminals must be identical to reduce the error caused by input bias current.

As should be expected, input bias current varies with temperature. Figure 2.6A shows a typical curve of this temperature dependence. If we always make sure that the resistances seen by both inputs are identical, bias-current changes with temperature should not cause problems.

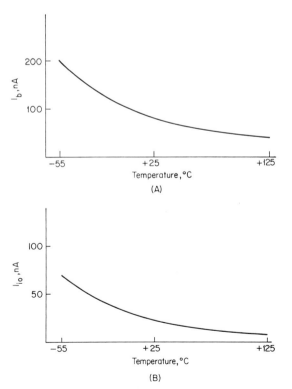

**Fig. 2.6** Typical variation of op amp input currents with temperature. (A) Input bias current $I_b$. (B) Input offset current $I_{ib}$.

## 2-8 FUNDAMENTALS OF CIRCUIT DESIGN USING OP AMPS

**INPUT OFFSET CURRENT** $I_{io}$  The two input transistors of any op amp will always require slightly different bias currents. The difference between these two currents is defined as the input offset current $I_{io}$. As might be expected, $I_{io}$ also varies with temperature. Figure 2.6B shows this variation with temperature for a typical op amp. Since the offset current may flow into either op amp input terminal, the resulting output-voltage error is

$$V_o = \pm I_{io} R_f$$

The bias-current compensation scheme presented in the last section (i.e., using $R_p$) does nothing to cancel the effects of $I_{io}$. We can compensate for any given $I_{io}$ by choosing an $R_p$ slightly larger or smaller than $R_1 R_f/(R_1 + R_f)$. However, at another temperature $I_{io}$ is different, and as a result the op amp output voltage will shift. Several circuits which cancel the effects of $I_{io}$ changes with temperature are shown in Fig. 2.7 (Ref. 2).

Figure 2.7A is a simple compensation scheme which is recommended for inverting-amplifier circuits. The current from the base of the PNP transistor is injected into the NPN input transistor inside the LM101. The 2N2605 is a

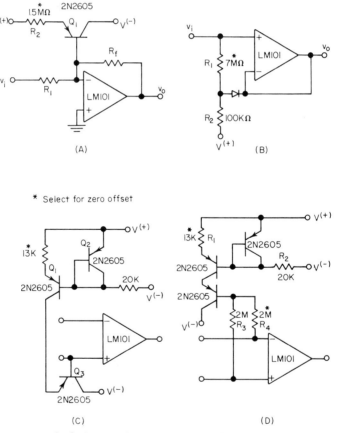

**Fig. 2.7** Various methods for canceling the effects of changes in $I_{io}$ with temperature. (A) Basic inverting amplifier. (B) Voltage follower. (C) For use if large common-mode range required. (D) Differential input compensation. (*National Semiconductor Corp.*)

silicon planar transistor which has nearly the same base-current characteristics over temperature as the NPN input transistors. The small difference between the devices is corrected by selecting $R_2$ for zero op amp output offset voltage. This circuit does not require the compensation resistor $R_p$.

For voltage-follower applications the circuit of Fig. 2.7B is often used. The compensation current comes through $R_1$, which is adjustable for each particular op amp. The diode regulates the compensation current so that compensation does not change with signal level. The input impedance of the voltage follower is not reduced with this circuit. For the LM101 shown, the typical input impedance is 1,000 MΩ.

Figure 2.7C provides even better compensation over temperature than Fig. 2.7B. It is also useful over a larger common-mode range. The emitter of $Q_3$ is fed from a current source ($Q_1$ and $Q_2$). This prevents the input-voltage level from changing the compensation current from the base of $Q_3$. Variations of input bias and offset currents with power-supply voltage are also reduced with this circuit.

The final circuit, shown in Fig. 2.7D, provides all the good features of Fig. 2.7C, except that compensation on both input terminals is now provided. $R_3$ and $R_4$ are selected for the different bias-current requirements on these two inputs. The circuit can be further optimized over temperature if the source resistances seen by the inputs are made equal (see Fig. 2.5). However, in applications where this is not possible, $R_3$ and $R_4$ can be selected for minimum output drift.

If one does not want to design external compensation circuits for the above op amp deficiencies, a higher-quality op amp may be used. Of course, one must be willing to pay the price for op amps which are optimized for low input voltage offset, low input bias current, and low input offset current. Three types of op amps which have low input offsets and temperature coefficients are FET input, varactor input, and chopper-stabilized op amps. These are available as discrete packaged op amps. The FET and chopper-stabilized types are also available in monolithic form in the TO-99 package.

**EQUIVALENT INPUT NOISE** $V_n$ and $I_n$  This parameter affects both the ac and dc characteristics of op amp circuits. Op amp specification sheets tabulate data on equivalent input noise from 0.01 Hz to over 1 MHz. Op amp noise is specified by use of an equivalent input-noise voltage and an equivalent input-noise current. The actual noise in the op amp is created in a number of places in the first few stages in the op amp. To simplify noise calculations, all these noise sources are assumed to be lumped into a single equivalent input-noise current and a single equivalent input-noise voltage. Figure 2.8 shows the circuit placement of these noise sources in front of a noiseless op amp.

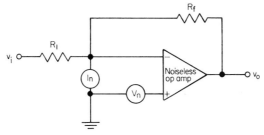

**Fig. 2.8**  Placement of the equivalent input-noise voltage and equivalent input-noise current generators in front of a noiseless op amp.

At least four different types of units are used to specify input noise. These units are:

Voltage:
  Volts/$\sqrt{\text{Hz}}$ rms
  Volts$^2$/Hz rms
  Volts peak-to-peak
  Noise figure in dB

Current:
  Amperes/$\sqrt{\text{Hz}}$ rms
  Amperes/Hz rms
  Amperes peak-to-peak
  Noise figure in dB

In this text we will use volts/$\sqrt{\text{Hz}}$ and amperes/$\sqrt{\text{Hz}}$, which are rms numbers quite easily correlated with simple measurements. Before we get into methods to optimize op amp circuits for low noise, we will work out methods to make conversions from the other three types of noise specification to the one we will use in this text. The volts/$\sqrt{\text{Hz}}$ and amperes/$\sqrt{\text{Hz}}$ terminology is used by a large number of op amp and transistor manufacturers. These units allow the easiest computation of circuit parameters which will minimize noise. When we say volts/$\sqrt{\text{Hz}}$ we mean the rms voltage over a 1-Hz bandwidth. This number must also be specified at some center frequency. Some texts call this the spot noise. It is not usually measured with instruments having a 1-Hz bandwidth, as this is usually too difficult. Rather, a more convenient bandwidth such as $\frac{1}{10}$ or $\frac{1}{100}$ the center operating frequency is used. The number obtained in this manner is then divided by the square root of the bandwidth. This converts the number to volts/$\sqrt{\text{Hz}}$.

As an example, suppose we measure the spot noise of some device at 10 kHz using an rms voltmeter having a bandwidth of 100 Hz. The equivalent input-voltage noise measures 300 nV. The spot noise is therefore

$$V_n = \frac{300 \text{ nV}}{\sqrt{100 \text{ Hz}}}$$

$$= \frac{300 \text{ nV}}{10 \sqrt{\text{Hz}}} = 30 \text{ nV}/\sqrt{\text{Hz}} \text{ at } 10 \text{ kHz}$$

The same type of calculation could be performed using data from a noise-current measurement.

Some data sheets provide noise data in units of volts$^2$/Hz or amperes$^2$/Hz. To obtain volts/$\sqrt{\text{Hz}}$ and amperes/$\sqrt{\text{Hz}}$ one needs merely to take the square root of these numbers. For example, the Fairchild $\mu$A741 has a noise voltage of $4 \times 10^{-16}$ volts$^2$/Hz at 10 kHz. The spot noise at 10 kHz is therefore $2 \times 10^{-8}$ volts/$\sqrt{\text{Hz}}$.

Very low frequency noise, such as that in the region from 0.01 to 1 Hz, is difficult to measure with an rms meter. One approach to this problem is to pass the signal through a low-pass filter having a 1-Hz upper cutoff frequency. The filter output is applied to an oscilloscope and the peak-to-peak excursions are estimated. The rms voltage is then 0.707 of one-half the measured peak-to-peak amplitude. The numbers obtained with these peak-to-peak measurements are only a rough estimate, since the amplitude of low-frequency noise in most devices increases as the frequency goes down. As a result, if a peak-to-peak estimate is once made, it will always be exceeded by a larger signal if one waits long enough. The correct peak-to-peak reading is that amplitude which is exceeded only 10 to 15 percent of the time (Ref. 3).

The fourth type of noise information often shown in data sheets is the "noise figure" expressed in dB. The noise figure is a measure of additional noise which is contributed by the amplifier above that noise already in the input signal. If the input signal is only that due to source resistor noise, its equivalent rms voltage is

$$V_R = \sqrt{4kTR_s} \qquad \text{V}/\sqrt{\text{Hz}} \qquad (2.7)$$

where $k$ = Boltzmann's constant = $1.374 \times 10^{-23}$ J/K
$T$ = temperature, K
$R_s$ = source resistance, $\Omega$

At room temperature the resistor noise is

$$V_R = 0.13 \sqrt{R_s} \qquad \text{nV}/\sqrt{\text{Hz}} \qquad (2.8)$$

If the amplifier contributes an equivalent input-voltage noise of $V_n$ and an equivalent input-current noise of $I_n$, the noise figure is defined as

$$\text{NF} = 10 \log_{10} \frac{V_n^2 + I_n^2 R_s^2 + 4kTR_s}{4kTR_s} \qquad \text{dB} \qquad (2.9)$$

To find $V_n$ and $I_n$ at any given frequency, we must use data-sheet curves, such as that shown in Fig. 2.9, which relate NF to $R_s$ and frequency. The equation must be solved for two NFs at the given frequency, since both $V_n$ and $I_n$ are unknowns.

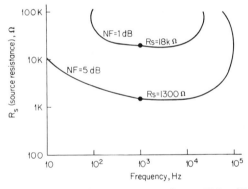

**Fig. 2.9** Narrow-band spot-noise-figure contours for $\mu$A725. (*Fairchild Semiconductor.*)

An example of the above procedure would be worthwhile at this point. Suppose we want to find $V_n$ and $I_n$ of the Fairchild $\mu$A725 at 1 kHz. Using Fig. 2.9, we note that NF = 1 dB if $f$ = 1 kHz and $R_s$ = 18 k$\Omega$. Likewise, NF = 5 dB if $f$ = 1 kHz and $R_s$ = 1,300 $\Omega$. These numbers are substituted into Eq. (2.9) for each $R_s$:

$$V_n^2 + I_n^2 (18{,}000)^2 = 4kT(18{,}000) \left[ \text{antilog}\left(\frac{1 \text{ dB}}{10}\right) - 1 \right] \qquad (2.10)$$

$$V_n^2 + I_n^2 (1{,}300)^2 = 4kT(1{,}300) \left[ \text{antilog}\left(\frac{5 \text{ dB}}{10}\right) - 1 \right] \qquad (2.11)$$

Subtracting the bottom equation from the top, we get

$$I_n[(18{,}000)^2 - (1{,}300)^2]$$

$$= 4kT\left\{ 18{,}000\left[ \text{antilog}\left(\frac{1}{10}\right) - 1 \right] - 1{,}300\left[ \text{antilog}\left(\frac{5}{10}\right) - 1 \right] \right\}$$

## 2-12 FUNDAMENTALS OF CIRCUIT DESIGN USING OP AMPS

Now we need only substitute in $k$ and $T$, then solve for $I_n$. (Note that the value of antilog $\frac{1}{10}$ is 1.259 and antilog $\frac{5}{10}$ is 3.163.) The final result becomes $I_n = 0.307$ pA/$\sqrt{\text{Hz}}$. This value of $I_n$ is substituted into either Eq. (2.10) or Eq. (2.11) to solve for $V_n$. After this calculation we get $V_n = 6.77$ nV/$\sqrt{\text{Hz}}$.

We will now show how to minimize the op amp output noise once we have curves of $V_n$ and $I_n$ vs. frequency. The output-voltage noise due to $V_n$ in the circuit of Fig. 2.8 is

$$V_{onv} = \frac{R_1 + R_f}{R_1} V_n \qquad (2.12)$$

or

$$V_{onv} = \frac{Z_1 + Z_f}{Z_1} V_n \qquad (2.13)$$

The relation $(Z_1 + Z_f)/Z_1$ is called the voltage-noise gain of the circuit. Note that the voltage-noise gain is larger than the gain for normal input voltages. For a noninverting amplifier the gains will be identical. The op amp output-voltage noise due to $I_n$ is

$$V_{oni} = R_f I_n \qquad (2.14)$$

or

$$V_{oni} = Z_f I_n \qquad (2.15)$$

The total output noise is the rms sum of $V_{onv}$ and $V_{oni}$.

$$V_{on} = \sqrt{V_{onv}^2 + V_{oni}^2}$$

Therefore the minimum value of $V_{on}$ is achieved when $V_{onv} = V_{oni}$. To satisfy this, we must set Eq. (2.12) equal to Eq. (2.14) [or Eq. (2.13) equal to Eq. (2.15)]. This results in the following:

$$\frac{R_1 + R_f}{R_1} V_n = R_f I_n$$

Rearranging this, we get

$$\frac{V_n}{I_n} = \frac{R_1 R_f}{R_1 + R_f} \qquad (2.16)$$

Minimum noise is therefore achieved when the parallel resistance of $R_1$ and $R_f$ is made equal to the ratio $V_n/I_n$. The latter ratio is appropriately called the noise resistance of the op amp.

**INPUT RESISTANCE $R_{id}$ and $R_{ic}$** These parameters have already been explored in detail in Sec. 1.6, where real and ideal op amps were compared. Equations were given which showed the effect of $R_{id}$ (differential input resistance) and $R_{ic}$ (common-mode input resistance) on the circuit gain $A_{vc}$. Equation (1.6) showed the effect on the inverting amplifier, and Eq. (1.7) was for the noninverting amplifier. The effect of $R_{ic}$ on circuit gain is so small that it is usually disregarded. The effect of $R_{id}$ is also extremely small and is neglected unless it is in the order of 10 k$\Omega$ or less. Since the cheapest monolithic op amps have an $R_{id}$ much greater than 10 k$\Omega$, the effect of input resistance can be neglected in all but very specialized applications.

**INPUT CAPACITANCE $C_{id}$ and $C_{ic}$** These two parameters are seldom stated on op amp data sheets. Their effect on the closed-loop gain of the inverting-amplifier configuration is negligible. Both $C_{id}$ and $C_{ic}$ have typical values of 1 to 2 pF and maximum values of 3 pF for monolithic op amps. The common-mode input capacitance $C_{ic}$, however, does have some deleterious effect on the noninverting amplifier at high frequencies. Since this type of amplifier is often driven by a high-impedance source, a capacitance to ground at the op amp positive input will attenuate high frequencies. The only way around this problem is to use careful layout procedures and to choose an op amp with low $C_{ic}$. (See Appendix III for fabrication procedures which minimize stray capacitances.) As we will find in Chap. 3, the input capacitances play a major role in the op amp loop stability. We will defer further comment on this subject until Chap. 3.

**OUTPUT RESISTANCE $R_o$** This parameter was discussed in Sec. 1.6. Equation (1.8) showed that the relationship between op amp output resistance $R_o$ and closed-loop output resistance $R_{out}$ is $R_{out} = R_o/\beta A_v$. At frequencies much lower than the loop-gain unity crossover the parameter $R_{out}$ is very small.

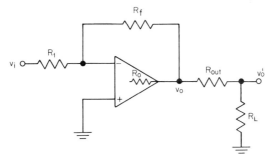

**Fig. 2.10** The closed-loop circuit output resistance $R_{out}$ is placed as shown to determine its effect on circuit performance.

However, as shown with Eq. (1.8), in the region of unity loop-gain crossover the op amp output resistance goes up and $\beta A_v$ drops. This tends to cause both stability and gain problems.

Solution to the feedback-stability problems caused by $R_o$ will be covered in Chap. 3. We will discuss the gain problem only briefly. As shown in Fig. 2.10, we can represent $R_{out}$ as a resistance in series with the load $R_L$. The actual circuit output voltage $v_o'$ will be slightly lower than $v_o$ because of the voltage divider formed by $R_{out}$ and $R_L$. Thus,

$$v_o' = \frac{R_L}{R_L + R_{out}} v_o \qquad (2.17)$$

At low frequencies, where $\beta A_v$ is large and $R_o$ is small, $R_{out}$ will be small. As an example, consider the 748 op amp. At frequencies up to 50 kHz, $R_o$ is approximately 70 Ω. At 50 kHz the open-loop gain is 200. If $\beta = \frac{1}{4}$, $\beta A_v = 50$ at 50 kHz. The output resistance of the circuit at this frequency is therefore [using Eq. (1.8)]

$$R_{out} = \frac{70}{50} = 1.4 \ \Omega$$

If $R_L = 1$ kΩ and $v_o = 10$ V, the output-voltage reduction due to $R_{out}$ is [using Eq. (2.17)]

$$v_o - v_o' = \left[1 - \frac{R_L}{R_L + R_{\text{out}}}\right] v_o$$

$$= \left[1 - \frac{1{,}000}{1{,}000 + 1.4}\right] 10 = 14 \text{ mV reduction}$$

At lower frequencies the voltage reduction will be even smaller. This error would probably go unnoticed. Now consider what happens at unity loop gain ($\beta A_v = 1$), where, at the same time $R_o$ has increased to 120 Ω (this happens at $f = 100$ kHz),

$$R_{\text{out}}(100 \text{ kHz}) = \frac{R_o}{\beta A_v} = \frac{120}{1} = 120 \text{ Ω}$$

The output-voltage reduction will be

$$v_o - v_o' = \left[1 - \frac{R_L}{R_L + R_{\text{out}}}\right] v_o$$

$$= \left[1 - \frac{1{,}000}{1{,}000 + 120}\right] 10 = 1.07 \text{ V}$$

The error in this case is more than 10 percent.

How do we reduce these errors caused by $R_o$? The most obvious suggestions are: (1) Do not require operation at frequencies where $\beta A_v = 10$ or less. (2) Keep $R_L$ large. (3) Place an emitter follower between the op amp and $R_L$. The last suggestion will make the load resistance $R_L$ seen by the op amp very large. Thus the voltage-divider action between $R_{\text{out}}$ and $R_L$ will be very small.

**OPEN-LOOP GAIN $A_v$ AND OPEN-LOOP DC GAIN $A_{vo}$** These parameters are also called the ac and dc differential gains, since they are the ratio of the op amp output voltage $v_o$ to the difference between the input terminals $v_p - v_n$. We discussed the degradation caused by a finite open-loop gain in Sec. 1.6. Equations (1.4) and (1.5) related closed-loop dc gain $A_{vco}$ to open-loop dc gain $A_{vo}$. We found that gain was reduced 1 percent or more (from the ideal gain) if $A_{vo}/A_{vco}$ was 100 or less. If we can guarantee $A_{vo}/A_{vco} > 100$ for all temperatures, then as $A_{vo}$ varies with temperature the circuit gain will be stable to within 1 percent. Likewise, if a 0.1 percent amplifier is required, we must guarantee that $A_{vo}/A_{vco} > 1{,}000$ under all conditions.

The degradation to $A_{vco}$ by a finite $A_{vo}$ follows the above rules only for frequencies up to the first pole of the op amp. Between the first-pole frequency and the second-pole frequency the true circuit gain is

$$A_{vc}(\text{inverting}) = \frac{-A_v R_f / R_1}{\sqrt{A_v^2 + (1 + R_f/R_1)^2}} \qquad (2.18)$$

and $$A_{vc}(\text{noninverting}) = \frac{+A_v R_f/(R_1 + R_f)}{\sqrt{A_v^2 + [1 + R_f/(R_1 + R_f)]^2}} \qquad (2.19)$$

These equations are the basis for Table 1.2 of percent gain error (ac only). Since $A_v$ is 90° out of phase with $A_{vc}$, the effect of $A_v$ on the accuracy and stability of $A_{vc}$ is much less than the effect of $A_{vo}$ on $A_{vco}$.

Suppose we wish to find the maximum frequency with which we can expect 1 percent accuracy for an X10 amplifier using the Fairchild μA741A. We first refer to the data sheet and determine the minimum open-loop gain and minimum unity-gain bandwidth at +25°C. As shown in Fig. 2.11, we use the above data to make a plot of $A_v$ as a function of frequency. The minimum value of $A_v$ at dc (i.e., $A_{vo}$) is given to be 50,000. We draw a horizontal line at that value. The unity-gain crossover frequency (called the bandwidth in this data sheet) is 0.44 MHz. We then draw a line having a slope of $-20$ dB/decade such that it passes through $A_v = 1$ at 0.44 MHz. This line is extended up and to the left until it intersects the $A_v = 50,000$ line. We define the resulting plot as the worst-case minimum curve of $A_v$ at $+25°C$.

Now, to determine the worst-case $A_v$ over temperature, we must assume two things: (1) The first pole at 8.5 Hz does not appreciably change frequency

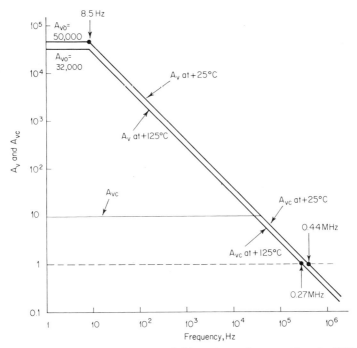

**Fig. 2.11** Curves of $A_v$ for the Fairchild $\mu$A741A taken at $+25$ and $+125°C$.

with temperature. (2) The slope of $A_v$ above the pole frequency remains at $-20$ dB/decade. We next examine the data sheet to find the minimum $A_{vo}$ over temperature. The test conditions for this $A_{vo}$ (and the $A_{vo}$ at $+25°C$) must not be too different from the planned use of the amplifier. Assuming the power supplies to be used are $\pm 20$ V, $R_L = 2$ k$\Omega$, and the output $= \pm 15$ V (peak-to-peak), the minimum $A_{vo}$ over temperature is 32,000 at $+125°C$. We draw a horizontal line at 32,000 on Fig. 2.11. At 8.5 Hz we change the slope to $-20$ dB/decade and note that the $A_v = 1$ intersection occurs at 0.27 MHz.

The closed-loop gain $A_{vc}$ curve is drawn as a straight horizontal line at $A_{vc} = 10$ until it intersects the two $A_v$ curves. We next determine the gain errors due to the lower $A_v$ curve. According to Table 1.1, the error will be less than $-0.1$ percent at dc, since $A_{vo}/A_{vco} = 3,200$ at dc. At 28 Hz the phase shift of $A_v$ will be almost $-90°$ relative to $A_{vc}$. At this frequency $A_v/A_{vc} = 1,000$; so the gain error according to Table 1.2 is $-5 \times 10^{-5}$ percent. Likewise, at 280 Hz the gain error is $-0.006$ percent, at 2,800 Hz the gain error is $-0.6$ percent, and at 28 kHz the gain error is $-33$ percent. We can reasonably assume this circuit will be better than 1 percent stable up to about 3,000 Hz and over the temperature range −55 to +125°C.

Our suggested methods to reduce the errors caused by changes in $A_v$ and $A_{vo}$ are:
1. Choose an op amp with a high dc gain and/or wide bandwidth.
2. Keep the temperature variations to a minimum.
3. Make sure the ratios $A_v/A_{vc}$ and $A_{vo}/A_{vco}$ are as high as possible over as many frequencies as possible.

**BANDWIDTH** $f_u, f_{cp}, f_f$   The bandwidth $f_u$ as defined in Chap. 1 is the frequency where the op amp gain is 1. It can also be determined by measuring the op amp risetime when the op amp is connected as a unity-gain noninverting amplifier. The bandwidth is then computed from

$$f_u = \frac{0.35}{t_r} \tag{2.20}$$

where $t_r$ is the 10 to 90 percent risetime.

We are usually more interested in the closed-loop bandwidth than in the open-loop bandwidth. Throughout the text we will use $f_{cp}$ as the frequency at which the closed-loop gain is down 3 dB. This is the dominant (first) pole frequency of the closed-loop circuit. Referring back to Fig. 2.11, the closed-loop bandwidth is the frequency where the open- and closed-loop curves intersect. Thus, in Fig. 2.9 the closed-loop bandwidth is 44 kHz at +25°C and 27 kHz at +125°C.

Some data sheets provide information on the change of closed-loop bandwidth with temperature. One must be careful, however, because these are usually typical data. The worst-case minimum closed-loop bandwidth, over temperature, may be much less than the data sheet implies. The method outlined above will probably give more accurate worst-case data.

When the circuit must supply large output-voltage swings, the maximum

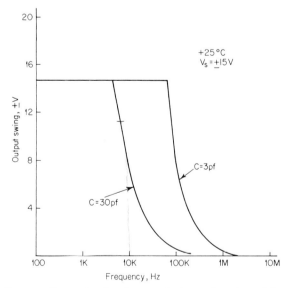

**Fig. 2.12** Full power (large-signal voltage swing) as a function of frequency for the LM101 op amp. (*National Semiconductors.*)

closed-loop bandwidth is much less than $f_u$. Depending on the size of the peak output voltage, the bandwidth may only be $1/10$ or $1/100$ of $f_u$. This high-level bandwidth is called $f_f$, since it is the maximum frequency at which full-power output response can be expected. Many data sheets provide curves showing the maximum peak-to-peak output voltage as a function of frequency. Figure 2.12 is a curve of this type for the LM101 op amp. The curve was obtained by noting the voltage at each frequency where ≤5 percent distortion occurs. If more than 5 percent distortion is acceptable, slightly higher voltages may be allowed.

The strong relationship between full-power response and the compensation capacitor is immediately obvious. Also, it should be noted that the curves flatten off rather abruptly at the top because the data were taken using ±15-V power supplies. Thus one would expect abrupt limiting as soon as the peak-to-peak amplitude approaches 30 V.

Bandwidth may be extended with the following methods:

1. Keep output amplitude low so that the full-power response curves are not approached.
2. Use the minimum compensation on the widest-bandwidth device.
3. Allow slightly more distortion in the output signal.

**SLEW RATE S**  As mentioned in Chap. 1, the maximum slew rate is the maximum rate at which an overdriven op amp can change its output voltage. This limiting action does not take place suddenly. It is observed to begin at one location on a sine wave as the frequency or amplitude is increased and then broaden out to include most of the sine wave. The exact nature of this complicated phenomenon depends on the type of op amp, the compensation used, and the load capacitance. We can best illustrate how slew rate depends on amplitude and frequency with a specific example. Figure 2.13 shows output waveforms from an LM101 with a sine-wave input. The circuit tested was the unity-gain noninverting type with a 30-pF compensation capacitor. The figure shows that it takes both high frequency and large voltage to cause maximum distortion. If the peak-to-peak voltage is small, the LM101 can provide distortionless gain up to 200 kHz. If the compensation capacitor is only 3 pF, distortionless gain is possible up to 1 MHz. Output waveforms up to 20 V or more are possible (for frequencies below 20 kHz).

Slew-rate limiting is characterized by a definite flattening on one portion of the sine wave. This flat portion is due to a constant-current source charging a capacitance. Precision ramp generators are built with the same principle. If the constant current is $I$, the slope of this slew-rate limit is

$$\frac{\Delta v}{\Delta t} = \frac{I}{C} \quad (2.21)$$

The capacitances and current generators causing the slew-rate limiting may be in several locations.

1. Often a compensation capacitor $C_c$ is placed between the collectors of the input differential stage. The slew-rate limit at this point is then

$$\frac{\Delta v}{\Delta t} = \frac{2I_c}{C_c}$$

where $I_c$ is the quiescent collector current of either input transistor.

2. If a large-load capacitor $C_L$ is connected to the op amp, the maximum slew rate is

$$\frac{\Delta v}{\Delta t} = \frac{I_o}{C_L}$$

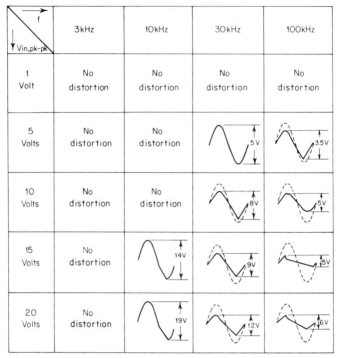

**Fig. 2.13** Output waveforms from an LM101 with a unity-gain noninverting configuration and 30-pF compensation. The circuit input was a sine wave in all cases. The dashed lines represent the ideal output waveform.

where $I_o$ is the maximum available op amp output current. The ultimate slew-rate limitation of the circuit will be the smallest $I/C$ ratio in the op amp.

If slew-rate limiting is a problem, the designer may consider the following suggestions:

1. Slew rate is higher for high-gain circuits. Perhaps the input signal can be reduced and the circuit gain increased.
2. The compensation-capacitor size may be too large. Some op amps have several methods for compensation from which to choose. Use of input lead-lag compensation mentioned in Chap. 3 is one possible suggestion for increasing slew rate.
3. If a large-load capacitance is the cause of slew-rate limiting, perhaps an emitter-follower buffer will help isolate the op amp from $C_L$.

Maximum slew rate $S_{max}$ is related to full-power response by

$$S_{max} = 2\pi f_f V_{pp} \tag{2.22}$$

where $f_f$ and $V_{pp}$ are the coordinates of a point on a curve such as Fig. 2.12. Suppose we wish to find the maximum slew rate of the LM101 if $V_{pp} = 10$ V and $C_f = 30$ pF. From Fig. 2.12 we note that $f_f = 8$ kHz at $V_{pp} = 10$ V and $C_f = 30$ pF. Thus,

$$S_{max} = 2\pi f_f V_{pp} = (6.28)(8 \times 10^3 \text{ Hz})(10 \text{ V})$$
$$= 5 \times 10^5 \text{ V/s} = 0.5 \text{ V}/\mu\text{s}$$

**COMMON-MODE REJECTION RATIO (CMRR)** Nearly all op amps have differential inputs. Many applications of op amps require both these differential inputs for proper operation. Some of these same applications require that any common voltage simultaneously applied to both inputs should not be amplified. This is not entirely possible in real op amps. This "common-mode" voltage always arrives at the output at some finite level. The common-mode rejection ratio (CMRR) is a measure of how much this common-mode signal is rejected relative to the desired differential-mode signal. CMRR is defined as

$$\text{CMRR} = \frac{A_{vo}}{A_{cmo}} \quad \text{at dc} \tag{2.23A}$$

or

$$\text{CMRR} = \frac{A_v}{A_{cm}} \quad \text{at any frequency} \tag{2.23B}$$

where  $A_{vo}$ = op amp differential gain at dc
$A_v$ = op amp differential gain as a function of frequency
$A_{cmo}$ = op amp common-mode gain at dc
$A_{cm}$ = op amp common-mode gain as a function of frequency

Figure 2.14 clarifies the definition of $A_v$ and $A_{cm}$ (or $A_{vo}$ and $A_{cmo}$, which are merely the dc components of $A_v$ and $A_{cm}$). Data sheets provide CMRR data in several forms. The most common form is merely one or two numbers stating the minimum and/or typical CMRR at dc. More useful data sheets provide curves showing minimum CMRR vs. frequency. A tabulated minimum CMRR at dc in conjunction with a typical CMRR vs. frequency could be used to make an approximate minimum CMRR vs. frequency plot.

Figure 2.15A shows a typical CMRR curve for the 101A op amp. The data sheet also states that the minimum CMRR at dc is 80 dB over the military temperature range (−55 to +125°C). The curve of CMRR vs. frequency (Fig. 2.15A) must accordingly be lowered by 23 dB at all frequencies if worst-case performance calculations are to be made. This is a rough approximation for worst-case minimum CMRR, but it will usually be satisfactory if no better data exist.

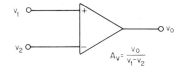

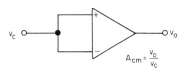

**Fig. 2.14** Definitions of op amp differential gain $A_v$ (usually called the open-loop gain) and common-mode gain $A_{cm}$.

Assume the designer knows the nature of his common-mode input voltage over frequency. A reasonable question to ask is: What will the output voltage be over frequency due to the common-mode input? To answer this question, we need plots of CMRR and $A_v$ which were both obtained under similar operating conditions. Figure 2.15A and B will be used. If both plots are in dB vs. frequency, the calculations will be simplified. Since CMRR = $A_v/A_{cm}$, $A_{cm} = A_v/$CMRR. Therefore

$$20 \log A_{cm} = 20 \log A_v - 20 \log \text{CMRR} \tag{2.24}$$

(See Appendix V for rules of logarithms.) The three terms are all expressed in dB. To obtain 20 log $A_{cm}$ (in dB) at a given frequency, we merely subtract 20 log CMRR (in dB) from 20 log $A_v$ (in dB) at that frequency.

Subtracting log terms is the same as dividing normal numbers (i.e., if we calculate $20 \log 1,000 - 20 \log 10$, we get $20 \times 3 - 20 \times 1 = 60 - 20 = 40 = 20 \log 100$. This is the same as saying $1,000/10 = 100$). The resulting plot of $20 \log A_{cm}$ is shown in Fig. 2.15C. To obtain the plot of output voltage vs. frequency due to a common-mode voltage, we must multiply $A_{cm}$ by the input common-mode voltage. Again, this is most easily done if both are in dB.

The circuit surrounding the op amp also has a CMRR of its own. In Chap. 9 we will show how to incorporate the op amp CMRR into the circuit CMRR to obtain the total CMRR. As it turns out, the total CMRR is always less than the circuit or op amp CMRR. We will take up ways to increase circuit CMRR in Chap. 9. The op amp CMRR can be optimized in the following ways:

1. Choose an op amp with a large minimum CMRR at dc.
2. Choose an op amp with the largest possible CMRR values over the same frequency range to be used in the circuit.

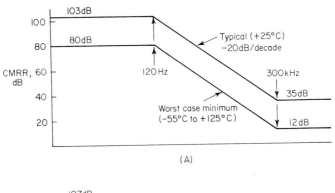

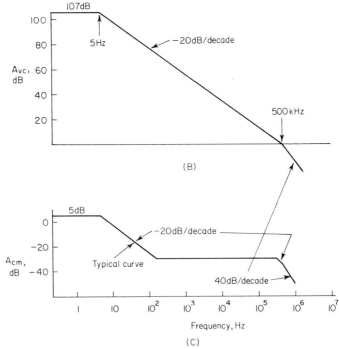

**Fig. 2.15** Curves of (A) CMRR, (B) $A_v$, and (C) $A_{cm}$ for the 101A op amp.

3. Measure each op amp for maximum CMRR before installation in the circuit (see Appendix IV).

4. Make sure circuit CMRR is at least ten times larger than op amp CMRR (see Chap. 9).

**POWER-SUPPLY REJECTION RATIO (PSRR)** This parameter is defined as the ratio of a change in input offset voltage to a change in power-supply voltage. The test is usually performed at some given frequency (60 Hz, 1 kHz, etc.); however, data sheets seldom state the frequency. The resulting number is expressed in $\mu$V/V or dB. When stated in dB, the number is actually negative, since the input offset change is much smaller than the power-supply change. A negative dB means the ratio is less than 1 (i.e., $-20$ dB $= 0.1$).

To determine the effect of PSRR on a given circuit, we can use the same type of calculation used to determine the effects of input offset voltage. As we recall, the op amp output voltage $V_o$ is related to the input offset voltage $V_{io}$ by the following:

$$V_o = \left(1 + \frac{R_f}{R_1}\right) V_{io} \qquad (2.25)$$

This expression holds for both inverting and noninverting amplifiers (see Sec. 2.2). From the definition of PSRR we have

$$\text{PSRR} = \frac{V_{io}}{V_s} \qquad (2.26)$$

where $V_{io}$ is an equivalent ac rms voltage across the op amp input terminals and $V_s$ is an ac rms voltage on both power-supply terminals (in phase). We can find the op amp output ripple voltage by combining Eqs. (2.25) and (2.26). The result is

$$V_o = \left(1 + \frac{R_f}{R_1}\right) V_s \times \text{PSRR} \qquad (2.27)$$

Suppose the power supplies have a 0.1-V rms ripple and the op amp has a PSRR of 20 $\mu$V/V ($-94$ dB). If the op amp is connected as an X1000 inverting amplifier, its output ripple will be

$$V_o(\text{rms}) = (1 + 1{,}000)(0.1 \text{ V rms})(20 \times 10^{-6} \text{V/V})$$
$$= 0.02 \text{ V rms}$$

Whether or not this disturbs the circuit function depends on the size of the real output signal and its required signal-to-noise ratio.

The effects of op amp CMRR can be minimized several ways:
1. Choose an op amp with a small CMRR.
2. Reduce the power-supply ripple with additional filtering.
3. Increase the input signal.
4. Decrease the circuit gain.

## 2.3 GENERAL METHOD TO COMPUTE $A_{vc}$

In Secs. 2.1 and 1.5 we state that the gain of a simple inverting amplifier is $-Z_f/Z_1$ (see Fig. 2.16A). Likewise, the gain of a simple noninverting amplifier is $1 + Z_f/Z_1$ (see Fig. 2.16B). These equations are true only if $Z_1$ and $Z_f$ are two-terminal devices (i.e., resistors, capacitors, inductors, diodes, etc.). If $Z_1$ and/or $Z_f$ are changed to three-terminal circuits with one terminal grounded as in Fig. 2.16C, the equations above do not apply. A more gen-

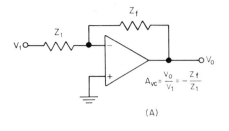

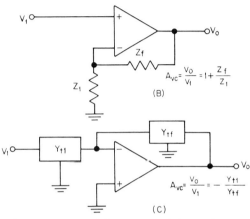

**Fig. 2.16** (A) Simple inverting-amplifier circuit. (B) Simple noninverting-amplifier circuit. (C) Complex inverting-amplifier circuit which requires use of Y parameters.

eral approach using Y parameters is required. We will now introduce Y parameters and then show how they are used in op amp circuits.

**Y PARAMETERS** We will start with the general three-terminal network shown in Fig. 2.17A. All three-terminal networks can be reduced to three blocks as shown, where each block can be an impedance Z or an admittance Y. The admittance has a real part G (the conductance) and an imaginary part B (the susceptance). Thus admittance is expressed as

$$Y = G + jB$$

just as impedance is expressed as

$$Z = R + jX$$

where R is resistance and X is reactance. We convert from one system to the other by

$$Z = \frac{1}{Y} \quad \text{or} \quad R + jX = \frac{1}{G + jB}$$

Individual components are expressed as follows:

Resistance: $\quad R = \dfrac{1}{G} \quad \text{or} \quad G = \dfrac{1}{R}$

Inductance: $\quad X = sL = \dfrac{1}{B} \quad \text{or} \quad B = \dfrac{1}{sL}$

Capacitance: $$X = \frac{1}{sC} = \frac{1}{B} \quad \text{or} \quad B = sC$$

By use of Fig. 2.17B and C we can obtain the four Y parameters. If the output is shorted to ground, we get

$$Y_{11} = \text{short-circuit input admittance} = \frac{I_1}{V_1}$$

$$Y_{21} = \text{short-circuit forward transfer admittance}$$
$$= \frac{I_2}{V_1}$$

If the input is then shorted to ground, the other two parameters are

$$Y_{12} = \text{short-circuit reverse transfer admittance}$$
$$= \frac{I_1}{V_2}$$

$$Y_{22} = \text{short-circuit output admittance}$$
$$= \frac{I_2}{V_2}$$

If the three-terminal circuit is made up of passive elements (resistors, capacitors, inductors, transformers, diodes, etc.), $Y_{12} = Y_{21}$. This greatly simplifies

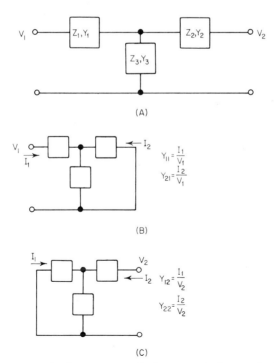

**Fig. 2.17** Determining the Y parameters of a three-terminal network. (A) The general three-terminal circuit put in simplified three-block form. (B) Computation of $Y_{11}$ and $Y_{21}$. (C) Computation of $Y_{12}$ and $Y_{22}$.

## 2-24 FUNDAMENTALS OF CIRCUIT DESIGN USING OP AMPS

the algebra in the following pages. We will simply call $Y_{12}$ and $Y_{21}$ the transfer admittance $Y_t$ in the following discussion.

It is a rather laborious task to compute or measure the short-circuit parameters of a network as a function of frequency. Therefore, we have provided the Y parameters for many common RC networks in Appendix VI. In the following pages we will show how to use the curves in Appendix VI to obtain the frequency response of almost any op amp circuit which uses RC networks.

**COMPUTING $A_{vc}$ WITH Y PARAMETERS** The most basic op amp circuit which requires the use of Y parameters is shown in Fig. 2.16C. As it turns out, we need to know only the transfer parameter $Y_t$ of the input and output networks. The closed-loop gain of the entire op amp circuit is

$$A_{vc} = \frac{V_2}{V_1} = -\frac{Y_{ti}}{Y_{tf}} \tag{2.28}$$

where $Y_{ti} = Y_{12} \, (= Y_{21})$ of the input network
$Y_{tf} = Y_{12} \, (= Y_{21})$ of the feedback network

The parameters $Y_{ti}$ and $Y_{tf}$ often contain many $sC$ terms which vary with frequency. To obtain the overall circuit response as a function of frequency, it is best to use dB plots. The circuit gain is then simply the $Y_{tf}$ plot (in dB) subtracted from the $Y_{ti}$ plot (in dB).

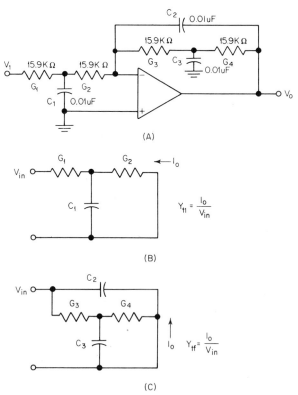

**Fig. 2.18** Calculation of an op amp circuit response using Y parameters. (A) The circuit. (B) Computing $Y_{ti}$ of input network. (C) Computing $Y_{tf}$ of feedback network.

GENERAL METHOD TO COMPUTE $A_{vc}$  2-25

An example would be worthwhile at this point. Given the circuit of Fig. 2.18A, what is the plot of voltage gain vs. frequency? The input network is composed of $G_1$, $G_2$, and $C_1$ and the feedback network is composed of $C_2$, $C_3$, $G_3$, and $G_4$.

The transfer admittance of the input network is found (see Fig. 2.18B) by grounding the right side of $G_2$. The current through $G_2$ is calculated for an input voltage on the left of $G_1$. The transfer admittance is

$$Y_{ti} = \frac{I_o}{V_{in}} = \frac{-G_1 G_2/C_1}{s + [(G_1 + G_2)/C_1]}$$

This transfer admittance is plotted in Fig. 2.19A. The transfer admittance for the feedback network is similarly found as shown in Fig. 2.18C. The result is

$$Y_{tf} = \frac{I_o}{V_{in}} = \frac{-C_2 \{s^2 + s\,[(G_3 + G_4)/C_3] + (G_3 G_4/C_2 C_3)\}}{s + [(G_3 + G_4)/C_3]}$$

This equation is plotted in Fig. 2.19B. If we (point-by-point) subtract $Y_{tf}$ (in dB) from $Y_{ti}$ (in dB), we get the circuit transfer function shown in Fig. 2.19C. The result is a two-pole low-pass filter with a corner frequency of 1 kHz.

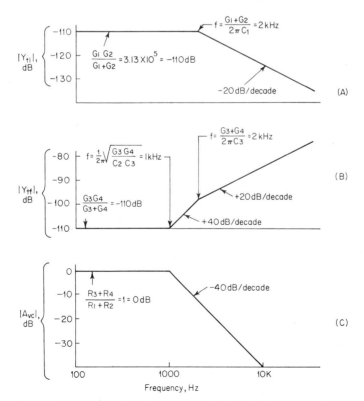

**Fig. 2.19** (A) The input transfer admittance. (B) The feedback transfer admittance. (C) The ratio $Y_{ti}/Y_{tf}$.

## REFERENCES

1. Giles, J. N.: "Fairchild Semiconductor Linear Integrated Circuits Applications Handbook," Fairchild Semiconductor, 1967.
2. Widlar, R. J.: "Drift Compensation Techniques for Integrated DC Amplifiers," National Semiconductor Corp., AN-3, 1967.
3. Smith, L., and D. H. Sheingold: Noise and Operational Amplifier Circuits, *Analog Dialogue*, Analog Devices, Inc., vol. 3, no. 1, March 1969.

Chapter **3**

# Feedback Stability

## 3.1 REVIEW OF FEEDBACK THEORY

Feedback of the circuit output to the circuit input is required for most op amp applications. The feedback network totally dominates the circuit characteristics. The op amp merely provides gain—usually orders of magnitude more than required. The feedback network determines what is done with all this gain. This feedback may be negative, which is usually done with the op amp negative (−) input. The feedback may also be positive by utilizing the op amp positive (+) input. Or a combination of positive and negative feedback is often used. Other variations are possible: If the feedback network has a phase shift, feedback to the negative input may become positive. Likewise, feedback to the positive input can become negative feedback under similar conditions. In this chapter we will discuss negative feedback and how it can cause circuit instability whenever it becomes positive feedback. Several things can cause this unwanted positive feedback. We call these various reasons for positive feedback "the seven causes of op amp instability." We will explore them in detail in this chapter. These seven causes may not be obvious causes of circuit instability to the reader at this point. By the end of this chapter, we hope the reader will be able to use these seven ideas to handle an op amp stability analysis of any complexity. The seven causes of instability are:

    1. Compensation recommended by op amp data sheet not used.
    2. Closed-loop gain too low for type (and amount) of compensation used.
    3. Excessive capacitive load on op amp.
    4. Incorrect phase lead/lag in feedback network.
    5. Excessive resistance between ground and op amp positive input.
    6. Excessive stray capacitance between op amp output and balance terminals.
    7. Inadequate power-supply bypassing.

The above seven causes of op amp instability fall in two categories: (1) design errors in the load and feedback circuit and (2) unexpected circuit elements. The first category is avoided by careful consideration of the first five causes of instability. The second category is handled by assuming worst-case circuit elements (such as stray capacitance at op amp input) and by using proved construction techniques. Both categories will be adequately covered in this chapter.

## 3-2 FEEDBACK STABILITY

We will first lay a firm foundation of feedback theory. However, we will discuss only that portion of the theory which relates to op amps. This approach greatly simplifies the analysis of op amp stability. Only voltage-mode transfer functions will therefore be required in this chapter. That is, the op amp transfer function will always be expressed as volts out/volts in. Likewise, the transfer function of the feedback network will be expressed as a voltage ratio. This ratio can be easily computed for feedback networks of almost any complexity, as we will show later in the chapter. As we showed in Sec. 2.3, a computation of the forward gain of complex op amp circuits requires the use of $Y$ parameters. This is not necessary for feedback analysis. We will do all our work with voltage ratios.

The most important parameter in an op amp stability analysis is *loop gain*. Loop gain is merely the product of the op amp gain and the feedback-network gain. If the true loop gain of a feedback circuit is known, the margin of stability of that circuit is easily determined. This is done by making plots of loop gain and its phase as a function of frequency. It takes only a few seconds to determine the circuit stability with these plots. If the plots show a marginal stability, the loop-gain equation will tell the designer which components are responsible. We will spend the rest of Sec. 3.1 showing how to develop the loop-gain equation, how to plot it, and how to determine stability margins from the plots. Section 3.2 will then describe in detail the various methods available to modify the loop gain to achieve stability.

Section 3.3 will analyze each of the seven causes of instability. Methods of overcoming each cause will be discussed in detail. Loop gain will again be the primary tool when discussing the seven causes. Then in Sec. 3.4 we will use the ideas developed in this chapter to perform a complete stability analysis of several typical circuits.

**RESULTS OF POSITIVE AND NEGATIVE FEEDBACK** It will be a good idea at this point to compare briefly positive with negative feedback. Sometimes the differences between them are so subtle that it is difficult at first glance to determine which type of feedback is used in a circuit. Most circuits are designed to use one type of feedback. A few are designed to use both types of feedback. In reality, all circuits have both types of feedback present at one frequency or another.

Positive feedback is used for circuits of the following types:
1. Generator (sine, square, pulse, sawtooth, etc.)
2. Bistable (flip-flop)
3. Comparator

Note that in the applications listed above, either no input is required to generate an output, or the output waveshape bears no resemblance to the input waveshape. This is the major characteristic of positive feedback. Electronically, the following happens in positive-feedback circuits: Some type of waveform on the input (noise, pulse, etc.) starts to be amplified by the op amp. The amplified waveform at the output of the op amp then passes back through the feedback network to the point where the input waveform is occurring. Since the feedback signal is the same polarity as the original signal, they add to each other and create an even larger waveform. If the positive feedback is dc-coupled to the input, the op amp is driven into saturation and may remain locked up. This locked-up condition can be changed only by removing power or driving the input with a large signal of the opposite polarity. If the positive feedback is not dc-coupled to the input, the circuit will oscillate at the frequency where the op amp gain times the feedback circuit gain equals 1.

Negative feedback is useful for the following reasons:
1. To widen amplifier bandwidth
2. To reduce amplifier distortion
3. To minimize phase shift and flatten frequency response
4. To minimize temperature-induced gain variations
5. To allow parts interchangeability without affecting circuit performance
6. To reduce or increase output resistance, depending on circuit configuration
7. To reduce or increase input resistance, depending on circuit configuration

In reviewing the above two lists, we note that (1) the main function of positive feedback is waveform generation, and (2) the main function of negative feedback is to allow accurate control of existing waveforms. Since these functions are so different, we cannot say one type of feedback has any advantage or disadvantage over the other. However, considered by themselves, each type of feedback has several disadvantages of which one must be aware.

Negative feedback has two problems:
1. All the advantages of negative feedback are obtained at the expense of gain. If a 1,000:1 improvement of some parameter (such as bandwidth or distortion) is required, the gain of the open-loop circuit must be reduced by 1,000 to obtain it. For this very reason, op amps typically have open-loop gains of 10,000 to over 1 million. We will discuss this limitation many times throughout this book.

2. Negative feedback sometimes becomes positive feedback at certain frequencies—thus causing the circuit to oscillate. This problem is the main topic of the present chapter. We will explore this problem from every angle. By the end of this chapter, the reader should know how to prevent the occurrence of positive feedback when negative feedback is desired.

Positive feedback also has the following potential drawback:
If positive feedback is used without negative feedback, the desired waveform generation may not be stable with temperature, load, aging, etc. The function of the circuit must not be dependent on the magnitude of op amp gain. This requirement will be discussed further in the chapters on oscillators and waveform generators.

**FIRST-CUT STABILITY ANALYSIS** If the designer does not have time for a complete stability analysis using the procedure developed in this chapter, the following general rule of thumb may be useful in simple designs: (1) As shown in Fig. 3.1, make a plot of op amp open-loop gain. (2) On the same sheet of paper, plot the circuit closed-loop gain. (3) Determine the rate of closure of the two plots at their point of intersection. (4) If this rate of closure is less than 40 dB/decade (12 dB/octave), the circuit will *probably* be stable.

This rule of thumb assumes several things: (1) The capacitive load $C_L$ on the op amp output terminal is small ($< 100$ pF in most cases). (2) Stray capacitance $C_S$ between the op amp input terminals is very small ($< 5$ pF). (3) The feedback resistor $R_f$ is small ($< 20$ k$\Omega$) if the stray input capacitance is 5 pF or more. Also, the rule of thumb is most accurate for high values of closed-loop gains ($A_{vc} > 10$). If the closed-loop gain approaches 1, it is best to keep stray and load capacitance small and to keep the rate of closure less than 40 dB/decade.

**DEVELOPMENT OF THE LOOP-GAIN EQUATION** An unstable feedback circuit is one which has 360° of phase lag around the loop. The first 180° is already accounted for by using the negative input terminal of the op amp. Feedback analysis is primarily concerned with causes of the second 180°. All plots, formulas, and discussions regarding feedback stability are centered on this second 180° phase lag. Thus, when we say the loop phase shift is 90°, it is actually $180 + 90 = 270°$. The low-frequency phase lag of the op amp output relative to the noninverting input is zero degrees. Likewise, relative to the inverting input the phase lag is 180°. Op amp data sheets always plot phase lag relative to the noninverting input. However, when working with negative feedback, the inverting input is always used. One should always remember to add 180° mentally to the data-sheet phase lag to obtain the actual phase lag. Otherwise, a reader surveying literature on the subject may be led to believe that a 180° phase shift of loop gain will cause oscillation.

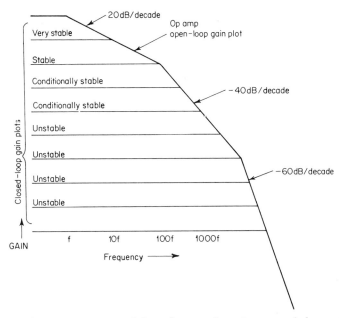

**Fig. 3.1** Finding approximate stability of circuit by using rate of closure between closed-loop gain and op amp open-loop gain.

In Chap. 1 we showed that the gain of an inverting amplifier is $-R_f/R_1$. We also showed that the gain of a noninverting amplifier is $1 + R_f/R_1$. These gain equations were derived using the ideal op amp properties:

1. Input resistance $= \infty$    $(R_{in} = \infty)$
2. Output resistance $= 0$    $(R_o = 0)$
3. Voltage gain $= \infty$    $(A_v = \infty)$
4. Bandwidth $= \infty$    $(f_{op1} = \infty)$
5. Offset $= 0$    $(V_{io} = I_b = I_{io} = 0)$

In Chap. 2 we computed the effect of nonideal op amp properties on amplifier performance. We will now determine the effect of nonideal properties on feedback stability. It will become apparent in the following discussion that as the ideal properties 1, 2, and 4 become more nonideal the circuit becomes more unstable.

REVIEW OF FEEDBACK THEORY   3-5

An op amp with infinite bandwidth will not oscillate when it is operated with purely resistive negative feedback. Infinite bandwidth means that the op amp has zero phase shift at all frequencies. A real op amp, however, has a −3-dB bandwidth at some finite frequency—usually in the 1-Hz to 1-MHz range. Real op amps thus have 90 or more degrees of phase lag at normal operating frequencies. Likewise, the feedback network and load circuit create additional phase lag. When the op amp phase lag is combined with these other phase lags, a potentially unstable circuit is possible. This instability will now be shown mathematically through use of the closed-loop gain equations derived in Chap. 2. These equations are repeated below:

$$A_{vc} = \frac{V_o}{V_i} = \frac{A_v}{1 + \beta A_v} \qquad (3.1)$$

for the noninverting amplifier, or

$$A_{vc} = \frac{V_o}{V_i} = \frac{A_v(\beta - 1)}{1 + \beta A_v} \qquad (3.2)$$

for the inverting amplifier, where
  $\beta$ = voltage transfer ratio of feedback network (function of frequency)
  $A_v$ = open-loop voltage gain of op amp (function of frequency). This is the gain relative to the positive input terminal.

These closed-loop gains become unstable ($A_{vc} = \infty$) whenever the denominator term equals zero. This occurs when $A_v = -1 = 1 \underline{/180°}$. Either $A_v$ or $\beta$ or both can cause this condition. The term $\beta A_v$ is called the loop gain, since it is the gain through the op amp back through the feedback network to the op amp input—thus completing a loop.

In both the inverting amplifier and the noninverting amplifier $\beta$ is defined as

$$\beta = \frac{R_1}{R_1 + R_f} \qquad (3.3)$$

Suppose $R_1$ and $R_f$ each become a two-terminal network of electronic components. $R_1$ then becomes $Z_{in}$ and $R_f$ becomes $Z_f$ as shown in Fig. 3.2A. $\beta$ is now defined as

$$\beta = \frac{Z_{in}}{Z_{in} + Z_f} \qquad (3.4)$$

Note that we show an impedance $Z_p$ from the op amp positive input to ground. This $Z_p$ is usually a resistor or a resistor bypassed with a capacitor. The function of this resistor was explained in Chap. 2. If $Z_p = Z_f/100$ or less at frequencies near unity loop gain, Eq. (3.4) will be unaffected. Otherwise $\beta$ must be computed with the general method to be outlined below.

Often $Z_{in}$ and/or $A_f$ are not two-terminal networks as shown in Fig. 3.2A but are three-terminal networks as shown in Fig. 3.2B. The simple formula for $\beta$ above cannot then be used. The procedure for obtaining $\beta$ in these complex cases will now be developed. The simple circuit shown in Fig. 3.2A is merely a special case of the method to be developed. This method of analysis will be most useful because it will allow one to incorporate load capacitance and input stray capacitance. Both these capacitances decrease amplifier stability.

In the following analysis we will compute the feedback voltage $V_a - V_b$ across the op amp input terminals relative to the op amp output voltage $V_o$.

## 3-6 FEEDBACK STABILITY

If $Z_{in}$ and $Z_f$ are each simple two-terminal networks as shown in Fig. 3.2A, and $Z_p = 0$, then

$$\beta = \frac{V_a - V_b}{V_o} = \frac{Z_{in}}{Z_{in} + Z_f} \quad (3.5)$$

This $\beta$ can be used to compute both closed-loop gain (see Chap. 2) and feedback stability. However, with a circuit like that in Fig. 3.2B we can still derive a $(V_a - V_b)/V_o$, but this ratio is not $\beta$. Instead, we will call this term the feedback factor $A_f$. We define

$$A_f = \frac{V_a - V_b}{V_o} \quad (3.6)$$

This term cannot be substituted into the closed-loop gain equation in place of $\beta$. The resulting calculated value for closed-loop gain $A_{vc}$ will sometimes be in error if this is tried. $A_f$ is used in place of $\beta$ only for determining the characteristics of the loop gain $A_f A_v$. When $A_f A_v = -1$, the circuit is unstable. As we will show in the following sections, the margin of stability can be determined by examining the gain and phase of $A_f A_v$.

The feedback factor $A_f$ is computed assuming no input to the circuit. The branch of the circuit containing the input-voltage source is accordingly replaced with the output impedance of the generator $Z_g$. Figure 3.2C shows how this is done. Note that the feedback loop has been spread out in a straight line for easier computation of $A_f$. The feedback factor is simply $V_5/V_2$.

Another simplification has been made to Fig. 3.2C to speed the calculation of $A_f$. The op amp input impedance $Z_{id}$ has changed places with $Z_p$. The

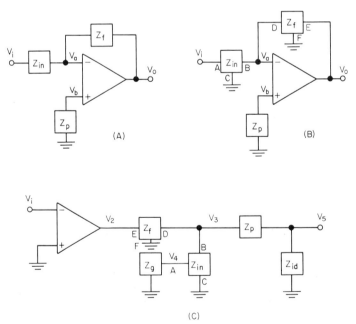

**Fig. 3.2** Rearrangement of op amp circuit to simplify calculation of $A_f$. (A) Simple circuit with two-terminal networks. (B) Complex circuit with three-terminal networks. (C) Loop opened up so that $A_f$ can be computed.

reason for this is as follows: Feedback-stability calculations require that we know the magnitude and phase of the op amp differential input voltage $V_a - V_b$. But $V_a - V_b$ is a fraction $Z_{id}/(Z_{id} + Z_p)$ of the voltage $V_a$. This fraction is unchanged if $Z_p$ and $Z_{id}$ change places. One may argue that the op amp is sensitive to $V_a$ and $V_b$ independently, thus invalidating the ideas above. Most op amps have a common-mode rejection ratio of 1,000 or more. Thus $V_a$ and/or $V_b$ have less than 0.001 the effect of $V_a - V_b$ on the op amp output. If $Z_{ib}$ and $Z_p$ exchange places, the effect on the op amp output will be almost negligible. This rearrangement makes it easier to visualize the computation of $V_5$ and $A_f$. $V_5$ is much easier to obtain if it is the result of the last voltage divider in a ladder network. If $Z_p$ and $Z_{id}$ had remained in their original locations, we would have had to find $V_a$ and $V_b$ separately, then determine $V_a - V_b$. We will now proceed to calculate $V_5$ to show how this simplified method operates.

Assume the circuit of Fig. 3.2C is composed of the components as shown in Fig. 3.3A. $Z_f$ is three resistors, $Z_{in}$ is three resistors, $Z_p$ is a single resistor, $R_8$ is the op amp input resistance, and $C$ is the op amp input capacitance. Networks of this type which are driven from one voltage source ($V_2$) are quite easy to simplify. The object is to simplify the feedback network until it looks like one of the classical $RC$ networks shown in Appendix VI.

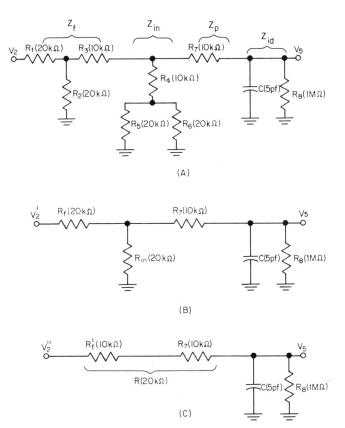

**Fig. 3.3** (A) Feedback network of Fig. 3.2C, where $Z_f$, $Z_{in}$, and $Z_p$ are assumed to be as shown. (B) Simplified network. (C) Ultimate simplification.

## 3-8  FEEDBACK STABILITY

The first simplification is to combine $R_1$, $R_2$, and $R_3$ into one equivalent resistor in series with a new voltage source $V_2'$. This is called a Thévenin equivalent circuit (see Ref. 1). The equivalent voltage replacing $V_2$ is

$$V_2' = \frac{R_2 V_2}{R_1 + R_2} = 0.5\, V_2 \tag{3.7}$$

which is merely the open-circuit output voltage of $V_2$, $R_1$, $R_2$, and $R_3$ (i.e., $R_3$ not connected to $R_4$ or $R_7$). The equivalent resistance in series with $V_2'$ is found by shorting $V_2$ to ground and finding the resistance from $D$, back through the $Z_f$ network to ground. This resistance is

$$R_f = R_3 + \frac{R_1 R_2}{R_1 + R_2}$$
$$= 10 + 10 = 20\text{ k}\Omega \tag{3.8}$$

The network $Z_{\text{in}}$ can be replaced by a single resistor to ground with a magnitude of

$$R_{\text{in}} = R_4 + \frac{R_5 R_6}{R_5 + R_6}$$
$$= 10 + 10 = 20\text{ k}\Omega \tag{3.9}$$

Figure 3.3B shows the resulting simplified network. One more step, however, is required to put the feedback network into a classical form. $R_{\text{in}}$, $R_f$, and $V_2'$ can be combined into another Thévenin equivalent circuit just as above. Accordingly,

$$V_2'' = \frac{V_2' R_{\text{in}}}{R_{\text{in}} + R_f} = \frac{20}{20 + 20} V_2' = 0.5\, V_2' \tag{3.10}$$

and

$$R_f' = \frac{R_{\text{in}} R_f}{R_{\text{in}} + R_f} = \frac{(20)(20)}{20 + 20} = 10\text{ k}\Omega \tag{3.11}$$

$V_2''$ is related to $V_2$ by substituting Eq. (3.7) into Eq. (3.10), resulting in

$$V_2'' = \frac{V_2 R_2 R_{\text{in}}}{(R_1 + R_2)(R_{\text{in}} + R_f)}$$
$$= \frac{(20)(20) V_2}{(40)(40)} = 0.25\, V_2 \tag{3.12}$$

$R_7$ and $R_f'$ are now added to make one equivalent feedback resistor $R$ (see Fig. 3.3C). The final simplified network is identical to one of the $RC$ networks shown in Appendix VI. The transfer function for this network is

$$\frac{V_5}{V_2''} = \frac{1/CR}{s + [(R + R_8)/CRR_8]} \tag{3.13}$$

Substituting Eq. (3.12) into Eq. (3.13) gives us

$$A_f = \frac{V_5}{V_2} = \frac{R_{\text{in}} R_2/CR(R_{\text{in}} + R_f)(R_1 + R_2)}{s + [(R + R_8)/CRR_8]}$$
$$= \frac{R_{\text{in}} R_2/2\pi CR(R_{\text{in}} + R_2)(R_1 + R_2)}{jf + [(R + R_8)/2\pi CRR_8]} \tag{3.14}$$

Substituting component values from previous calculations into the above equation, we get

$$A_f = \frac{(20)(20)/2\pi(5 \times 10^{-12})(20)(40)(40)}{jf + [(20 + 1,000)/2\pi(5 \times 10^{-12})(20)(1,000)]}$$

$$= \frac{4 \times 10^5}{jf + 1.6 \times 10^6}$$

The gain and phase plots of the above transfer function are shown in Fig. 3.4.

Loop gain $A_f A_v$ is obtained by multiplying the $A_f$ found above with the $A_v$ given in the op amp data sheet.

Assume the op amp has a dc open-loop gain of $10^5$ and pole frequencies of 10 and $10^6$ Hz. The transfer function of the op amp is

$$A_v = \frac{(2\pi)^2 \times 10^{12}}{(s + 20\pi)(s + 2\pi \times 10^6)}$$

$$= \frac{10^{12}}{(jf + 10)(jf + 10^6)}$$

Figure 3.4B shows the gain and phase plots of this op amp. The product of op amp gain and the feedback factor is thus

$$A_f A_v = \frac{4 \times 10^5}{jf + 1.6 \times 10^6} \frac{10^{12}}{(jf + 10)(jf + 10^6)}$$

$$= \frac{4 \times 10^{17}}{(jf + 10)(jf + 10^6)(jf + 1.6 \times 10^6)}$$

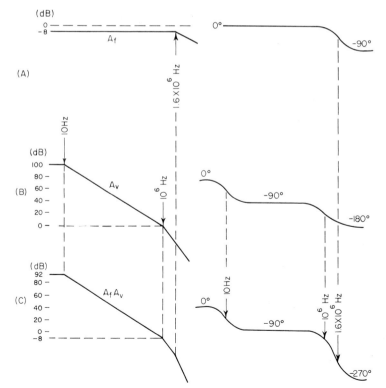

**Fig. 3.4** Bode plots of gain and phase. (A) Feedback network $A_f$. (B) Op amp $A_v$. (C) Loop gain $A_f A_v$.

The product $A_f A_v$ is shown in Fig. 3.4C. Note that if the gain plots are all in dB, each vertical coordinate in Fig. 3.4C is merely the sum of corresponding points at the same frequency in Fig. 3.4A and B. Likewise, phase in Fig. 3.4C is the sum of phase in Fig. 3.4A and B at each frequency.

The example just outlined shows how the loop gain of fairly complex op amp circuits can be quite easily obtained. By utilizing the loop-gain equation in conjunction with gain and phase plots, computational errors can be minimized. One should never rely on just the equations or the plots. It is also helpful to relate specific components to specific poles in the loop gain. This will be helpful in the following sections, where we will find that poles are the prime cause of instability. If it is known that a given resistor or capacitor is the cause of a pole at a critical frequency, often the size of these components can be adjusted to stabilize the circuit.

The preceding example could have had complex $RC$ networks instead of all the resistors shown. The method of computing $A_f$ would still be the same.

**GAIN MARGIN AND PHASE MARGIN** Now that we have a simplified method for computing loop gain, we will put it to use determining stability. We will first state several rules of thumb which will greatly simplify the calculations of this section.

RULE 1: For the gain plot use linear coordinates for gain and logarithmic coordinates for frequency. The gain, however, should be expressed in dB, thus making it a logarithmic plot. This will allow us to multiply two gain plots by simply adding the dB at each frequency.

RULE 2: Make all plots using the Bode approximations. After the final loop-gain and -phase plots are determined, convert the approximate plots to actual plots using the method discussed in Chap. 2.

We now define gain margin and phase margin:

PHASE MARGIN $\phi_m$: Using the loop-gain plot, determine the frequency $f_g$ at which the gain equals 1. Then using the loop-phase plot, at the frequency $f_g$, measure the number of degrees that the loop phase is above $-180°$. This number is the phase margin $\phi_m$.

GAIN MARGIN $A_m$: Using the loop-phase plot, determine the frequency $f_\pi$ at which the phase lag equals $180°$. Then using the loop-gain plot, measure the number of dB that the loop gain is below zero dB at the frequency $f_\pi$. This number is the gain margin $A_m$.

If the loop-gain and -phase plots are accurately constructed, $A_m$ and $\phi_m$ can be quickly determined. It is best to determine phase margin first, since it is sometimes difficult to determine gain margin. This is so because some op amp data sheets do not show the phase lag passing through $-180°$. The loop phase also may not pass through $-180°$ at frequencies of interest. In these cases the phase margin itself will suffice to give a number for stability margin.

What values of gain and phase margins are required in a feedback circuit? Ideally we want a gain margin of 40 or more dB and a phase margin of $180°$. This set of conditions is impossible to obtain; so a compromise is usually accepted. A reasonable phase margin used by circuit designers is $45°$. Likewise a gain margin of 10 dB is acceptable. These margins will result in a closed-loop gain having a small-signal step-response overshoot of approximately 20 percent. A $\phi_m$ of $45°$ also means that the closed-loop frequency response will have less than 3 dB of peaking. If analysis shows that the worst-case minimum phase margin is $45°$, the design is adequate. Some designers use $45°$ as nominal and $30°$ as the worst-case minimum. If this is a true worst case, which considers all possible circuit-degradation factors, then $30°$ worst case is satisfactory.

In the last paragraph we hinted that gain and phase margins are related to closed-loop transient response and closed-loop frequency response. The closed-loop responses are handy tools with which to determine stability margin, since loop-gain characteristics are sometimes very difficult to measure. We will explore this further in Sec. 3.5.

## 3.2 COMPENSATION CIRCUITS

If the calculation of gain and phase margins indicates poor stability, compensation can restore stability. In this section we will review the popular compensation schemes seen on op amp data sheets. Since these data sheets seldom tell the user what the compensation scheme is doing to loop gain, we will discuss that aspect extensively. The exact effect of each compensation scheme on gain and phase margins will be another goal of this section.

**LAG COMPENSATION** This type of compensation is also known as dominant-pole compensation. The dominant pole in a loop-gain plot is the lowest-frequency pole. Consider the loop-gain plot shown in Fig. 3.5, which has two poles. Plot A is the original loop gain, which has the dominant pole $f^A_{lp1}$ at 100 kHz and a second pole $f^A_{lp2}$ at 10 MHz. The plot shows that the second pole occurs at about one-third the frequency of unity-gain crossover. The dominant pole causes a phase lag of 90° for all frequencies above ten times $f^A_{lp1}$. At unity-gain crossover, or three times $f^A_{lp2}$, another lag of 72° takes place

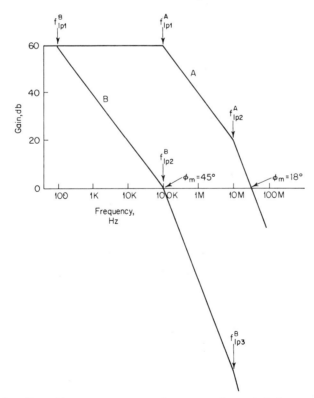

**Fig. 3.5** The effect of lag compensation on loop gain. Curve A: before compensation. Curve B: after lag compensation.

owing to $f_{lp2}^A$, for a total lag of 162° at that point. The resulting phase margin is 18° and the gain margin is 20 dB. This circuit would be very close to instability, especially if component values changed slightly because of temperature changes.

Lag compensation can be used to produce the loop-gain curve B shown in Fig. 3.5. A new dominant pole is added to the loop response at a frequency sufficiently low that the old dominant pole occurs near unity gain. As noted in the figure, the phase margin becomes 45°. This is always the result if one can force the second pole of loop gain to occur at unity-gain crossover.

At least three methods are commonly utilized to create a simple lag as noted above. Figure 3.6 shows these three most common types. The method shown in Fig. 3.6A can be implemented on any op amp, even if a special lag terminal is not available. It does have the disadvantage, however, of requiring fairly large capacitors. The dominant-pole frequency is found from

$$f_p = \frac{1}{2\pi R_o C_c} \tag{3.15}$$

where
$R_o$ = output resistance of op amp
$C_c$ = compensation capacitance

Many op amps have an output resistance of approximately 100 Ω. If a 100-Hz pole is required as shown in Fig. 3.5,

$$f_p = \frac{1}{6.28 \times 100 C_c} = 100 \text{ Hz}$$

or
$$C_c = 15.9 \ \mu\text{F}$$

Obviously, this is a difficult way to lag-compensate an op amp. An unpolarized capacitor is required for this application, and the physical size of an un-

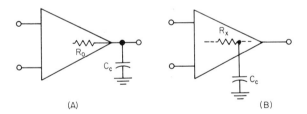

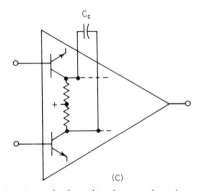

**Fig. 3.6** Three circuit methods utilized to produce lag compensation.

polarized 15.9-$\mu$F capacitor would be unreasonably large. If a resistor of, say, 10,000 $\Omega$ is placed in series with the op amp output, the required $C_c$ is only 0.159 $\mu$F. This will not increase the circuit output impedance if the op amp output can still be used as the circuit output terminal. The feedback circuitry, however, must be in series with the lag network. In some applications this will not be possible, so the networks shown in Fig. 3.6B or C may be required.

Some op amps have a lag-compensation terminal available for creating a low-frequency dominant pole. The internal circuitry of this type of op amp is generally as shown in Fig. 3.6B. The output resistance of the op amp at this point is $R_x$.

If this resistance $R_x$ is quite large, the external capacitor can be reasonably small. The output resistance of the op amp output terminal is unaffected by $R_x$ unless $R_x$ is after the last stage.

Figure 3.6C shows another way to achieve lag compensation. Two op amp terminals are required in this method. The output impedance of the circuit at these terminals is at least several thousand ohms. The required compensation capacitor will therefore be less than 0.1 $\mu$F in most cases.

Lag compensation has several disadvantages of which the designer must be aware. First, this method of assuring stability severely decreases the op amp available bandwidth. It is recommended only for dc or low-frequency applications. Second, the capacitor required to make the circuit unconditionally stable is often too large physically.

**LEAD COMPENSATION**  The main cause of instability problems is the excess number of poles in the loop gain. We must make sure that only one or two poles occur before the loop-gain frequency response decreases to unity. Preferably, the second pole occurs right at unity-loop-gain crossover. This gives us a phase margin of exactly 45°.

Lead compensation can create a zero in the loop gain at any desired frequency. If the loop gain has three poles before unity-gain crossover, this zero can cancel one of the poles. The net result is two poles before unity crossover. Any of the three poles can be canceled, although best results are usually achieved by eliminating the second pole. The third pole can then be made to coincide with unity-gain crossover, resulting in a 45° phase margin. With this technique, the circuit bandwidth may be increased by at least several octaves. If the new second pole occurs at unity-gain crossover, we can also connect the circuit up as a unity-gain amplifier without worrying about instability. As we shall see in Sec. 3.3, unity-gain amplifiers are difficult to construct using op amps which have two or more poles before unity-loop-gain crossover. Even lead compensation will not work in some cases, as we shall see below.

Lead compensation is implemented as shown in Fig. 3.7A. A small capacitor $C_f$ is placed in parallel with the feedback capacitor $R_f$. No special op amp compensation terminals are required. The zero frequency is

$$f_z = \frac{1}{2\pi R_f C_f} \tag{3.16}$$

This compensation capacitor also inserts another pole into the loop gain at a frequency of

$$f_p = \frac{(R_1 + R_f)f_z}{R_1} = A_{vc}f_z \tag{3.17}$$

where $A_{vc}$ is the noninverting circuit gain.

## 3-14 FEEDBACK STABILITY

If the circuit gain is low, $f_p$ is not large with respect to $f_z$. In this case the improvement in stability may not be worth the effort, since $f_z$ almost cancels $f_p$ and the second op amp pole effectively moves up slightly to a new value $f_p'$. This is shown graphically in Fig. 3.7B for a circuit gain of

$$\frac{R_1 + R_f}{R_1} = 2$$

Also shown in Fig. 3.7B is the ideal case where $f_p \gg f_z$. In this case the phase margin before compensation was less than zero. After compensation the phase margin became 45°.

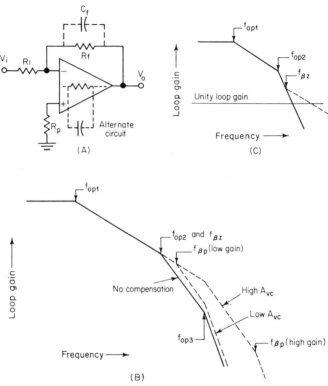

**Fig. 3.7** (A) Circuit with lead compensation. (B) Bode loop gain plot of circuit A before compensation and after compensation (shown for high and low $A_{vc}$). (C) Bode loop gain where we set $f_z > f_{op2}$.

It is possible in many cases, if the required circuit gain $A_{vc}$ is high, to set $f_z > f_{op2}$ by several octaves. This will increase the bandwidth even further than indicated above. It all depends on how close $f_{op3}$ is to $f_{op2}$. If they are widely separated and $f_{op3}$ does not interfere with the zero $f_z$ we can set $f_z > f_{op2}$. The main criterion to watch for is that the slope of loop gain is $-20$ dB/decade as it goes through unity gain. This slope must have been established for several octaves before passing through unity gain. Figure 3.7C shows the Bode plots of this type of lead compensation.

If lead compensation is used to cancel out a pole caused by op amp input capacitance, a new pole is not created along with the zero. The reader should compare the circuits in Appendix VI. The simple lead network with a resistive load has both a zero and a pole. When the resistive load is shunted with stray capacitance and we set $R_1 C_s = R_f C_f$, the pole and zero cancel. The net effect is an all-pass transfer function with no poles and no zeros.

**LEAD-LAG COMPENSATION** Design flexibility is enhanced if the designer can create a pole and a zero in the loop gain. Any of the lag circuits of Fig. 3.6 can be used in conjunction with the lead circuit shown in Fig. 3.7 to manipulate the loop-gain curve. Note, however, that the lag circuits create only one pole but the lead circuit creates both a pole and a zero. Several better ways to implement lead-lag compensation will be shown below. These methods

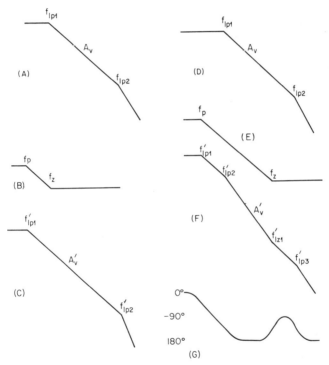

**Fig. 3.8** Two possible ways in which independent lead and lag networks can increase stability. $(A,B,C)$ Standard method. $(D,E,F)$ Wide-bandwidth method. $(G)$ Phase of $A'_v$ for wide-bandwidth case.

create exactly one pole and one zero. However, the pole is not independent of the zero. If one needs total independence of the pole and zero, the circuits of Fig. 3.6 and 3.7 must be used.

Figure 3.8 shows two of the possible ways lead and lag networks can be used to increase stability of a feedback circuit. In part $A$ of the figure a loop-gain curve with a phase margin near zero is shown. A lead-lag network with a pole having a frequency lower than the loop-gain dominant pole $f_{lp1}$ is shown in $B$. Figure 3.8$C$ shows the effect on loop gain when the lead-lag compensation is included. The compensation network zero $f_z$ is chosen such that it exactly cancels the original dominant pole $f_{lp1}$. The compensation pole $f_p$ ideally is placed slightly below $f_{lp1}$ so that the second loop-gain pole occurs right at unity-loop-gain crossover (or slightly higher in frequency).

Since the lead-lag compensation pole and zero are not independent, it may not be possible to achieve both the main two objectives: pole-zero cancellation and $f_{lp2}^1$ at unity-gain crossover. However, we do have some design latitude on both these objectives. If we can allow the phase margin to be as low as 27°, we can let $f_{lp2}^1$ occur up to an octave before unity-loop-gain crossover. We can also allow $f_{lp2}^1$ to occur anywhere after unity-loop-gain crossover, but this is wasteful of the circuit capabilities.

The placement of the zero also has some latitude. It does not have to exactly cancel the original dominant pole. Figure 3.8D, E, and F shows a case where $f_z \gg f_{lp1}$. In this case closed-loop gains having bandwidths several octaves higher than in Fig. 3.8C are possible. The main thing one must watch for is that the zero $f_z = f_{z1}'$ is at least two or three octaves lower in frequency than unity-loop-gain crossover. It is a good practice, in complicated cases like this, to make phase plots for loop gain (before and after compensation). Precise values for both gain and phase margins can then be determined. The phase plot of $A_v$ shown in Fig. 3.8F is plotted in Fig. 3.8G. The phase margin for this particular compensation example is 39°. The gain margin is undefined because the phase lag never goes below $-180°$. The phase margin before compensation was zero, since the second-pole $f_{lp2}$ occurred $1\frac{1}{2}$ orders of magnitude below the unity-loop-gain crossover frequency.

The three most common circuits which are used for lead-lag compensation are shown in Fig. 3.9. Each of these methods has advantages and disadvantages when compared with the other two. We will find that circuit B has the best overall flexibility and performance.

The circuit shown in Fig. 3.9A is a common method of compensation. The equations for the compensation pole and zero are

$$f_z = \frac{1}{2\pi R_c C_c} \tag{3.18}$$

$$f_p = \frac{1}{2\pi (R_x + R_c) C_c} \tag{3.19}$$

The major disadvantages of this method are: (1) one or two special compensation terminals are required on the op amp; (2) the pole and zero are not independent; and (3) the resistance $R_x$ is not usually specified on the data sheet. The value of $R_x$ can be computed by solving Eq. (3.19) for $R_x$:

$$R_x = \frac{1}{2\pi f_p C_c} - R_c \tag{3.20}$$

and correlating this with curves of frequency response on the data sheet. For example, the $\mu$A709 data sheet indicates that $f_p = 1$ kHz if $R_c = 1,500$ $\Omega$ and $C_c = 100$ pF. Computing $R_x$, we get

$$R_x = \frac{1}{6.28 \times 10^3 \times 10^{-10}} - 1,500$$
$$= 1.59 \text{ M}\Omega$$

Figure 3.9B shows a lead-lag-compensation circuit which requires no special op amp terminals. This method is called input lead-lag compensation. The zero frequency in this case is

$$f_z = \frac{1}{2\pi R_c C_c} \tag{3.21}$$

Note that this is exactly the same zero-frequency formula as the first lead-lag circuit. The pole, however, for input lead-lag compensation is

$$f_p = \frac{R_1 + R_f}{2\pi C_c(R_1 R_f + R_1 R_c + R_1 R_p + R_c R_f + R_p R_f)} \quad (3.22)$$

We observe that the above pole is independent of any internal op amp resistance $R_x$. This will help stabilize the pole frequency over temperature if high-quality external components are used. This means that the phase margin will be more stable over temperature compared with a compensation method which relies on an internal $R_x$.

Equation (3.22) indicates that $f_p$ depends on $R_1$, $R_f$, $R_p$, $R_c$, and $C_c$. The zero $f_z$ depends only on $R_c$ and $C_c$. Thus $f_p$ can be made nearly independent of $f_z$. $R_1$, $R_f$, and $R_p$ are determined by the closed-loop-gain requirements. The closed-loop gain is equal to $R_f/R_1$. $R_p$ is then computed from $R_p =$

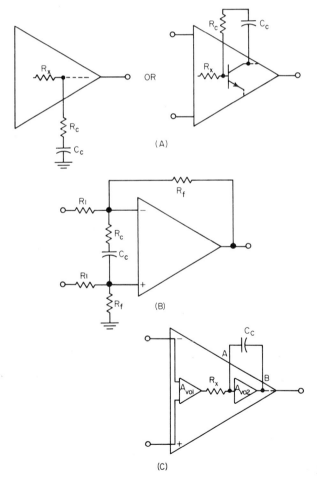

**Fig. 3.9** Three typical lead-lag-compensated op amp circuits. (A) Lead-lag compensation of an internal op amp node. (B) Input lead-lag compensation. (C) Miller-effect lead-lag compensation.

$R_f R_1/(R_f + R_1)$. $R_f$ and $R_1$ can usually be varied over a wide range as long as the ratio $R_f/R_1$ is equal to the required closed-loop gain.

Op amp slewing rate depends on the location of the compensation network. The higher the signal level at the point of compensation, the more severely slewing rate is affected. If one is handling fairly large signal levels in the circuit and does not want slew-rate limiting to become worse than the basic op amp, input lead-lag compensation is recommended. At no point in the op amp circuit is the signal level lower than across the input terminals. This is an ideal place for compensation when slewing rate is a design concern.

The last type of lead-lag compensation we will discuss is often called Miller-effect compensation. Figure 3.9C shows the general type of circuit involved in this method. Leads are brought out of the op amp from the input and output of an internal stage. This is often just the base and collector leads of a single common-emitter stage inside the op amp. Only one small capacitor $C_c$ across these terminals is required to produce lead-lag compensation. The zero created exactly cancels out the dominant-pole $f_{lp1}$ of the loop-gain function. The new pole created is at a frequency

$$f_p = \frac{1}{2\pi R_x C_c A_{vo2}} \qquad (3.23)$$

where  $R_x$ = output resistance at op amp at terminal A
$A_{vo2}$ = voltage gain of op amp from terminal A to terminal B

This method of compensation also gets rid of the second-pole $f_{lp2}$ of the loop gain. A mathematical analysis of the effect of $C_c$ on loop gain shows that $f_{lp2}$ is not canceled, but it shifts down in frequency and becomes the new $f_p$.

The main problem with Miller-effect compensation is the calculation of $f_p$. Data sheets do not give values for $R_x$ and $A_{vo2}$. However, curves of open-loop gain before and after compensation are available which allow us to determine stability margins. One could even compute the product $R_x A_{vo2}$ from these data, but this is not as important as the effect on loop gain. Thus the curves are usually sufficient.

A primary advantage of Miller-effect compensation is the fact that two poles are canceled out and one new pole created. The other two lead-lag-compensation schemes created one pole and one zero. Miller-effect compensation would therefore be useful to stabilize wide-bandwidth op amps which have three or more poles above unity gain. Since the new dominant-pole $f_p$ depends on two op amp parameters (both of which are temperature-dependent), one must choose $C_c$ so that for the worst-case temperature extremes an adequate phase margin is realized. Data sheets almost never give the user information regarding effects of compensation over temperature. When using any compensation scheme which depends on internal op amp parameters, a totally different approach to stability analysis must be used. The designer must either measure stability of all production circuits using the procedure outlined in Sec. 3.5 or must design a circuit with a very large phase margin.

## 3.3 THE SEVEN MAJOR CAUSES OF OP AMP INSTABILITY

The frequency-compensation techniques developed in the last section are usually sufficient to design stable op amp circuits. We also include this section and Sec. 3.4 for those circuit designs which still have instability problems. These sections can be used by the more thorough circuit designers

## THE SEVEN MAJOR CAUSES OF OP AMP INSTABILITY

who wish to consider every angle of stability analysis. Only in this way can one be sure the circuit will be stable under the worst set of conditions. This approach may be too expensive for some design tasks but is mandatory for others.

There are many ways to list the causes of op amp instability. The seven causes presented here are grouped so that they are a practical system for op amp stability analysis. The designer does not have to follow the order presented here, since any one of the seven causes can be the main cause of instability in a particular design. However, the order presented here closely parallels the normal design steps of an op amp circuit. Design and stability analysis should proceed in parallel.

The seven causes of instability are as follows:
1. Compensation recommended by op amp data sheet not used.
2. Closed-loop gain too low for type (and amount) of compensation used.
3. Excessive capacitive load on op amp.
4. Incorrect phase lead/lag in feedback network.
5. Excessive resistance between ground and op amp positive input.
6. Excessive stray capacitance between op amp output and balance terminals.
7. Inadequate power-supply bypassing.

We will consider each of the seven separately in the pages to follow. There will occasionally be some overlapping among the first five, since these can all be analyzed using loop gain. Experienced designers, or those having computer-aided design facilities available, can handle the first five all at once with one large loop-gain formula. In this text, however, we will emphasize taking one at a time, since this gives us better visibility of the problem.

**FIRST CAUSE OF INSTABILITY — COMPENSATION RECOMMENDED BY DATA SHEET NOT USED** This should be the first item to check in troubleshooting an unstable op amp circuit. A manufacturer's recommended compensation is the result of extensive testing with a large number of op amps. It should work for most applications if reasonable layout and bypassing are provided. If anything other than the standard inverting and noninverting amplifier is being designed, however, special precautions must be taken. The rest of this section will consider some of these special precautions. For the present, let us discuss further the compensation recommended on op amp data sheets.

The frequency-compensation networks given by a manufacturer are designed to provide stability for the full production distribution of their device. They usually have taken into account the unit-to-unit variation of open-loop gain, output resistance of compensation terminals, and phase shift at unity-gain crossover, and also the effects of temperature and power-supply voltage on these parameters. It would be too conservative to compensate for the worst-case variations of all op amp parameters at once, since some of them effectively cancel out the other's effect on stability. For example, in some op amps when the gain drops with reduced supply voltage, the compensation-terminal output resistance also goes down. According to Eq. (3.15) (lag compensation) or Eq. (3.19) (lead-lag compensation), this will increase the frequency of the compensated loop-gain dominant pole. The phase-margin change will therefore be minimal, even though two parameters affecting it have changed substantially.

If a dc or low-frequency circuit is being designed, it might be a good idea to overcompensate. This is one way to be absolutely sure the full production run of circuits will be stable. This sort of "brute-force" approach works best for lag compensation and is summarized in the following rule of thumb: A

**3-20 FEEDBACK STABILITY**

tenfold increase in compensation capacitance will provide ten times more stability and ten times less bandwidth.

A corollary to the above rule of thumb might state: Compensation less than the data-sheet recommendation must only be done if one computes (and plots) the loop gain for the worst-case combination of variables. If the phase margin is still greater than 30°, using less than recommended compensation might be acceptable.

**SECOND CAUSE OF INSTABILITY—CLOSED-LOOP GAIN TOO LOW FOR TYPE (AND AMOUNT) OF COMPENSATION USED**  We have mentioned several times that some op amps cannot be operated in the unity-gain configuration unless they have been compensated in a special manner. We will now explore this problem in detail and outline a method which will allow any op amp to be operated with any closed-loop gain.

This problem will be analyzed first graphically and then mathematically. In Sec. 3.1 we stated a rule of thumb which allowed a first-cut stability analysis to be performed using plots of closed-loop gain and op amp open-loop

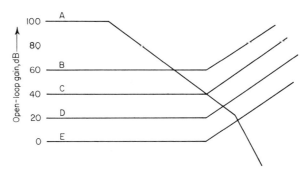

**Fig. 3.10**  Determining phase margin by using open- and closed-loop-gain curves.

gain. Now that we have discussed loop gain, phase margin, and compensation circuits, we can expand our use of that rule of thumb to determine minimum allowable gain for a circuit. The rule of thumb stated that if the rate of closure between plots of open-loop and closed-loop gains was 40 dB/decade or less, the circuit would probably be stable. In place of closed-loop gain we can use $1/A_f$, since this simplifies the procedure. However, as we recall from earlier discussions, we should not use $1/A_f$ actually to determine the closed-loop circuit gain. This will be a substitution which is completely accurate for stability analysis but is sometimes wrong when used to determine closed-loop gain.

Fig. 3.10 will be used to show how the graphical construction technique operates. Curve A is a typical op amp open-loop-gain plot. So that we can easily relate stability to closed-loop gain in this example, we will assume $A_f = \beta$. This is true for simple inverting and noninverting amplifiers which contain only two-terminal input and feedback networks (see Sec. 2.5). B, C, D, and E are plots of $1/A_f$ corresponding to closed-loop gains of 1,000(60 dB), 100(40 dB), 10(20 dB), and 1(zero dB), respectively. For curve B the rate of closure is 20 dB/decade, which results in an absolutely stable circuit. Since the zero frequency in $1/A_f$ is an order of magnitude above the point of intersection, the circuit has a phase margin of approximately 90°. Curve C intersects at 20 dB/decade. However, the zero of $1/A_f$ occurs at the same point.

The resulting phase margin is 45°. Curve $D$ intersects at 40 dB/decade, which puts it on the threshold of oscillation. Its phase margin is very close to zero, since $1/A_f$ occurs an order of magnitude lower in frequency than the intersection. Curve $E$ intersects $A$ at 60 dB/decade and represents an unstable circuit. The phase margin in this case is less than zero degrees.

It is apparent from Fig. 3.10 that instability increases as closed-loop gain is reduced. The real cause of this instability is the zero in $1/A_f$ and the second pole of $A_v$. If the second pole of $A_v$ is canceled with compensation, the circuit could be stable down to a 20-dB closed-loop gain. As it exists, a minimum of 40-dB closed-loop gain is recommended. If additional compensation is utilized to increase the zero frequency of $1/A_f$ by one or two orders of magnitude, the circuit could be made stable down to unity closed-loop gain.

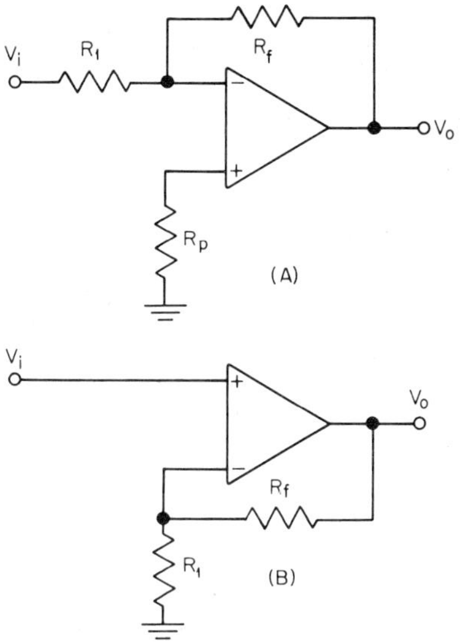

**Fig. 3.11** Standard amplifier circuits. (A) Inverting amplifier. (B) Noninverting amplifier.

We will now explore the mathematics of the foregoing comments and work out a method to compute accurately the minimum stable closed-loop gain. The formulas shown will apply only to the standard inverting and noninverting amplifier circuits. Determining minimum closed-loop gain for more complex circuits can be done using a similar approach. The closed-loop-gain formula of an inverting amplifier (see Fig. 3.11) is

$$A_{vc} = \frac{R_f}{R_1 + R_f} \frac{-A_v}{1 + [R_1 A_v/(R_1 + R_f)]} \quad (3.24)$$

For the noninverting amplifier the formula is

$$A_{vc} = \frac{A_v}{1 + [R_1 A_v/(R_1 + R_f)]} \quad (3.25)$$

In both these equations $A_v$ is a positive real number at low frequencies. That is, $A_v$ is referred to the positive input terminal of the op amp.

In Eq. (3.24) or (3.25), sustained oscillation is likely if

$$\frac{R_1 A_v}{R_1 + R_f} = -1 = 1\,\underline{/180°} \quad (3.26)$$

From the op amp data sheet determine the op amp gain at the frequency where its phase lag is 180°. Call this gain $A_\pi$. Oscillation now occurs when

$$\frac{R_1 A_\pi}{R_1 + R_f} = -1 = 1\,\underline{/180°} \quad (3.27)$$

From Chap. 1 we recall that

$$A_{vc} = \frac{R_1 + R_f}{R_1} \quad \text{(noninverting)} \quad (3.28)$$

and

$$A_{vc} = \frac{R_1 + R_f}{R_1} - 1 \quad \text{(inverting)} \quad (3.29)$$

Therefore, the minimum closed-loop gains which assure stability ($\phi_m \geq 0$) are

$$A_{vc}(\min) = A_\pi \qquad \text{(noninverting)} \qquad (3.30)$$

$$A_{vc}(\min) = A_\pi - 1 \qquad \text{(inverting)} \qquad (3.31)$$

If a phase margin $\phi_m$ greater than zero is required, one must determine the new gain $A_\pi$ at the frequency where the phase lag is $\phi_m°$ less than 180°. This new $A_\pi$ is substituted into Eqs. (3.30) and (3.31) to determine minimum possible closed-loop gain.

Now that we have both graphical and mathematical methods to determine minimum closed-loop gain, let us discuss ways to lower this minimum. In examining Fig. 3.10, we note that two changes can be made which will lower the minimum closed-loop gain: (1) The zero in $1/A_f$ (pole in $A_f$) must increase in frequency or be eliminated. (2) The second pole of $A_v$ must be increased in frequency or eliminated. Recalling all we learned in Sec. 3.2, the way to accomplish the two objectives above is to compensate. Depending on the bandwidth required, we can use lag, lead, or lead-lag compensation. The reader is referred to Sec. 3.2 for ways to accomplish this goal. In Sec. 3.4 we will give specific examples of this procedure.

**THIRD CAUSE OF INSTABILITY—EXCESSIVE CAPACITIVE LOAD ON OP AMP** All operational amplifiers have a finite output resistance $R_o$. This resistance is a function of frequency and increases quite rapidly as the op amp gain approaches unity. As output resistance is usually defined, it is effectively in series with the output terminal. If a load capacitor $C_L$ is connected from the output terminal to ground, a lag network is created. This lag network is in series with the feedback network; so it causes another pole in the loop gain. If the pole frequency of this lag network is near gain crossover, the phase margin of the circuit will be reduced. In practice it is best to keep this pole frequency at least two to ten times larger than the loop-gain crossover frequency. One method of doing this is by controlling the size of $C_L$.

Suppose the maximum output resistance $R_o$ is 500 Ω at 1 MHz. If 1 MHz is also the loop-gain crossover frequency and the phase margin is 20°, the phase lag caused by $R_o$ and $C_L$ must be not more than a few degrees at 1 MHz. If we choose $C_L$ such that the pole frequency of $R_o$ and $C_L$ is 10 MHz, then the phase shift they cause at 1 MHz will be 6° (see the phase plot of a simple lag network in Chap. 2). Solving for $C_L$,

$$C_L = \frac{1}{2\pi f R_o} = \frac{1}{(6.28)10^7 \times 500}$$

$$= 32 \text{ pF} \qquad \text{(not very much!)} \qquad (3.32)$$

Thus op amps with small phase margins and a large output resistance will oscillate if they drive much of a capacitive load. For refined calculations we must also include the effect of load resistance $R_L$ and feedback resistance $R_f$. The real pole caused by $C_L$ must be calculated using $R_o$ in parallel with $R_L$ and $R_f$. Usually $R_o$ is much smaller than the other two; so $R_L$ and $R_f$ can be neglected.

As might be expected, a large capacitive load also decreases circuit bandwidth. Figure 3.12 shows a family of curves which indicate how increasing $C_L$ lowers bandwidth as it increases instability. The peaking of the closed-loop response is an indicator of instability. In Sec. 3.5 we will show graphically the relationship between closed-loop peaking and phase margin.

If $C_L$ cannot be reduced, a circuit modification as shown in Fig. 3.13 will handle any size of capacitive load. The mathematics of this circuit is very

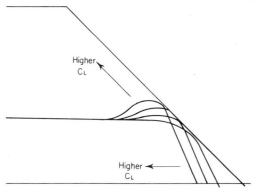

**Fig. 3.12** Curves showing how increased load capacitance affects both instability and bandwidth.

complex, but we will show a simplified method to determine component sizes which will get us started. Then the stability test procedure to be shown in Sec. 3.5 can be used to place the phase margin accurately at any desired value.

In Fig. 3.13, the lag network composed of $R_o$, $R_2$, and $C_L$ creates a pole in the loop gain. The frequency of this pole is approximately

$$f_p = \frac{1}{2\pi(R_o + R_2)C_L} \tag{3.33}$$

The components $C_f$ and $R_f$ likewise form a lead network. This places a zero in the loop gain at an approximate frequency of

$$f_z = \frac{1}{2\pi R_f C_f} \tag{3.34}$$

The object is to make the zero cancel out the pole. Setting $f_p = f_z$, we find that the compensation capacitor $C_f$ required is

$$C_f = \frac{C_L(R_o + R_2)}{R_f} \tag{3.35}$$

In practice $R_2$ is chosen to be 50 to 500 Ω. Higher values will tend to increase the closed-loop output resistance.

This method of compensating for large $C_L$ has one drawback. It will work

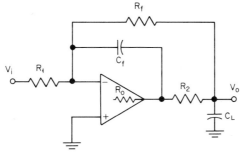

**Fig. 3.13** Op amp circuit which will handle large capacitive loads without reducing phase margin.

satisfactorily only for op amps which are stable at unity gain. This can be achieved by using lag compensation, lead-lag compensation, or an op amp with this inherent capability. Many op amps, such as the 741, are stable at unity gain. These types of op amps are identified by the fact that the second pole occurs at unity-gain crossover.

**FOURTH CAUSE OF INSTABILITY—INCORRECT PHASE LEAD/LAG IN FEEDBACK NETWORK**
Many op amp circuits possess a phase lead or lag in the feedback network even though the designer did not plan it that way. Phase lead can be caused by stray capacitance across the feedback resistor. This type of stray capacitance can also occur in printed-circuit boards if the copper trace from the op amp output is too close to the input, or vice versa.

Phase lag in the feedback network is troublesome in amplifiers which require a large feedback resistor—i.e., FET op amps, high-gain–high-input-resistance inverting amplifiers, etc. In these cases the input resistor is large. The feedback resistor is even larger if the circuit is to have gain. The feedback resistor and input capacitance form a 90° lag network. The corner frequency of this lag network must be at least ten times the unity-loop-gain frequency in order to not subtract more than 6° from the phase margin.

Since unwanted phase lag in the feedback network is most likely to cause instability, we will concentrate on it. There are actually three stray capacitances which must be considered at the op amp input terminals. Figure 3.14A shows that we are considering the stray capacitances from the input terminals to ground ($C_1$ and $C_3$) and also between the input terminals ($C_2$). The feedback factor $A_f$ is approximately

$$A_f = \frac{R_p C_3 [s + (1/R_p C_3)]/R_f^2 C_1 (C_2 + C_3)}{[s + (1/R_f C_1)]\{s + [1/R_p(C_2 + C_3)]\}}$$

$$= \frac{R_p C_3 [jf + (1/2\pi R_p C_3)]/R_f^2 C_1 (C_2 + C_3)}{2\pi [jf + (1/2\pi R_f C_1)]\{jf + [1/2\pi R_p(C_2 + C_3)]\}} \quad (3.36)$$

where
$$R_p = \frac{R_1 R_f}{R_1 + R_f}$$

We note that these stray capacitances have caused two poles and a zero in the feedback factor. We can make several comments regarding Eq. (3.36) if we assume some component values such as $C_1 = C_2 = C_3$ and $R_f = 3R_p$ (closed-loop gain $R_f/R_1 = 2$). In this case the two poles will be one-half and one-third the frequency of the zero. Since the poles occur first, and there are two of them, $A_f$ will have a substantial phase lag at all frequencies near these poles and the zero.

A solution to this excessive phase lag is to create another zero in $A_f$. This can be accomplished in several ways. One approach is lead compensation. If we put a small capacitor $C_f$ around the feedback resistor, we get

$$A_f = \frac{C_3 C_f [jf + (1/2\pi R_p C_3)][jf + (1/2\pi R_f C_f)]}{(C_1 + C_f)(C_2 + C_3)\{jf + [1/2\pi R_p(C_1 + C_f)]\}\{jf + [1/2\pi R_p(C_2 + C_3)]\}} \quad (3.37)$$

All these poles and zeros are fairly close together in frequency; so pole-zero cancellation is possible. However, the required capacitance of $C_f$ is very small. For example, if $R_1 = 1$ k$\Omega$, $R_f = 10$ k$\Omega$, $C_1 = C_2 = C_3 = 5$ pF, then the required $C_f$ is in the range of 0.5 to 0.9 pF. These two values depend on which pole we attempt to cancel out. In either case, stray capacitance around the feedback resistor will be perhaps half of any $C_f$ we attempt to install. The phase-margin testing procedure outlined in Sec. 3.5 is required after using this type of compensation.

Another approach to the problem is to use input lead-lag compensation. Utilizing the circuit of Fig. 3.14B, we get

$$A_f = \frac{[R_1R_c/(R_1R_f + R_1R_c + R_1R_p + R_fR_c + R_fR_p)][jf + (1/2\pi R_c C_c)]}{jf + [(R_1 + R_f)/2\pi C_c(R_1R_f + R_1R_c + R_1R_p + R_fR_c + R_fR_p)]} \quad (3.38)$$

The following assumptions were made to derive Eq. (3.38):

1. The effects of stray capacitances can be neglected, since the unity-loop-gain crossover is one or two orders of magnitude lower after compensation (see Fig. 3.8). This is possible only if $R_f$ is not too large. Check to see

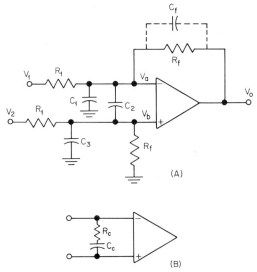

**Fig. 3.14** Stray capacitances at input of op amp which can cause unwanted phase lag. Stray-capacitance cancellation is possible using lead compensation A or lead-lag compensation B.

that $f = 1/2\pi R_f C_s$ (where $C_s$ = a typical stray capacitance) is at least several octaves above the new unity-loop-gain crossover frequency.

2. Op amp input resistance is much larger than $R_1$ and $R_f$.

**FIFTH CAUSE OF INSTABILITY—EXCESSIVE RESISTANCE BETWEEN GROUND AND OP AMP POSITIVE INPUT** A closer examination of the mathematics which derived Eq. (3.36) reveals another possible pole in the loop gain. This pole is caused by a special set of conditions which only some designers will have. If the op amp has a fairly low input resistance and the input resistor is large (> 10 kΩ), this extra pole may occur near unity-loop-gain crossover. The problem is compounded because op amp input resistance decreases an order of magnitude or more at frequencies near its unity-gain crossover. It is not worthwhile to expand Eq. (3.36) here to show the effect of low $R_i$ and high $R_p$, since the mathematics gets too involved. We will merely outline the steps required to analyze and solve this potential problem.

## 3-26  FEEDBACK STABILITY

The problem may cause excessive lag in loop gain if:
 1. Op amp input resistance $R_i$ is ten times $R_f$ or smaller at op amp unity-gain crossover.
 2. $R_1$ is large (> 10 k$\Omega$).

The solution to this problem is simple—bypass the positive op amp input to ground with a capacitor $C$:

$$C \geq \frac{10}{2\pi f_u R_p} \qquad (3.39)$$

where $f_u$ is the op amp unity-gain-crossover frequency, and $R_p = R_1 R_f/(R_1 + R_f)$.

Some designers may wish to calculate the effect of $R_p$ on loop gain. Referring to Fig. 3.3 and the accompanying text, we note that $R_p$ can be lumped in with other parts in the computation of $A_f$. This procedure can also be used to determine the effectiveness of the bypass capacitor across $R_p$.

**SIXTH CAUSE OF INSTABILITY—EXCESSIVE STRAY CAPACITANCE BETWEEN OP AMP OUTPUT AND BALANCE TERMINALS**  Many op amps have one of their balance terminals right next to the output terminal. For example, the 101 output terminal is pin 6 and one of the balance terminals is pin 5 (TO-99 package). This particular balance terminal has a signal which is of the same polarity as the output terminal. If any excessive stray capacitance exists between these terminals, a slight amount of positive feedback is possible. As this stray capacitance gets larger, the loop phase margin is lowered. Experiments with the 101, for instance, revealed the following: Starting with a compensation which produces a phase margin of 45°, different small capacitances were placed between pins 5 and 6. It was found that $3/4$ pF reduced the phase margin to 34°, 1 pF to 32°, and $1\frac{1}{4}$ pF to 30°. These are substantial reductions to the phase margin from such an unlikely source.

Stray capacitance between balance and output terminals usually occurs because the designer thinks the balance terminal is a dc-adjustment terminal. The balance wires (or traces on a printed-circuit board) may run 6 or 8 inches to a potentiometer which is used for the dc balance. The output terminal may follow the same route. Under these conditions several picofarads stray capacitance between terminals is possible.

The balance terminals must be treated like any wire in the signal path. Wire lengths should be kept short, and their placement relative to other wires should be carefully studied before artwork on a board is firmed up. Multilayer boards with ground planes between layers of signal traces are recommended. Some manufacturers recommend placing a 0.1-$\mu$F capacitor between balance terminals when they are used. If the balance terminals are not required, they can be tied together.

**SEVENTH CAUSE OF INSTABILITY—INADEQUATE POWER-SUPPLY BYPASSING**  This problem is seldom handled with feedback analysis, although it can be done if sufficient data are available. Few designers resort to an analysis, however, since a few rules of thumb are sufficient for all but the very exotic applications of op amps. These rules are:
 1. Bypass the positive and negative power supplies with capacitors to ground.
 2. These bypass capacitors can be 0.01- to 0.1-$\mu$F ceramic capacitors in most op amp ($f_u \approx 1$ MHz) applications.
 3. High-speed (wide-bandwidth) op amp circuits and those using feedforward compensation should be bypassed with low-inductance capacitors at each op amp on both supply leads.

4. Bypass at least once per card (both + and − supplies) or every 3 to 5 op amps, whichever comes first.
5. Extra bypassing is required for op amps driving high-capacitance loads.
6. A large tantalum capacitor ($\approx 22\ \mu F$) on the negative supply to ground is recommended at least once per board. This is not usually required for the positive supply.
7. High-current buffers driven by op amps require additional bypassing so that transients will not get back to the op amp. At least 0.1 $\mu F$ ceramic per 50 mA is recommended.

## 3.4 FEEDBACK-STABILITY-DESIGN EXAMPLES

We will now present the stability-design portion of several op amp circuit designs. The discussion will be centered around the first five causes of instability and the various compensation schemes presented in Sec. 3.2. The last two causes of instability are not easily subject to analysis and rely mainly on rules of thumb based on experience.

**FIRST FEEDBACK-STABILITY EXAMPLE — NONINVERTING AMPLIFIER WITH GAIN OF 10**
As our first example, let us design a circuit with a gain of 10 and using the 101A op amp. We will determine the widest possible bandwidth using standard compensation techniques shown in this chapter. The main stability criterion we will require is that a phase margin of 45° is guaranteed at room temperature using typical 101A parameters. A true worst-case design might allow the phase margin to be only 30° when all parameters go to their worst-case limits simultaneously. We will not go into that depth in this example.

A good starting point is to make plots of the 101A open-loop gain and phase using minimum recommended compensation. Figure 3.15A and B shows these plots. For those interested in detailed mathematical calculations to supplement the graphical approach used here, we find the op amp gain equation as follows:

1. The two poles give us denominator terms such as $jf + 40$ and $jf + 2 \times 10^6$.
2. Multiply the numerator by these pole frequencies.
3. Multiply the numerator by the op amp dc gain ($1.6 \times 10^5$).
4. The op amp gain as a function of frequency becomes

$$A_v = \frac{40 \times 2 \times 10^6 \times 1.6 \times 10^5}{(jf + 40)(jf + 2 \times 10^6)}$$

$$= \frac{1.28 \times 10^{13}}{(jf + 40)(jf + 2 \times 10^6)}$$

The first-cut circuit is shown in Fig. 3.16A. Minimum compensation requires a 3-pF capacitor between the compensation terminals. Referring to Sec. 3.2, we note that this is lead-lag compensation of the type shown in Fig. 3.9C. The pole defined by Eq. (3.23) is at 40 Hz. It need not be computed, since the data-sheet plot for minimum compensation is given.

Let us assume the circuit must have an input impedance of 100 k$\Omega$. The input resistor $R_1$ must then be 100 k$\Omega$. For a gain of 10 the required feedback resistor is 1 M$\Omega$. According to Eq. (3.5) the feedback factor is

$$\beta = A_f = \frac{R_1}{R_1 + R_f} = \frac{10^5}{10^5 + 10^6} = 0.0909 = -20.8 \text{ dB}$$

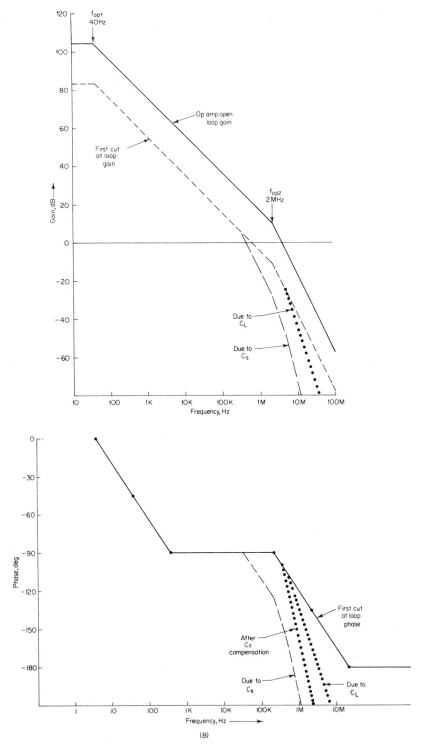

**Fig. 3.15** (A) LM101A open-loop gain and loop gain. (B) LM101A open-loop phase and loop phase.

To obtain the loop gain we must subtract 20.8 dB from all points on the op amp gain plot. Figure 3.15A shows the result in short dashed lines. The unity-loop-gain crossover frequency is 600 kHz. The phase lag at this frequency is 110°, which makes a phase margin of $180° - 110° = 70°$. The amplifier appears to be very stable indeed.

Let us add a few practical considerations to the circuit and see how much the phase margin is reduced. First we consider the effect of load capacitance. The op amp data sheet shows an approximate output resistance $R_o$ of 100 Ω near unity-loop-gain crossover. We will assume a load capacitance

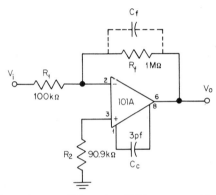

**Fig. 3.16** Inverting amplifier using 101A op amp with a gain of 10.

$C_L$ of 300 pF. Referring to Eq. (3.32), we see that because of $R_o$ and $C_L$ a new pole is created at a frequency of

$$f_p = \frac{1}{2\pi R_o C_L} = \frac{1}{6.28 \times 10^2 \times 3 \times 10^{-10}} = 5.5 \text{ MHz}$$

This pole is added to the loop-gain plot in Figure 3.15A. No change in unity-loop-gain crossover frequency is observed. However, at that frequency the phase plot in Fig. 3.15B shows that approximately 5° has been subtracted from the phase margin. This makes the phase margin 65°, which is still very good. The transfer function for $R_o$ and $C_L$ is $A_L = 5 \times 10^6/(jf + 5 \times 10^6)$. This term will be incorporated into the final loop-gain equation after all other factors in the loop gain are computed.

Next we compute the effects of stray capacitance across the op amp input terminals. Assume the combination $C_s$ of op amp input capacitance and stray capacitance across the input adds up to 5 pF. The circuit composed of $R_1$, $R_f$, and $C_s$ (with $V_i = 0$) perfectly matches one of the classical circuits shown in Appendix VI. The feedback factor for this circuit is

$$A_f = \frac{1/2\pi R_f C_s}{jf + [(R_1 + R_f)/2\pi R_1 R_f C_s]}$$

$$= \frac{1/6.28 \times 10^6 \times 5 \times 10^{-12}}{jf + [(10^5 + 10^6)/6.28 \times 10^5 \times 10^6 \times 5 \times 10^{-12}]}$$

$$= \frac{31,800}{jf + 350,100}$$

This new pole in the loop gain is shown in Fig. 3.15A. The loop gain now becomes unity at about 400 kHz instead of the original 600 kHz. The phase plot is also drastically changed, and at 400 kHz the loop phase lag is 150°. The phase margin is therefore 180° − 150° = 30°. This phase margin is 15° less than the goal stated in the opening paragraph of this section. We must carefully examine the reasons for not achieving this goal and work out a way to overcome this potential instability problem.

The capacitance across the op amp input in combination with the 1-M$\Omega$ feedback resistor are the real cause of the problem. We could assume a smaller capacitance, but a value lower than 4 pF is not recommended. The data sheets of many op amps specify a 3-pF maximum input capacitance. The printed-circuit board must be extremely well laid out to keep stray capacitance below 1 pF. For these reasons 5 pF is a reasonably conservative value for $C_s$.

The feedback resistor is the next likely candidate in our attempt to restore stability. If $R_f$ is reduced to 500 k$\Omega$ (and $R_1$ is reduced to 50 k$\Omega$), the pole caused by $R_f$ and $C_s$ increases to 600 kHz. The unity-loop-gain crossover then returns to 600 kHz. The loop-phase plot also moves slightly to the right. However, a close examination reveals that the phase margin is still approximately 30°. In addition to this problem, the circuit input impedance has been lowered to 50 k$\Omega$.

Additional compensation can restore the phase margin to 45°. The only problem is to decide which method is optimum. We will briefly tabulate several options and then try the one which looks most promising.

1. Lead compensation: If we add a small capacitor across $R_f$, a zero is added to the loop gain. This zero is calculated so that it cancels the pole caused by $R_f$ and $C_s$. The closed-loop bandwidth will remain essentially unchanged.

2. Input lead-lag compensation: By inserting a series $RC$ circuit across the op amp input terminals, we can create both a pole and a zero in the loop gain. The pole is at a lower frequency than the zero, as shown in Fig. 3.8E. The zero must be placed at a frequency several octaves below the pole caused by $C_s$ and $R_f$. Likewise, the new compensation pole must be at least several octaves lower in frequency than the above zero. This increases phase margin but reduces closed-loop bandwidth.

3. Increase $C_c$ compensation: If $C_c$ is increased, the dominant pole of loop gain is reduced in frequency. This makes the second pole occur closer to unity gain, causing the phase margin to increase. The closed-loop bandwidth is also reduced.

Of these three options, only the first does not decrease the closed-loop bandwidth. We will therefore use lead compensation to stabilize the amplifier. The instability is caused by the 350-kHz pole owing to $C_s$ and $R_f$. The zero from the lead compensation must be placed at this same frequency. Since $R_f$ and $f_z$ are fixed, we solve for $C_f$:

$$C_f = \frac{1}{2\pi f_z R_f} = \frac{1}{6.28 \times 3.5 \times 10^5 \times 10^6} = 1.6 \text{ pF}$$

The plot of loop gain will now look identical to the plot before $C_s$ was considered. The new value of $A_f$ is again

$$A_f = \frac{R_1}{R_1 + R_f} = \frac{10^4}{10^4 + 10^5} = 0.0909$$

The phase margin is restored to 65°, showing that the 1.6-pF capacitor is sufficient to restore stability.

We must still consider the fifth cause of instability, namely, the resistor $R_2$ between the op amp positive input and ground. The size of this resistor is

90.9 kΩ. In most cases if $R_2$ is below 10 kΩ, the designer need not worry about it. If $R_2$ is too large, either it can be bypassed or $C_f$ can be made slightly larger. In the generation of the total loop-gain equation, $R_2$ merely adds to $R_f$ (see Sec. 3.1 under Development of the Loop-Gain Equation). The new value for $A_f$ (assuming $R_2$ is unbypassed) is

$$A_f(\text{final}) = \frac{10^5}{10^5 + 1.09 \times 10^6} = 0.084$$

A recalculation of the pole caused by $C_s$ shows it is now 347.7 kHz. Since the pole changed only 3 kHz, it is unlikely $C_f$ will need to be changed.

We can now compute the final loop-gain equation. It will be the product of op amp gain, the transfer function $A_L$ caused by $R_o$ and $C_L$, and $A_f$.

$$\text{Loop gain} = A_v A_f A_L$$

$$= \frac{1.28 \times 10^{13}}{(jf + 40)(jf + 2 \times 10^6)} \frac{0.084(jf + 3.5 \times 10^5)}{jf + 3.5 \times 10^5} \frac{5 \times 10^6}{jf + 5 \times 10^6}$$

$$= \frac{5.38 \times 10^{18}}{(jf + 40)(jf + 2 \times 10^6)(jf + 5 \times 10^6)}$$

**SECOND FEEDBACK-STABILITY EXAMPLE – WIDE-BANDWIDTH UNITY-GAIN INVERTING AMPLIFIER** The procedures presented in this chapter are especially useful for difficult design problems. As an example of a difficult design problem we will look at the stability of a unity-gain amplifier capable of 10-MHz operation. The circuit input resistance is to be 10,000 Ω. After data sheets of several op amps are examined, the μA702 (Fairchild) is chosen. Its data sheet shows that if compensation is used as shown in Fig. 3.17, a unity-gain 20-MHz circuit is possible. This is a good margin above the required 10-MHz bandwidth.

We start by making accurate plots of open-loop gain and phase for the μA702. Figure 3.18A shows the gain plot obtained from the data sheet. This device has three poles before unity-gain crossover, and their estimated locations are 0.7, 4, and 40 MHz. Figure 3.18B shows how we can graphically obtain the phase plot when it is not shown on the data sheet. The dashed lines show the 90° lag plot for each pole. The −45°

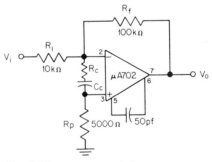

**Fig. 3.17** Recommended compensation for μA702 if unity-gain 10-MHz operation is required.

point for each of these plots correspond to the pole center frequency. The slope of the phase plots is −45° per decade for one decade above and below the pole frequency. The slope is zero elsewhere. The phase plot for the op amp is drawn by merely adding the contributions from each pole at each frequency.

We next determine the effect of the 50-pF lead-compensation capacitor on the op amp gain plot. According to the data sheet this capacitor cancels out the second pole $f_{op2}$. The resultant open-loop gain with only this capacitor installed is shown in Fig. 3.18A (dashed lines). Referring to Eq. (3.31), we note that the minimum closed-loop gain (for $\phi_m = 45°$) is now $A_\pi - 1 =$ 32 dB − 1 = 39.8 − 1 = 38.8, whereas before compensation the minimum

**3-32** FEEDBACK STABILITY

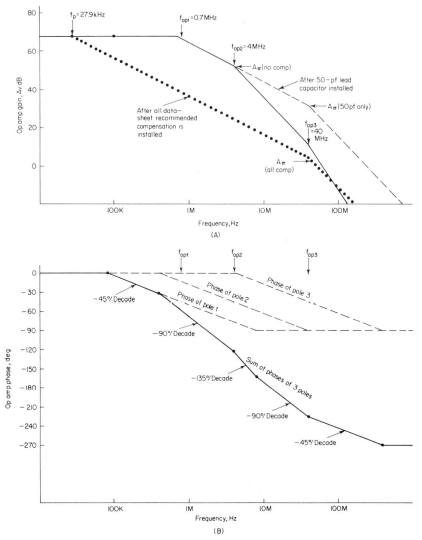

**Fig. 3.18** (A) Compensating the $\mu$A702 gain. (B) Compensating the $\mu$A702 phase.

gain was $A_\pi - 1 = 52$ dB $- 1 = 398 - 1 = 397$. This is getting closer to our goal, but more compensation is required to allow unity closed-loop gain.

Figure 3.17 shows that the data sheet also recommends input lead-lag compensation. We will now compute and plot the effect of this additional compensation on open-loop gain. Equation (3.21) is used to compute the zero frequency.

$$f_z = \frac{1}{2\pi R_c C_c} = \frac{1}{(6.28)(200)(10^{-9})} = 795 \text{ kHz}$$

This is very close to the open-loop dominant pole of 700 kHz; so the two will cancel. The new dominant-pole frequency is found by using Eq. (3.22).

$$f_p = \frac{R_1 + R_f}{2\pi C_c(R_1R_f + R_1R_c + R_1R_p + R_cR_f + R_pR_f)}$$

$$= \frac{10^4 + 10^4}{6.28 \times 10^{-9}(10^4 \times 10^4 + 10^4 \times 200 + 10^4 \times 500 + 200 \times 10^4 + 500 \times 10^4)}$$

$$= 27.9 \text{ kHz}$$

The effects of this new pole and zero are plotted on Fig. 3.18A (dotted line). The second pole now occurs close to unity-gain crossover, making $A_\pi = 4$ dB. The minimum gain for a phase margin of 45° is now $A\pi - 1 = 4$ dB $- 1 = 1.58 - 1 = 0.58$, which is less than unity. It appears at this point that the compensation recommended by the data sheet is adequate. Next we look into the third, fourth, and fifth causes of instability before the stability analysis is complete.

We will now determine the load capacitance, which will decrease the phase margin by, say, 6°. Before we do this, however, we must find the approximate frequency of unity-loop-gain crossover. This requires us to estimate loop gain. Assume for the present that the feedback network $R_1$ and $R_f$ has no stray capacitance and hence no phase lag. The feedback factor $A_f$ is therefore $R_1/(R_1 + R_f) = 10^4/2 \times 10^4 = 0.5 = -6$ dB. The loop gain is $A_f$ times the op amp open-loop gain $A_v$. This means that we merely subtract 6 dB from all points on the totally compensated plot shown in Fig. 3.18A. This new plot is shown in Fig. 3.19A. The corresponding plot of loop phase is shown in Fig. 3.19B. The second pole now occurs below unity gain, which will give us good loop stability. Since the loop-gain crossover is at 32 MHz, the phase margin is $180 - 130 = 50°$. This is close to the $\phi_m = 45°$ determined above.

An accurate plot of a pole (see Chap. 1) shows a phase lag of 6° at a frequency ten times lower than the pole. If we want the load capacitance $C_L$ to diminish the phase margin by no more than 6°, we must make [using Eq. (3.32)]

$$C_L \leq \frac{1}{2\pi(10f_g)R_o}$$

where
$f_g$ = unity-loop-gain crossover frequency
$R_o$ = op amp output resistance = 200 Ω

In other words, the pole caused by $R_o$ and $C_L$ must be at least ten times the frequency of unity-loop-gain crossover to cause less than 6° reduction in phase margin. This would reduce the phase margin to $50 - 6 = 44°$. Calculating $C_L$, we get

$$C_L = \frac{1}{6.28 \times 10 \times 3 \times 10^7 \times 200} = 2.65 \text{ pF}$$

This is such a small load capacitance that we should probably take a different approach. Suppose our load is guaranteed to be less than 25 pF. What reduction in phase margin will this cause? Again using Eq. (3.32), we get

$$f_p = \frac{1}{2\pi R_o C_L} = \frac{1}{6.28 \times 200 \times 25 \times 10^{-12}} = 32 \text{ MHz}$$

If we plot this additional pole on Fig. 3.19A, we note that unity-gain crossover is still at 32 MHz. However, the phase plot from about 4 MHz and higher is picking up the lag due to two poles. By the time 32 MHz is reached, the phase margin is only about 5°. This is a very poor phase margin.

## 3-34 FEEDBACK STABILITY

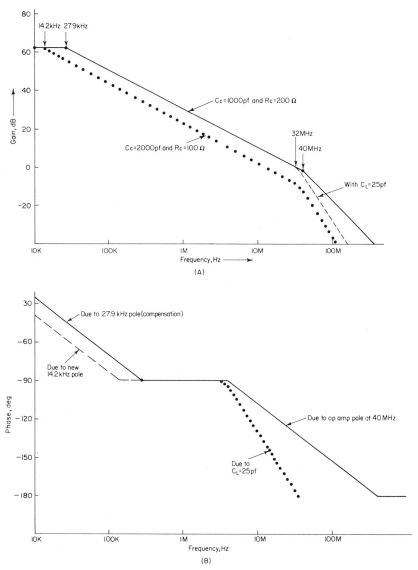

**Fig. 3.19** (A) Final compensation for $\mu$A702 gain. (B) Final compensation of $\mu$A702 phase.

We could specify that $C_L$ shall never go above 5 or 10 pF, but that severely restricts the utility of the amplifier. Perhaps we can increase $\phi_m$ by reducing the closed-loop bandwidth to approximately 10 MHz. We cannot merely increase $C_c$ to reduce bandwidth in Fig. 3.18A, since $C_c$ will reduce the frequency of the compensation pole and zero. The resultant phase margin would not change. We must find a way of reducing the pole frequency by 2 without appreciably affecting the zero frequency. This can be done by doubling $C_c$ and dividing $R_c$ by 2. The zero frequency remains unchanged while the new pole becomes [see Eq. (3.22)]

$$f_p = \frac{10^4 + 10^4}{6.28 \times 2 \times 10^{-9}(10^4 \times 10^4 + 10^4 \times 10^2 + 10^4 \times 500 + 10^2 \times 10^4 + 500 \times 10^4)}$$

$$= 14.2 \text{ kHz}$$

The new loop-gain plot is shown with a dotted line in Fig. 3.19A. The unity-loop-gain crossover frequency becomes 13 MHz. The phase plot of Fig. 3.19B is changed only for frequencies near the new $f_p$. The phase lag near unity-loop-gain crossover is not affected. Therefore, the phase margin is increased to a value of $180 - 137 = 43°$.

The effects of stray capacitance across the op amp input terminals will now be considered. In previous sections we showed that this stray capacitance $C_s$ in combination with the feedback resistor creates another pole at a frequency of $f_p = 1/2\pi R_f C_s$. With input lead-lag compensation this pole is no longer present. The impedance of $R_c$ and $C_c$ is so low at frequencies near loop-gain crossover that any stray capacitance up to about 30 pF is swamped out. The effects of $R_p$ on loop gain are also minimized for this same reason. The fourth and fifth causes of instability need not be considered further.

## 3.5 FEEDBACK-STABILITY MEASUREMENTS

A good designer never relies on a paper design of a circuit but instead always does extensive testing of all circuit parameters and if time is available, does many of these tests with worst-case environments and/or part-parameter variations. Feedback stability is probably the most important circuit parameter. It can totally disable an op amp circuit for quite a few reasons. Some of these reasons are quite straightforward and can be quickly analyzed. Other reasons for instability may be difficult to predict and also difficult to analyze. It is therefore mandatory that a good op amp circuit designer know at least two or three stability-testing methods. The number of methods he uses for any particular circuit will depend on the time available and the degree to which the designer wants to understand the circuit. We will present three methods of measuring stability in this section.

**LOOP-GAIN MEASUREMENT METHOD** A designer who wants to verify each loop-gain calculation can measure the transfer function of each portion of the loop. Several of these measurements are quite difficult, but the effort always results in greater understanding of circuit operation.

Most op amp oscillation problems occur at frequencies between 10 kHz and 10 MHz. Some wide-bandwidth op amps have problems up to 100 MHz. The equipment required for measuring transfer functions affecting loop gain is as follows:

1. Sine-wave generator capable of 1 kHz to 10 MHz (or more). This generator should have a constant output amplitude over frequency.

2. Dual-beam or dual-trace oscilloscope. This oscilloscope should have the capability of showing the precise phase relationship between two inputs. That is, both traces should be triggered by one input. A low-capacitance (< 10-pF) probe should be used.

The op amp open-loop gain as a function of frequency can be measured by several methods. Op amp manufacturers have published many circuits for performing this test in various application notes. This is a difficult test at frequencies below 100Hz, since the low-frequency op amp gain is so high. For this reason the gain test is usually restricted to frequencies above 1 kHz.

The change in loop gain due to a load capacitance can be determined by running the above test with a load capacitance installed. The op amp output resistance changes substantially at frequencies near its unity-gain crossover. Therefore, this test should not be made with a simulated $R_o$ but must include the actual $R_o$ of the op amp. The transfer function due to $R_o$ and $C_L$ can be obtained by subtracting (on a dB scale) the op amp gain plot from the op amp plus $C_L$ gain plot. Be sure to include the oscilloscope probe capacitance in the value recorded for $C_L$.

The transfer function for the feedback network must be made very carefully. The phase lag under investigation may be caused by only a few picofarads of stray capacitance $C_s$. The oscilloscope probe may have more input capacitance $C_i$ than the stray capacitance being simulated. If one has an oscilloscope probe with the same input capacitance as the $C_s$ to be simulated, the probe can be used in place of the simulated $C_s$. If the probe capacitance is higher than $C_s$, the probe effects must be subtracted from the transfer-function plot. Assume that the measured transfer-function pole frequency is $f_p = 1/2\pi R_f C$, where $R_f$ is the feedback capacitor and $C = C_s + C_i$. If we know $f_p$, $R_f$, $C_s$, and $C_i$, the pole frequency $f'_p$ without $C_i$ is computed from $f'_p = 1/2\pi R_f C_s$, where $C_s = 1/(2\pi f_p R_f) - C_i$. It is a good idea to run this test on the circuit board using the actual $R_f$, $C_s$, and op amp. The op amp should have power applied, but the side of $R_f$ normally connected to the op amp output should be disconnected. The generator then drives $R_f$ at this point and the low-capacitance probe is attached across the op amp input. If too much difficulty is experienced with this test, the probe should be connected to the op amp output and the op amp plot subtracted from the results of this test (on a dB scale).

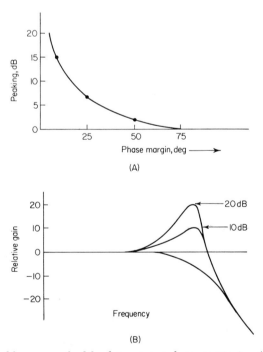

**Fig. 3.20** Closed-loop ac method for determining phase margin $\phi_m$. (A) Phase margin as a function of closed-loop peaking. (B) Frequency plots showing how peaking is defined.

**CLOSED-LOOP AC METHOD** Closed-loop measurements are much easier to perform than open-loop measurements. The only problem with closed-loop measurements is that one cannot isolate a problem down to a specific part very easily. The ease of a closed-loop measurement, however, more than outweighs this small disadvantage.

The ac method requires a sine-wave generator and an oscilloscope. The generator must be capable of frequencies at least twice the closed-loop bandwidth. Likewise, the oscilloscope must not appreciably attenuate frequencies of the same magnitude. The method will work for either inverting- or noninverting-amplifier configurations.

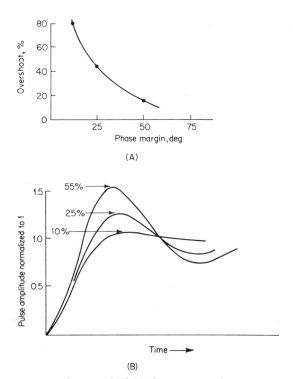

**Fig. 3.21** Transient-overshoot method for determining phase margin $\phi_m$. (A) Curve showing approximate relationship of $\phi_m$ to percent overshoot. (B) Overshoot curves showing how percent overshoot is defined.

The closed-loop ac method gives us approximate phase margin once we determine the magnitude of closed-loop peaking. Figure 3.20A shows the approximate phase margin as a function of closed-loop ac peaking. Figure 3.20B shows how closed-loop peaking is defined. Each curve shown represents a different damping factor $\zeta$. The frequency $f_n$ is the frequency at which the circuit will oscillate if $\zeta = 0$ (i.e., if the phase margin $\phi_m = 0$). This family of curves are graphical representations of the formula

$$|A_{vc}| = \frac{A_{vo}}{\sqrt{[1 - (f^2/f_n^2)]^2 + (4\zeta^2 f^2/f_n^2)}} \qquad (3.40)$$

**3-38 FEEDBACK STABILITY**

where $|A_{vc}|$ = magnitude of closed-loop gain
$A_{vo}$ = op amp open-loop gain at dc
$f_n$ = frequency of peaking or frequency of oscillation if $\phi_m = 0$
$\zeta$ = damping factor

Use of Fig. 3.20A to determine phase margin has several limitations:
1. The frequency of the op amp dominant pole must be more than a decade lower than the intersection frequency of closed- and open-loop plots.
2. The third pole of op amp open-loop gain (if any) must be at least two octaves higher in frequency than the second pole.

If these conditions are not met, the ac method can still be used to observe relative stability. The relationship between $\phi_m$ and closed-loop peaking will not be right, but one can still determine roughly how close the circuit is to instability. It is simple to add $C_s$ and $C_L$ and watch the peaking grow until the circuit finally breaks out into oscillation.

**TRANSIENT-RESPONSE METHOD** The fastest method for determining phase margin is by observing the transient overshoot. Because this method is quick and easy to set up, it is best suited for production-line testing of circuit stability. This method has the same two limitations as the ac method; it is accurate only for two-pole response and the poles must be widely separated.

The relationship between transient overshoot and phase margin is shown in Fig. 3.21A. In Fig. 3.21B we show how overshoot is defined. Suppose we let the steady-state value of the output be 100. The percent overshoot is merely the difference between the peak value and 100.

Transient-overshoot measurements must be made at a low level. The peak output voltage should be less than 100 or 200 mV to avoid slew-rate limiting.

The transient-response method is ideal for experimentally determining such things as maximum $C_L$, maximum $R_f$, minimum gain, and $C_s$ associated with different printed-circuit layouts.

## REFERENCES

1. Millman, J., and H. Taub: "Pulse, Digital, and Switching Waveforms," McGraw-Hill Book Company, New York, 1965.
2. Tobey, G. E., J. G. Graeme, and L. P. Huelsman: "Operational Amplifiers — Design and Applications," McGraw-Hill Book Company, New York, 1971.
3. Millman, J., and C. C. Halkias: "Integrated Electronics," McGraw-Hill Book Company, New York, 1972.
4. "Linear Integrated Circuits Applications Handbook," Fairchild Semiconductor, Mountain View, 1967.
5. "Linear Integrated Circuits Data Catalog," Fairchild Semiconductor, Mountain View, 1973.
6. "Feedback and Control Systems — Theory and Problems" (Schaum's Outline Series), Schaum Publishing Co., New York, 1967.

Chapter **4**

# Amplifiers

**INTRODUCTION**

Amplifiers are the most widely utilized application for operational amplifiers. We will cover eight of these applications in this chapter. Differential-mode applications will be explored separately, in detail, in Chap. 9, as they are such an important class of amplifiers. Since the inverting amplifier is the fundamental building block of most analog systems, it will be covered first with a numerical example worked out in detail. Other basic circuits such as (1) the noninverting amplifier, (2) the current amplifier, (3) the current-to-voltage amplifier (or the transresistance amplifier), and (4) the voltage-to-current amplifier (or the transconductance amplifier) will be presented with a large number of design equations.

At the end of the chapter several other types of amplifiers will be briefly presented along with several of their design equations. These include ac-coupled inverting amplifier, charge-sensitive amplifier, and summing amplifier.

## 4.1 BASIC INVERTING AMPLIFIER

**ALTERNATE NAMES** Inverting-mode amplifier, inverting configuration, phase inverter, inverter.

**EXPLANATION OF OPERATION** In Fig. 4.1A, assume $v_i$ starts to drive $v_n$ (i.e., $v_n - v_p$) in the same direction as $v_i$. The op amp output $A_{vo}(v_p - v_n)$ will then be driven in a direction opposite to that of $v_i$. This will cause a feedback current through $R_f$ which attempts to drive $v_n - v_p$ back toward zero. If $A_{vo}$ is very large, $v_n - v_p$ is driven to nearly zero and the current into the op amp negative input terminal also approaches zero. All the input current through $R_1$ must therefore flow through $R_f$. The current through $R_p$ also approaches zero as $A_{vo}$ becomes very large. The voltage at $v_n$ can thus be assumed equal to zero, which makes computation of the circuit gain quite easy. Since the same current $i_i$ flows in $R_1$ and $R_f$,

$$i_i = \frac{v_i}{R_1} = -\frac{v_o}{R_f}$$

The closed-loop-circuit voltage gain at dc becomes

## 4-2 AMPLIFIERS

$$A_{vco} = \frac{v_o}{v_i} = -\frac{R_f}{R_1}$$

In the design equations and design procedure to follow we will show the effect of various op amp parameters on the accuracy and stability of $A_{vco}$. Likewise, we will examine the closed-loop voltage gain as a function of frequency $A_{vc}$. The effects of nonideal op amp parameters on $A_{vc}$ will also be summarized.

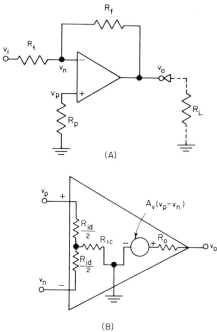

**Fig. 4.1** (A) Basic inverting-amplifier circuit. (B) Definition of input/output resistances and voltage gain.

### DESIGN PARAMETERS

| Parameter | Description |
|---|---|
| $A_v$ | Open-loop voltage gain of op amp as a function of frequency |
| $A_{vo}$ | Open-loop dc voltage gain of op amp ($A_{vo}$ may be substituted for $A_v$ in any equation if dc characteristics are wanted) |
| $A_{vc}$ | Closed-loop voltage gain of circuit as a function of frequency |
| $A_{vco}$ | Closed-loop dc voltage gain of circuit |
| $\beta$ | Voltage feedback ratio of $R_1$ and $R_f$, i.e., $\beta = R_1/(R_1 + R_f)$ |
| $f_{cp}$ | First-pole frequency of circuit, i.e., the $-3$-dB bandwidth |
| $f_{op}$ | First-pole frequency of op amp |
| $I_b$ | Input bias current of op amp |
| $I_{io}$ | Input offset current of op amp |
| $I_n$ | Equivalent input noise current of op amp |
| $R_1$ | Input resistor |
| $R_f$ | Feedback resistor |
| $R_{ic}$ | Common-mode input resistance of op amp |

## BASIC INVERTING AMPLIFIER 4-3

## DESIGN PARAMETERS (Continued)

| Parameter | Description |
|---|---|
| $R_{id}$ | Differential input resistance of op amp (Fig. 4.1B) |
| $R_{in}$ | Input resistance of circuit |
| $R_L$ | Load resistance of circuit |
| $R_o$ | Output resistance of op amp |
| $R_{out}$ | Output resistance of circuit |
| $t_r$ | Risetime of circuit (10 to 90%) |
| $V_{io}$ | Input offset voltage of op amp |
| $V_n$ | Equivalent input noise voltage of op amp |
| $V_{on}$ | Output noise voltage of circuit |
| $v_o$ | Output voltage of circuit |

## DESIGN EQUATIONS

| Eq. No. | Description | Equation |
|---|---|---|
| 1 | Closed-loop voltage gain assuming ideal op amp parameters | $A_{vc} = -\dfrac{R_f}{R_1}$ |
| 2 | Closed-loop voltage gain if finite op amp gain $A_v$ is included | $A_{vc} = \dfrac{-R_f/R_1}{1 + 1/\beta A_v}$ where $\beta = \dfrac{R_1}{R_1 + R_f}$ |
| 3 | Closed-loop voltage gain if differential input resistance $R_{id}$ is included ($A_v$ must also be included) | $A_{vc} = \dfrac{-R_f/R_1}{1 + 1/\beta A_v + 2R_f/A_v R_{id}}$ |
| 4 | Closed-loop voltage gain if the op amp output resistance $R_o$ is included ($A_v$ must also be included) | $A_{vc} = \dfrac{-R_f/R_1}{1 + (R_f + R_o)/\beta A_v R_f}$ |
| 5 | Size of $R_f$ for minimum gain error due to $A_v$, $R_{id}$, and $R_o$ | $R_f(\text{opt}) = \left(\dfrac{R_{id} R_o}{2\beta}\right)^{1/2}$ |
| 6 | Input resistance of circuit assuming ideal op amp parameters | $R_{in} = R_1$ |
| 7 | Input resistance of circuit assuming finite $A_{vo}$ | $R_{in} = R_1\left(1 + \dfrac{R_f}{A_{vo} R_1}\right)$ |
| 8 | Output resistance of circuit assuming ideal op amp parameters | $R_{out} = 0$ |
| 9 | Output resistance of circuit assuming finite op amp output resistance $R_o$ and finite $A_v$ | $R_{out} = \dfrac{R_o}{1 + \beta A_v}$ |
| 10 | Bandwidth of circuit assuming bandwidth ($-3$ dB) of op amp is at $f_{op}$ ($f_{op}$ = first pole of the op amp) | $f_{cp} = \dfrac{f_{op} A_{vo} R_1}{R_f}$ |
| 11 | Small-signal risetime of circuit (10 to 90%) | $t_r = \dfrac{0.35\, R_f}{f_{op} A_{vo} R_1}$ |
| 12 | Output dc voltage change due to input offset-voltage change of op amp (assuming $I_b$ and $I_{io} = 0$) | $\Delta V_o = \pm\, \Delta V_{io}\, \dfrac{R_1 + R_f}{R_1}$ |

## AMPLIFIERS

### DESIGN EQUATIONS (Continued)

| Eq. No. | Description | Equation |
|---|---|---|
| 13 | Output dc voltage due to input bias current of op amp assuming $R_p = 0$ and $V_{io} = 0$ | $V_o = I_b R_f$ |
| 14 | Output dc voltage change due to input offset-current change of op amp assuming $R_p = \dfrac{R_1 R_f}{R_1 + R_f}$ and $V_{io} = 0$ | $\Delta V_o = \pm \Delta I_{io} R_f$ |
| 15 | Output-noise voltage due to an equivalent op amp input noise voltage in volts/$\sqrt{\text{Hz}}$ | $V_{on} = V_n \left(1 + \dfrac{R_f}{R_1}\right)$ V/$\sqrt{\text{Hz}}$ |
| 16 | Output-noise voltage due to both equivalent op amp input-voltage noise and current noise (V/$\sqrt{\text{Hz}}$ and A/$\sqrt{\text{Hz}}$ or V²/Hz and A²/Hz) | $V_{on} = \left[V_n^2\left(1 + \dfrac{R_f}{R_1}\right)^2 + I_n^2 R_f^2\right]^{1/2}$ |
| 17 | Optimum value for $R_p$ to minimize output-voltage offset due to $I_b$ | $R_p = \dfrac{R_1 R_f}{R_1 + R_f}$ |

### DESIGN PROCEDURE

The approach one takes in designing an inverting amplifier depends on the application of the circuit. Many designs will merely require usage of Eqs. 1, 6, and 9 (voltage gain and input/output resistances). High-frequency applications may require a trade-off between Eqs. 1 (voltage gain) and 10 (closed-loop bandwidth). Low-level precision dc amplifiers may require compromises among Eqs. 2, 3, 4, 5, 7, 9, 12, 13, 14, 16, and 17. This last category would be the most difficult, since there are many conflicting requirements in the above list of equations. For example, Eq. 5 indicates that an optimum $R_f$ can be chosen which reduces the effect of changes in $A_v$, $R_{id}$, and $R_o$ on the circuit. Contrary to this, Eqs. 7, 12, 13, 14, and 16 imply that $R_f$ should be as small as practical. The final choice for $R_f$ in this case would probably be simplified if a few extra dollars were invested in a good-quality instrumentation op amp.

In the design steps to follow, we will assume that the op amp type is already chosen and that both the closed-loop dc gain $A_{vco}$ and the minimum input resistance $R_{in}$ are specified.

The variation of dc closed-loop gain $A_{vco}$ as a result of variations in open-loop dc gain $A_{vo}$, differential input resistance $R_{id}$, $R_1$, $R_f$, and output resistance $R_o$ will be calculated. The drift of the circuit output dc voltage as a result of drifts in input offset voltage $V_{io}$ and input offset current $I_{io}$ will be determined. The small-signal risetime, bandwidth, and output noise will also be computed.

### DESIGN STEPS

*Step 1.* Compute the optimum $R_f$ with Eq. 5:

$$R_f(\text{opt}) = \left(\frac{R_{id} R_o}{2\beta}\right)^{1/2}$$

where $\beta = 1/(1 - A_{vco})$ may be used as a first cut.

*Step 2.* Determine the size of $R_1$ using Eq. 1 and the results of step 1 above:

$$R_1 = \frac{-R_f}{A_{vco}}$$

If this computed $R_1$ is less than the minimum specified circuit input resistance, a compromise among Eqs. 1, 5, and 6 may be required. Assuming Eqs. 1 and 6 are more important, we will calculate the errors caused by a nonoptimum $R_f$ in the following steps. This approach means we let $R_1 = R_{in}$ and $R_f = -A_{vco} R_1$.

*Step 3.* Given the range of temperatures within which the circuit is to be operated, the variations in $A_{vo}$, $R_{id}$, $R_1$, $R_f$, $V_{io}$, and $I_{io}$ are determined from the op amp data sheet.

*Step 4.* Assuming a given range of frequencies must be amplified with minimum error and noise, the following parameters as a function of frequency are also found from the data sheet. $A_v$, $R_o$, $V_n$, and $I_n$.

*Step 5.* Compute the variations in closed-loop dc voltage gain $A_{vco}$ using data from step 3 and Eqs. 2, 3, and 4. Repeat this step using $A_v$ at selected frequencies of interest.

*Step 6.* Determine the circuit output resistance $R_{out}$ at selected frequencies using Eq. 9. Compute the reduction of $A_{vc}$ using $R_{out}$ and $R_L$ with voltage-divider theory. The effective voltage gain $A_{vc}$ with $R_L$ attached is

$$A_{vc} \text{ (with } R_L) = \frac{A_{vc}(\text{no load})R_L}{R_L + R_{out}}$$

*Step 7.* Determine the true input resistance by usage of Eq. 7. If the input resistance is critical and this calculation reveals a design deficiency, $R_1$ may need to be increased. Steps 2, 5, 6, and 7 will then need to be repeated.

*Step 8.* Compute the variation in $A_{vco}$ due to resistance changes in $R_1$ and $R_f$. Since $A_{vc} = -R_f/R_1$, a $\pm 1$ percent change in either $R_f$ or $R_1$ will result in a $\pm 1$ percent change in $A_{vc}$. If $R_1$ increases 1 percent as $R_f$ decreases 1 percent, $A_{vc}$ will decrease 2 percent. Resistor variations are unpredictable with some types, and the $\pm$ sign must be used (see Appendix III).

*Step 9.* Compute $R_p$ according to Eq. 17.

*Step 10.* If $R_p$ has been determined according to Eq. 17, Eq. 13 will not need to be calculated. The output offset drift voltage will be controlled only by $\Delta V_{io}$ and $\Delta I_{io}$. Determine the total output-offset-voltage change from Eqs. 12 and 14. If the temperature-dependent drifts of $V_{io}$ and $I_{io}$ are known in both magnitude and direction, the actual $\Delta V_o$ can be determined. However, in most cases $\pm$ symbols should be used, since either input terminal of the op amp can require more bias voltage or more bias current than the other. The output dc voltage error caused by the value of $V_{io}$ at $+25°C$ can be nulled out using special terminals on most op amps.

*Step 11.* Use Eq. 10 to compute the small-signal bandwidth and Eq. 11 for the small-signal risetime.

*Step 12.* If the equivalent input-noise current and noise voltage is available as a function of frequency, the output-noise voltage as a function of frequency can be determined. Equation 16 requires that $V_n$ be in units of $V/\sqrt{Hz}$ or $V^2/Hz$ and $I_n$ be in units of $A/\sqrt{Hz}$ or $A^2/Hz$. This equation will need to be solved at 5 or 10 frequencies using $V_n$ and $I_n$ data at these same frequencies.

**EXAMPLE OF INVERTING-AMPLIFIER DESIGN** An actual design of an inverting amplifier with a gain of $-100$ will now be presented.

## 4-6 AMPLIFIERS

*Design Requirements*
$A_{vco} = -100$
$R_{in} \geq 1{,}000 \ \Omega$
Op amp = $\mu$A741A (Fairchild)
$R_L \geq 2{,}000 \ \Omega$

*Device Data* ($-55$ to $+125°C$ and $\pm 20$ V supply voltages)
$V_{io} = \pm 0.8$ mV typical, $\pm 4.0$ mV worst case ($-55$ or $+125°C$)
$\Delta V_{io} = \pm 15 \ \mu V/°C$ maximum
$I_{io} = \pm 3.0$ nA typical, $\pm 70$ nA worst case ($-55°C$)
$\Delta I_{io} = \pm 0.5$ nA/°C maximum
$R_{id} = 6 \ M\Omega$ typical, $0.5 \ M\Omega$ worst case ($-55°C$)
$A_{vo} = 5 \times 10^4$ minimum at $+25°C$ and $3.2 \times 10^4$ minimum over temperature ($-55$ to $+125°C$)
$f_{op} = 8$ Hz typical, 6 Hz at $+125°C$
$R_o = 70 \ \Omega$ dc to 10 kHz, $90 \ \Omega$ at 100 kHz, and $280 \ \Omega$ at 1 MHz
$V_n = 5 \times 10^{-15} \ V^2/Hz$ at 10 Hz, $10^{-15} \ V^2/Hz$ at 100 Hz, $5 \times 10^{-16} \ V^2/Hz$ from 1 to 100 kHz
$I_n = 5 \times 10^{-23} \ A^2/Hz$ at 10 Hz, $5 \times 10^{-24} \ A^2/Hz$ at 100 Hz, $8 \times 10^{-25} \ A^2/Hz$ at 1 kHz, $3 \times 10^{-25} \ A^2/Hz$ from 10 to 100 kHz
$\Delta R_1 = \pm 100$ ppm/°C (i.e., $\pm 10^{-4}$ change in 1°C)
$\Delta R_f = \pm 100$ ppm/°C

*Step 1.* The value of $\beta$, the voltage transfer ratio of the feedback network, is

$$\beta = \frac{1}{1 - A_{vco}} = \frac{1}{1 + 100} = 0.0099$$

The optimum $R_f$ can now be computed from

$$R_f(\text{opt}) = \left(\frac{R_{id} R_o}{2\beta}\right)^{1/2}$$

$$= \left(\frac{6 \times 10^6 \times 70}{2 \times 0.0099}\right)^{1/2}$$

$$= 145{,}600 \ \Omega$$

*Step 2.* The size of $R_1$ becomes

$$R_1 = \frac{-R_f(\text{opt})}{A_{vco}} = \frac{-1.456 \times 10^5}{-100} = 1{,}456 \ \Omega$$

This resistance also satisfies the $1{,}000 \ \Omega$ minimum-input-resistance requirement.

*Step 3.* The variations of the following parameters over the temperature range of $-55$ to $+125°C$ are (using data at $+25°C$ as a reference):

| Parameter | $-55°C$ | $+25°C$ | $+125°C$ |
|---|---|---|---|
| $A_{vo}$ | $3.2 \times 10^4$ | $5 \times 10^4$ | |
| $R_{id}$ | $0.5 \times 10^6$ | $6 \times 10^6$ | |
| $\Delta V_{io}$ | $\pm 2$ mV | $\pm 0.8$ mV | $\pm 2.3$ mV |
| $\Delta I_{io}$ | $\pm 43$ nA | $\pm 3$ nA | $\pm 53$ nA |
| $R_1$ | $1{,}456 \pm 11.6 \ \Omega$ | $1{,}456 \ \Omega$ | $1{,}456 \pm 14.6 \ \Omega$ |
| $R_f$ | $145{,}600 \pm 1{,}160 \ \Omega$ | $145{,}600 \ \Omega$ | $145{,}600 \pm 1{,}456 \ \Omega$ |

(The above are from the Fairchild data sheet.)

## BASIC INVERTING AMPLIFIER   4-7

*Step 4.* The variations of the following parameters over the frequency range of dc to 1 MHz are:

| Parameter | dc | 10 Hz | 100 Hz | 1 kHz | 10 kHz | 100 kHz | 1 MHz |
|---|---|---|---|---|---|---|---|
| $A_v$ | $5 \times 10^4$ | $4 \times 10^4$ | 4,000 | 400 | 40 | 4 | 0.4 |
| $R_o$ | 70 | 70 | 70 | 70 | 70 | 90 | 280 |
| $V_n^2$, V²/Hz | | $5 \times 10^{-15}$ | $10^{-15}$ | $5 \times 10^{-16}$ | $5 \times 10^{-16}$ | $5 \times 10^{-16}$ | |
| $I_n^2$, A²/Hz | | $5 \times 10^{-23}$ | $5 \times 10^{-24}$ | $8 \times 10^{-25}$ | $3 \times 10^{-25}$ | $3 \times 10^{-25}$ | |

(The above is from the Fairchild data sheet.)

*Step 5.* The ideal gain should be

$$A_{vco}(\text{ideal}) = -\frac{R_f}{R_1}$$

$$= -\frac{145,600}{1,456} = -100$$

Since the open-loop dc gain $A_{vo}$ is only $5 \times 10^4$ worst-case minimum at +25°C, the actual gain will be (assuming ideal values for $R_1$ and $R_f$)

$$A_{vco}(A_{vo} \neq \infty) = \frac{-R_f/R_1}{1 + 1/\beta A_{vo}} = \frac{-145,600/1,456}{1 + 1/0.0099 \times 5 \times 10^4}$$

$$= -99.798387$$

If $A_{vo}$ decreases to $3.2 \times 10^4$ at −55°C, the dc closed-loop gain is reduced to

$$A_{vco}(A_{vo} = 3.2 \times 10^4) = \frac{-145,600/1,456}{1 + 1/0.0099 \times 3.2 \times 10^4}$$

$$= -99.685336$$

This gain is approximately 0.1 percent lower than the closed-loop gain at room temperature. The degradation caused by the typical input resistance $R_{id}$ at +25°C is

$$A_{vco}(R_{id} = 6 \text{ M}\Omega) = \frac{-R_f/R_1}{1 + 1/\beta A_{vo} + 2R_f/A_{vo}R_{id}}$$

$$= \frac{-145,600/1,456}{1 + 1/0.0099 \times 5 \times 10^4 + 2 \times 145,600/5 \times 10^4 \times 6 \times 10^6}$$

$$= -99.798290$$

An input resistance of 6 MΩ is obviously not an error source of concern. If the worst-case minimum $R_{id}$ of 0.5 MΩ is used, we get

$$A_{vco}(R_{id} = 0.5 \text{ M}\Omega) = \frac{-145,600/1,456}{1 + 1/0.0099 \times 5 \times 10^4 + 2 \times 145,600/5 \times 10^4 \times 0.5 \times 10^6}$$

$$= -99.797227$$

This is still a very small error when compared with the effects of a finite $A_{vo}$. The change of $R_{id}$ over the worst-case temperature range will affect the closed-loop gain by approximately 0.001 percent.

The effects of a finite $R_o$ in conjunction with $R_f$ are determined by Eq. 4.

$$A_{vco}(R_o = 70\ \Omega) = \frac{-R_f/R_1}{1 + (R_f + R_o)/\beta A_{vo}R_f}$$

$$= \frac{-145{,}600/1{,}456}{1 + (145{,}600 + 70)/0.0099 \times 5 \times 10^4 \times 145{,}600}$$

$$= -99.79829$$

This gain is approximately the same as that computed when only the finite $A_{vo}$ was considered.

The effects of output resistance at frequencies above the first op amp pole ($\approx 8$ Hz) must be determined with caution. Since $A_v$ lags $A_{vc}$ 90° for these frequencies, Eq. 4 must take this into account. This is done as follows (at 1 kHz where $A_v = -j400$):

$$A_{vc}(f = 1\ \text{kHz},\ R_o = 70\ \Omega) = \frac{-100}{1 + (145{,}600 + 70)/0.0099 \times (-j400) \times 145{,}600}$$

$$= \frac{-100}{1 + j0.2526} = \frac{-100}{1.0314\ \underline{/14°}}$$

$$= 96.95\ \underline{/-14°}$$

The closed-loop gain has a slight phase lag at 1 kHz. If we had not considered the phase shift of $A_v$, we would have mistakenly computed $A_{vc}$ as follows:

$$A_{vc} = \frac{-100}{1 + (145{,}600 + 70)/0.0099 \times 400 \times 145{,}600} = 79.83$$

Suppose one wishes to compute the reduction of $A_{vco}$ due to two causes such as $R_o = 70\ \Omega$ (at dc) and $R_{id} = 0.5$ MΩ. This is done as follows:

$$A_{vco} = -100\ \frac{99.79829}{100}\ \frac{99.797227}{100} = -99.595926$$

*Step 6.* As in the previous step, the phase shift of $A_v$ must be considered if the op amp first-pole frequency is exceeded.

$$R_{\text{out}}(\text{dc}) = \frac{R_o}{1 + \beta A_v}$$

$$= \frac{70}{1 + 0.0099 \times 5 \times 10^4} = 0.1411\ \Omega$$

$$R_{\text{out}}(1\ \text{kHz}) = \frac{70}{1 + 0.0099(-j400)}$$

$$= \frac{70}{\sqrt{1^2 + (-3.96)^2}} = \frac{70}{\sqrt{1 + 15.68}}$$

$$= \frac{70}{4.084\ \underline{/-76°}} = 17.14\ \underline{/76°}\ \Omega$$

$$R_{\text{out}}(100\ \text{kHz}) = \frac{90}{1 + 0.0099(-j4)}$$

$$= \frac{90}{\sqrt{1^2 + (-0.0396)^2}} = \frac{90}{\sqrt{1 + 0.001568}}$$

$$= \frac{90}{1.00078} = 89.93$$

The reduction in closed-loop gain $A_{vc}$ due to interaction between $R_{out}$ and $R_L$ is computed by use of voltage-divider theory. If $R_{out} = 17.14\,\Omega$ (1 kHz) and $R_L = 2\,k\Omega$, the effective closed-loop gain $A_{vc}$ is

$$A_{vc}(\text{with } R_L) = \frac{A_{vco}(\text{no load})R_L}{R_L + R_{out}}$$

$$= \frac{-96.95 \times 2{,}000}{2{,}000 + 17.14} = -96.12$$

*Step 7.* The true input resistance at dc is

$$R_{in} = R_1\left(1 + \frac{R_f}{A_{vo}R_1}\right) = 1{,}456\left(1 + \frac{145{,}600}{5 \times 10^4 \times 1{,}456}\right)$$

$$= 1{,}459\,\Omega$$

*Step 8.* If we use $+25°C$ as a reference, the closed-loop-gain variation due to resistor changes over temperature is (+25 to +125°C)

$$A_{vco} = -\frac{R_f \pm \Delta R_f}{R_1 \pm \Delta R_1}$$

$$= -\frac{145{,}600 \pm 145{,}600 \times 10^{-4} \times 100}{1{,}456 \pm 1{,}456 \times 10^{-4} \times 100}$$

$$= -\frac{145{,}600 \pm 1{,}456}{1{,}456 \pm 14.56} = -98.02 \text{ to } -102.02$$

*Step 9.* The required value of $R_p$ is

$$R_p = \frac{R_1 R_f}{R_1 + R_f} = \frac{(1{,}456)(145{,}600)}{1{,}456 + 145{,}600} = 1{,}442\,\Omega$$

*Step 10.* The change in dc output voltage due to the change of $V_{io}$ over 100°C is (+25 to +125°C)

$$\Delta V_o = \pm V_{io}\frac{R_1 + R_f}{R_1}$$

$$= \pm \frac{15 \times 10^{-6} \times 100°C}{°C} \frac{1{,}456 + 145{,}600}{1{,}456} = \pm 0.1515\,V$$

The change in dc output voltage due to the change of $I_{io}$ over 100°C is

$$\Delta V_o = \pm \Delta I_{io} R_f = \pm \frac{0.5 \times 10^{-9} \times 100°C \times 145{,}600}{°C} = 7.28\,mV$$

*Step 11.* The small-signal bandwidth is

$$f_{cp} = \frac{f_{op} A_{vo} R_1}{R_f}$$

$$= \frac{(8\,Hz)\,5 \times 10^4\,(1{,}456)}{145{,}600} = 4\,kHz$$

The small-signal risetime is

$$t_r = \frac{0.35\,R_f}{f_{op} A_{vo} R_1} = \frac{0.35\,(145{,}600)}{8 \times 5 \times 10^4 \times 1{,}456}$$

$$= 87.5\,\mu s$$

*Step 12.* The output noise at 10 Hz is computed from

$$V_{on}(10 \text{ Hz}) = \left[ V_n^2\left(1 + \frac{R_f}{R_1}\right)^2 + I_n^2 R_f^2 \right]^{1/2}$$

$$= \left[ 5 \times 10^{-15}\left(1 + \frac{145{,}600}{1{,}456}\right)^2 + 5 \times 10^{-23}(145{,}600)^2 \right]^{1/2}$$

$$= 7.2 \ \mu\text{V rms at 10 Hz}$$

At 100 Hz, 1 kHz, 10 kHz, and 100 kHz the output noise is

$$V_{on}(100 \text{ Hz}) = [10^{-15}(101)^2 + 5 \times 10^{-24}(145{,}600)^2]^{1/2}$$
$$= 3.2 \ \mu\text{V rms at 100 Hz}$$

$$V_{on}(1 \text{ kHz}) = [5 \times 10^{-16}(101)^2 + 8 \times 10^{-25}(145{,}600)^2]^{1/2}$$
$$= 2.3 \ \mu\text{V rms at 1 kHz}$$

$$V_{on}(10 \text{ kHz}) = [5 \times 10^{-16}(101)^2 + 3 \times 10^{-25}(145{,}600)^2]^{1/2}$$
$$= 2.3 \ \mu\text{V rms at 10 kHz}$$

$$V_{on}(100 \text{ kHz}) = [5 \times 10^{-16}(101)^2 + 3 \times 10^{-25}(145{,}600)^2]^{1/2}$$
$$= 2.3 \ \mu\text{V rms at 100 kHz}$$

## REFERENCES

1. Niu, G.: Gain-Error Nomograms for Op Amps, *EEE*, February 1967, p. 104.
2. Moschytz, G. S.: The Operational Amplifier in Linear Active Networks, *IEEE Spectrum*, January 1970, p. 42.
3. Tobey, G. E., J. G. Graeme, and L. P. Huelsman: "Operational Amplifiers—Design and Applications," pp. 427–436, McGraw-Hill Book Company, New York, 1971.

## 4.2 BASIC NONINVERTING AMPLIFIER

**ALTERNATE NAMES** Voltage follower, noninverting configuration.

**EXPLANATION OF OPERATION** The operation of this circuit is similar to that of the basic inverting amplifier except for the following (see Fig. 4.2):

1. The input signal is applied directly to the noninverting input, thereby making the output in phase with the input.
2. The resistor $R_1$ is connected to ground instead of being connected to the input voltage.
3. The closed-loop voltage gain for the noninverting amplifier is

$$A_{vc} = \frac{v_o}{v_i} = 1 + \frac{R_f}{R_1}$$

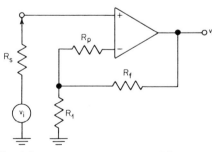

**Fig. 4.2** Basic noninverting-amplifier circuit.

## BASIC NONINVERTING AMPLIFIER

### DESIGN PARAMETERS

These will be identical to those of the inverting amplifier except that $R_1$ is no longer the input resistor.

### DESIGN EQUATIONS

| Eq. No. | Description | Equation |
|---|---|---|
| 1 | Closed-loop voltage gain assuming ideal op amp parameters | $A_{vc} = 1 + \dfrac{R_f}{R_1}$ |
| 2 | Closed-loop voltage gain if finite op amp gain $A_v$ is included | $A_{vc} = \dfrac{1 + R_f/R_1}{1 + 1/\beta A_v}$ |
| 3 | Closed-loop voltage gain if differential input resistance $R_{id}$ and common-mode input resistance $R_{ic}$ are included ($A_v$ must also be included) | $A_{vc} = \dfrac{1 + R_f/R_1}{1 + 1/\beta A_v + 2R_f/A_v R_{id}}$ |
| 4 | Closed-loop voltage gain if the op amp output resistance $R_o$ is included ($A_v$ must also be included) | $A_{vc} = \dfrac{1 + R_f/R_1}{1 + (R_1 + R_f + R_o)/A_v R_1}$ |
| 5 | Size of $R_f$ for minimum gain error due to $A_v$, $R_{id}$, and $R_o$ | $R_f(\text{opt}) = \left(\dfrac{R_{id} R_o R_f}{2R_1}\right)^{1/2}$ |
| 6 | Input resistance of circuit assuming ideal op amp parameters | $R_{in} = \infty$ |
| 7 | Input resistance of circuit assuming finite $A_v$, finite $R_{id}$, and and nonzero $R_o$ | $R_{in} = \dfrac{\beta A_v R_{id}{}^2 R_f}{(R_f + R_o)(R_{id} + 2\beta R_f)}$ $\approx \beta A_v R_{id}$ |
| 8 | Output resistance of circuit assuming ideal op amp parameters | $R_{out} = 0$ |
| 9 | Output resistance of circuit assuming finite op amp output resistance $R_o$ and finite $A_v$ | $R_{out} = \dfrac{R_o(R_f + R_o)(R_{id} + 2\beta R_f)}{\beta A_v R_f R_{id}}$ $\approx \dfrac{R_o}{\beta A_v}$ |
| 10 | Bandwidth of circuit assuming bandwidth (−3 dB) of op amp is at $f_{op}$ ($f_{op}$ = first pole of the op amp) | $f_{cp} = \dfrac{f_{op} A_{vo} R_1}{R_1 + R_f}$ |
| 11 | Small-signal risetime of circuit (10 to 90%) | $t_r = \dfrac{0.35\,(R_1 + R_f)}{f_{op} A_{vo} R_1}$ |
| 12 | Output dc-voltage change due to an input offset-voltage change of op amp (assuming $I_b$ and $I_{io} = 0$) | $\Delta V_o = \pm\,\Delta V_{io}\,\dfrac{R_1 + R_f}{R_1}$ |
| 13 | Output dc voltage due to input bias current of op amp assuming $R_p = 0$ and $V_{io} = 0$ | $V_o = I_b R_f$ |
| 14 | Output dc-voltage change due to an input-offset-current change of op amp assuming $R_p = \dfrac{R_1 R_f}{R_1 + R_f}$ and $V_{io} = 0$ | $\Delta V_o = \pm\,\Delta I_{io} R_f$ |

## DESIGN EQUATIONS (Continued)

| Eq. No. | Description | Equation |
|---|---|---|
| 15 | Output-noise voltage due to an equivalent op amp input-noise voltage in $V/\sqrt{Hz}$ | $V_{on} = V_n \left(1 + \dfrac{R_f}{R_1}\right) V/\sqrt{Hz}$ |
| 16 | Output-noise voltage due to both equivalent op amp input-noise voltage and current ($V/\sqrt{Hz}$ and $A/\sqrt{Hz}$ or $V^2/Hz$) | $V_{on} = \left[ V_n^2 \left(1 + \dfrac{R_f}{R_1}\right)^2 + I_n^2 R_f^2 \right]^{1/2}$ |
| 17 | Optimum value for $R_p$ to minimize output offset voltage due to $I_b$ | $R_p = R_s - \dfrac{R_1 R_f}{R_1 + R_f}$<br>If this is negative, place $R_p$ in series with noninverting input and let<br>$R_p = \dfrac{R_1 R_f}{R_1 + R_f} - R_s$ |

## REFERENCES

1. Moschytz, G. S.: The Operational Amplifier in Linear Active Networks, *IEEE Spectrum*, January 1970, p. 42.
2. Tobey, G. E., J. G. Graeme, and L. P. Huelsman: "Operational Amplifiers—Design and Applications," pp. 427–436, McGraw-Hill Book Company, New York, 1971.

## 4.3 CURRENT AMPLIFIER

**ALTERNATE NAMES** Current-to-current converter, impedance transformer.

**EXPLANATION OF OPERATION** This circuit supplies an output current $i_o$ through the load resistor $R_L$ which is proportional to the input current $i_i$. The output current is independent of the resistance of $R_L$ over a specified range of $R_L$. Current amplifiers, ideally, have zero input resistance and infinite output resistance.

The circuit operates as follows: According to Chap. 2, we can (for a first approximation) assume that (1) the voltage across the op amp input terminals is zero, and (2) neither op amp input terminal draws any current. Using these two statements, we conclude that

$$v_i = 0$$

and

$$v_x = -i_i R_f$$

Using basic circuit theory, we also note that

$$i_o = i_s + i_i$$

and

$$v_x = -i_s R_s$$

Combining the last three equations, we find that the current gain is

$$A_{ic} = \frac{i_o}{i_i} = 1 + \frac{R_f}{R_s}$$

Note that the direction of $i_o$ is into the op amp if $i_i$ is as shown.

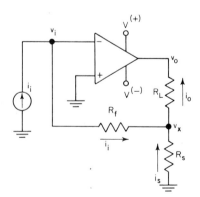

**Fig. 4.3** Current amplifier.

## DESIGN EQUATIONS

| Eq. No. | Description | Equation |
|---|---|---|
| 1 | Closed-loop current gain assuming ideal op amp parameters | $A_{ic} = \dfrac{i_o}{i_i} = 1 + \dfrac{R_f}{R_s}$ |
| 2 | Closed-loop current gain if finite op amp gain $A_v$ is included | $A_{ic} = \dfrac{1 + R_f/R_s}{1 + 1/\beta A_v}$ where $\beta = \dfrac{R_s}{R_s + R_L}$ |
| 3 | Closed-loop current gain if $A_v$, the op amp output resistance $R_o$, and $R_L$ are included | $A_{ic} = \dfrac{(R_s + R_f)A_v}{R_L + R_o + R_s(1 + A_v)}$ |
| 4 | Closed-loop current gain if $A_v$, $R_o$, $R_L$, and the op amp input resistance $R_{id}$ are included | $A_{ic} = \dfrac{R_{id}[(R_f + R_s)A_v + R_s]}{(R_{id} + R_f)(R_s + R_L + R_o) + R_s(R_L + R_o) + R_s R_{id} A_v}$ |
| 5 | Closed-loop input resistance considering $A_v$ as the only nonideal parameter | $R_{in} = \dfrac{R_f}{1 + A_v}$ |
| 6 | Closed-loop input resistance if $A_v$, $R_{id}$, $R_L$, and $R_o$ are considered | $R_{in} = \dfrac{R_{id}[R_f(R_s + R_L + R_o) + R_s(R_L + R_o)]}{(R_{id} + R_f)(R_s + R_L + R_o) + R_s(R_L + R_o) + R_s R_{id} A_v}$ |
| 7 | Closed-loop output resistance considering $A_v$ as the only nonideal parameter | $R_{out} = R_s(1 + A_v)$ |
| 8 | Closed-loop output resistance considering $A_v$, $R_{id}$, and $R_L$ | $R_{out} = \dfrac{(R_f + R_s)(R_o + R_s)[(R_{id} + R_f)(R_s + R_o) + R_s R_o + R_s R_{id} A_v]}{(R_{id} + R_s + R_s)(R_f R_s + R_f R_o + R_s R_o)}$ |
| 9 | Range of $R_L$ resistance over which Eq. 1 is true ($i_i < 0$) | $0 < R_L < \dfrac{|V^+| - 3}{|i_i(\max)|(1 + R_f/R_s)}$ |
| 10 | Range of $R_L$ resistance over which Eq. 1 is true ($i_i > 0$) | $0 < R_L < \dfrac{|V^-| - 3}{|i_i(\max)|(1 + R_f/R_s)}$ |

## DESIGN PARAMETERS

| Parameter | Description |
|---|---|
| $A_{ic}$ | Current gain of circuit |
| $A_v$ | Open-loop voltage gain of op amp as a function of frequency |
| $A_{vo}$ | Open-loop dc voltage gain of op amp ($A_{vo}$ may be substituted for $A_v$ in an equation if dc characteristics are wanted) |
| $A_{vc}$ | Closed-loop voltage gain of op amp as a function of frequency |
| $A_{vco}$ | Closed-loop dc voltage gain of op amp |
| $\beta$ | Voltage feedback ratio of $R_s$ and $R_L$ |
| $i_i$ | Input current |
| $i_o$ | Output current |
| $R_{id}$ | Differential input resistance of op amp |
| $R_{in}$ | Input resistance of circuit |
| $R_o$ | Output resistance of op amp |
| $R_{out}$ | Output resistance of circuit |
| $V^{(+)}$ | Positive supply voltage |
| $V^{(-)}$ | Negative supply voltage |

## REFERENCE

1. Nieu, G.: Op Amps Act as Universal Gain Elements, *Electron. Des.*, Jan. 18, 1968, p. 78.

## 4.4 TRANSRESISTANCE AMPLIFIER

**ALTERNATE NAMES** Current-to-voltage converter, transimpedance amplifier, I-to-V converter, photodiode amplifier.

**EXPLANATION OF OPERATION** A transresistance amplifier behaves as if it were a resistor with power gain. It provides an output voltage which is proportional to the input current. The proportionality constant is the feedback resistor $R_f$ such that $v_o = -i_i R_f$. This circuit is characterized by zero input resistance and zero output resistance if the op amp is ideal. Only one resistor ($R_f$) is required for this circuit. The maximum size of this resistor (i.e., the circuit gain) is limited only by the output-voltage capability of the op amp and the size of input current.

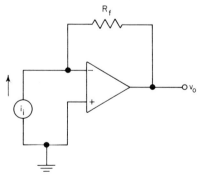

**Fig. 4.4** Transresistance amplifier.

## DESIGN PARAMETERS

| Parameter | Description |
|---|---|
| $A_{rc}$ | Closed-loop transresistance of circuit |
| $A_v$ | Open-loop voltage gain of op amp as a function of frequency (for dc performance $A_{ro}$ may be substituted for $A_v$) |
| $R_f$ | Feedback resistor |
| $R_{id}$ | Differential input resistance of op amp |
| $R_{in}$ | Input resistance of circuit |
| $R_o$ | Output resistance of op amp |
| $R_{out}$ | Output resistance of circuit |
| $v_o$ | Output voltage |

## DESIGN EQUATIONS

| Eq. No. | Description | Equation |
|---|---|---|
| 1 | Closed-loop transresistance assuming ideal op amp parameters | $A_{rc} = \dfrac{v_o}{i_i} = -R_f$ |
| 2 | Closed-loop transresistance if $A_v$ and the op amp input resistance $R_{id}$ are considered | $A_{rc} = \dfrac{-R_f R_{id} A_v}{R_f + R_{id}(1 + A_v)}$ |
| 3 | Closed-loop transresistance if $A_v$, $R_{id}$, and the op amp output resistance $R_o$ are considered | $A_{rc} = \dfrac{-R_{id}(R_f + R_{id})(R_f + R_o)A_v - R_o(R_{id} + R_f + R_o)}{(R_{id} + R_f + R_o)[R_o + R_f + R_{id}(1 + A_v)]}$ |
| 4 | Input resistance of circuit assuming ideal op amp parameters | $R_{in} = 0$ |
| 5 | Input resistance of circuit assuming a finite $A_v$ | $R_{in} = \dfrac{R_f}{1 + A_v}$ |
| 6 | Input resistance of circuit assuming finite $A_v$, $R_{id}$, and $R_o$ | $R_{in} = \dfrac{R_{id}(R_o + R_f)}{R_o + R_f + R_{id}(1 + A_v)}$ |
| 7 | Output resistance of circuit assuming ideal op amp parameters | $R_{out} = 0$ |
| 8 | Output resistance of circuit assuming a finite $A_v$ | $R_{out} = \dfrac{R_o}{1 + A_v}$ |
| 9 | Output resistance of circuit assuming finite $A_v$, $R_{id}$, and $R_o$ | $R_{out} = \dfrac{R_o(R_f + R_{id})}{R_o + R_f + R_{id}(1 + A_v)}$ |

## REFERENCE

1. Nieu, G.: Op Amps Act as Universal Gain Elements, *Electron. Des.*, Jan. 18, 1968, p. 78.

## 4.5 TRANSCONDUCTANCE AMPLIFIER

See also Current Regulators in Chap. 24.

**ALTERNATE NAMES** Voltage-to-current converter, V-to-I converter, transadmittance amplifier, current source, controlled-current source.

**EXPLANATION OF OPERATION** This circuit provides a current through $R_L$ which is proportional to the input voltage $v_i$. The output current $i_o$ is sensed by the sense resistor $R_s$. The resulting sense voltage is fed back in series with the input voltage $v_i$. The circuit closely resembles the noninverting amplifier shown in Fig. 4.2 except that in this case $R_s$ is usually quite small. The voltage at $v_o$ is

$$v_o = \frac{v_i(R_s + R_L)}{R_s}$$

The current through $R_L$ is

$$i_o = \frac{v_o}{R_s + R_L}$$

Combining these equations, we get

$$i_o = \frac{v_i}{R_s}$$

and the output current is seen to be independent of $R_L$.

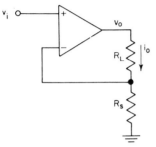

**Fig. 4.5** The transconductance amplifier with a floating load.

This circuit is often used for a current regulator where $v_i$ is held constant. We will cover this subject in detail in Chap. 24.

### DESIGN PARAMETERS

| Parameter | Description |
| --- | --- |
| $A_{gc}$ | Closed-loop transconductance of circuit |
| $A_v$ | Open-loop voltage gain of op amp as a function of frequency (for dc performance $A_{vo}$ may be substituted for $A_v$) |
| $\beta$ | Transfer function of feedback network |
| $i_o$ | Output current of circuit |
| $R_g$ | Internal resistance of voltage-source driving circuit |
| $R_{id}$ | Differential input resistance of op amp |
| $R_{in}$ | Input resistance of circuit |
| $R_L$ | Load resistance |
| $R_o$ | Output resistance of op amp |
| $R_{out}$ | Output resistance of circuit |
| $R_s$ | Output-current-sensing resistor |
| $v_i$ | Input voltage to circuit |

## DESIGN EQUATIONS

| Eq. No. | Description | Equation |
|---|---|---|
| 1 | Closed-loop transconductance assuming ideal op amp parameters | $A_{gc} = \dfrac{i_o}{v_i} = \dfrac{1}{R_s}$ |
| 2 | Closed-loop transconductance if finite op amp gain $A_v$ is considered | $A_{gc} = \dfrac{1/R_s}{1 + 1/\beta A_v}$ where $\beta = \dfrac{R}{R_s + R_L + R_o}$ |
| 3 | Closed-loop transconductance if $A_v$ and the op amp input resistance $R_{id}$ are considered | $A_{gc} = \dfrac{R_{id}A_v - R_s}{R_L(R_s + R_{id} + R_g) + R_s(R_{id} + R_g) + R_s R_{id} A_v}$ |
| 4 | Closed-loop transconductance if $A_v$, $R_{id}$, and the op amp output resistance $R_o$ are considered | $A_{gc} = \dfrac{R_{id}A_v - R_s}{(R_o + R_L)(R_s + R_{id} + R_g) + R_s(R_{id} + R_g) + R_s R_{id} A_v}$ |
| 5 | Input resistance of circuit assuming ideal op amp parameters | $R_{in} = \infty$ |
| 6 | Input resistance of circuit assuming finite $A_v$, $R_o$, and $R_{id}$ | $R_{in} = R_{id}(1 + \beta A_v)$ |
| 7 | Input resistance of circuit considering total effect of $A_v$, $R_o$, $R_{id}$, $R_o$, and $R_L$ | $R_{in} = R_{id} + R_g + \dfrac{R_s(R_o + R_L + R_{id}A_v)}{R_s + R_L + R_o}$ |
| 8 | Output resistance of circuit assuming ideal op amp parameters | $R_{out} = \infty$ |
| 9 | Output resistance of circuit assuming finite $A_v$ | $R_{out} = R_s(1 + A_v)$ |
| 10 | Output resistance of circuit considering all factors | $R_{out} = R_o + \dfrac{R_s[R_g + R_{id}(1 + A_v)]}{R_s + R_{id} + R_g}$ |

## REFERENCE

1. Nieu, G.: Op Amps Act as Universal Gain Elements, *Electron. Des.*, Jan. 18, 1968, p. 78.

## 4.6 AC-COUPLED INVERTING AMPLIFIER

**ALTERNATE NAMES** Capacitor-coupled amplifier, dc-isolated amplifier.

**EXPLANATION OF OPERATION** If dc isolation of stages is required, an input and/or output isolation capacitor may be used. (See Fig. 4.6.) DC biasing

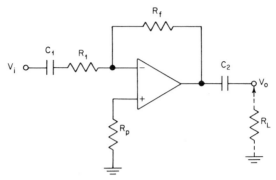

**Fig. 4.6** AC-coupled inverting amplifier.

to the inverting input is obtained through the feedback resistor. The size of $R_p$ is approximately equal to $R_f$, since $R_1$ is isolated from the circuit.

This type of circuit is also useful for shaping of the frequency characteristics. As will be shown in the design equations, the lower cutoff frequency depends on $R_1$ and $C_1$ or $R_L$ and $C_2$ (whichever frequency is lowest). The upper cutoff frequency depends on the op amp.

### DESIGN EQUATIONS

| Eq. No. | Description | Equation |
|---|---|---|
| 1 | Closed-loop voltage gain of circuit | $A_{vc} = \dfrac{v_o}{v_1} = -\dfrac{R_f}{R_1} \dfrac{s}{s + 1/R_1 C_1}$ |
| 2 | Lower cutoff frequency (−3 dB) | $f = \dfrac{1}{2\pi R_1 C_1}$ or $\dfrac{1}{2\pi R_L C_2}$ (whichever is highest) |
| 3 | Output offset voltage assuming $R_p = 0$ and $C_2$ removed | $\Delta V_o = \pm V_{io} + I_b R_f$ |
| 4 | Nominal size of $R_p$ to minimize output offset assuming $C_2$ not used | $R_p \approx R_f$ |
| 5 | Input impedance | $Z_{in} = R_1 + \dfrac{1}{sC_1}$ |
| 6 | Output impedance | $Z_{out} \approx \dfrac{R_o}{\beta A_v} + \dfrac{1}{sC_2}$ |
|   |   | where $\beta = \dfrac{R_1 + 1/sC_1}{R_1 + R_f + 1/sC_1}$ |

### REFERENCE

1. Tobey, G. E., J. G. Graeme, and L. P. Huelsman: "Operational Amplifiers—Design and Application," p. 222, McGraw-Hill Book Company, New York, 1971.

### 4.7 CHARGE-SENSITIVE AMPLIFIER

**ALTERNATE NAMES** Charge-to-voltage converter, capacitive transducer amplifier.

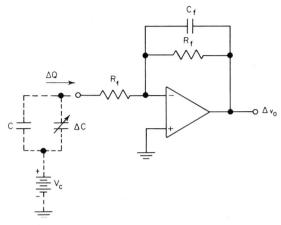

**Fig. 4.7** Charge-sensitive amplifier.

**EXPLANATION OF OPERATION** Many high-impedance transducers such as proportional counters, capacitance microphones, and some accelerometers require an amplifier which converts a transfer of charge into a change of voltage. The voltage across the transducer is usually held constant. We can therefore assume that either the capacitance of the transducer changes by $\Delta C$ or a charge of $\Delta Q$ is emitted from the transducer. The equation relating these two phenomena is

$$\Delta Q = V_c \Delta C$$

where  $\Delta Q$ = quantity of charge transferred
 $\Delta C$ = change in capacitance
 $V_c$ = constant voltage across the transducer

The capacitance microphone actually has a changing capacitance which varies in proportion to the acoustical input. The proportional counter (used to detect x-rays) puts out a quantity of charge $\Delta Q$ in response to each detected x-ray.

The amplifier output-voltage change is

$$\Delta v_o = \frac{-V_c \Delta C}{C_f}$$

if the transducer is of the capacitive type. For charge-emitting transducers the amplifier output-voltage change is

$$\Delta v_o = \frac{-\Delta Q}{C_f}$$

since $\Delta Q = V_c \Delta C$.

The lower cutoff frequency (−3 dB) of this circuit is

$$f_{cp1} = \frac{1}{2\pi R_f C_f}$$

The upper cutoff frequency (−3 dB) is

$$f_{cp2} = \frac{1}{2\pi R_1 C}$$

## 4-20 AMPLIFIERS

The resistor $R_f$ is required only to discharge $C$ and $C_f$. Without its presence these capacitors would build up a charge when input signals are present. If the leakage paths around $C$ and $C_f$ are too small, this bias could gradually run the op amp into saturation.

### DESIGN EQUATIONS

| Eq. No. | Description | Equation |
|---|---|---|
| 1 | Midband gain for an input-capacitance change of $\Delta C$ | $\dfrac{\Delta v_o}{\Delta C} = -\dfrac{V_c}{C_f}$ |
| 2 | Midband gain for an input charge of $\Delta Q$ | $\dfrac{\Delta v_o}{\Delta Q} = -\dfrac{1}{C_f}$ |
| 3 | Lower cutoff frequency (−3 dB) | $f_{cp1} = \dfrac{1}{2\pi R_f C_f}$ |
| 4 | Upper cutoff frequency (−3 dB) | $f_{cp2} = \dfrac{1}{2\pi R_1 C}$ |

### REFERENCES

1. Tobey, G. E., J. G. Graeme, and L. P. Huelsman: "Operational Amplifiers – Design and Applications," p. 233, McGraw-Hill Book Company, New York, 1971.
2. Stout, D. F.: A Low Noise Charge Sensitive Video Preamplifier for Use with the SEC Camera Tube, Space Programs Summary, Jet Propulsion Laboratory, 37–54, vol. III, p. 137.

## 4.8 SUMMING AMPLIFIER

**ALTERNATE NAMES** Adding/subtracting amplifier, weighted summing amplifier, adder, subtractor.

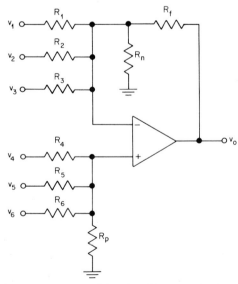

**Fig. 4.8** Summing amplifier for addition and subtraction.

**EXPLANATION OF OPERATION** The negative feedback used in the basic inverting and noninverting-amplifier circuits tends to drive the two op amp input terminals to the same voltage. This makes summation of currents at both input terminals possible without interaction between input branches. Thus, one op amp can be used for both addition and subtraction of a large number of voltages. These may be dc or ac voltages (or both).

This circuit must be designed in the following sequence:

1. Select the value of the feedback resistor $R_f$. Its maximum size is determined by the allowable output-voltage offset using $\Delta V_o = \pm I_{io} R_f$, where $I_{io}$ is the op amp input offset current.
2. Select resistors $R_1$ through $R_6$ as if all input signals were to be inverted.
3. Calculate the parallel value of $R_1$, $R_2$, $R_3$, and $R_f$. Call this $R_A$.
4. Calculate the parallel value of $R_4$, $R_5$, and $R_6$. Call this $R_B$.
5. If $R_A > R_B$, do not use $R_p$. Choose $R_n$ such that $R_A = R_B$, where $R_n$ is now included in the parallel-resistance calculation for $R_A$.
6. If $R_B > R_A$, do not use $R_n$. Choose $R_p$ such that $R_A = R_B$, where $R_p$ is now included in the parallel-resistance calculation for $R_B$.

## DESIGN EQUATIONS

| Eq. No. | Description | Equation |
|---|---|---|
| 1 | Output voltage of circuit | $v_o = A_{vc1} v_1 + A_{vc2} v_2 + A_{vc3} v_3 + A_{vc4} v_4 + A_{vc5} v_5 + A_{vc6} v_6$ |
| 2 | Voltage gain for inverting input voltages ($v_1$, $v_2$, and $v_3$) | $A_{vc1} = \dfrac{v_o}{v_1} = -\dfrac{R_f}{R_1}$ <br><br> $A_{vc2} = \dfrac{v_o}{v_2} = -\dfrac{R_f}{R_2}$ <br> etc. |
| 3 | Voltage gain for noninverting input voltages ($v_4$, $v_5$, and $v_6$) | $A_{vc4} = \dfrac{R_C}{R_C + R_4}\left(1 + \dfrac{R_f}{R_x}\right)$ <br><br> where $R_C$ = parallel resistance of $R_5$, $R_6$, and $R_p$ <br><br> $R_x$ = parallel resistance of $R_1$, $R_2$, $R_3$, and $R_n$ <br><br> $A_{vc5} = \dfrac{R_D}{R_D + R_5}\left(1 + \dfrac{R_f}{R_x}\right)$ <br><br> where $R_D$ = parallel resistance of $R_4$, $R_6$, and $R_p$ <br> etc. |
| 4 | Output offset voltage due to input offset current | $\Delta V_o = \pm I_{io} R_f$ |
| 5 | Output offset voltage due to input offset voltage | $\Delta V_o = \pm V_{io}\left(1 + \dfrac{R_f}{R_x}\right)$ |

## REFERENCES

1. Kostanty, R. G.: Doubling Op Amp Summing Power, *Electronics*, Feb. 14, 1972, p. 73.
2. Barber, J. C.: Mix Various Signals by a Simple Method, *Electron. Des.*, Mar. 1, 1968, p. 90.

Chapter 5

# Comparators

## INTRODUCTION

This class of circuits is used to convert analog signals into bilevel signals. This is done by comparing an input signal with a reference voltage. Whenever the signal changes from less than the reference to greater than the reference (or vice versa), the output voltage of the comparator abruptly changes state. The character of this change of state is subject to the designer's choice. It can be TTL compatible or ECL compatible or can possess a wide set of limits such as $\pm 15$ V. Likewise, the input characteristics of the comparator can take on many possible forms. The voltage comparison can take place at zero voltage or any $\pm$ voltage through manipulation of resistor networks.

In this chapter we will provide design information on the eight basic comparator types:
1. Inverting zero-crossing detector
2. Noninverting zero-crossing detector
3. Inverting zero-crossing detector with hysteresis
4. Noninverting zero-crossing detector with hysteresis
5. Inverting level detector
6. Noninverting level detector
7. Inverting level detector with hysteresis
8. Noninverting level detector with hysteresis

The seventh (and eighth) type is the most mature of the list. We will therefore provide a complete design procedure and example for this type only. For the preceding six we will provide the explanation of operation and design equations.

## 5.1 ZERO-CROSSING DETECTOR

**ALTERNATE NAMES** Zero-crossing comparator, zero-level detector, Schmitt trigger.

**EXPLANATION OF OPERATION** We will begin this discussion with the inverting zero-crossing detector. Afterward we will compare these results with the noninverting zero-crossing detector.

**Inverting zero-crossing detector** A zero-crossing detector determines if an input voltage is greater than zero or less than zero. In response to this deter-

## 5-2 COMPARATORS

mination, the output voltage can assume only two possible states. The output assumes the positive state if $v_i < 0$ and the negative state if $v_i > 0$. As shown in Fig. 5.1A, the magnitudes of the positive and negative output voltages are determined only by zener diodes $Z_1$ and $Z_2$. If $v_i < 0$, $v_o = V_{Z1}$ and if $v_i > 0$, $v_o = -V_{Z2}$. Figure 5.1B illustrates the approximate input-output transfer function of the circuit.

Several sources of error in this circuit should be recognized. The errors due to the op amp input currents will be discussed first. These include the input bias current $I_b$ and the input offset current $I_{io}$. Ideally, the output voltage should switch whenever $v_i$ passes through zero volts. When the output switches, this means that the current direction through the zener diodes is reversing. At this instant the output voltage is passing through zero voltage. When $v_o = 0$, we must have $i_i = I_b$. We conclude that the

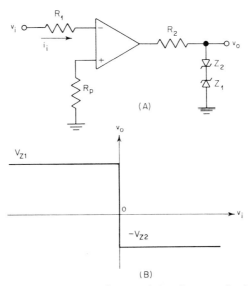

**Fig. 5.1** Inverting zero-crossing detector (A) and its transfer function (B).

circuit does not switch until $v_i = i_i R_1 = I_b R_1$. The op amp input bias current therefore causes an error in switching at zero input voltage. This error is minimized by installing an $R_p$ as shown such that $R_p = R_1$. The only error remaining from input bias current will be due to the difference in bias currents into the two op amp inputs—i.e., the input offset current $I_{io}$. Since $I_{io} \ll I_b$, if we make $R_p$ adjustable (from $\frac{1}{4}R_1$ to $4R_1$), $R_p$ can be set so that switching occurs at zero input voltage.

The op amp input offset voltage $V_{io}$ causes a switching error in $v_i$ equal in magnitude to $V_{io}$. In the worst case this must be added to the errors caused by $I_b$ and $I_{io}$. $I_{io}$ and $V_{io}$ can be of either polarity with respect to $I_b$. Assuming all errors are stacked up in the same direction, the worst-case offset at $v_i$ is $V_{\text{off}} = V_{io} + I_{io}R_1$ if $R_1 = R_p$ and $V_{\text{off}} = V_{io} + I_b |(R_1 - R_p)|$ if $R_1 \neq R_p$. The null terminals (and $R_p$) can be utilized to cancel most of $V_{\text{off}}$ at room temperature. Since $V_{io}$, $I_b$, and $I_{io}$ are time- and temperature-sensitive, $V_{\text{off}}$ will assume nonzero values at later times and/or other temperatures. One should realize, therefore, that the zero-crossing detector can be made to switch when $v_i$ is exactly equal to zero at one temperature only. At a later time and/or temperature, when $I_b$, $V_{io}$, and $I_{io}$ have changed, the switching will take place at a voltage slightly different from zero.

Figure 5.2 shows both typical and ideal voltage-transfer functions of an op amp. The fact that these transfer functions differ creates another error source for the zero-crossing detector. The total voltage swing at the op amp output, i.e., $V^{(+)}-3$ to $V^{(-)}+3$ (or vice versa), can take place only if the input voltage $v_i$ has changed by more than

$$\Delta v_i(\min) = \frac{V^{(+)}-V^{(-)}-6}{A_{vo}}$$

where $A_{vo}$ is the (large-signal) dc voltage gain of the op amp. The output-switching time also gets very long as the $\Delta v_i(\min)$ above is approached. The input voltage change must usually exceed $\Delta v_i(\min)$ by a factor of 10 to 100 in

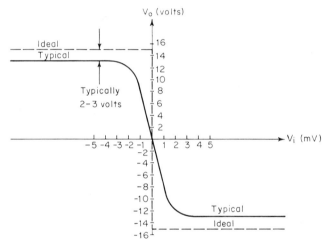

**Fig. 5.2** Typical and ideal voltage-transfer curves for an op amp (from negative input to output).

order to achieve the maximum output-switching speed (i.e., the maximum slew rate of the op amp).

The basic zero-crossing detector has a major drawback because of a phenomenon called chatter. If the input voltage has noise in the order of $\Delta v_i(\min)$, then as $v_i$ goes through zero volts the output may switch states several times before the final decision is made. This can be avoided only by filtering out the noise, using a lower-gain op amp, or using hysteresis (to be discussed later in this chapter). Hysteresis will also speed up the circuit-switching time. However, the switching speed cannot be made faster then the open-loop slew rate of the op amp.

**Noninverting zero-crossing detector** This circuit, shown in Fig. 5.3A, merely reverses the use of $R_1$ and $R_p$ as shown in Fig. 5.1A. The performance of the noninverting circuit is identical to that of the inverting circuit except that the transfer function (Fig. 5.3B) is rotated about the vertical axis.

## 5-4 COMPARATORS

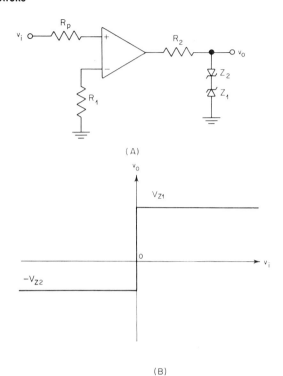

**Fig. 5.3** Noninverting zero-crossing detector (A) and its transfer function (B).

## DESIGN PARAMETERS

| Parameter | Description |
|---|---|
| $A_{vo}$ | Voltage gain of op amp at dc |
| $I_b$ | Input bias current of op amp |
| $I_{io}$ | Input offset current of op amp |
| $R_1$ | Resistor attached to op amp inverting input |
| $R_2$ | Resistor which establishes correct current in $Z_1$ and $Z_2$ |
| $R_{id}$ | Differential input resistance of op amp |
| $R_{in}$ | Input resistance of circuit |
| $R_{out}$ | Output resistance of circuit |
| $R_p$ | Resistor used to nullify the effects of $I_b$ |
| $R_{Z1}$ | Dynamic resistance of $Z_1$ |
| $R_{Z2}$ | Dynamic resistance of $Z_2$ |
| $v_i$ | Input voltage to circuit |
| $\Delta v_i(\min)$ | Minimum input-voltage change which may cause a full output change |
| $V^{(+)}$ | Positive supply voltage |
| $V^{(-)}$ | Negative supply voltage |
| $V_{io}$ | Input offset voltage of op amp |
| $v_o$ | Circuit output voltage |
| $V_{off}$ | Error in trip voltage of circuit caused by nonideal op amp input parameters |
| $V_{Z1}$ | Breakdown voltage of $Z_1$, plus the forward breakdown voltage of $Z_2$ |
| $V_{Z2}$ | Breakdown voltage of $Z_2$, plus the forward breakdown voltage of $Z_1$ |

## DESIGN EQUATIONS

| Eq. No. | Description | Equation |
|---|---|---|
| 1 | Output voltage when $v_i < 0$ assuming ideal op amp parameters and square zener characteristics | INVERTING<br>$v_o = V_{Z1}$<br>NONINVERTING<br>$v_o = -V_{Z2}$ |
| 2 | Output voltage when $v_i > 0$ assuming ideal op amp parameters and square zener characteristics | INVERTING<br>$v_o = -V_{Z2}$<br>NONINVERTING<br>$v_o = V_{Z1}$<br>NOTE: Zener voltages are dependent on their bias current. For this circuit the bias current depends on the size of $R_2$. |
| 3 | Maximum offset (from zero) of $v_i$ trip point considering op amp input parameters and $R_1 = R_p$ | $V_{\text{off}} = \pm(V_{io} + I_{io}R_1)$ |
| 4 | Maximum offset (from zero) of $v_i$ trip point considering op amp input parameters and $R_1 \neq R_p$ | $V_{\text{off}} = \pm(V_{io} + I_b|(R_1 - R_p)|)$ |
| 5 | Minimum change in $v_i$ required to provide full-magnitude output change of state | $\Delta v_i(\min) = \dfrac{V^{(+)} - V^{(-)} - 6}{A_{vo}}$ |
| 6 | Optimum $R_1$ source resistance if effects of changes in $V_{io}$ and $I_b$ with temperature are to be minimized | $R_1 = \dfrac{\Delta V_{io}/\Delta T}{\Delta I_b/\Delta T}$ |
| 7 | Input resistance of circuit | $R_{\text{in}} = R_1 + R_p + R_{id}$ |
| 8 | Output resistance of circuit | $R_{\text{out}} = R_{Z1}$ (positive output) or $R_{Z2}$ (negative output) |
| 9 | Optimum size for $R_2$ | The resistance of $R_2$ is chosen such that the zener diodes are operated at their recommended current levels |

## 5.2 ZERO-CROSSING DETECTOR WITH HYSTERESIS

**ALTERNATE NAMES** Zero-crossing comparator, zero-level comparator, regenerative comparator, Schmitt trigger, bilevel latch, latching comparator, latching zero-crossing detector.

**EXPLANATION OF OPERATION** The operator of this circuit is almost identical to that of the basic zero-crossing detector except that hysteresis is now included.
**Inverting zero-crossing detector with hysteresis** Hysteresis is provided by merely adding $R_f$ to the circuit. Since $R_f$ is connected from the op amp output to the noninverting input, it provides a small amount of positive feedback. The effect of this type of positive feedback is best explained by comparing Fig. 5.4B with Fig. 5.1B. Hysteresis causes the Z curve of Fig.

## 5-6 COMPARATORS

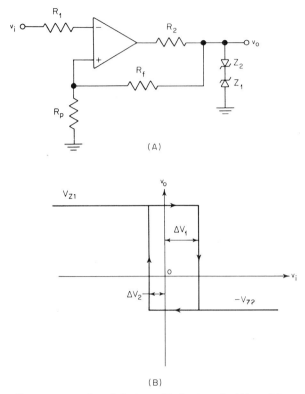

**Fig. 5.4** Inverting zero-crossing detector with hysteresis (A) and its transfer function (B).

5.1B to widen out into the boxed Z shape of Fig. 5.4B. If the box width reduces to zero width, i.e., if $R_f = \infty$, the circuit and transfer function become identical to that of Fig. 5.1. Note that $v_o$ vs. $v_i$ always travels clockwise around the box. If $v_i$ is less than zero and becoming more positive, it has to cross zero and rise to $R_p V_{Z1}/(R_p + R_f)$ before the output switches states. The arrows in Fig. 5.4B show that this is the only path by which this change of state can occur. After the transition has taken place, the output cannot return to the positive state unless $v_i$ has a negative noise spike of at least

$$v_i(\text{noise, peak}) = \frac{R_p(V_{Z1} + V_{Z2})}{R_p + R_f}$$

The hysteresis circuit therefore provides noise immunity and prevents the output from chattering between states as $v_i$ passes through zero. The only disadvantage to this circuit is that larger voltage excursions are required to initiate a change of states. The hysteresis also makes a substantial improvement in output-switching speed. The maximum switching speed, however, is equal to the maximum slew rate of the op amp.

**Noninverting zero-crossing detector with hysteresis** As with the basic zero-cross detector, the noninverting circuit is identical to the inverting circuit except that the transfer function is rotated about the vertical axis.

## ZERO-CROSSING DETECTOR WITH HYSTERESIS 5-7

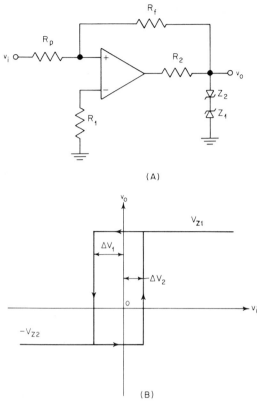

**Fig. 5.5** Noninverting zero-crossing detector with hysteresis (A) and its transfer function (B).

## DESIGN PARAMETERS

| Parameter | Description |
|---|---|
| $A_{vo}$ | Op amp voltage gain at dc |
| $I_b$ | Input bias current of op amp |
| $I_{io}$ | Input offset current of op amp |
| $R_1$ | Resistor for inverting input |
| $R_2$ | Resistor to set current level in $Z_1$ and $Z_2$ |
| $R_f$ | Feedback resistor to establish hysteresis |
| $R_{id}$ | Differential input resistance of op amp |
| $R_{in}$ | Input resistance of circuit |
| $R_{out}$ | Output resistance of circuit |
| $R_p$ | Part of hysteresis feedback circuit |
| $R_{Z1}$ | Dynamic resistance of zener $Z_1$ |
| $R_{Z2}$ | Dynamic resistance of zener $Z_2$ |
| $\Delta V_1$ | Portion of hysteresis loop caused by $V_{Z1}$ |
| $\Delta V_2$ | Portion of hysteresis loop caused by $V_{Z2}$ |
| $v_i$ | Input voltage to circuit |
| $\Delta v_i(\min)$ | Minimum input-voltage change which may cause a full output change |
| $V^{(+)}$ | Positive power-supply voltage |
| $V^{(-)}$ | Negative power-supply voltage |

## 5-8 COMPARATORS

### DESIGN PARAMETERS (Continued)

| Parameter | Description |
|---|---|
| $V_{io}$ | Input offset voltage of op amp |
| $v_o$ | Output voltage of circuit |
| $V_{off}$ | Error in trip voltage due to nonideal op amp input parameters |
| $V_{Z1}$ | Breakdown voltage of $Z_1$, plus the forward breakdown voltage of $Z_2$ |
| $V_{Z2}$ | Breakdown voltage of $Z_2$, plus the forward breakdown voltage of $Z_1$ |

### DESIGN EQUATIONS

| Eq. No. | Description | Equation |
|---|---|---|
| 1 | Value of positive output voltage | $v_o = V_{Z1}$ |
| 2 | Value of negative output voltage | $v_o = -V_{Z2}$ |
| 3 | Ideal positive trip point for $v_i$ assuming ideal op amp and square zener characteristics | INVERTING $$\Delta V_1 = \frac{R_p V_{Z1}}{R_p + R_f}$$ NONINVERTING $$\Delta V_2 = \frac{R_p V_{Z2}}{R_p + R_f}$$ |
| 4 | Ideal negative trip point for $v_i$ assuming ideal op amp and square zener characteristics | INVERTING $$\Delta V_2 = -\frac{R_p V_{Z2}}{R_p + R_f}$$ NONINVERTING $$\Delta V_1 = -\frac{R_p V_{Z1}}{R_p + R_f}$$ |
| 5 | Maximum error in above trip points considering op amp input parameters and assuming $R_1 = R_p$ and $R_f \gg R_p$ | $V_{off} = \pm (V_{io} + I_{io} R_1)$ |
| 6 | Maximum error in above trip points considering op amp input parameters and assuming $R_1 \neq R_p$ and $R_f \gg R_p$ | $V_{off} = \pm [V_{io} + I_b \mid (R_1 - R_p) \mid]$ |
| 7 | Minimum change in $v_i$ required to provide full-magnitude output change of state | $\Delta v_i \approx 0$ since positive feedback makes forward gain approach $\infty$ |
| 8 | Optimum $R_1$ (source + input resistor) if effects of $V_{io}$ and $I_b$ over temperature are to be minimized | $R_1 = \dfrac{\Delta V_{io}/\Delta T}{\Delta I_b/\Delta T}$ |
| 9 | Input resistance of circuit | INVERTING $R_{in} \approx R_1 + R_p + R_{id}$ NONINVERTING $R_{in} \approx R_p + \dfrac{R_f R_{id}}{R_f + R_{id}}$ |
| 10 | Output resistance of circuit | $R_{out} = R_{Z1}$ (positive output) or $= R_{Z2}$ (negative output) |
| 11 | Optimum size for $R_2$ | $R_2$ is chosen to provide recommended current through zener diodes |

## 5.3 LEVEL DETECTOR

**ALTERNATE NAMES**  Schmitt trigger, level comparator.

**EXPLANATION OF OPERATION**  The operation of this circuit is similar to that of the zero-crossing detector except that the resistor ($R_1$ or $R_p$) which is normally grounded is returned to a reference voltage $V_R$. This change makes the output voltage change states whenever the input voltage passes through $V_R$ rather than zero. $V_R$ can be positive or negative, or it may be a variable which varies according to some system function.

**Inverting level detector**  A level detector determines if an input voltage is greater or less than a reference voltage. In response to this determination, the level-detector output voltage can assume only two possible states. The output assumes the positive state if $v_i < V_R$ and the negative state if $v_i > V_R$. Figure 5.6A shows the inverting level-detector circuit, and Fig. 5.6B indicates several of the possible transfer functions. Note that $V_R$ can be positive (solid line) or negative (dashed line). The two output voltage levels are determined only by $Z_1$ and $Z_2$.

Errors due to $I_b$, $I_{io}$, and $V_{io}$ are similar to those of the zero-crossing detector. They are minimized by making $R_p \approx R_1$. The total output swing is still $V_{Z1} + V_{Z2}$ as before.

This circuit will also tend to chatter at the instant of state changing if $v_i$ has noise larger than

$$v_i(\min) = \frac{V_{Z1} + V_{Z2}}{A_{vo}}$$

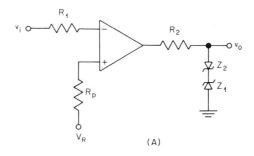

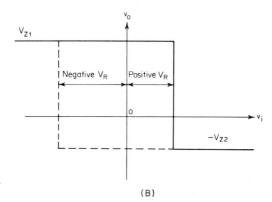

**Fig. 5.6**  Inverting level detector (A) and its transfer function (B).

## 5-10 COMPARATORS

This chattering can be avoided by noise filtering, reducing $A_{vo}$ (using a different type of op amp), or using hysteresis as shown in the next section.

**Noninverting level detector** Figure 5.7 shows the noninverting level detector and its transfer function. The output voltage in this circuit assumes the positive state $V_{Z1}$ if $v_i > V_R$ and the negative state $-V_{Z2}$ if $v_i < V_R$. As before, $V_R$ can be positive, negative, or variable. Bias errors are reduced by incorporating $R_p$. Noise problems are reduced by filtering, lowering $A_{vo}$, or incorporating hysteresis.

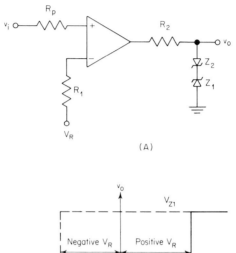

**Fig. 5.7** Noninverting level detector (A) and its transfer function (B).

### DESIGN PARAMETERS

| Parameter | Description |
|---|---|
| $A_{vo}$ | Op amp voltage gain at dc |
| $I_b$ | Input bias current of op amp |
| $I_{io}$ | Input offset current of op amp |
| $R_1$ | Resistor attached to op amp inverting input |
| $R_2$ | Resistor which establishes correct current in $Z_1$ and $Z_2$ |
| $R_{id}$ | Differential input resistance of op amp |
| $R_{in}$ | Input resistance of circuit |
| $R_p$ | Resistor used to nullify the effects of $I_b$ |
| $R_{Z1}$ | Dynamic resistance of zener diode $Z_1$ |
| $R_{Z2}$ | Dynamic resistance of zener diode $Z_2$ |
| $v_i$ | Input voltage to circuit |
| $\Delta v_i$ (min) | Minimum input-voltage change which may cause a full output change |
| $V^{(+)}$ | Positive power-supply voltage |

## DESIGN PARAMETERS (Continued)

| Parameter | Description |
|---|---|
| $V^{(-)}$ | Negative power-supply voltage |
| $V_{io}$ | Input offset voltage of op amp |
| $v_o$ | Output voltage of circuit |
| $V_{\text{off}}$ | Error in trip voltage due to nonideal op amp input parameters |
| $V_R$ | Reference voltage used to establish trip point |
| $V_{Z1}$ | Breakdown voltage of $Z_1$, plus the forward breakdown voltage of $Z_2$ |
| $V_{Z2}$ | Breakdown voltage of $Z_2$, plus the forward breakdown voltage of $Z_1$ |

## DESIGN EQUATIONS

| Eq. No. | Description | Equation |
|---|---|---|
| 1 | Output voltage when $v_i < V_R$, assuming ideal op amp parameters and square zener characteristics | INVERTING $v_o = V_{Z1}$ <br> NONINVERTING $v_o = -V_{Z2}$ |
| 2 | Output voltage when $v_i > V_R$, assuming ideal op amp parameters and square zener characteristics | INVERTING $v_o = -V_{Z2}$ <br> NONINVERTING $v_o = V_{Z1}$ |
| 3 | Maximum deviation from $V_R$ of $v_i$ trip point considering op amp input parameters and $R_1 = R_p$ | $V_{\text{off}} = \pm (V_{io} + I_{io}R_1)$ |
| 4 | Maximum deviation from $V_R$ of $v_i$ trip point considering op amp input parameters and $R_1 \neq R_p$ | $V_{\text{off}} = \pm [V_{io} + I_b \mid (R_1 - R_p) \mid]$ |
| 5 | Minimum change in $v_i$ required to provide full-magnitude output change of state | $\Delta v_i(\min) = \dfrac{V_{Z1} + V_{Z2}}{A_{vo}}$ |
| 6 | Optimum source resistance $R_1$ if effects of changes in $V_{io}$ and $I_b$ with temperature are to be minimized | $R_1 = \dfrac{\Delta V_{io}/\Delta T}{\Delta I_b/\Delta T}$ |
| 7 | Input resistance of circuit | $R_{\text{in}} = R_1 + R_p + R_{id}$ |
| 8 | Output resistance of circuit | $R_{\text{out}} = R_{Z1}$ (positive output) or $R_{Z2}$ (negative output) |
| 9 | Optimum size for $R_2$ | $R_2$ is chosen to provide the recommended bias current through the zener diodes |

## 5.4 LEVEL DETECTOR WITH HYSTERESIS

**ALTERNATE NAMES** Latching comparator, latching Schmitt trigger, latching level detector, regenerative comparator, regenerative level detector.

**EXPLANATION OF OPERATION** A level detector with hysteresis is the most versatile and useful of the comparator circuits discussed in this chapter. It can be designed to change output states whenever the input voltage passes through any selected reference voltage. The noise immunity can be tailored to each application by choosing the amount of hysteresis. The absolute voltages of the two output states are selected by using two appropriate zener diodes.

**Inverting level detector with hysteresis**  A level detector determines if an input voltage $v_i$ is above or below a reference voltage $V_R$. In response to this determination, the output voltage will assume one of two possible states. Referring to Fig. 5.8B, the output-voltage states are $+V_{Z1}$ if $v_i < (V_R R_f - R_p V_{Z2})/(R_p + R_f)$ and $-V_{Z2}$ if $v_i > (V_R R_f + R_p V_{Z1})/(R_p + R_f)$. The actual

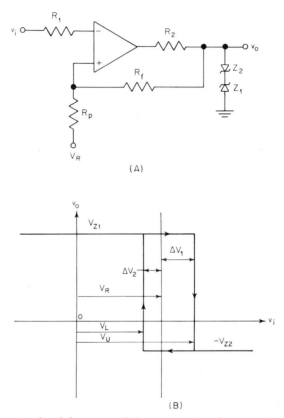

**Fig. 5.8**  Inverting level detector with hysteresis (A) and its transfer function (B).

reference voltage is therefore $V_R R_f/(R_p + R_f)$ instead of $V_R$. In practice, however, since only small amounts of hysteresis are required, we usually make $R_p \ll R_f$. If $R_f = 100 R_p$, the actual reference voltage is only about 1 percent from $V_R$.

The hysteresis voltage above $V_R$ is

$$\Delta V_1 = \frac{R_p V_{Z1}}{R_p + R_f}$$

and the hysteresis voltage below $V_R$ is

$$\Delta V_2 = \frac{R_p V_{Z2}}{R_p + R_f}$$

The two trip points are therefore approximately $V_R + \Delta V_1$ and $V_R - \Delta V_2$.

Op amp input currents will cause another error in the trip points. Input bias current $I_b$ flowing through $R_1$ will make $v_i$ differ from the actual voltage at the op amp inverting input $v_n$. These two voltages will differ by $v_i - v_n = I_b R_1$. If $R_1$ is made equal to $R_f$ in parallel with $R_p$, the effects of the bias current into the op amp input terminals will cancel and this error source will be minimized. The difference between the two input currents, i.e., the input offset current $I_{io}$, will be the only remaining current error. The error caused by $I_{io}$ can be minimized by making $R_1$ adjustable. Since $I_b$ may be larger in either op amp terminal, $R_1$ should be adjustable from above to below the value computed above.

The op amp input offset voltage causes another error in the trip point equal in magnitude to the offset voltage. In the worst case this must be added to the offset voltages due to $I_b$ and $I_{io}$. $I_{io}$ and $V_{io}$ can be of either polarity with respect to $I_b$. Assuming all errors are stacked up in the same direction, the worst-case offset at $v_i$ is

$$V_{\text{off}} = V_{io} + I_{io} R_1$$

if $R_1 = R_p$, and

$$V_{\text{off}} = V_{io} + I_b \mid (R_1 - R_p) \mid$$

if $R_1 \neq R_p$. The op amp null terminals and the adjustable $R_1$ can be utilized to cancel most of $V_{\text{off}}$ at room temperature. Since $V_{io}$, $I_b$, and $I_{io}$ are all time- and temperature-sensitive, $V_{\text{off}}$ will assume nonzero values at later times and/or other temperatures. The level detector will therefore switch when $v_i = V_R + \Delta V_1$ or $v_i = V_R - \Delta V_2$ at only one temperature. At later times and/or temperatures the trip points will differ from these by a small voltage.

Since this circuit has positive feedback, the output changes abruptly whenever the trip points are exceeded. The slew rate will be limited to the maximum slew rate of the op amp with the compensation used. Once the transition between states has taken place, noise will not cause the output to chatter between states as long as the noise in $v_i$ is less than

$$\Delta v_i (\min) = \Delta V_1 + \Delta V_2$$

**Noninverting level detector with hysteresis**  This circuit is similar to the inverting level detector with hysteresis. The resistors $R_1$ and $R_p$ have been interchanged, which reverses the transfer function (compare Fig. 5.8B with Fig. 5.9B). This also causes the trip voltages to be slightly different, since the positive feedback current is now summed with the input current. In the inverting circuit the positive feedback current was summed with current from the reference voltage $V_R$.

## DESIGN PARAMETERS

| Parameter | Description |
|---|---|
| $A_{vo}$ | Op amp voltage gain at dc |
| $I_b$ | Input bias current of op amp |
| $I_{io}$ | Input offset current of op amp |
| $R_1$ | Resistor attached to op amp inverting input |
| $R_2$ | Resistor which establishes correct current in $Z_1$ and $Z_2$ |
| $R_f$ | Feedback resistor which establishes hysteresis |
| $R_{id}$ | Differential input resistance of op amp |
| $R_{in}$ | Input resistance of circuit |
| $R_{out}$ | Output resistance of circuit |

## DESIGN PARAMETERS (Continued)

| Parameter | Description |
|---|---|
| $R_p$ | Part of hysteresis feedback circuit |
| $R_{Z1}$ | Dynamic resistance of zener diode $Z_1$ |
| $R_{Z2}$ | Dynamic resistance of zener diode $Z_2$ |
| $\Delta V_1$ | Portion of hysteresis loop caused by $V_{Z1}$ |
| $\Delta V_2$ | Portion of hysteresis loop caused by $V_{Z2}$ |
| $v_i$ | Input voltage to circuit |
| $\Delta v_i(\min)$ | Minimum input-voltage change which may cause a full output-voltage change |
| $V^{(+)}$ | Positive power-supply voltage |
| $V^{(-)}$ | Negative power-supply voltage |
| $V_{io}$ | Input offset voltage of op amp |
| $V_L$ | Lower trip voltage |
| $v_o$ | Output voltage of circuit |
| $V_{\text{off}}$ | Error in trip voltage due to nonideal op amp input parameters |
| $V_R$ | Reference voltage used to establish trip voltages |
| $V_U$ | Upper trip voltage |
| $V_{Z1}$ | Breakdown voltage of $Z_1$, plus the forward breakdown voltage of $Z_2$ |
| $V_{Z2}$ | Breakdown voltage of $Z_2$, plus the forward breakdown voltage of $Z_1$ |

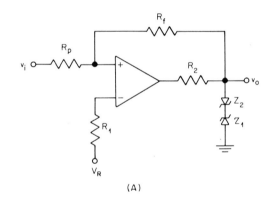

(A)

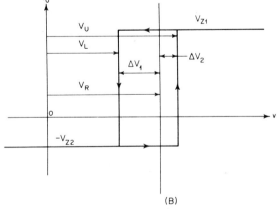

(B)

**Fig. 5.9** Noninverting level detector with hysteresis (A) and its transfer function (B).

## DESIGN EQUATIONS

| Eq. No. | Description | Equation |
|---|---|---|
| 1 | Value of positive output voltage | $v_o = V_{Z1}$ |
| 2 | Value of negative output voltage | $v_o = -V_{Z2}$ |
| 3 | Upper trip voltage for $v_i$, assuming ideal op amp parameters and square zener characteristics | INVERTING $$V_U = \frac{V_R R_f + R_p V_{Z1}}{R_f + R_p}$$ NONINVERTING $$V_U = \frac{V_R(R_p + R_f) + R_p V_{Z2}}{R_f}$$ |
| 4 | Lower trip voltage for $v_i$, assuming ideal op amp parameters and square zener characteristics | INVERTING $$V_L = \frac{V_R R_f - R_p V_{Z2}}{R_f + R_p}$$ NONINVERTING $$V_L = \frac{V_R(R_p + R_f) - R_p V_{Z1}}{R_f}$$ |
| 5 | Width of hysteresis loop | $\Delta V_1 + \Delta V_2 = \dfrac{(V_{Z1} + V_{Z2})R_p}{R_p + R_f}$ |
| 6 | Maximum error in input trip point due to op amp input errors if $R_1 = R_p R_f/(R_p + R_f)$ | $V_{\text{off}} = \pm(V_{io} + I_{io} R_1)$ |
| 7 | Maximum error in input trip point due to op amp input errors if $R_1 \ne R_p R_f/(R_p + R_f)$ | $V_{\text{off}} = \pm\left(V_{io} + I_b \left| R_1 - \dfrac{R_p R_f}{R_p + R_f} \right|\right)$ |
| 8 | Minimum change in $v_i$ required to provide full-magnitude output change of state | $\Delta v_i \approx 0$ since positive feedback makes forward gain approach $\infty$ |
| 9 | Optimum $R_1$ if effects of $V_{io}$ and $I_b$ over temperature are to be minimized | $R_1 = \dfrac{\Delta V_{io}/\Delta T}{\Delta I_b/\Delta T}$ |
| 10 | Input resistance of circuit | INVERTING $R_{\text{in}} \approx R_1 + R_p + R_{id}$ NONINVERTING $R_i \approx R_p + \dfrac{R_f R_{id}}{R_f + R_{id}}$ |
| 11 | Output resistance of circuit | $R_{\text{out}} = R_{Z1}$ (positive output) or $R_{Z2}$ (negative output) |
| 12 | Optimum size relationship between $R_1$ and $R_p$ (it is best to leave $R_1$ adjustable) | $R_1 = \dfrac{R_p R_f}{R_p + R_f}$ |

## DESIGN PROCEDURE

One must first establish which performance requirements are most important before design can proceed. Often, different performance requirements result in conflicts among the various design equations. It is best to write down the performance requirements in a list in order of descending priority. Only in this way can rational trade-offs be performed when conflicts among various design equations take place.

## 5-16  COMPARATORS

For this design procedure we will assume the descending priority of performance requirements are as follows:
  1. The nominal reference voltage $(V_U + V_L)/2$ is specified. Its stability over a given range of temperatures is to be calculated.
  2. The noise level in $v_i$ is specified, which implies a minimum size for the hysteresis loop $(\Delta V_1 + \Delta V_2)$.
  3. The two output levels are given.
  4. The input resistance is to be chosen for maximum trip-voltage stability.
  5. The output resistance is to be calculated.
  6. The minimum allowable supply voltages are to be calculated.

### DESIGN STEPS FOR INVERTING LEVEL DETECTOR WITH HYSTERESIS

*Step 1.* Select diodes $Z_1$ and $Z_2$ which are approximately 0.6 V below the specified upper and lower output voltages (see Eqs. 1 and 2). Choose $R_2$ so that the zener diodes operate at the current (and voltages) specified on their data sheets. Make sure the chosen current level is comfortably below the minimum short-circuit output current of the op amp.

*Step 2.* Compute a nominal value for $R_1$ using Eq. 9. $R_1 = (\Delta V_{io}/\Delta T)/(\Delta I_b/\Delta T)$. Use the same $\Delta T$ in both the denominator and numerator.

*Step 3.* Choose a hysteresis-loop size $V_U - V_L$ which is larger than the peak-to-peak noise in $v_i$. An $R_f$ which satisfies Eqs. 3, 4, and 12 is found from

$$R_f = \frac{R_1(V_{Z1} + V_{Z2})}{V_U - V_L}$$

*Step 4.* Compute the following:

$$R_p = \frac{R_1 R_f}{R_f - R_1}$$

*Step 5.* Using the specified $V_R$, find the upper trip voltage from Eq. 3:

$$V_U = \frac{V_R R_f + R_p V_{Z1}}{R_f + R_p}$$

Find the lower trip voltage from Eq. 4:

$$V_L = \frac{V_R R_f - R_p V_{Z2}}{R_f + R_p}$$

Verify that $V_U - V_L$ is correct as chosen in step 3.

*Step 6.* Compute the maximum error in trip voltages $V_U$ and $V_L$ by using Eq. 6:

$$V_{\text{off}} = \pm (V_{io} + I_{io} R_1)$$

*Step 7.* Compute the input resistance of the circuit.

$$V_{\text{in}} \approx R_1 + R_p + R_{id}$$

*Step 8.* Compute the output resistance of the circuit using the dynamic resistances of the zener diodes at their operating points.

$$R_{\text{out}} = R_{Z1} \text{ (positive output) or } R_{Z2} \text{ (negative output)}$$

**DESIGN EXAMPLE**  An example of an inverting level detector with hysteresis will be described below. The 741 type of op amp will be used, although this device is not recommended for fast level detectors.

## LEVEL DETECTOR WITH HYSTERESIS    5-17

*Design Requirements*

$$\frac{V_U + V_L}{2} = +4 \text{ V (approximately)}$$

$v_i$ (noise $< 0.1$ V peak-to-peak
$V_{Z1} = +10$ V ($\pm 0.5$ V)
$-V_{Z2} = -4$ V ($\pm 0.5$ V)

*Device Data*
*741*

$V_{io}$ ($+25°$C) $= 3$ mV max

$\dfrac{\Delta V_{io}}{\Delta T}$ ($+ 25°$C) $= 15$ $\mu$V/°C

$I_{io}$ ($+ 25°$C) $= 30$ nA max

$\dfrac{\Delta I_b}{\Delta T}$ ($+ 25°$C) $= 1.3$ nA/°C

$R_{id}$(min) $- 3 \times 10^5$ $\Omega$
Minimum short-circuit output current $= 10$ mA

*Zener Diodes*

$V_{Z1}$ (1N5240) $= 9.4$ V at 3 mA ($+ 0.6$ V forward)
$V_{Z2}$ (1N703) $= 3.4$ V at 5 mA ($+ 0.6$ V forward)
$R_{Z1}$ (at 3 mA) $= 10$ $\Omega$
$R_{Z2}$ (at 5 mA) $= 80$ $\Omega$

NOTE: The zener voltages do not have to be operated at their "data sheet" test currents. In operation at lower currents, the zener voltage will decrease and the dynamic resistance will increase. Preliminary testing of the zener diodes on a transistor/diode curve tracer will allow the designer to determine the voltage and dynamic resistance at any operating current.

*Step 1.* Trial-and-error calculation of zener currents and voltages is required, since both zener currents pass through the same resistor $R_2$. Assume that $\pm 20$ V is used for the op amp supply voltages. If we also assume that 3 V (maximum) is lost in the 741 because of saturation, the op amp output voltage will switch between $\pm 17$ V. The current through $R_2$ will be approximately

$$I_{Z1} = \frac{17\text{V} - 10\text{V}}{R_2} = \frac{7\text{ V}}{R_2}$$

during the time the op amp output is $+17$ V. While the op amp output voltage is $-17$ V, the current through $R_2$ will be approximately

$$I_2 = \frac{17\text{V} - 4\text{V}}{R_2} = \frac{11\text{ V}}{R_2}$$

The current through $R_2$ when the output voltage is negative will therefore be $^{11}/_7$ larger than that when the output voltage is positive. Consequently, it is recommended that the low-voltage zener be a higher-test-current device than the higher-voltage zener. For $Z_2$ we will use an IN5240 (10 V at 20 mA but 9.4 V at 3 mA according to the curve tracer). The ratio 5 mA/3 mA $= 1.66$ is reasonably close to the ratio of currents through $R_2$ ($^{11}/_7 = 1.57$). If we choose $I_1$ to be 3 mA,

$$R_2 = \frac{7\text{ V}}{3\text{ mA}} = 2{,}300 \text{ }\Omega$$

## 5-18 COMPARATORS

The current $I_2$ will be approximately

$$I_2 = \frac{11 \text{ V}}{2{,}300 \text{ }\Omega} = 4.8 \text{ mA}$$

*Step 2*

$$R_1 = \frac{\Delta V_{io}/\Delta T}{\Delta I_b/\Delta T} = \frac{15 \text{ }\mu\text{V}/°\text{C}}{1.3 \text{ nA}/°\text{C}} = 11{,}500 \text{ }\Omega$$

*Step 3.* Since the peak-to-peak noise in $v_i$ is less than 0.1 V, we will make the hysteresis-loop size 0.2 V. Thus we have

$$R_f = \frac{R_1(V_{Z1} + V_{Z2})}{V_U - V_L}$$

$$= \frac{11{,}500(9.4 + 3.4 + 1.2)}{0.2} = 805 \text{ k}\Omega$$

*Step 4.* $R_p$ is found from

$$R_p = \frac{R_1 R_f}{R_f - R_1}$$

$$= \frac{11{,}500(805{,}000)}{805{,}000 - 11{,}500} = 11{,}667 \text{ }\Omega$$

*Step 5.* The upper trip voltage is

$$V_U = \frac{V_r R_f + R_p V_{Z1}}{R_f + R_p}$$

$$= \frac{4 \times 805{,}000 + 11{,}667(9.4 + 0.6)}{805{,}000 + 11{,}667} = 4.0857 \text{ V}$$

The lower trip voltage is

$$V_L = \frac{V_R R_f - R_p V_{Z2}}{R_f + R_p}$$

$$= \frac{4 \times 805{,}000 - 11{,}667(3.4 + 0.6)}{805{,}000 + 11{,}667} = 3.8857$$

The actual hysteresis width will be

$$V_U - V_L = 4.0857 - 3.8857 = 0.20 \text{ V}$$

*Step 6.* The maximum error in the trip voltages $V_U$ and $V_L$ at +25°C will be

$$V_{\text{off}} = \pm(V_{io} + I_{io}R_1)$$
$$= \pm(3 \times 10^{-3} + 3 \times 10^{-8} \times 11{,}500) = \pm 6.45 \text{ mV}$$

*Step 7.* The circuit input resistance is

$$R_{\text{in}} \approx R_1 + R_p + R_{id}$$
$$\approx 11{,}500 + 11{,}667 + 300{,}000 \approx 323{,}167 \text{ }\Omega$$

*Step 8.* The circuit output resistance for positive output voltages is 10 $\Omega$ ($R_{Z1}$). For negative output voltages the output resistance is 80 $\Omega$ ($R_{Z2}$).

Chapter **6**

# Converters

## INTRODUCTION

This chapter will be confined to two types of converters, analog-to-digital converters and digital-to-analog converters. The analog-to-digital converter is commonly called the A/D converter, or ADC for brevity. The digital-to-analog converter is likewise called the D/A converter or DAC. Since these are rather complex circuits, we will restrict ourselves to one approach for each type.

## 6.1 DUAL-SLOPE A/D CONVERTER

**ALTERNATE NAMES** Analog-to-digital converter, ADC, A-D converter, dual-ramp A/D converter.

**EXPLANATION OF OPERATION** Figure 6.1 shows an ADC designed to operate on positive input voltages. An inverter in front of the $v_i$ terminal is required for negative input voltages. Likewise, the reference voltage $V_R$ must be negative. Sequencing of the logic is automatic. The digital equivalent of $v_i$ will be repeatedly determined once per cycle. This is the familiar technique used in digital voltmeters. The cycle time may range from seconds down to milliseconds, depending principally on the speed of the op amps.

The following explanation of operation assumes an 8-bit ADC. The design equations, however, are general enough to handle word lengths other than 8.

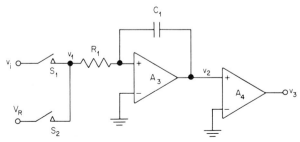

**Fig. 6.1** Simplified analog portion of dual-slope ADC.

## 6-2 CONVERTERS

This ADC actually has four modes of operation. These modes last for durations of $T_1$, $T_2$, $T_3$, and $T_4$ s. They are known as:

$T_1$: clear-pulse mode
$T_2$: Input-voltage integration mode
$T_3$: Reference-voltage integration mode
$T_4$: Standby mode when automatic zeroing may take place

Briefly, the following events occur during these modes (refer to Figs. 6.1 to 6.3):

$T_1$: This pulse clears the 7493s in preparation for a new cycle. Both $S_1$ and $S_2$ are off. FF1 and 2 are set to $Q = 1$.

$T_2$: $S_1$ is on and $S_2$ is off. During this *fixed* length of time the input voltage $v_i$ is integrated. Comparator $A_4$ changes state at the beginning of $T_2$ as $v_2$ passes through zero.

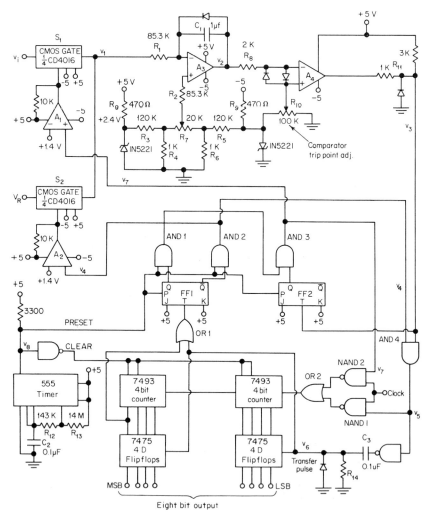

**Fig. 6.2** An 8-bit dual-slope ADC.

$T_3$: $S_2$ is on and $S_1$ is off. During this *variable* length of time the negative reference voltage drives the integrator output back toward zero. At the end of $T_3$, when $v_2$ passes through zero, the comparator again changes state.

$T_4$: $S_1$ and $S_2$ are both off. This standby time can be used for automatic zeroing. This type of circuitry is quite complex and will not be covered here.

During $T_3$ an 8-bit counter accumulates pulses from a clock. At the end of $T_3$ the clock is turned off and the counter contents are transferred to an 8-bit output register. The binary number in this register is proportional to time $T_3$.

Assume that $v_i$ is constant during $T_2$ (this can be done with a sample-and-hold circuit). The integrator output voltage at the end of $T_2$ is $v_2(t_2) = -v_i T_2/R_1 C_1$. During $T_3$ this voltage is returned to zero using $V_R$; so we also have $v_2(t_3) = v_2(t_2) - V_R T_3/R_1 C_1$. But $v_2(t_3) = 0$; so the following is true:

$$-\frac{v_i T_2}{R_1 C_1} = \frac{V_R T_3}{R_1 C_1}$$

or
$$T_3 = \frac{v_i T_2}{-V_R} \quad \text{(where } V_R \text{ is negative)}$$

The comparator may change states at some nonzero input voltage. This will not create an error, since the first integration ($T_2$) is in the opposite direction of the second integration ($T_3$) and the errors cancel. The error sources which must be considered are the input offset voltage and current of the integrator. These are partially canceled out (at one temperature) with the resistor network shown connected to the noninverting input of $A_3$.

Details of circuit operation are rather lengthy, and reference to the timing chart (Fig. 6.3) will be necessary. This type of ADC is automatic in operation

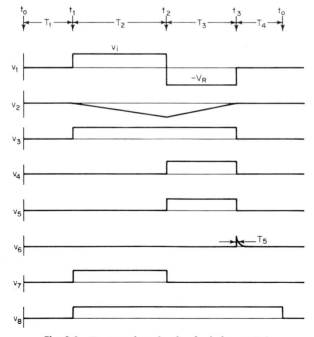

**Fig. 6.3** Timing chart for the dual-slope ADC.

and performs one analog-to-digital conversion each $T_1 + T_2 + T_3 + T_4$ s. We assume a starting point for our discussion at the instant $T_1$ starts. Pulse $T_1$ sets both flip-flops to $Q = 1$ and clears the eight-stage counter to zero. It is assumed all zeroing and drift compensation is completed prior to $T_1$. $T_1$ is generated by the 555 timer and may have a duration of 1 to 10 ms. The spacing between $T_1$ pulses should be at least greater than $T_2 + T_3$ + any zeroing time required.

AND 1 and AND 2 prevent the output of FF1 from being used until the end of $T_1$. After $T_1$ is over, AND 1 and AND 3 pass a HIGH $v_7$ to $A_1$ which turns on the CMOS gate $S_1$. $A_3$ is then allowed to integrate $v_i$ for a period of time $T_2$. As $v_2$ begins to integrate downward, $A_4$ switches $v_3$ to the HIGH state. This signal passes over to AND 4 but cannot pass farther, since $v_4$ is LOW. Meanwhile, $v_7$ is allowing NAND 2 to transfer clock pulses to the 8-bit counter. After 256 pulses have been accumulated, the last stage of the counter toggles FF1 through OR 1. This terminates the integration of $v_i$.

At this time $T_2$ ends and $T_3$ starts. $S_1$ is turned off and $S_2$ is turned on. This action applies a negative $V_R$ to the integrator, resulting in a $v_2$ which ramps upward. During this time the eight-stage counter receives pulses through NAND 1, since $v_3$, $v_4$, and $v_5$ are all HIGH. When $v_2$ passes through zero, $v_3$ and $v_5$ go LOW, and the counter stops. At this same instant a $v_6$ pulse is generated which transfers the counter contents to some other registers (such as a pair of 7475s). This 8-bit word will remain in the holding registers until the same time in the next cycle. As the transfer takes place, a trigger is sent over to FF1 to switch $v_4$ to the LOW state. At the end of $T_3$ the trailing edge of the $v_3$ pulse also toggles FF2. This also keeps $v_7$ at the LOW state. During $T_4$ both $v_4$ and $v_7$ remain LOW. Automatic zeroing can be incorporated during this time to offset any integrator drift which has occurred since the start of $T_1$.

### DESIGN PARAMETERS

| Parameter | Description |
| --- | --- |
| ADC | Analog-to-digital converter |
| $A_1$ | Comparator used to drive the CMOS gate $S_1$ |
| $A_2$ | Comparator used to drive the CMOS gate $S_2$ |
| $A_3$ | High-quality op amp connected as an integrator |
| $A_4$ | Comparator used to sense the position of $v_2$ |
| $C_1$ | Determines integration time |
| $f_c$ | Clock frequency |
| $I_{io}$ | $A_3$ input offset current |
| $N$ | Number of stages in binary counter |
| $R_1$ | Part of integrator time constant |
| $R_{S1}, R_{S2}$ | ON resistances of $S_1$ and $S_2$ |
| $S_1$ to $S_2$ | CMOS switches |
| $t_o$ to $t_3$ | Time of different events as shown in Fig. 6.3 |
| $T_1$ | Time of clearing and presetting mode |
| $T_2$ | Time for integration of input voltage (fixed duration) |
| $T_3$ | Time to integrate $v_2$ back to zero (variable duration) |
| $T_4$ | Standby mode |
| $T_c$ | Time of one clock cycle ($T_c = 1/f_c$) |
| $v_1$ to $v_8$ | Voltage waveforms as shown in Figs. 6.1 to 6.3 |
| $v_i$ | Input voltage to ADC |
| $V_{io}$ | $A_3$ input offset voltage |
| $V_R$ | Negative reference voltage |

## DESIGN EQUATIONS

| Eq. No. | Description | Equation |
|---|---|---|
| 1 | Integrator output voltage $v_2$ at the end of $T_2$ | $v_2(t_2) = -\dfrac{v_i T_2}{R_1 C_1}$ |
| 2 | Integrator output voltage $v_2$ at the end of $T_3$ | $v_2(t_3) = v_2(t_2) - \dfrac{V_R T_3}{R_1 C_1}$ |
| 3 | Pulse width $T_1$ | $T_1 \approx 0.7\, R_{12}\, C_2$ |
| 4 | Cycle time $T_1 + T_2 + T_3 + T_4$ | $T_1 + T_2 + T_3 + T_4 \approx 0.7(R_{12} + R_{13})C_2$ |
| 5 | Length of time $T_2$ | $T_2 = 2^N\, T_c$ |
| 6 | Length of time $T_3$ | $T_3 = \dfrac{-v_i T_2}{V_R}$ |
| 7 | Optimum $v_6$ pulse width | $T_5 \approx R_{14} C_3 \ll T_c$ |
| 8 | Drift of $v_2$ during $T_2 + T_3$ integration period | $\Delta v_2 = \dfrac{\pm(T_2 + T_3)[V_{io}(1 + R_1 C_1) + R_1 I_{io}]}{R_1 C_1}$ |
| 9 | Fractional error in $v_i$ conversion due to ON resistances of $S_1$ and $S_2$ | $\Delta v_i(\text{max}) = \dfrac{|R_{s1} - R_{s2}|\ \text{max}}{R_1}$ |

## DESIGN PROCEDURE

The design of the analog portion of this circuit is straightforward, since it contains only an integrator, three comparators, and two CMOS switches. To simplify this presentation, no error-correction circuits other than the integrator compensation circuit are shown. Many zeroing and compensation circuits are seen in the ADC literature.

The prime trade-offs to be considered are the cycle time vs. speed of devices and the number of output bits vs. offset errors. The dual-slope ADC is normally used in low-speed, high-accuracy systems. In this procedure we will assume, therefore, that the cycle time $T_1 + T_2 + T_3 + T_4$ is fixed. The offset errors are mostly removed by the integrator compensation-resistor network. Higher-accuracy systems would require many more gates and logic than shown here.

## DESIGN STEPS

*Step 1.* Determine $R_{12}$ and $C_2$ with Eq. 3. The time $T_1$ is chosen to be 1 to 10 ms, which is much more than adequate to set FF1 and FF2 and clear the counter.

*Step 2.* After a total cycle time $T_1 + T_2 + T_3 + T_4$ is chosen, $R_{13}$ is computed using Eq. 4. $R_{13}$ can be made adjustable if a variable cycle time is required. The minimum $R_{13}$, however, is constrained by the maximum $T_2$ and $T_3$ expected in the following steps.

*Step 3.* Calculate the time $T_2$ using Eq. 5. Ascertain that this gives reasonable values for $R_1$ and $C_1$ using Eq. 1. This is done as follows: First compute $T_2$ using $N$, the number of conversion bits required, and $T_c$, the period of one clock cycle. Let $v_i$ be the largest expected input voltage and let $|v_2(t_2)|$ be at least 2 V less than $|V^{(-)}|$. The $R_1 C_1$ product resulting from using Eqs. 1 and 5 may be quite large if long integration times are desired. Large values of $C_1$ may result in capacitor leakage currents approaching the input bias current of $A_3$. The zero-adjust circuit will make up for some of this error at one temperature. Large values of $R_1$ will cause an equivalent input-voltage offset due to the input bias current. Again, the zero-adjust circuit will probably cancel out much of this error, but caution is still advisable.

*Step 4.* Use Eq. 6 to determine the required reference voltage $V_R$. Let $T_3 = T_2$ as computed in step 3. Again, use $v_i(\text{max})$ for the calculation.

*Step 5.* Select resistor ratios $R_3/R_4$ and $R_5/R_6$ which will provide at least several times $\pm(V_{io} + I_b R_1)$ at each end of $R_7$. Let $R_7$ be at least ten times $R_4 + R_6$. Set $R_2 = R_1$ so that bias currents into $A_3$ will have a first-order cancellation. $R_7$ can be used to trim any remaining offsets.

*Step 6.* $R_8$ is chosen to be greater than the manufacturer's recommended minimum-load resistor for $A_3$.

*Step 7.* $R_{14}$ is selected so that it will sink the input current of OR 1 and the 7475s without producing more than 1 V.

*Step 8.* The maximum drift in $v_2$ at the end of the $T_2$ to $T_3$ integration period is found from Eq. 8.

**EXAMPLE OF AN ADC DESIGN** We will numerically illustrate the design procedure by designing a basic 8-bit ADC.

*Design Requirements*
$v_i(\text{max}) = +1$ V
$V_R = -1$ V
$f_c = 1.0$ kHz ($T_c = 10^{-3}$ s)
$N = 8$
Cycle time = 1 s
$T_1 = 10$ ms
$v_2(t_2) = -3$ V maximum
$V^{(\pm)} = \pm 5$ V

*Device Data* ($-55$ to $+125°$C)
$V_{io} = \pm 5$ mV
$I_b = 50$ nA       } HA2700 op amp
$I_{io} = \pm 30$ nA
$R_L(\text{nominal}) = 2{,}000\ \Omega$
$S_1$ to $S_2$ ON resistance range: 100 to 1,000 $\Omega$

*Step 1.* We have chosen $T_1 = 10$ ms so

$$R_{12}C_2 = \frac{T_1}{0.7} = \frac{0.01}{0.7} = 0.0143$$

Let us choose a convenient value for $C_2$ such as 0.1 $\mu$F. Thus,

$$R_{12} = \frac{0.0143}{C_2} = \frac{0.0143}{10^{-7}} = 143\ \text{k}\Omega$$

*Step 2.* The total cycle time is determined principally from $R_{13}$ and $C_2$. Since $C_2$ is already found,

$$R_{13} = \frac{T_1 + T_2 + T_3 + T_4}{0.7\ C_2} - R_{12}$$

$$= \frac{1}{0.7\ (10^{-7})} - 143{,}000 = 14\ \text{M}\Omega$$

*Step 3.* Using Eq. 5, we get

$$T_2 = 2^N T_c = 2^8 \times 10^{-3} = 256\ \text{ms}$$

From a slight rearrangement of Eq. 1 we get

$$R_1 C_1 = \frac{T_2 v_i(\max)}{|v_2(\max)|} = \frac{(0.256) \times 1}{|-3|} = 0.0853$$

If we choose $C_1 = 1\ \mu\text{F}$,

$$R_1 = \frac{0.0853}{C_1} = \frac{0.0853}{10^{-6}} = 85.3\ \text{k}\Omega$$

*Step 4.* Using Eq. 6 we get

$$V_R = \frac{-v_i T_2}{T_3} = \frac{-v_i T_2}{T_2} = -v_i = -1\ \text{V}$$

*Step 5.* The maximum input offset over temperature will be

$$V_{io}(\text{equivalent}) = \pm(V_{io} + I_b R_1) = \pm(5 \times 10^{-3} + 5 \times 10^{-8} \times 8.53 \times 10^4)$$
$$= \pm 12\ \text{mV}$$

We need to generate at least $\pm 12$ mV from $\pm 2.4$-V sources. If a maximum offset adjustment of $\pm 20$ mV is assumed, we need

$$\frac{R_4}{R_3 + R_4} = \frac{R_6}{R_5 + R_6} = \frac{20\ \text{mV}}{2.4\ \text{V}}$$

This can be satisfied by $R_4 = R_6 = 1{,}000\ \Omega$ and $R_3 = R_5 = 120\ \text{k}\Omega$. $R_7$ can be a 20-k$\Omega$ potentiometer. First-order cancellation of input offsets is achieved by letting $R_2 = R_1 = 85.3\ \text{k}\Omega$.

*Step 6.* As per the manufacturer's recommendation we choose $R_8 = 2{,}000\ \Omega$.

*Step 7.* If OR 1 is a 7432 and the 4-bit registers are 7475s, $R_{14}$ must sink 10 mA while $v_6$ is in the LOW state. This current must not produce more than 1 V across $R_{14}$. We therefore require that

$$R_{14} \leq 1\ \text{V}/10\ \text{mA} = 100\ \Omega$$

We will assume $R_{14} = 100\ \Omega$. This will make the transfer pulse width

$$V_6(\text{pulse width}) \approx R_{14} C_3 \approx 100 \times 10^{-7} \approx 10\ \mu\text{s}$$

This is a sufficient trigger pulse width for any TTL device.

*Step 8.* If an active integrator drift-compensation circuit is not used, the maximum error of $v_2$ at the end of each cycle may be found from Eq. 8:

$$\Delta v_2(\max) = \frac{\pm(T_2 + T_3)\ [V_{io}(1 + R_1 C_1) + R_1 I_{io}]}{R_1 C_1}$$

$$= \frac{\pm(2 \times 0.256)\ [5 \times 10^3(1 + 0.0853) + 85{,}300 \times 3 \times 10^{-8}]}{0.0853}$$

$$= \pm 48\ \text{mV}$$

Voltage $v_2$ integrates down to $-3$ V for a maximum $v_i$ of 1 V. Thus, a 48-mV error in $v_2$ is equivalent to a $48/3 = 16$-mV error in the measurement of $v_i$. This is significantly larger than the basic A/D error caused by converting $v_i$ into 256 discrete digital output numbers (this is called quantization error). The quantization error of an 8-bit ADC operating on a $v_i(\max)$ of 1 V is 1 V/256 = 3.9 mV. However, this effective 16-mV error in $v_i$ occurs only in the worst case over the full $-55$ to $+125°$C temperature range.

**6-8 CONVERTERS**

## REFERENCES

1. Goldberg, H. S.: Three-Phase A/D Conversion Has High Accuracy and Low Cost, *EDN*, Jan. 20, 1973, p. 82.
2. Tobey, G. E., J. G. Graeme, and L. P. Huelsman: "Operational Amplifiers – Design and Applications," p. 346, McGraw-Hill Book Company, New York, 1971.

## 6.2 DIGITAL-TO-ANALOG CONVERTER

**ALTERNATE NAMES** D/A converter, DAC, unipolar D/A converter.

**EXPLANATION OF OPERATION** The $R$-$2R$ ladder method of D/A conversion has the advantage of requiring only two resistance values in the resistor network. Converters which utilize ladders having binary-weighted resistors result in a large spread of required resistor values. For example, a 10-bit D/A converter would require the largest ladder resistor to have 1,024 times the resistance of the lowest resistor. Converter errors may occur with large resistance values because of op amp input bias current. Low resistor values create possible error sources because the resistance of the switch (op amp voltage follower) for each binary bit becomes appreciable in comparison with that of the resistor. The $R$-$2R$ ladder overcomes both these problems by utilizing mid-range resistors. We will present equations which give the errors caused by resistors (in the ladder network) which are too large or too small.

The operation of this D/A converter is best understood by first assuming that one bit of the input digital word is OFF and all other bits are ON. After $v_o$ is determined for each bit, one at a time, superposition is used to determine $v_o$ for any arbitrary combination of input bits. Suppose first that bit 1, the most significant bit (MSB), is OFF ($v_1 = 0$) and all other bits are ON ($v_2$ through $v_8 = +5$ V). If $G1$ is an inverting gate, such as 1/6 of a 74CO4 CMOS hex inverter, then $v_9$ will be OFF when $v_1$ is ON. The output of $A_1$, $v_{10}$ will also be ON. A CMOS gate driving a voltage follower produces precisely +5.00 V when $v_{10}$ is ON and 0.00 V when $v_{10}$ is OFF. The error is less than 1 mV if selected 74CO4 devices are used. Of course, the +5.00 V supply for these gates must be regulated as shown in Fig. 6.4. The voltage at $v_{11}$ is determined by noting that this type of ladder network has a resistance to ground of exactly $2R$ when looking left or right of any node. $A_1$ is driving $v_{11}$ with a $2R$ source resistance; so the voltage at $v_{11}$ is precisely 5/3 V when $v_{10} = 5.00$ V.

By similar reasoning, if $v_2$ is OFF and all other inputs are ON, $v_{14} = 5/3$ V. Transferring $v_{14}$ over to $v_{11}$ results in 5/6 V at $v_{11}$. This is one-half that achieved at $v_{11}$ when $v_1$ was OFF and all other inputs were ON. Going down the ladder, we find that each bit, by itself, contributes one-half the voltage to $v_{11}$ of the bit to its right. If all inputs are OFF, the voltage at $v_{11}$ is

$$v_{11}(\max) = +\frac{5}{3} + \frac{5}{6} + \frac{5}{12} + \frac{5}{24} + \frac{5}{48} + \frac{5}{96} + \frac{5}{192} + \frac{5}{384} = 3.320 \text{ V}$$

The $A_9$ circuit has a gain of

$$A_{vc} = 1 + \frac{R_8}{R_7}$$

In practice $R_7$ is usually twice $R_8$, giving $A_{vc} = 1.5$. In this case we get $v_o = 3.320 \times 1.5 = 4.9536$ V. The full-scale trim pot $R_3$ can be adjusted so that the output voltage is exactly 5.000 V when all inputs $v_1$ to $v_8$ are OFF. The zero trim pot $R_5$ is set so that $v_o = 0.000$ when $v_1$ to $v_8$ are all ON. This

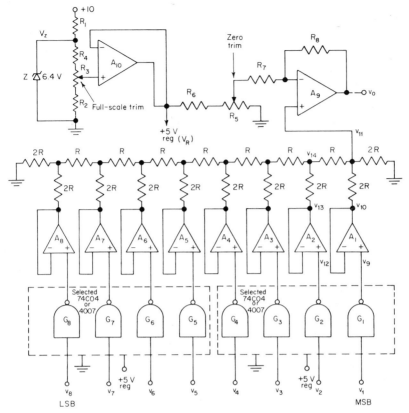

**Fig. 6.4** A high-precision R-2R D/A converter.

zero trim may not be possible with some types of op amps. In these cases a negative supply voltage (−5 or −10) may be required for $A_9$.

The error sources in this circuit are the resistor ladder, input offset drifts of $A_9$, and the stability of Z. The feedback network resistors of $A_9$, the zero trim pot, and the full-scale trim pot must also be considered. All resistors mentioned above should be of the low-drift metal-film type if 8 bits or more of D/A conversion is required. $A_9$ should be a low-offset, low-drift op amp. The R and 2R resistors should be kept under 100 kΩ so that $A_9$ bias currents do not develop appreciable errors. Lastly, the reference diode Z should be sufficiently stable for the number of bits chosen in this DAC.

## DESIGN PARAMETERS

| Parameter | Description |
|---|---|
| $A_1$ to $A_8$ | Voltage followers which provide nearly zero loading of the CMOS gates and drive the resistor network with a nearly zero source resistance |
| $A_9$ | Inverting amplifier which provides nearly zero output resistance |
| $A_{10}$ | Voltage follower to provide low-resistance source of +5-V reference |
| $A_v$ | Op amp open-loop gain |
| DAC | Digital-to-analog converter |

## DESIGN PARAMETERS (Continued)

| Parameter | Description |
|---|---|
| $G_1$ to $G_8$ | CMOS gates (selected for low saturation voltages) |
| $I_{io}$ | Input offset current of $A_9$ |
| LSB, MSB | Least significant and most significant bit of input digital word |
| N | Number of conversion bits |
| R | Standard resistance used in precision resistor network. This value is chosen such that a reasonable compromise between the errors calculated in Eqs. 4 and 5 is achieved |
| $R_o$ | Output resistance of op amps $A_1$ to $A_8$ (closed-loop) |
| $R_{out}$ | Open-loop output resistance of op amps |
| $v_o$ | Output voltage of DAC |
| $v_1$ to $v_{14}$ | Voltages as indicated in Fig. 6.4 |
| $V_{io}$ | Input offset voltage of $A_9$ |
| $V_R$ | Reference voltage |
| $V_z$ | Zener breakdown voltage |
| Z | Reference diode |

## DESIGN EQUATIONS

| Eq. no. | Description | Equation |
|---|---|---|
| 1 | Output voltage as a function of inputs | $v_o = \dfrac{\bar{v}_1}{2} + \dfrac{\bar{v}_2}{4} + \cdots + \dfrac{\bar{v}_N}{2^N}$ |
| 2 | Voltage gain from $v_{11}$ to $v_o$ | $A_{vc} = 1 + \dfrac{R_8}{R_7}$ |
| 3 | Output resistance of $A_1$ to $A_8$ (closed-loop) | $R_o = \dfrac{R_{out}}{1 + A_v}$ |
| 4 | Worst-case error in $v_{10}$ due to the change in output resistance of $A_1$ (important if R is low) | $\Delta v_{10}(\text{max}) = \dfrac{\Delta R_o(\text{max})}{3R}$<br>NOTE: This implies that $3R$ must be at least $2^N$ times larger than $\Delta R_o(\text{max})$ in order to maintain accuracy |
| 5 | Worst-case error in $v_o$ due to changes in $V_{io}$ and $I_{io}$ of $A_9$ (important if R is high) | $\Delta v_o = \pm \left[\left(1 + \dfrac{R_8}{R_7}\right)V_{io} + R_8 I_{io}\right]$ |
| 6 | Resistor values, $R_1$ | $R_1 \approx \dfrac{V^{(+)} - V_z}{I_z}$ |
| 7 | $R_T$ | $R_T = R_2 + R_3 + R_4 > 10\,R_1$ |
| 8 | $R_2$ | $R_2 = \dfrac{R_T[V_R - 5V_{io}(\text{max})]}{V_z}$ |
| 9 | $R_3$ | $R_3 = \dfrac{10 V_{io} R_T}{V_z}$ |
| 10 | $R_4$ | $R_4 = R_T - R_2 - R_3$ |
| 11 | $R_5$ | $R_5 = 100$ to $1{,}000\ \Omega$ |
| 12 | $R_6$ | $R_6 = \dfrac{R_5 V_R}{10 V_{io}(\text{max})}$ |
| 13 | $R_7$ | $R_7 = \dfrac{2 R A_{vc}}{3(A_{vc} - 1)}$ |
| 14 | $R_8$ | $R_8 = R_7(A_{vc} - 1)$ |
| 15 | R | $R \gg 2^N R_o$ |

## REFERENCES

1. Widlar, R. J.: A Fast Integrated Voltage Follower with Low Input Current, National Semiconductor Application Note AN-5, May 1969.
2. Tobey, G. E., J. G. Graeme, and L. P. Huelsman: "Operational Amplifiers—Design and Applications," p. 335, McGraw-Hill Book Company, New York, 1971.

Chapter **7**

# Demodulators and Discriminators

### INTRODUCTION

Signal-processing circuits are often called upon to extract one type of information from a composite waveform. This process is called demodulation, since it is the inverse to modulation. Traditionally, some demodulator circuits are called discriminators, and these two names are often used interchangeably. We will present design equations for three types of demodulators in this chapter. For AM demodulation we choose the synchronous demodulator, since the standard continuous AM demodulator, or precision rectifier, is presented in Chap. 16. The circuit to be shown for FM demodulation is a precision-type circuit which requires no tuned circuits. Pulse-width demodulation (or discrimination) is the last subject to be covered.

### 7.1 SYNCHRONOUS AM DEMODULATOR

**ALTERNATE NAMES** Phase-sensitive demodulator, suppressed-carrier AM demodulator, synchronous-switching demodulator, phase-sensitive detector, lock-in amplifier, synchronous detector.

**EXPLANATION OF OPERATION** In addition to all the above names given to this circuit, it can also be considered to be a synchronized full-wave rectifier. If the input signal $v_i$ is exactly in phase with the carrier signal $v_c$, the output will look like a full-wave rectified waveform. As shown in Fig. 7.2, the average value of the full-wave rectified output is the parameter that is maximized. If $v_i$ is 90° out of phase, the average value of the output is zero. Likewise, most random noise will not be synchronous with $v_c$ and will be greatly attenuated in this circuit. Synchronous demodulators are often used in low-noise systems because of this noise-reducing property.

The $A_1$ stage of Fig. 7.1 is an inverting amplifier with a gain of $-R_2/R_1$. Thus $-v_i R_2/R_1$ appears at the left side of switch $S$. This signal receives further gain of $-R_7/R_6$ during the time when $S$ is on. Through this route the signal receives a total gain of $R_2 R_7/R_1 R_6$. To this must be added the signal at $v_o$ coming directly through $R_5$ with a gain of $-R_7/R_5$. During the time $S$ is on we have

7-1

## 7-2 DEMODULATORS AND DISCRIMINATORS

$$v_o(S \text{ on}) = v_i \left(\frac{R_2 R_7}{R_1 R_6} - \frac{R_7}{R_5}\right)$$

While S is off we have

$$v_o(S \text{ off}) = v_i \left(-\frac{R_7}{R_5}\right)$$

To provide a symmetrical full-wave rectified signal at $v_o$ with minimum ripple (when $v_i$ and $v_c$ are in phase), the gain term $R_2 R_7 / R_1 R_6$ must be twice the size of $R_7/R_5$. Under these conditions the maximum output voltage is

$$v_o = |v_i| \frac{R_7}{R_5}$$

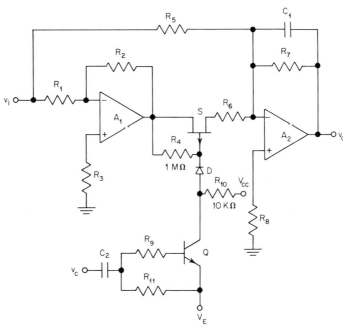

**Fig. 7.1** Synchronous AM demodulator.

The dc output voltage as a function of phase angle $\phi$ between $v_i$ and $v_c$ is (assuming $v_i$ is a sinusoid)

$$v_o(\text{dc}) = 0.637 \, |v_i(\text{peak})| \frac{R_7}{R_5} \cos \phi$$

The 0.637 constant is the average value of a half sine wave. If $v_i$ is another type of waveform, another appropriate constant will be needed. The $A_2$ stage is an integrator with dual inputs (through $R_5$ and $R_6$). The degree of ripple in $v_o$ can be controlled with $C_1$. If this capacitor is not too large, no integrator reset is needed. A compromise between output ripple and drift must sometimes be made. The peak-to-peak output-ripple voltage when $v_i$ and $v_c$ are in phase is

$$v_o(\text{peak-to-peak}) \leq \frac{v_i(\text{peak-to-peak})}{2\pi R_5 C_1 f_c}$$

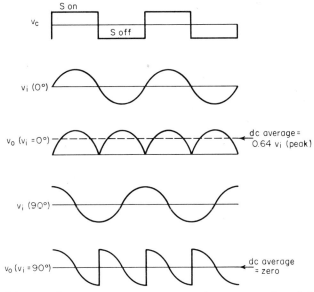

**Fig. 7.2** Waveforms at various points in synchronous AM demodulator.

## DESIGN PARAMETERS

| Parameter | Description |
|---|---|
| $A_{vc}$ | Closed-loop gain of entire circuit |
| $\beta$ | Current gain of $Q$ |
| $C_1$ | Capacitor which controls magnitude of ripple in $v_o$ |
| $C_2$ | Coupling capacitor for $v_c$ |
| $D$ | Switching diode |
| $f_c$ | Chopper (or carrier) frequency into $Q$ |
| $I_{co}$ | Collector-base leakage current of $Q$ |
| $\phi$ | Phase difference between $v_i$ and $v_c$ |
| $Q$ | Drive transistor for FET |
| $R_1, R_2$ | Gain-determining resistors for $A_1$ |
| $R_5, R_6, R_7$ | Gain-determining resistors for $A_2$ |
| $R_3, R_8$ | Used to null effects of input bias currents in $A_1$ and $A_2$ |
| $R_4$ | Causes the FET source-to-gate voltage to return quickly to zero bias when $Q$ is turned on and $D$ is reverse-biased |
| $R_9, R_{10}$ | Resistors used to set proper current levels in $Q$ |
| $S$ | FET switch (low ON resistance desirable) |
| $V_{cc}$ | Collector supply voltage for $Q$ |
| $v_c$ | Chopper (or carrier) input rectangular waveform |
| $v_i$ | Input signal to be demodulated |
| $v_i(0°)$ | Input signal exactly in phase with $v_c$ |
| $v_i(90°)$ | Input-signal phase leading the phase of $v_c$ by 90° |
| $v_o$ | Circuit output signal |
| $V_p$ | FET pinch-off voltage |

## 7-4 DEMODULATORS AND DISCRIMINATORS

### DESIGN EQUATIONS

| Eq. No. | Description | Equation |
|---|---|---|
| 1 | Voltage gain while S is on | $A_{vc} = \dfrac{v_o}{v_i} = \dfrac{R_2 R_7}{R_1 R_6} - \dfrac{R_7}{R_5}$ |
| 2 | Voltage gain while S is off | $A_{vc} = \dfrac{v_o}{v_i} = -\dfrac{R_7}{R_5}$ |
| 3 | Relationship between gain-determining resistors to provide minimum unfiltered output ripple | $R_2 R_5 = 2\, R_1 R_6$ |
| 4 | Average dc output voltage as a function of phase angle between $v_i$ and $v_c$ (assuming $v_i$ is a sinusoid) | $v_o(\text{dc}) = 0.637\, |v_i(\text{peak})|\, \dfrac{R_7}{R_5} \cos \phi$ |
| 5 | Peak-to-peak output-ripple voltage when $v_i$ and $v_c$ are in phase | $v_o(\text{peak-to-peak}) \leqq \dfrac{v_i(\text{peak-to-peak})}{2\pi R_5 C_1 f_c}$ |
| 6 | Optimum size of $R_3$ | $R_3 = \dfrac{R_1 R_2}{R_1 + R_2}$ |
| 7 | Optimum size of $R_8$ | $R_8 = \dfrac{1}{1/R_5 + 1/R_6 + 1/R_7}$ |
| 8 | Minimum required $V_{cc}$ | $V_{cc} > v_i(\text{peak}) + V_p + 2$ |
| 9 | Maximum required $V_E$ | $V_E < v_i(\text{peak}) - 2$ |
| 10 | Optimum size for $R_9$ | $R_9 = \dfrac{v_c(\text{peak})\beta R_{10}}{V_{cc} + |V_E|}$ |
| 11 | Optimum size for $R_{11}$ | $R_{11} < \dfrac{1}{10 I_{co}(\text{max})} - R_9$ |
| 12 | Minimum size for $C_2$ | $C_2 > \dfrac{3(R_9 + R_{11})}{f_c R_9 R_{11}}$ |

### REFERENCES

1. Bergersen, T. B.: Field Effect Transistors in Chopper and Analog Switching Circuits, Motorola Semiconductor Products, Inc., Application Note AN-220, 1966.
2. Tobey, G. E., J. G. Graeme, and L. P. Huelsman: "Operational Amplifiers—Design and Applications," p. 413, McGraw-Hill Book Company, New York, 1971.
3. Lloyd, A. G.: Phase Detector/Modulator Operates from DC to 30 kHz, *Electron. Des.*, June 10, 1971, p. 82.

### 7.2 FM DEMODULATOR

**ALTERNATE NAMES** Time-averaging FM demodulator, FM discriminator, FM detector, pulse-counting FM demodulator, frequency meter.

**EXPLANATION OF OPERATION** An FM demodulator of this type does not require tuned circuits. Thus the circuit will not drift out of tune as it ages. This is very important in applications where unattended operation for long periods of time is mandatory. Basically, the demodulator shown in Fig. 7.3 is composed of four simple circuits discussed in other chapters of this book. First, a zero-crossing detector (see Chap. 5) changes the input signal into rectangular waveforms. This signal, as shown in Fig. 7.4, is differentiated in the second circuit (see Chap. 15). Then the negative (or positive) pulses from the differentiator are removed with a precision rectifier (see Chap. 16). A simple passive rectifier circuit can be used if dc stability of the demodulator transfer function is not important. Fourth, a low-pass filter (or integrator) produces an output waveform which is equal to the average value of the positive (or negative) pulses. This filter controls the magnitude of ripple present in the output signal. The low-pass filter circuit is easily combined with the precision rectifier by merely installing the capacitor $C_4$. If separate low-pass filtering is desired, Chap. 10 should be consulted.

As shown in Fig. 7.3, $A_1$ is connected as an inverting zero-crossing detector with hysteresis (see Sec. 5.2). The peak-to-peak hysteresis is chosen so that it is at least an order of magnitude larger than the peak-to-peak noise present

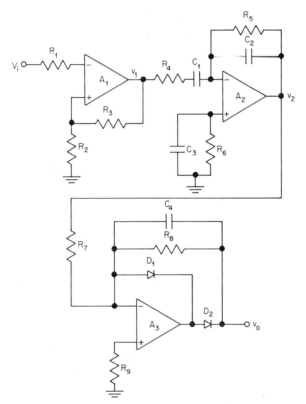

**Fig. 7.3** A time-averaging FM demodulator.

in $V_i$. $A_1$ can be a comparator type of op amp if high speed is important. The slew rate of $A_1$ over the peak-to-peak limits of $v_1$ must be substantially faster than the time of one cycle of $V_i$. Otherwise $v_1$ will be triangular in shape instead of rectangular.

The design of the $A_2$ stage should proceed as outlined in Sec. 15.1. $R_4$ and $C_2$ are required to guarantee feedback stability. $R_6$ and $C_3$ help reduce bias drift and noise problems. $C_1$ and $R_5$ determine the actual differentiation properties of this circuit.

The transfer function of the differentiator is

$$\frac{v_2}{v_1} = \frac{R_4 C_1 \exp(-t/R_5 C_2) - R_5 C_2 \exp(-t/R_4 C_1)}{R_4 C_1 - R_5 C_2}$$

## 7-6 DEMODULATORS AND DISCRIMINATORS

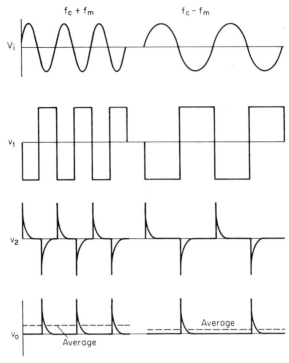

**Fig. 7.4** Waveforms at various locations in Fig. 7.3.

$C_1$ and $R_5$ must be sized according to the constraints imposed by the following:

$$v_2(\text{max}) = R_5 C_1 \left.\frac{dv_1}{dt}\right|_{\text{max}} = R_5 C_1 S(\text{max})$$

where

$$\left.\frac{dv_1}{dt}\right|_{\text{max}} = S(\text{max}) = \text{slew rate of } A_1$$

Also, the time constant $C_2 R_5$ must be approximately

$$R_5 C_2 \approx \frac{1}{5 f_c(\text{min})}$$

where $f_c(\text{min})$ is the minimum carrier frequency expected. This equality guarantees that each $v_2$ pulse will return to the baseline before the next pulse occurs. If the tail on each pulse gets too long, the amplitude of $v_2$ is reduced. This reduces the overall demodulation sensitivity of the circuit. If $C_2 R_5$ is too short, the average rectified signal at $v_o$ will be too small, and again, sensitivity will be reduced.

The last stage of the circuit can be designed for a positive or negative output as shown in Chap. 16. The filter capacitor is chosen such that the carrier frequency $f_c$ ripple is greatly attenuated compared with the FM modulation frequency $f_m$. According to Chap. 16 we must have

$$\frac{1}{2\pi f_c R_8} \ll C_4 \ll \frac{1}{2\pi f_m R_8}$$

FM DEMODULATOR 7-7

If the modulation is not present, the quiescent output voltage $v_o$ will be

$$v_o(\text{quiescent}) = \frac{0.45\, v_2(\text{rms})R_8}{R_7}$$

## DESIGN PARAMETERS

| Parameter | Description |
|---|---|
| $A_1$ | Op amp for zero-crossing-detector stage |
| $A_2$ | Op amp for differentiator stage |
| $A_3$ | Op amp for rectifier-filter stage |
| $C_1$ | Primary differentiation capacitor |
| $C_2$ | Provides feedback stability in $A_2$ |
| $C_3$ | Provides feedback stability and reduces equivalent input noise of $A_2$ |
| $C_4$ | Filter capacitor for $A_3$ |
| $D_1, D_2$ | Provides precision rectification in conjunction with $A_3$ |
| $f_c$ | Input carrier frequency |
| $f_m$ | Modulation frequency |
| $R_1$ | Protects $A_1$ from large $V_i$ transients |
| $R_2, R_3$ | Establishes hysteresis in zero-crossing detector |
| $R_4$ | Provides feedback stability in $A_2$ |
| $R_5$ | Primary differentiation resistor |
| $R_6$ | Used to cancel effect of op amp input bias current |
| $R_7, R_8$ | Establishes gain of rectifier-filter stage |
| $S$ | Slew rate of $A_1$ |
| $V_i$ | Input carrier containing FM |
| $V_n$(peak-to-peak) | Peak-to-peak noise in $V_i$ |
| $v_1$ | Rectangular waveform produced by $A_1$ |
| $v_2$ | Differentiated $v_1$ |
| $v_o$ | Modulation signal extracted from $V_i$ |

## DESIGN EQUATIONS

| Eq. No. | Description | Equation |
|---|---|---|
| 1 | Transfer function of differentiation stage | $\dfrac{v_2}{v_1} = \dfrac{R_4C_1\exp(-t/R_5C_2) - R_5C_2\exp(-t/R_4C_1)}{R_4C_1 - R_5C_2}$ |
| 2 | Maximum allowable slew rate of input stage | $S(\text{max}) = \dfrac{v_2(\text{max})}{R_5C_1}$ |
| 3 | Nominal value for time constant of pulse tail | $R_5C_2 \approx \dfrac{1}{5f_c(\text{min})}$ |
| 4 | Range of $C_4$ so that $f_c$ ripple is minimized and $f_m$ modulation is maximized | $\dfrac{1}{2\pi f_c R_8} \ll C_4 \ll \dfrac{1}{2\pi f_m R_8}$ |
| 5 | Quiescent output voltage with no FM modulation present | $v_o(\text{quiescent}) = \dfrac{0.45\, v_2(\text{rms})R_8}{R_7}$ |
| 6 | Resistor values | $R_1 = \dfrac{R_2 R_3}{R_2 + R_3}$<br>$R_2 = 1$ to $10$ k$\Omega$ |

| Eq. No. | Description | Equation |
|---|---|---|
| | | $R_3 = R_2 \left[ \dfrac{V_1(\text{max})}{10 V_n(\text{peak-to-peak})} - 1 \right]$ |
| | | $R_4, R_5, R_6$: See steps 1 through 9 of Sec. 15.1 |
| | | $R_7 = 2$ to $10$ k$\Omega$ |
| | | $R_8 = 10$ to $100$ k$\Omega$ (depends on size of $v_o$ required) |
| | | $R_9 = \dfrac{R_7 R_8}{R_7 + R_8}$ |
| 7 | Capacitor values | $C_1, C_2, C_3$: See steps 1 through 9 of Sec. 15.1 |
| | | $C_4$: See steps 5 through 7 in Sec. 16.2 |

## REFERENCES

1. Whittington, K. R.: Simple F-M Demodulator for Audio Frequencies, *Electronics*, Nov. 30, 1962, p. 89.
2. Tobey, G. E., J. G. Graeme, and L. P. Huelsman: "Operational Amplifiers—Design and Applications," p. 419, McGraw-Hill Book Company, New York, 1971.

## 7.3 PULSE-WIDTH DISCRIMINATOR

**ALTERNATE NAMES**  Pulse-width demodulator, pulse-width detector, pulse catcher.

**EXPLANATION OF OPERATION**  In communication systems a need often arises to extract pulses of a given width from a line containing a multitude of waveforms. The circuit described here generates an output pulse if and only if a pulse is received which is a specified width ± some given tolerance. This circuit allows the designer to choose both the pulse width and the tolerance. Referring to Fig. 7.5, this pulse-width demodulator operates as follows:

1. The input pulse is first "squared up" using a level detector with hysteresis. This circuit changes all input waveforms into pulses of uniform amplitude and rejects all noise below a specified threshold. This is further clarified in Fig. 7.6, where pulses of various lengths, amplitude, and noise content are shown.
2. The input pulse, of unknown length $T_x$, is then applied to an integrator-level detector circuit, and also to one input of a NOR gate.
3. The integrator-level detector will trip if $T_x$ is longer than the required pulse $T_p$ minus a specified tolerance. Thus $v_3$ will go low if $T_x > T_p - \Delta T_p$.
4. When $v_3$ goes low, it causes the $A_3$ single shot to generate a pulse having a width $T_t = 2 \Delta T_p$. This pulse goes to the other input of the NOR gate.
5. If the input pulse $T_x$ ends during the time $T_t$ is present, a pulse will appear at $v_5$. If $T_x$ ends before $T_t$ starts or after $T_t$ ends, no pulse will occur at $v_5$.
6. The pulse width at $v_5$ will depend on the width of each $T_x$ pulse and the width of $\Delta T_p$. Another single shot, $A_4$, is therefore added so that uniform output pulses are produced.

The first level detector $A_1$ is designed according to rules outlined in Sec. 5.4. In some applications a zero-crossing detector with hysteresis may be more useful (Sec. 5.2). Section 20.3 provides all the necessary design in-

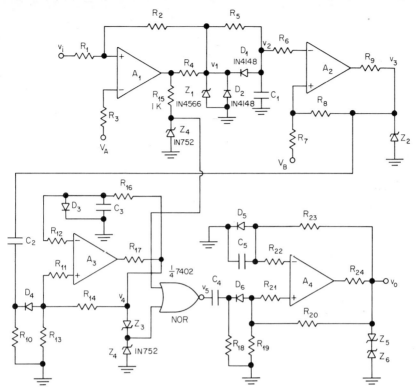

**Fig. 7.5** Pulse-width discriminator for positive-going pulses.

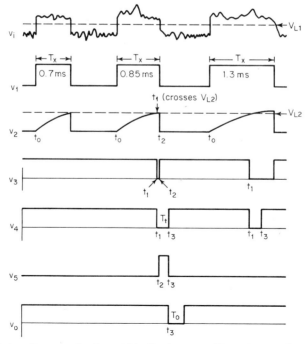

**Fig. 7.6** Timing diagram of pulse-width discriminator ($T_p = 1.0$ ms and $T_t = 0.4$ ms is assumed).

7-9

## 7-10 DEMODULATORS AND DISCRIMINATORS

formation for the two one-shot circuits. If a positive output pulse $T_o$ is required, an op amp or transistor inverter may follow the $A_4$ circuit.

The integrator is merely a passive circuit composed of $R_5$ and $C_1$. These should be low-drift (and low-leakage) parts if precision pulse-width discrimination is required. $C_1$ is discharged through $D_1$ at the termination of $T_x$.

Suppose one wants to make the pulse width $T_p$ and tolerance $\pm T_p$ adjustable. $R_5$ is the best candidate for a $T_p - \Delta T_p$ adjustment. $R_{16}$ is the best choice for a potentiometer to adjust $2\Delta T_p$.

### DESIGN PARAMETERS

| Parameter | Description |
|---|---|
| $\Delta I_{b1}/\Delta T$ | Change in input bias current as a function of temperature for $A_1$ |
| $\Delta I_{b2}/\Delta T$ | Change in input bias current as a function of temperature for $A_2$ |
| $I_{Z1}$ to $I_{Z6}$ | Optimum zener currents for $Z_1$ to $Z_6$ |
| $S_3$ | Slew rate of $A_3$ |
| $S_4$ | Slew rate of $A_4$ |
| $T_o$ | Output pulse width |
| $T_p$ | Nominal acceptable input pulse width |
| $T_p \pm \Delta T_p$ | Range of input pulse widths which cause circuit to generate an output pulse |
| $T_t$ | Tolerance of acceptable pulse widths ($T_t = 2\Delta T_p$) |
| $T_x$ | Input pulse of unknown width |
| $v_1$ to $v_5$ | Nonlinear waveforms at various points in circuit |
| $\Delta v_1$ | Peak-to-peak hysteresis for first level detector |
| $\Delta v_2$ | Peak-to-peak hysteresis for second level detector |
| $\Delta v_5$ | Change of $v_5$ as the NOR gate output changes from high to low |
| $v_i$ | Input voltage to circuit |
| $v_o$ | Output voltage from circuit |
| $V_{D1}$ | Forward breakdown voltage of $D_1$ |
| $V_{DZ2}$ | Forward breakdown voltage of $Z_2$ |
| $V_{L1}$ | Trip voltage for first level detector |
| $V_{L2}$ | Trip voltage for second level detector |
| $V_{Z1}$ to $V_{Z6}$ | Optimum zener breakdown voltages of $Z_1$ to $Z_6$. Note that calculations using $V_{Z3}$ to $V_{Z6}$ must include the forward breakdown voltage of the other zener in series with them |
| $\Delta V_{io1}/\Delta T$ | Change of $A_1$ input offset voltage with temperature |
| $\Delta V_{io2}/\Delta T$ | Change of $A_2$ input offset voltage with temperature |
| $V^{(+)}$ | Positive power-supply voltage |
| $V^{(-)}$ | Negative power-supply voltage |

### DESIGN EQUATIONS

| Eq. No. | Description | Equation |
|---|---|---|
| 1 | Required pulse width minus tolerance | $T_p - \Delta T_p = R_5 C_1 \ln\left(\dfrac{V_{Z1}}{V_{Z1} - V_{L2}}\right)$ where $V_{L2}$ = second-level-detector trip point |
| 2 | Pulse width of $T_t$ | $T_t = 2\Delta T_p = 0.8\, R_{16} C_3$ if $R_{13} = R_{14}$ and $V_{Z4} > 3$ V |
| 3 | Output pulse width | $T_o = 0.8\, R_{23} C_5$ if $R_{19} = R_{20}$ and $V_{Z6} > 3$ V |

# PULSE-WIDTH DISCRIMINATOR

| Eq. No. | Description | Equation |
|---|---|---|
| 4 | Pulse width at $v_5$ | $t_3 - t_2 = T_p + \Delta T_p - T_x$ (only if $T_x$ is in the range $T_p \pm \Delta T_p$, otherwise $t_3 - t_2 = 0$) |
| 5 | Trip voltage for first level detector | $V_{L1} = \dfrac{V_A(R_1 + R_2) + R_1 V_{D1}}{R_2}$ |
| 6 | Peak-to-peak hysteresis for first level detector | $\Delta v_1 = \dfrac{(V_{Z1} + V_{D1})R_1}{R_1 + R_2}$ |
| 7 | Trip voltage for second level detector | $V_{L2} = \dfrac{V_B R_8 + V_{Z2} R_7}{R_7 + R_8}$ |
| 8 | Peak-to-peak hysteresis for second level detector | $\Delta v_2 = \dfrac{(V_{Z2} + V_{DZ2})R_7}{R_7 + R_8}$ |
| 9 | Optimum $R_3$ if effects of $V_{io}$ and $I_b$ (of $A_1$) over temperature are to be minimized | $R_3 = \dfrac{\Delta V_{io1}/\Delta T}{\Delta I_{b1}/\Delta T}$ |
| 10 | Optimum $R_6$ if effects of $V_{io}$ and $I_b$ (of $A_2$) over temperature are to be minimized | $R_6 = \dfrac{\Delta V_{io2}/\Delta T}{\Delta I_{b2}/\Delta T}$ |
| 11 | Optimum relationship among $R_1$, $R_2$, and $R_3$ | $R_3 = \dfrac{R_1 R_2}{R_1 + R_2}$ |
| 12 | Optimum relationship among $R_6$, $R_7$, and $R_8$ | $R_6 = \dfrac{R_7 R_8}{R_7 + R_8}$ |
| 13 | Optimum resistance of $R_4$ | $R_4 = \dfrac{V^{(+)} - V_{Z1}}{I_{Z1}}$ |

## DESIGN PROCEDURE

As with any circuit-design task, the designer must first ask: What are the constant and variable parameters in this circuit? We will assume that the constants of this design task are $T_p$, $\pm \Delta T_p$, the minimum full-width voltage of $T_x$, the peak-to-peak noise in $v_i$, and the characteristics of the output pulse. The other parameters will depend on these constants, as shown in Eqs. 1 to 8.

## DESIGN STEPS

*Step 1.* Compute a nominal value for $R_3$ using Eq. 9.

$$R_3 = \frac{\Delta V_{io1}/\Delta T}{\Delta I_{b1}/\Delta T}$$

Use the same $\Delta T$ in both the numerator and denominator. Do not allow $R_3$ to go above 20 k$\Omega$ (bipolar op amps).

*Step 2.* Choose a highly stable regulator diode for $Z_1$. Otherwise, the voltage charging the $R_5$-$C_1$ integrator will vary with temperature and $T_p - \Delta T_p$ will be unstable.

*Step 3.* Choose a hysteresis width $\Delta v_1$ which is 1 to 10 percent of the trip level $V_{L1}$. At this point Eqs. 6 and 11 can be combined to determine $R_1$:

$$R_1 = \frac{R_3(V_{Z1} + V_{D1})}{V_{Z1} - \Delta v_1}$$

*Step 4.* Use Eq. 11 to calculate

$$R_2 = \frac{R_1 R_3}{R_1 - R_3}$$

*Step 5.* Equation 5 is now rearranged to determine the required bias voltage $V_A$:

$$V_A = \frac{R_2 V_{L1} - R_1 V_{D1}}{R_1 + R_2}$$

*Step 6.* Calculate a resistance for $R_4$ using Eq. 13 which will drive the optimum current through $Z_1$. This current will be found on the $Z_1$ data sheet. Make sure that $R_4 \ll R_1 + R_2$ so that $R_4$ will drive the correct current into $Z_1$. This inequality should be satisfied by a factor of at least 100.

*Step 7.* Compute a nominal value for $R_6$ using Eq. 10:

$$R_6 = \frac{\Delta V_{io2}/\Delta T}{\Delta I_{b2}/\Delta T}$$

Use the same $\Delta T$ in numerator and denominator. Keep $R_6 < 20$ k$\Omega$ for bipolar op amps.

*Step 8.* $Z_2$ does not need to be a voltage-regulator diode, since $A_2$ is merely providing a trigger pulse for $A_3$. The nominal value for $R_9$ is calculated as follows:

$$R_9 = \frac{V^{(+)} - V_{Z2}}{I_{Z2}}$$

It is recommended that $V_{Z2} \approx V_{Z3}$ so that $V_3$ will adequately trigger the $T_t$ single shot $A_3$. As will be shown in step 13, for TTL logic circuits, $V_{Z2} \approx 5$ V.

*Step 9.* Choose a hysteresis width $\Delta v_2$ which is 1 to 10 percent of the trip level $V_{L2}$. Equations 8 and 12 are combined to provide a method to find $R_7$:

$$R_7 = \frac{R_6(V_{Z2} + V_{DZ2})}{V_{Z2} + V_{DZ2} - \Delta v_2}$$

*Step 10.* Use Eq. 12 to calculate

$$R_8 = \frac{R_6 R_7}{R_7 - R_6}$$

*Step 11.* Equation 7 is now rearranged to determine the required bias voltage $V_B$:

$$V_B = \frac{V_{L2}(R_7 + R_8) - V_{Z2} R_7}{R_8}$$

*Step 12.* Equation 1 is rearranged so that $R_5 C_1$ can be computed:

$$R_5 C_1 = \frac{T_p - \Delta T_p}{\ln[V_{Z1}/(V_{Z1} - V_{L2})]}$$

Individual values for $R_5$ and $C_1$ are determined using the following ideas:

1. $R_5$ must not load down the regulated voltage established by $Z_1$. Otherwise Eq. 1 will not be true. If possible, let $R_5$ be 100 or 1,000 times larger than $R_4$ to overcome this problem.

2. $C_1$ should be a high-quality low-leakage capacitor if stability of

$T_p - \Delta T_p$ over temperature is required. It is recommended that $C_1$ not exceed 0.1 $\mu$F.

*Step 13.* Choose $Z_4$ to be compatible with the voltage requirements of the NOR gate. For TTL logic a 4.8- to 5.1-V zener diode should be adequate. $R_{17}$ is chosen to provide the zener current, the current through $R_{14}$ to $R_{16}$, and the current into the digital device. If the digital device is TTL, it will draw no current when $v_4$ is high. Thus, if $R_{14}$ and $R_{16}$ are 10 to 100 times larger than $R_{17}$, their loading effect on $V_{Z3}$ will not be worth considering. The nominal value for $R_{17}$ is

$$R_{17} = \frac{V^{(+)} - V_{Z4}}{I_{Z4}}$$

*Step 14.* Find the resistances of $R_{13}$ and $R_{14}$ as follows:

$$R_{13} = R_{14} = \frac{100 V_{Z4}}{I_{Z4}}$$

*Step 15.* Let $R_{11} = R_{12} = 10$ k$\Omega$ if a bipolar monolithic op amp is used for $A_3$. Set $R_{10} = 10 R_{13}$.

*Step 16.* Make sure that

$$\frac{V_{Z3} + V_{Z4}}{2 S_3} < T_t$$

If this inequality is not true, $A_3$ may be too slow ($S_3$ too low) for this application.

*Step 17.* Calculate $C_2 = T_t/R_{10}$. If $R_{10}C_2$ is too small, triggering will not occur. If $R_{10}C_2$ is too large, multiple output pulses will occur for each input pulse.

*Step 18.* Set $R_{16} = 2 R_{13}$. Compute $C_3$ as follows:

$$C_3 = \frac{T_t}{0.8 R_{16}}$$

*Step 19.* Choose $Z_5$ and $Z_6$ according to the output-pulse requirements of the discriminator. $R_{24}$ is computed as follows:

$$R_{24} = \frac{V^{(+)} - V_{Z6}}{I_{Z6}}$$

*Step 20.* Determine the resistances of $R_{19}$ and $R_{20}$ from

$$R_{19} = \frac{50 \Delta v_5}{I_{Z6}}$$

$$R_{20} = \frac{100 V_{Z6}}{I_{Z6}} - R_{19}$$

*Step 21.* Let $R_{21} = R_{22} = 10$ k$\Omega$ if a bipolar monolithic op amp is used for $A_4$. Set $R_{18} = 10 R_{19}$.

*Step 22.* Make sure that

$$\frac{V_{Z5} + V_{Z6}}{2 S_4} < T_o$$

If this inequality is not true, $A_4$ may be too slow ($S_4$ too low) for this application.

*Step 23.* Calculate $C_4 = T_o/R_{18}$. Multiple triggering or lack of triggering will result unless $C_4$ is properly sized.

## 7-14 DEMODULATORS AND DISCRIMINATORS

*Step 24.* Set $R_{23} = 2R_{19}$. Compute $C_5$ as follows:

$$C_5 = \frac{T_o}{0.8 \, R_{23}}$$

**DESIGN EXAMPLE** As a numerical illustration we will design a discriminator for extracting 10-ms pulses from a noisy signal. The results of tests on an actual circuit built according to the design steps confirm the validity of this example.

*Design Requirements*

$T_p = 10$ ms with +5-V amplitude and 10-ms spacing between pulses
$T_t$: 20 percent of $T_p$ or 2 ms
$v_i = \pm 5$-V composite waveform with 1-V noise (peak-to-peak)
$V^{(+)} = +10$ V and $V^{(-)} = -10$ V
$T_o$: 5 ms, +5.5 V standby, −5.5 V during pulse

*Device Data*

$\frac{\Delta V_{io1}}{\Delta T} = \frac{\Delta V_{io2}}{\Delta T} = 5$ mV (−55 to +125°C)

$\frac{\Delta I_{b1}}{\Delta T} = \frac{\Delta I_{b2}}{\Delta T} = 400$ nA (−55 to +125°C)

$I_{Z1} = 0.5$ mA (1N4566, 6.4 V)
$I_{Z2} = I_{Z3} = I_{Z4} = I_{Z5} = 5$ mA (1N705A, 4.8 V)
$S_3 = S_4 = 0.5$ V/$\mu$s

*Step 1.* Equation 9 produces

$$R_3 = \frac{\Delta V_{io1}/\Delta T}{\Delta I_{b1}/\Delta T}$$

$$= \frac{5 \times 10^{-3}}{4 \times 10^{-7}} = 12{,}500 \, \Omega$$

*Step 2.* A 1N4566 reference diode (6.4 V) is chosen for $Z_1$. Its temperature coefficient is 0.005%/°C; so it will contribute an error to $T_p$ which can be calculated using Eq. 1.

*Step 3.* The trip level $V_{L1}$ of the first level detector is chosen to be 3 V. This level is halfway between the peak noise and the peak voltage of the input pulses $T_r$. The hysteresis width $\Delta v_1$ is chosen to be 1 percent of 3 V, or 30 mV. $R_1$ is now computed:

$$R_1 = \frac{R_3(V_{Z1} + V_{D1})}{V_{Z1} - \Delta v_1} = \frac{(12{,}500)(6.4 + 0.7)}{6.4 - 0.03} = 13{,}932 \, \Omega$$

*Step 4.* Equation 11 now provides us with $R_2$:

$$R_2 = \frac{R_1 R_3}{R_1 - R_3} = \frac{(13{,}932)(12{,}500)}{13{,}932 - 12{,}500} = 121{,}613 \, \Omega$$

*Step 5.* The bias voltage $V_A$ is now computed:

$$V_A = \frac{R_2 V_{L1} - R_1 V_{D1}}{R_1 + R_2} = \frac{(121{,}613)3 - 13{,}932(0.7)}{13{,}932 + 121{,}613} = 2.62 \text{ V}$$

*Step 6.* The required $R_4$ is

$$R_4 = \frac{V^{(+)} - V_{Z1}}{I_{Z1}} = \frac{10 - 6.4}{0.5 \times 10^{-3}} = 7{,}200 \, \Omega$$

$R_4$ is 17 times less than $R_2 = 121{,}613\ \Omega$. This will make $I_Z$ lower than the required 0.5 mA by 0.047 mA. If $R_4$ is lowered to

$$R_4 = \frac{V^{(+)} - V_{Z1}}{I_Z + V_{Z1}/(R_1 + R_2)} = \frac{10 - 6.4}{0.5 \times 10^{-3} + 6.4/(13{,}932 + 121{,}613)} = 6{,}579\ \Omega$$

then $I_{Z1}$ will remain at 0.5 mA.

*Step 7.* $R_6$ is found from Eq. 10:

$$R_6 = \frac{\Delta V_{io2}/\Delta T}{\Delta I_{b2}/\Delta T} = 12{,}500\ \Omega \quad \text{(same as } R_3\text{)}$$

*Step 8.* $R_9$ is computed as follows:

$$R_9 = \frac{V^{(+)} - V_{Z2}}{I_{Z2}} = \frac{10 - 4.8}{5 \times 10^{-3}} = 1{,}040\ \Omega$$

*Step 9.* Let $V_{L2} = 3$ V. This is halfway up the $R_5C_1$ charging curve which levels off at $v_1(\max) = 6.4$ V. If $V_{L2}$ is more than 80 percent of $v_1(\max)$, small variations in $R_5$ and $C_1$ will begin to have an appreciable effect on stability. Let $\Delta v_2 = 0.1\ V_{L2} = 0.1(3) = 300$ mV. $R_7$ is now found from

$$R_7 = \frac{R_6(V_{Z2} + V_{DZ2})}{V_{Z2} + V_{DZ2} - \Delta v_2} = \frac{12{,}500(4.8 + 0.7)}{4.8 + 0.7 - 0.3} = 13{,}221\ \Omega$$

*Step 10.* Equation 12 produces

$$R_8 = \frac{R_6 R_7}{R_7 - R_6} = \frac{(12{,}500)(13{,}221)}{13{,}221 - 12{,}500} = 229.2\ \text{k}\Omega$$

*Step 11.* The bias voltage for the second level detector is

$$V_B = \frac{V_{L2}(R_7 + R_8) - V_{Z2}R_7}{R_8} = \frac{3(13{,}221 + 229{,}200) - 4.8(13{,}221)}{229{,}200} = 2.90\ \text{V}$$

*Step 12.* The $R_5C_1$ product is computed from Eq. 1:

$$R_5 C_1 = \frac{T_p - \Delta T_p}{\ln\ [V_{Z1}/(V_{Z1} - V_{L2})]} = \frac{0.01 - 0.001}{\ln\ [6.4/(6.4 - 3)]} = 0.01423$$

As a first approach, let $R_5 = 100\ R_4 = 100\ (6{,}579) = 657{,}900\ \Omega$. The resulting $C_1$ is

$$C_1 = \frac{0.01423}{R_5} = \frac{0.01423}{657{,}900} = 0.0216\ \mu\text{F}$$

If $C_1$ is readjusted up to a standard value of 0.022 $\mu$F,

$$R_5 = \frac{0.01423}{C_1} = \frac{0.01423}{2.2 \times 10^{-8}} = 646{,}800\ \Omega$$

*Step 13.* $Z_4$ is chosen to be compatible with TTL logic. The calculated $R_{17}$ is

$$R_{17} = \frac{V^{(+)} - V_{Z4}}{I_{Z4}} = \frac{10 - 4.8}{5 \times 10^{-3}} = 1{,}040\ \Omega$$

*Step 14.* The $R_{13} - R_{14}$ voltage divider is computed from

$$R_{13} = R_{14} = \frac{100\ V_{Z4}}{I_{Z4}} = \frac{100(4.8)}{5 \times 10^{-3}} = 96{,}000\ \Omega$$

*Step 15.* Assuming $A_3$ is a 741 type of op amp, we can let $R_{11} = R_{12} = 10$ k$\Omega$. We also compute $R_{10} = 10\ R_{13} = 10(96{,}000) = 960$ k$\Omega$.

## 7-16 DEMODULATORS AND DISCRIMINATORS

*Step 16.* Is the $A_3$ op amp slew rate sufficiently fast? Check the following:

$$\frac{V_{Z3} + V_{Z4}}{2 S_3} \stackrel{?}{<} T_t$$

$$\frac{4.8 + 4.8}{2(0.5 \times 10^6)} \stackrel{?}{<} 2 \times 10^{-3}$$

$$9.6 \times 10^{-6} < 2 \times 10^{-3}$$

The inequality is satisfied.

*Step 17.* $C_2$ is found from

$$C_2 = \frac{T_t}{R_{10}} = \frac{2 \times 10^{-3}}{9.6 \times 10^5}$$

$$= 2.08 \times 10^{-9} = 2{,}080 \text{ pF}$$

*Step 18.* $R_{16}$ and $C_3$ are determined.

$$R_{16} = 2 R_{13} = 2(96{,}000) = 192 \text{ k}\Omega$$

$$C_3 = \frac{T_t}{0.8 R_{16}} = \frac{2 \times 10^{-3}}{0.8(192{,}000)} = 0.013 \text{ }\mu\text{F}$$

*Step 19.* $R_{25}$ is found from

$$R_{25} = \frac{V^{(+)} - V_{Z6}}{I_{Z6}} = \frac{10 - (4.8 + 0.7)}{5 \times 10^{-3}} = 900 \text{ }\Omega$$

*Step 20.* We next compute

$$R_{19} = \frac{50 \Delta V_5}{I_{Z6}} = \frac{50(2)}{5 \times 10^{-3}} = 20{,}000 \text{ }\Omega$$

$$R_{20} = \frac{100 V_{Z6}}{I_{Z6}} - R_{19} = \frac{100(5.5)}{0.005} - 20{,}000 = 90{,}000 \text{ }\Omega$$

*Step 21.* Assuming $A_4$ is a 741 op amp, we let $R_{21} = R_{22} = 10$ k$\Omega$. We next compute $R_{18} = 10 R_{19} = 10(20{,}000) = 200$ k$\Omega$.

*Step 22.* The slew rate of $A_4$ is checked against the requirement:

$$\frac{V_{Z5} + V_{Z6}}{2 S_4} \stackrel{?}{<} T_o$$

$$\frac{5.5 + 5.5}{2(0.5 \times 10^6)} \stackrel{?}{<} 5 \times 10^{-3}$$

$$11 \times 10^{-6} < 5 \times 10^{-3}$$

The inequality is satisfied.

*Step 23.* The required $C_4$ is

$$C_4 = \frac{T_o}{R_{18}} = \frac{5 \times 10^{-3}}{200{,}000} = 0.025 \text{ }\mu\text{F}$$

*Step 24.* $R_{23}$ and $C_5$ are the last values to be determined:

$$R_{23} = 2 R_{19} = 2(20{,}000) = 40 \text{ k}\Omega$$

$$C_5 = \frac{T_o}{0.8 R_{23}} = \frac{5 \times 10^{-3}}{0.8(40{,}000)} = 0.156 \text{ }\mu\text{F}$$

## REFERENCE

1. Benson, R. A., Jr., and F. M. Cancillier: Pulse Width Discriminator Uses Unijunction Transistor, *Electron. Des.*, Mar. 1, 1968, p. 96.

# Chapter 8

# Detectors

## INTRODUCTION

In this chapter we will discuss two types of detectors. The first circuit, called the peak detector, determines the peak amplitude of a waveform during a given period of time. The second circuit, unrelated to the first, detects the phase difference between two input signals.

## 8.1 POSITIVE-PEAK DETECTOR

**ALTERNATE NAMES** Peak holder, peak-signal tracker.

**EXPLANATION OF OPERATION** The circuitry of a peak detector can be arranged for positive- or negative-peak detection. For each of these cases the output can be made positive or negative. The circuit we have chosen to discuss, in Fig. 8.1, selects positive peaks and produces a negative output.

Peak detectors track the input signal and hold the output at the highest

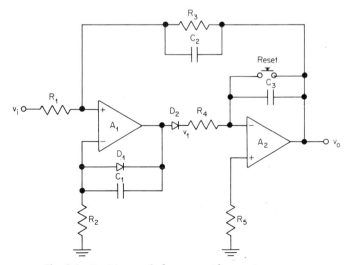

**Fig. 8.1** Positive-peak detector with negative output.

8-1

## 8-2 DETECTORS

peak found since operation of the reset switch. They continuously compare the input waveform with the stored peak value to determine if the stored value must be updated. This is graphically illustrated in Fig. 8.2. A peak detector may be thought of as a type of sample/hold circuit. It samples and holds the peak value of the largest peak in a given measurement interval. This is extremely useful in applications where widely spaced transients in a system must be measured.

This peak detector is actually a combination of two circuits described in other chapters of this handbook. The circuit of $A_1$ is similar to the precision rectifier discussed in Chap. 16. However, the feedback resistor $R_f$ and ca-

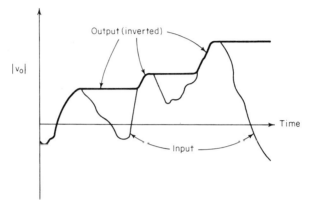

**Fig. 8.2** Input and output waveforms of positive-peak detector.

pacitor $C_f$ of Fig. 16.5 have been replaced by an active feedback network, namely, $R_3$, $R_4$, $R_5$, $C_2$, $C_3$, and $A_2$. The $A_2$ circuit is merely a fast integrator (see Chap. 15 for the rules of integrator design).

The circuit gain is defined as the ratio of peak output voltage to peak input voltage. In terms of circuit components the gain is

$$A_{vc} = \frac{v_o(\text{peak})}{v_i(\text{peak})} = -\frac{R_3}{R_1}$$

After a peak is stored on $C_3$, diode $D_2$ is reverse-biased for all succeeding lower amplitudes. This actually opens the feedback loop. The $A_1$ output will then try to saturate with negative $v_i$. Diode $D_1$ prevents this by holding the $A_1$ output near $-0.7$ V if $v_i$ becomes negative.

This circuit may be unstable with some types of op amps because of the large phase shift around the loop (see Chap. 3). The gain $A_{vc}$ must be critically damped or overdamped to prevent overshoot. An overshoot may be interpreted as a maximum peak, so caution in the feedback design is recommended. $C_1$ and $C_2$ are two possible compensation capacitors. Since the size of these capacitors is critically dependent on the types of op amp, an experimental approach is recommended: Start with $C_1 = C_2 = 5\,\text{pF}$ and work up or down from that value while observing the overshoot in $V_1$ with $v_i = $ a step function.

If the peak must be stored for long periods of time, $A_2$ should be an FET input op amp. $C_3$ should also be a low-leakage capacitor. The bias current of $A_2$ and the leakage current of $C_3$ will produce a peak-hold error of

$$\Delta v_o = \frac{I \times \text{hold time}}{C_3}$$

where $I$ is the sum of $A_2$ input bias current and $C_3$ leakage current.

## DESIGN PARAMETERS

| Parameter | Description |
|---|---|
| $A_{vc}$ | Voltage gain of entire circuit until first peak is reached. Afterward this is the ratio of the output voltage to the maximum input peak |
| $C_1, C_2$ | Compensation capacitors for feedback stability |
| $C_3$ | Integrating capacitor which holds peak output voltage |
| $D_1$ | Diode to prevent $A_1$ negative saturation during negative input voltages |
| $I_b$ | Input bias current of $A_2$ |
| $I_c$ | Leakage current of $C_3$ |
| $I_{o2}(\max)$ | Maximum output current of $A_2$ |
| $R_1$ | Determines gain of circuit along with $R_3$ |
| $R_2$ | Used to cancel most of the offset caused by input bias current of $A_1$ |
| $R_3$ | Determines gain of circuit along with $R_1$ |
| $R_4$ | Determines speed of response of circuit |
| $R_5$ | Used to cancel most of the integrator error caused by the input bias current of $A_2$ |
| $S_2$ | Slew rate of $A_2$ |
| $t_r$ | Approximate speed of response of circuit |
| $\Delta T$ | Sampling time of circuit, i.e., from reset to reset |
| $v_i$ | Input voltage to circuit |
| $v_1$ | Output of rectifier circuit |
| $v_o$ | Value of peak voltage determined during $\Delta T$ |
| $\Delta v_o$ | Error in $v_o$ due to $I_b$ and $I_c$ |

## DESIGN EQUATIONS

| Eq. No. | Description | Equation |
|---|---|---|
| 1 | Voltage gain of circuit (i.e., peak output to peak input) | $A_{vc} = \dfrac{v_o(\text{peak})}{v_i(\text{peak})} = -\dfrac{R_3}{R_1}$ |
| 2 | Approximate risetime of integrator (circuit cannot accurately respond to peaks having risetimes faster than this) | $t_r \approx R_4 C_3$<br>NOTE: This assumes the slew rate limit $S_2$ of $A_2$ is not exceeded<br>$[S_2 = I_{o2}(\max)/C_3]$ |
| 3 | Optimum value for $R_2$ | $R_2 = \dfrac{R_1 R_3}{R_1 + R_3}$ |
| 4 | Optimum value for $R_5$ | $R_5 = R_4$ |
| 5 | Error in stored peak value of $v_o$ due to $I_b$ and $I_c$ | $\Delta v_o = \dfrac{(I_b + I_c)\,\Delta T}{C_3}$ |

## REFERENCES

1. "Applications Manual for Computing Amplifiers," p. 88, George A. Philbrick Researches, Inc., 1966.
2. Tobey, G. E., J. G. Graeme, and L. P. Huelsman: "Operational Amplifiers—Design and Applications," p. 357, McGraw-Hill Book Company, New York, 1971.

8-4 DETECTORS

## 8.2 PHASE DETECTOR

**ALTERNATE NAMES** Phase-difference detector, phase-shift detector, phase-error detector, phase-to-dc converter.

**EXPLANATION OF OPERATION** Many systems require a measurement of the phase difference between two signals of the same frequency. The circuit shown in Fig. 8.3 will make this measurement accurately even if the input amplitudes are much different. The output $v_o$ will be zero if the phase difference between $v_A$ and $v_B$ is zero. If the phase of $v_B$ leads the phase of $v_A$, $v_o$ will be positive. The dc output voltage $v_o$ will vary linearly from zero to $+V_{Z2}$ as $\phi_B - \phi_A$ varies from 0 to 180°. Likewise, $v_o$ will vary linearly from zero to $-V_{Z1}$ as $\phi_B - \phi_A$ varies from 0 to $-180°$. This is shown for several cases in Fig. 8.4.

The circuits of $A_1$ and $A_2$ are zero-crossing detectors with hysteresis (see Chap. 5). $A_2$ is of the inverting type so that $v_2$ lags $v_B$ by 180°. Because of the high gain of these zero-crossing detectors, $v_1$ and $v_2$ are rectangular waveforms.

$C_1$, $R_7$ and $C_2$, $R_8$ are differentiation networks. $D_1$ and $D_2$ select the positive pulses resulting from this differentiation. The pulses from $D_1$ make the flip-flop circuit $A_3$ go into the low state such that $v_5 = -V_{Z1}$. (See Chap. 20 for a discussion of flip-flops.) Pulses from $D_2$ cause the flip-flop to go into

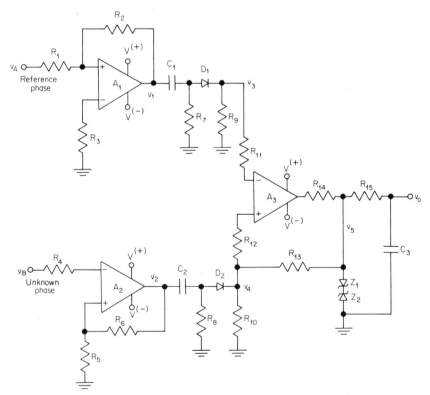

**Fig. 8.3** Phase detector utilizing zero-crossing detectors, differentiators, and an op amp flip-flop.

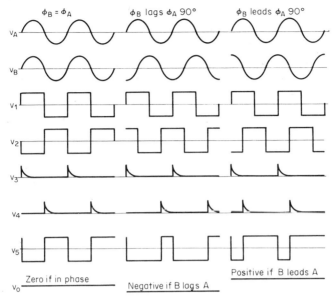

**Fig. 8.4** Waveforms at various locations in Fig. 8.3.

the other state so that $v_5 = V_{Z2}$. If $v_A$ is exactly in phase with $v_B$, the flip-flop will spend equal amounts of time in the high and low states. The voltage at $v_o$ will therefore be zero.

If the phase of $v_B$ leads the phase of $v_A$, the flip-flop will spend more time in the high state and $v_o$ will be positive. Likewise, if the phase of $v_B$ lags that of $v_A$, $v_o$ will be negative. The scale factor, i.e., volts/degree, is set only by the choice of $V_{Z1}$ and $V_{Z2}$. If these are identical diodes, the scale factor is

$$v_o = \frac{(\phi_B - \phi_A) V_{Z1}}{180} \frac{\text{volts}}{\text{degree}}$$

The range of frequencies over which accurate performance can be guaranteed depends on different factors at the low and high ends of the spectrum. At low frequencies the risetimes of $v_1$ and $v_2$ may not be fast enough to transfer adequate trigger pulses through the differentiation networks to the flip-flop. Also, the output filter $R_{15}$ and $C_3$ becomes less efficient at low frequencies. These deficiencies result in a $v_o$ which is noisy or temporarily saturated at $\pm V_{Z1}$.

At high frequencies the slew-rate limits of $A_1$ and $A_2$ start to reduce the peak-to-peak amplitude of $v_1$ and $v_2$. This will cause the trigger pulses $v_3$ and $v_4$ to diminish in amplitude also until the flip-flop no longer triggers. This must not be allowed to occur, since the flip-flop will again hang up in one state or the other.

## DESIGN PARAMETERS

| Parameter | Description |
| --- | --- |
| $A_1, A_2$ | Op amps used as zero-crossing detectors. Note that $A_2$ is inverting |
| $C_1, C_2$ | Differentiation capacitors |
| $f_c$ | Frequency of $v_A$ and $v_B$ |

## 8-6 DETECTORS

| Parameter | Description |
|---|---|
| $I_{Z1}, I_{Z2}$ | Nominal current through $Z_1$ and $Z_2$ |
| $\phi_A$ | Phase of $v_A$ (assumed zero) |
| $\phi_B$ | Phase of $v_B$ relative to $v_A$ |
| $R_1, R_2$ | Determines level of hysteresis in $A_1$ |
| $R_3$ | Prevents input bias current of $A_1$ from causing a nonzero crossover in $A_1$ circuit |
| $R_4$ | Prevents input bias current of $A_2$ from causing a nonzero crossover in $A_2$ circuit |
| $R_5, R_6$ | Determines level of hysteresis in $A_2$ |
| $R_7, R_8$ | Differentiation resistors |
| $R_{10}, R_{13}$ | Determines level of hysteresis in flip-flop |
| $R_{11}, R_{12}$ | Prevents excessive input currents in $A_3$ |
| $R_{14}$ | Controls current levels in $Z_1$ and $Z_2$ |
| $R_{15}$ | Determines ripple of $f_c$ in $v_o$ |
| $R_{iA}$ | Input resistance of circuit at $v_A$ input |
| $R_{iB}$ | Input resistance of circuit at $v_B$ input |
| $R_{i2}$ | Input resistance of $A_2$ |
| $R_{o1}$ | Output resistance of $A_1$ |
| $R_{o2}$ | Output resistance of $A_2$ |
| $T_r$ | Time for output voltage to settle to within 1% of final value |
| $v_A$ | Reference input signal |
| $v_B$ | Input whose phase is to be determined |
| $\Delta v_A(\min)$ | Peak-to-peak noise immunity of $A_1$ circuit |
| $\Delta v_B(\min)$ | Peak-to-peak noise immunity of $A_2$ circuit |
| $v_o$ | Output voltage of circuit which is proportional to phase difference between $v_A$ and $v_B$ |
| $V_{Z1}$ | Zener voltage of $Z_1$ plus the forward drop of $Z_2$ |
| $V_{Z2}$ | Zener voltage of $Z_2$ plus the forward drop of $Z_1$ |
| $V^{(+)}, V^{(-)}$ | Positive and negative power-supply voltages |

### DESIGN EQUATIONS

| Eq. No. | Description | Equation |
|---|---|---|
| 1 | Output voltage as a function of the phase difference between $v_A$ and $v_B$ (assuming $V_{Z1} = V_{Z2}$) | $v_o = \dfrac{(\phi_B - \phi_A) V_{Z1}}{180} \dfrac{\text{volts}}{\text{degree}}$ |
| 2 | Maximum allowed time constant of differentiation networks | $R_7 C_1 = R_8 C_2 < \dfrac{1}{2 f_c}$ |
| 3 | Input resistance at $A$ input | $R_{iA} = R_1 + R_2$ |
| 4 | Input resistance at $B$ input | $R_{iB} = R_4 + R_{i2}$ |
| 5 | Peak-to-peak noise immunity (hysteresis) of $A_1$ circuit | $\Delta v_A(\min) = \dfrac{[v_1(\max) - v_1(\min)] R_1}{R_1 + R_2}$ |
| 6 | Peak-to-peak noise immunity (hysteresis) of $A_2$ circuit | $\Delta v_B(\min) = \dfrac{[v_2(\max) - v_2(\min)] R_5}{R_5 + R_6}$ |
| 7 | Optimum sizes for differentiation resistors $R_7$ and $R_8$ | $5R_{o1} < R_7 < 0.1 R_9$ and $5R_{o2} < R_8 < \dfrac{0.1(R_{10} R_{13})}{R_{10} + R_{13}}$<br>NOTE: These resistors must satisfy Eq. 2 also |

| Eq. No. | Description | Equation |
|---|---|---|
| 8 | Standby value of $v_4$ | $v_4(\text{standby, high}) = \dfrac{V_{Z2}R_{10}}{R_{10} + R_{13}}$ when $v_5$ is high and $v_4(\text{standby, low}) = \dfrac{-V_{Z1}(R_8 \| R_{10})}{(R_8 \| R_{10}) + R_{13}}$ when $v_5$ is low |
| 9 | Required time constant of output network to achieve 1% ripple at $f_c(\text{min})$ | $R_{15}C_3 > \dfrac{100}{2\pi f_c(\text{min})}$ |
| 10 | Optimum size for $R_3$ | $R_3 = \dfrac{R_1 R_2}{R_1 + R_2}$ |
| 11 | Optimum size for $R_4$ | $R_4 = \dfrac{R_5 R_6}{R_5 + R_6}$ |

## DESIGN PROCEDURE

Assume the input voltage and its frequency are given. Second, the output-voltage limits for a $\pm 180°$ phase difference are specified. The maximum allowable ripple in the output is often of prime importance. We will now indicate a set of recommended design steps assuming the previous requirements are given.

## DESIGN STEPS

*Step 1.* Choose $V^{(+)}$ and $V^{(-)}$ to be compatible with the specified input- and output-voltage levels. $A_1$ and $A_2$ may be comparators operated between $V^{(+)}$ and ground, since $v_1$ and $v_2$ merely need to be squared-up versions of $v_A$ and $v_B$.

*Step 2.* Choose $Z_1$ $(=Z_2)$ so that Eq. 1 provides the required transfer function.

*Step 3.* Assume $v_1(\text{max}) = v_2(\text{max}) = V^{(+)} - 3$ V and $v_1(\text{min}) = v_2(\text{min}) = -V^{(-)} + 3$ V. In accordance with Eq. 3, we next choose $R_1 + R_2$ to be equal to the required input resistance $R_{iA}$ at $v_A$. For symmetry we let $R_5 + R_6 = R_1 + R_2$. Equation 5 is rearranged to compute $R_1$:

$$R_1 = \frac{\Delta v_A(\text{min})R_{iA}}{v_1(\text{max}) - v_1(\text{min})}$$

$R_2$ is then found from $R_2 = R_{iA} - R_1$.

*Step 4.* Equations 3 and 6 are combined with $R_1 + R_2 = R_{iA} = R_5 + R_6$ to determine $R_5$:

$$R_5 = \frac{\Delta v_B(\text{min})R_{iA}}{v_2(\text{max}) - v_2(\text{min})}$$

$R_6$ is then found from $R_6 = R_{iA} - R_5$.

*Step 5.* Compute $R_3$ and $R_4$ from Eqs. 10 and 11:

$$R_3 = \frac{R_1 R_2}{R_1 + R_2}$$

$$R_4 = \frac{R_5 R_6}{R_5 + R_6}$$

*Step 6.* Assume the $A_1$ and $A_2$ output resistances $R_{o1}$ and $R_{o2}$ are the maximum values given in their data sheets. This usually occurs at a high frequency near the unity-gain crossover frequency. Set $R_7 = 10R_{o1}(\text{max})$ and $R_8 = 10R_{o2}(\text{max})$. $C_1$ and $C_2$ are now computed with the aid of Eq. 2:

$$C_1 = \frac{1}{2f_c R_7}$$

$$C_2 = \frac{1}{2f_c R_8}$$

*Step 7.* Let $R_9 = 10R_7$ and $R_{10} = 10R_8$. Resistor $R_{13}$ is computed so that the trigger pulses at $v_3$ and $v_4$ are several times larger than the $v_4$ standby voltage. When $v_5$ is high, $v_4$ is computed from Eq. 8 (first part)

$$v_4(\text{standby}) = \frac{+V_{Z2}R_{10}}{R_{10} + R_{13}}$$

Resistor $R_{13}$ must be computed so that $v_1(\text{max}) - v_1(\text{min})$, the pulse amplitude at $v_1$ and $v_3$, is twice the amplitude of $v_4$ (standby). This assures us that $v_3$ will trigger the flip-flop from the high state to the low state. Compute $R_{13}$ from

$$R_{13} = \frac{R_{10}(2V_{Z2} - v_1(\text{max}) + v_1(\text{min}))}{v_1(\text{max}) - v_1(\text{min})}$$

*Step 8.* When $v_5$ is in the low state, $v_4$ is also in the low state. However, $D_2$ is forward-biased in this case, and $R_8$ and $R_{10}$ appear to be in parallel. As a consequence we have

$$|v_4(\text{standby, low})| < |v_4(\text{standby, high})|$$

Triggering the flip-flop from the low state to the high state will therefore be easier than the triggering from high to low described in step 7. If we have designed the circuit such that $A_1 = A_2$, $R_{o1} = R_{o2}$, $C_1 = C_2$, $R_7 = R_8$, and $R_9 = R_{10}$, triggering in both directions will be assured. If these five equalities are not true, $v_4(\text{standby, low})$ may need to be adjusted using $R_8$, $R_{10}$, or $R_{13}$. Triggering is assured only if

$$|v_2(\text{max}) - v_2(\text{min})| \geq |2v_4(\text{standby, low})|$$

*Step 9.* Resistor $R_{14}$ is now sized so that it provides the correct current through $Z_1$ and $Z_2$. If $Z_1$ and $Z_2$ are identical,

$$R_{14} = \frac{V^{(+)} - 3 - V_{Z2}}{I_{Z2}}$$

*Step 10.* The time constant $R_{15}C_3$ can be selected only after a trade-off analysis is made.

1. $R_{15}C_3$ must be large enough to diminish the $f_c$ ripple in $v_o$ to a reasonably low level.
2. $R_{15}C_3$ must not be too large or $v_o$ will have a response time which is too slow.

These two statements are summarized in the following opposing requirements:

$$R_{15}C_3 \approx \frac{100}{2\pi f_c} \quad \text{for } \approx 1\% \ f_c \text{ ripple in } v_o$$

$$R_{15}C_3 \approx \frac{T_r}{5} \quad \text{for a } v_o \text{ within 1\% of final value in time } T_r$$

PHASE DETECTOR 8-9

The output filter is a simple one-stage $RC$ filter. At frequencies above $f = 1/2\pi RC$ the filter attenuates by a factor of 10 for each factor of 10 increase in frequency. If we want an attenuation of 100 at $f_c$, we need $R_{15}C_3 = 100/2\pi f_c$ as stated above.

**EXAMPLE OF PHASE-DETECTOR DESIGN** We will numerically illustrate the preceding design steps with a 1,000-Hz phase-detector design. Experimental data from an actual circuit have confirmed the validity of the design steps.

*Design Requirements*
$R_{iA} = R_{iB} > 5,000 \; \Omega$
$f_c = 1,000 \; Hz$
$\Delta v_A(\min) = \Delta v_B(\min) = 0.1 \; V$
$v_A = $ 10-V peak-to-peak sine wave
$v_B = $ 10-V peak-to-peak sine wave with phase varying from $-180$ to $+180°$ with respect to $v_A$
Phase-to-voltage scale factor $= 4 \; V/180°$
$T_r < 1 \; s$

*Device Data*
$R_{O1}(\max) = R_{O2}(\max) = 200 \; \Omega$ at 1 MHz
$I_{Z1} \approx 5$ mA to develop 3.3 V using a 1N703A zener diode

   Step 1.  $v_A$ and $v_B$ have $\pm 5$-V peak values and $v_O$ has $\pm 4$-V peaks. We can safely allow $V^{(+)} = 5 \; V$ and $V^{(-)} = -5 \; V$ for $A_1$ and $A_2$. $A_3$ requires slightly larger power-supply voltages if the full $v_O(\max) = \pm 4 \; V$ is required.
   Step 2.  If we choose 1N703A zener diodes, we can expect zener voltages of 3.4 V $\pm 5$ percent at $I_Z = 5$ mA. We must add the zener forward breakdown voltage of 0.7 V to obtain the actual limits for $v_o$. The resultant voltage becomes $\pm(3.4 + 0.7) = \pm 4.1 \; V$. However, these are 5 percent diodes, so the voltage may come in at anything between $\pm(3.23 + 0.7) = \pm 3.93$ and $\pm(3.57 + 0.7) = \pm 4.27 \; V$. $R_{14}$ can be trimmed as a last step to achieve the required $v_o = \pm 4.0 \; V$.
   Step 3.  The peak-to-peak square waves at $v_1$ and $v_2$ are

$$v_1(\text{peak-to-peak}) = v_1(\max) - v_1(\min) = V^{(+)} - 3 + V^{(-)} - 3$$
$$= 5 - 3 + 5 - 3 = 4 \; V$$

$$v_2(\text{peak-to-peak}) = v_2(\max) - v_2(\min) = V^{(+)} - 3 + V^{(-)} - 3$$
$$= 10 - 6 = 4 \; V$$

If $R_1 + R_2 = R_5 + R_6 = 10 \; k\Omega$, the required minimum input resistance $R_{iA}$ at $v_A$ will be satisfied. At $v_B$ the input resistance $R_{iB}$ is much higher, since $R_4$ is in series with the op amp input resistance (usually $>1 \; M\Omega$). $R_1$ and $R_2$ are computed from

$$R_1 = \frac{\Delta v_A(\min) R_{iA}}{v_1(\max) - v_1(\min)} = \frac{0.1(10^4)}{4} = 250 \; \Omega$$

$$R_2 = R_{iA} - R_1 = 10,000 - 250 = 9,750 \; \Omega$$

   Step 4.  $R_5$ and $R_6$ are computed:

$$R_5 = \frac{\Delta v_B(\min) R_{iA}}{v_2(\max) - v_2(\min)} = \frac{0.1(10^4)}{4} = 250 \; \Omega$$

$$R_6 = R_{iA} - R_5 = 10,000 - 250 = 9,750 \; \Omega$$

## 8-10 DETECTORS

*Step 5.* $R_3$ and $R_4$ are computed:

$$R_3 = \frac{R_1 R_2}{R_1 + R_2} = \frac{250(9,750)}{250 + 9,750} = 244 \; \Omega$$

$$R_4 = \frac{R_5 R_6}{R_5 + R_6} = \frac{250(9,750)}{250 + 9,750} = 244 \; \Omega$$

*Step 6.* The differentiation networks are next designed:

$$R_7 = 10R_{o1}(\text{max}) = 10(200) = 2,000 \; \Omega$$
$$R_8 = 10R_{o2}(\text{max}) = 10(200) = 2,000 \; \Omega$$
$$C_1 = \frac{1}{2f_c R_7} = \frac{1}{2(1,000)2,000} = 0.25 \; \mu\text{F}$$
$$C_2 = \frac{1}{2f_c R_8} = \frac{1}{2(1,000)2,000} = 0.25 \; \mu\text{F}$$

*Step 7.* $R_9$, $R_{10}$, and $R_{13}$ are computed:

$$R_9 = 10R_7 = 10(2,000) = 20 \; \text{k}\Omega$$
$$R_{10} = 10R_8 = 10(2,000) = 20 \; \text{k}\Omega$$
$$R_{13} = \frac{R_{10}[2V_{z2} - v_i(\text{max}) + v_i(\text{min})]}{v_i(\text{max}) - v_i(\text{min})}$$
$$= \frac{(20,000)(2 \times 4 - 4)}{4} = 20 \; \text{k}\Omega$$

*Step 8.* The two inputs have been symmetrical up to this point; so triggering of the flip-flop in both directions is assured.

*Step 9.* The first-cut size for $R_{14}$ is computed from

$$R_{14} = \frac{V^{(+)} - 3 - V_{Z2}}{I_{Z2}} = \frac{10 - 3 - 4}{0.005} = 600 \; \Omega$$

This resistor could be replaced with a 1-k$\Omega$ pot having a title of VOLTS/DEGREE TRIM (range is limited).

*Step 10.* Suppose we size $R_{15}C_3$ so that $T_r = 0.1$ s. Since the output ripple is not specified, we will be satisfied with whatever the computations tell us. First we solve

$$R_{15}C_3 = \frac{T_r}{5} = \frac{0.1}{5} = 0.02 \; \text{s}$$

The corner frequency for this filter is

$$f = \frac{1}{2\pi R_{15}C_3} = \frac{1}{6.28(0.02)} = 7.96 \; \text{Hz}$$

Since 1,000 Hz is 125.7 times larger than 7.96 Hz, the ripple at $v_o$ will be 125.7 times less than the ripple at $v_5$. If we choose $C_3 = 1 \; \mu\text{F}$,

$$R_{15} = \frac{T_r}{5C_3} = \frac{0.1}{5 \times 10^{-6}} = 20 \; \text{k}\Omega$$

Laboratory verification of the above circuit showed the transfer function to be within 2 percent of that predicted. The offset-adjustment terminals were found to be quite useful on $A_1$ and $A_2$, since the precise trip level in each stage had a significant effect on the transfer function.

## REFERENCE

1. Woodbury, J. R.: Measuring Phase with Transistor Flip-Flops, *Electronics*, Sept. 22, 1961, p. 56.

Chapter **9**

# Differential Amplifiers

## INTRODUCTION

The differential-input–single-ended-output amplifier is one of the most versatile circuits available. It is required in innumerable applications where low-level transducer signals must be converted to a higher power level. Often the transducer has a high common-mode voltage (i.e., both sides of the transducer operating at some appreciable voltage off ground). This common-mode voltage must be rejected, and only the differential signal across the transducer must be amplified.

We will present here design equations for the two most popular differential amplifiers using op amps. The first circuit emphasizes simplicity and is the basic type of differential amplifier using a single op amp. It has the drawback of requiring tightly matched resistors to keep the common-mode gain low. The second circuit requires three op amps but results in a high-quality instrumentation amplifier.

## 9.1 BASIC DIFFERENTIAL AMPLIFIER

**ALTERNATE NAMES** Differential-input amplifier, low-cost instrumentation amplifier, differential dc amplifier, difference amplifier, error amplifier, data amplifier, transducer amplifier.

**EXPLANATION OF OPERATION** This circuit is the lowest-cost approach to a differential amplifier. It provides an output voltage which is proportional to the difference of two voltage signals. It is most often used in dc and low-frequency applications. However, frequencies up to several kHz are possible with slightly degraded performance. It also has a fairly low input impedance. For optimum differential-amplifier performance, the three op amp circuit in the following section is recommended.

Differential gain $A_d$ is defined as $A_d = v_o/(v_2 - v_1)$. If this circuit is perfectly balanced (i.e., if $R_3/R_1 = R_4/R_2$), the differential gain is

$$A_d = \frac{v_o}{v_2 - v_1} = \frac{R_3}{R_1} = \frac{R_4}{R_2}$$

The common-mode gain of a differential amplifier is defined as

## 9-2 DIFFERENTIAL AMPLIFIERS

$$A_c = \frac{2v_o}{v_1 + v_2} = \frac{v_o}{\text{average of input voltages}} = \frac{v_o}{v_{ic}}$$

If the amplifier is perfectly balanced and the op amp has no common-mode gain, the circuit will have no common-mode gain. This is a desirable feature but is difficult to achieve with this basic circuit.

If the circuit is not perfectly balanced and/or the op amp has a finite common-mode gain, the output voltage will be $v_2 - v_1$ times the differential gain plus $(v_2 + v_1)/2$ times the common-mode gain.

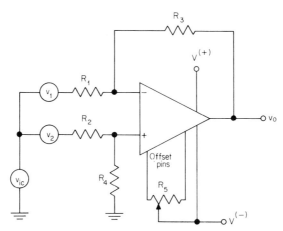

**Fig. 9.1** Basic differential amplifier.

Op amp data sheets do not specify common-mode gain. Instead, the common-mode rejection ratio (CMRR) is given, where

$$\text{CMRR} = \frac{\text{differential gain}}{\text{common-mode gain}} = \frac{A_d}{A_c}$$

If the circuit is perfectly balanced and CMRR $\neq \infty$, the common-mode gain due to the op amp only is

$$A_{co} = \frac{v_o}{v_{ic}} = \frac{R_3^2}{R_1(R_1 + R_3) \times \text{CMRR}}$$

If the op amp CMRR $= \infty$ but the circuit is not perfectly balanced, the common-mode gain due to the circuit unbalance only is

$$A_{cc} = \frac{v_o}{v_{ic}} = \frac{R_4 R_1 - R_2 R_3}{R_1 (R_2 + R_4)}$$

Common-mode rejection (CMR) is often used on data sheets. By convention CMR is given in dB and is defined as

$$\text{CMR} = 20 \log |\text{CMRR}|$$

The CMR of most op amps is 60 dB or higher.

The circuit output voltage due to $A_{co}$ times $v_{ic}$ is independent of the output voltage due to $A_{cc}$ times $v_{ic}$. If we consider both common-mode gains simultaneously, the output voltage is

$$v_o = A_{co} v_{ic} + A_{cc} v_{ic} = v_{ic}(A_{co} + A_{cc})$$

BASIC DIFFERENTIAL AMPLIFIER 9-3

The CMRR for both the circuit and op amp combined becomes

$$\text{CMRR}_c = \frac{\text{differential gain}}{\text{total common-mode gain}}$$

$$= \frac{A_d}{v_o/v_{ic}} = \frac{A_d}{A_{co} + A_{cc}}$$

The dc stability of $v_o$ over time and temperature often requires careful attention in differential-amplifier design. Many applications of this circuit are for low-level transducers which require a large dc gain. The offset of $v_o$ (and the drift of this offset) caused by $I_b$, $I_{io}$, and $V_{io}$ must be recognized. If $R_3/R_1 = R_4/R_2$, the $I_b$ term will cause no problem and the output offset is

$$\Delta V_o = \pm \frac{V_{io}(R_1 + R_3)}{R_1} \pm I_{io}R_3$$

Methods to reduce this offset (and its drift over time and temperature) are suggested in Chap. 2.

## DESIGN PARAMETERS

| Parameter | Description |
|---|---|
| $A_c$ | Common-mode gain of a differential amplifier circuit |
| $A_{cc}$ | Common-mode gain due to resistor mismatching |
| $A_{co}$ | Common-mode gain due to finite CMRR of op amp |
| $A_d$ | Differential voltage gain of circuit |
| $A_v$ | Open-loop gain of op amp (varies with frequency) |
| $A_{vo}$ | Open-loop dc gain of op amp |
| CMR | Op amp common-mode rejection = 20 log\|CMRR\| |
| CMRR | Common-mode rejection ratio of op amp. This is defined as the ratio of op amp differential gain to op amp common-mode gain. CMRR varies with frequency |
| $\text{CMRR}_c$ | Common-mode rejection ratio of entire circuit after both $A_{cc}$ and $A_{co}$ have been considered |
| $f_{\max}$ | Maximum frequency at which high-accuracy performance can be achieved |
| $f_{cm}$ | Frequency at which the minimum op amp CMRR is less than that required |
| $f_{cp}$ | Frequency at which the circuit gain is down 3 dB from a dc gain of $A_d$ (i.e., first-pole frequency of closed loop) |
| $f_u$ | Gain crossover frequency of op amp |
| $I_{io}$ | Input offset current of op amp. This parameter varies with temperature |
| $R_1, R_2, R_3, R_4$ | Matched set of resistors which determine gain of circuit |
| $R_5$ | Offset-adjustment potentiometer. The actual connection scheme varies with the type of op amp |
| $R_{inc}$ | Common-mode input resistance of circuit, i.e., between either $v_1$ or $v_2$ and ground |
| $R_{ind}$ | Differential input resistance of circuit, i.e., between terminals $v_1$ and $v_2$ |
| $R_o$ | Op amp output resistance |
| $v_1, v_2, v_o, v_{ic}$ | Voltages as shown in schematic. These may be dc and/or ac voltages |
| $\Delta V_o$ | Output offset error due to input voltage and current offsets |
| $V_{ic}$ | Input offset voltage of op amp. This parameter varies with temperature |

## 9-4 DIFFERENTIAL AMPLIFIERS

### DESIGN EQUATIONS

| Eq. No. | Description | Equation |
|---|---|---|
| 1 | Differential gain if $\dfrac{R_3}{R_1} = \dfrac{R_4}{R_2}$ | $A_d = \dfrac{v_o}{v_2 - v_1} = \dfrac{R_3}{R_1} = \dfrac{R_4}{R_2}$<br>NOTE: Generator output resistances must be included in $R_1$ and/or $R_2$ |
| 2 | Common-mode gain due to mismatch of $\dfrac{R_3}{R_1} = \dfrac{R_4}{R_2}$ | $A_{cc} = \dfrac{v_o}{v_{ic}} = \dfrac{R_4 R_1 - R_2 R_3}{R_1(R_2 + R_4)}$ |
| 3 | Differential gain if $\dfrac{R_3}{R_1} = \dfrac{R_4}{R_2}$<br>CMRR $= \infty$<br>$A_v \neq \infty$<br><br>(use $A_{vo}$ in place of $A_v$ for dc computations) | $A_d = \dfrac{v_o}{v_2 - v_1} = \dfrac{A_v R_3}{R_1 A_v + R_3}$ |
| 4 | Common-mode gain if op amp CMRR $\neq \infty$ | $A_{co} = \dfrac{v_o}{v_{ic}} = \dfrac{R_3^2}{R_1(R_1 + R_3)\mathrm{CMRR}}$ |
| 5 | Differential input resistance | $R_{ind} = R_1 + R_2$ |
| 6 | Common-mode input resistance | $R_{inc} = \dfrac{R_1 + R_2}{2}$ |
| 7 | Output offset voltage | $\Delta V_o = \pm \dfrac{V_{io}(R_1 + R_3)}{R_1} \pm I_{io} R_3$ |

### DESIGN PROCEDURE

The most common requirements for differential amplifiers, in order of importance, are as follows:
  1. A high and stable differential gain $A_d$ over time, temperature, and frequency
  2. Low common-mode gain $A_c$ relative to $A_d$
  3. Low output offset $\Delta V_o$ due to $I_b$, $I_{io}$, and $V_{io}$
  4. A high input resistance

### DESIGN STEPS

*Step 1.* Is the op amp open-loop gain $A_v$ at least 100 times larger than the required differential gain $A_d$ at all frequencies and temperatures of interest? If no, then $A_d$ may not be stable with temperature and power-supply voltage variations at those frequencies where $A_v < 100\, A_d$. Call the upper frequency at which this is true $f_{max}$. (See Fig. 9.2.)
*Step 2.* Set $R_3/R_1 = A_d$ and $R_4/R_2 = A_d$.
*Step 3.* Set $R_1 = R_2 = R_{ind}/2$.
*Step 4.* NOTE: $R_{inc}$ will now $= R_1$.
*Step 5.* Set $R_3 = A_d R_1$ and $R_4 = A_d R_1$.
*Step 6.* Is the output voltage offset $\Delta V_o = [V_{io}(R_1 + R_3)/R_1] + I_{io} R_3$ too large? If yes, lower the resistance of resistors $R_1$, $R_2$, $R_3$, and $R_4$ until $\Delta V_o$(off-

set) is satisfactory. Note, however, that this lowers both the differential- and common-mode input resistances. A compromise must usually be made.

*Step 7.* Compute the common-mode gain resulting from mismatch of $R_3/R_1 = R_4/R_2$:

$$A_{cc} = \frac{v_o}{v_{ic}} = \frac{R_4 R_1 - R_2 R_3}{R_1(R_2 + R_4)}$$

*Step 8.* A curve showing minimum CMRR of the op amp as a function of frequency must be found. However, data sheets usually give only tabular values for minimum and typical dc CMRR. Some data sheets also provide typical CMRR as a function of frequency. If the three pieces of information above are available, a plot of minimum CMRR as a function of frequency can

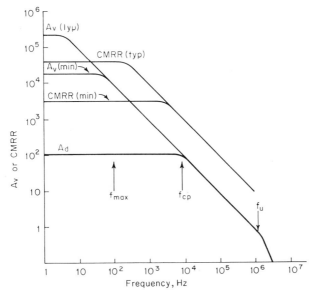

**Fig. 9.2** Curves of CMRR (typical and minimum) as a function of frequency. Curves of $A_v$ (typical and minimum) as a function of frequency. Curve of $A_d$ (typical and minimum) as a function of frequency.

be constructed as shown in Fig. 9.2. This new curve tells us the frequency at which minimum worst-case CMRR begins to degrade circuit performance. This frequency we call $f_{cm}$. Common-mode errors will get worse above $f_{cm}$.

*Step 9.* Compute the common-mode gain resulting from the op amp CMRR.

$$A_{co} = \frac{R_3^2}{R_1(R_1 + R_3)\text{CMRR}}$$

If $A_d = 100$ and the minimum op amp CMRR is 60 dB (1,000), $A_{co} = 9.9 \times 10^{-2}$. This can be reduced by decreasing $A_d$ or using a better op amp.

*Step 10.* Compute the common-mode rejection ratio of the entire circuit.

$$\text{CMRR}_c = \frac{A_d}{A_{cc} + A_{co}}$$

If $A_d = 100$, $A_{cc} = 3.95 \times 10^{-3}$, and $A_{co} = 9.9 \times 10^{-2}$ as computed above, then $CMRR_c = 971 = 59.7$ dB.

Step 11. Feedback stability must be examined using the seven causes of op amp instability listed in Chap. 3.

Step 12. If $A_{vo}$ is quite variable owing to temperature changes or power-supply changes, Eq. 3 must be solved for each value of $A_{vo}$ to determine its effect on differential gain. The effect of $A_v$ on $A_d$ at frequencies where $A_v$ is falling at $-20$ dB/decade must be handled differently. In this region of frequencies, $A_v$ has a 90° phase lag relative to $A_d$. Equation 3 must be modified to the following:

$$A_d = \frac{A_v R_3}{\sqrt{R_1^2 A_v^2 + R_3^2}}$$

## DESIGN EXAMPLE

*Design Requirements*
$R_{ind} = 10$ k$\Omega$
$R_{inc} = 5$ k$\Omega$
$A_d = 100$
$\Delta V_o < 0.1$ V
$CMRR_c > 60$ dB (1,000)
$f_{max} > 100$ Hz

*Device Data (741 op amp)*
$A_{vo} = 25{,}000$ if $V^{(\pm)} = \pm 2$ V
$A_{vo} = 250{,}000$ if $V^{(\pm)} = \pm 20$ V
CMRR (min) = 70 dB (3,160)
$V_{io}$ (max) = 7.5 mV (0 to 70°C range)
$I_{io}$ (max) = 300 nA (0 to 70°C range)
$\phi_m$ (open-loop) = 80°
Closed-loop bandwidth (normalized to 1 MHz at +25°C) = 1.12 at $-55$°C and 0.8 at $+125$°C.

Step 1. Start the design process by examining the 741 curve showing open-loop gain as a function of frequency. As shown in Fig. 9.2, a heavy line is drawn along the gain = 100 line until it intersects with the open-loop curve. The frequency of this intersection is $f_{cp}$. The heavy line then follows the open-loop curve down to below unity gain. The heavy line is the differential gain $A_d$. Since the heavy line is more than 40 dB below $A_v$ only from dc up to 100 Hz, $f_{max} = 100$ Hz. This particular differential amplifier will have highly stable $A_d$ only below $f_{max}$. It will be usable from $f_{max}$ to $f_{cp}$. However, temperature and power-supply-voltage variations will have a slight effect on its performance. We will compute these variations in step 12 below.

Step 2. $R_3/R_1 = R_4/R_2 = 100$.
Step 3. $R_1 = R_2 = R_{ind}/2 = 5{,}000$ $\Omega$ (use 5,110-$\Omega$ metal film).
Step 4. $R_{inc} = R_1 = 5{,}110$ $\Omega$. This is satisfactory.
Step 5. $R_3 = R_4 = A_d R_1 = 511$ k$\Omega$ (metal film).

Step 6. $\Delta V_o = \dfrac{V_{io}(R_1 + R_3)}{R_1} + I_{io} R_3$

$= \dfrac{(0.0075)(516{,}110)}{5{,}110} + (3 \times 10^{-7})(5.11 \times 10^5)$

$= 0.7575 + 0.1533 = 0.9108$ V, maximum worst case

**BASIC DIFFERENTIAL AMPLIFIER 9-7**

NOTE: With the 741 this offset can be nulled out using the offset terminals. The variations of $V_{io}$ and $I_{io}$ over temperature, however, will create a finite $\Delta V_o$ at other temperatures.

*Step 7.* Assume $R_1$, $R_4$ are both 0.1 percent high and $R_2$, $R_3$ are both 0.1 percent low in resistance.

$$A_{cc} = \frac{R_4 R_1 - R_2 R_3}{R_1(R_2 + R_4)}$$

$$= \frac{(511{,}511)(5{,}115.11) - (5{,}104.89)(510{,}489)}{5{,}115.11(5{,}104.89 + 511{,}511)} = 0.00395$$

*Step 8.* Assume operation of the differential amplifier is only for frequencies below 100 Hz where CMRR is always greater than 70 dB (70 dB = 3,160).

*Step 9.* $A_{co} = \dfrac{R_3^2}{R_1(R_1 + R_3)\text{CMRR}}$

$$= \frac{(511{,}000)^2}{5{,}110(5{,}110 + 511{,}000)(3{,}160)} = 0.0313$$

*Step 10.* $\text{CMRR}_c = \dfrac{100}{0.00395 + 0.0313} = 2{,}836.9 = 69.1 \text{ dB}$

*Step 11.* The 741 is internally compensated and has an open-loop phase margin of 80°. It is therefore unlikely that instability causes numbered 1, 2, or 4 will be applicable (see Chap. 3). The other instability causes will be briefly discussed.

3. The 741 has a maximum output resistance $R_o$ of 300 Ω at 1 mHz. We must make sure that capacitive loading does not create a pole near gain crossover—otherwise the 80° phase margin is reduced. If the pole is at a frequency ten times gain crossover, it will reduce the phase margin only 6°. The load capacitance to do this is

$$C_L(\text{max}) = \frac{1}{2\pi(10 f_u)R_o} = \frac{1}{(6.28)(10^7)(300)} = 53 \text{ pF}$$

5. The resistance between ground and the noninverting terminal is the parallel combination of 5,110 Ω and 511 kΩ. No problem should be experienced with a resistance this low.

6. Careful board layout should be used which minimizes the capacitance between the 741 offset and output terminals.

7. In most applications, 0.1-$\mu$F ceramic capacitors from $V^{(+)}$ to ground and $V^{(-)}$ to ground for each four or five op amps is adequate to prevent unwanted feedback.

*Step 12.* The changes of dc differential gain resulting from variations of open-loop gain are computed. While one does not expect $V^{(\pm)}$ to vary from ±2 to ±20 V, at least this calculation will give an indication of the sensitivity of $A_d$ to power-supply variations.

$$A_d(A_{vo} = 250 \text{ k, dc}) = \frac{A_{vo} R_3}{R_1 A_{vo} + R_3}$$

$$= \frac{250{,}000 \times 511{,}000}{5{,}110(250{,}000) + 510{,}000} = 99.9596$$

$$A_d(A_{vo} = 25 \text{ k, dc}) = \frac{25{,}000 \times 511{,}000}{5{,}110(25{,}000) + 510{,}000} = 99.6016$$

## 9-8 DIFFERENTIAL AMPLIFIERS

This calculation shows that the differential gain changes only 0.36 percent while the open-loop gain is reduced 90 percent.

At frequencies between $f_{op}$ and $f_{cp}$ the gain variations in $A_d$ must be computed using

$$A_d = \frac{A_v R_3}{\sqrt{R_1^2 A_v^2 + R_3^2}}$$

Assume we want to determine the circuit performance up to 2 kHz. The open-loop variations of $A_v$ at 2 kHz are a little more difficult to obtain from a data sheet. The 741 data sheet states that the closed-loop bandwidth varies from 10 kHz at 25°C to 11.2 kHz at −55°C to 8 kHz at 125°C. Since the open-loop-gain curve is decreasing at 20 dB/decade at these frequencies, the open-loop gain is changing at the same rate as the closed-loop bandwidth. The nominal $A_v$ at 2 kHz is 500. At −55°C, $A_v$ is (1.12)500 = 560 and at 125°C, $A_v$ is (0.8)500 = 400. Substituting these into the above equation,

$$A_d(\text{nominal, 2 kHz}) = \frac{500 \times 511{,}000}{\sqrt{5{,}110^2 \times 500^2 + 511{,}000^2}} = 98.05783$$

$$A_d(-55°C, \text{2 kHz}) = \frac{560 \times 511{,}000}{\sqrt{5{,}110^2 \times 560^2 + 511{,}000^2}} = 98.442473$$

$$A_d(125°C, \text{2 kHz}) = \frac{400 \times 511{,}000}{\sqrt{5{,}110^2 \times 400^2 + 511{,}000^2}} = 97.014167$$

These calculations show that the 2-kHz differential gain changes by +0.4 or −1 percent over the military temperature range (−55 to +125°C).

### REFERENCES

1. Cate, Tom: Op Amps or Instrumentation Amplifiers? *EEE*, August 1970, p. 52.
2. Schick, Larry L.: Linear Circuit Applications of Operational Amplifiers, *IEEE Spectrum*, April 1970, p. 36.
3. Barna, A.: "Operational Amplifiers," John Wiley & Sons, Inc., New York, 1971.

### 9.2 INSTRUMENTATION AMPLIFIER

The basic differential amplifier described in Sec. 9.1 has many limitations. Its input resistance is quite low, whereas differential-amplifier applications often require very high input resistances. High values of differential gain require a large feedback resistor which causes excess dc output offset due to the op amp input offset current. If both high gain and high input resistance are required, the problem is doubly compounded because the input resistor must be very large and the feedback resistor must be much larger than the input resistor. Since it is difficult to match high-megohm resistors, the CMRR due to resistor mismatching will suffer. Clearly, these four parameters ($R_{in}$, $A_d$, $CMRR_c$, and $\Delta V_o$) interact and a reasonable compromise cannot always be achieved with the basic differential amplifier.

The instrumentation-quality differential amplifier described here overcomes some of the above limitations. It does so, however, at the expense of requiring three op amps. With the availability of two, three, or four op amps in a package this is no longer of concern. The major problem in using multiple packaged op amps is matching the two input op amps. This is often needed to reduce the drift of input offset voltage.

The instrumentation amplifier shown in Fig. 9.3 is essentially the basic differential amplifier of Fig. 9.1 with noninverting amplifiers attached to each input. The output voltages from stages $A_1$ and $A_2$ are

$$v_3 = \left(1 + \frac{R_2}{R_1}\right)v_1 - \frac{R_2}{R_1}v_2 + v_{ic}$$

$$v_4 = \left(1 + \frac{R_3}{R_1}\right)v_2 - \frac{R_3}{R_1}v_1 + v_{ic}$$

where $v_{ic}$ is the common-mode input voltage $[v_{ic} = (v_1 + v_2)/2]$. If the output stage is perfectly balanced, i.e., if $R_6/R_4 = R_7/R_5$,

$$v_o = \frac{R_6}{R_4}(v_4 - v_3) = \frac{R_6(R_1 + R_2 + R_3)(v_2 - v_1)}{R_1 R_4}$$

If we let $R_2 = R_3$ and $R_4 = R_5 = R_6 = R_7$,

$$A_d = \frac{v_o}{v_2 - v_1} = 1 + \frac{2R_2}{R_1}$$

The input stages $A_1$ and $A_2$ can be designed for high gain without causing excessive dc offset. The output stage $A_3$ can use small resistors to minimize dc offset.

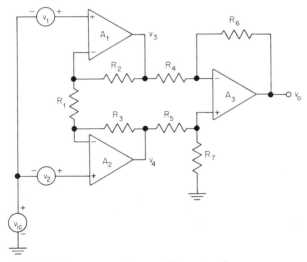

**Fig. 9.3** Instrumentation amplifier using three op amps.

As with the basic differential amplifier, the CMRR of the circuit depends on the CMRR of $A_3$ and how perfectly $R_6/R_4 = R_7/R_5$. A mismatch of $R_1$, $R_2$, or $R_3$ merely affects the differential gain $[A_d = v_o/(v_2 - v_1)]$ and not the common-mode gain $(A_c = v_o/v_{ic})$. The input impedances of the noninverting inputs are typically greater than $10^{10}$ $\Omega$ while the impedances driving $R_4$ and $R_5$ are nearly zero. The gain of this entire circuit is usually controlled only with adjustments to $R_1$.

## DESIGN PARAMETERS

| Parameter | Description |
|---|---|
| $A_{cc}$ | Common-mode gain of circuit |
| $A_d$ | Differential gain of circuit |
| CMRR | Common-mode rejection ratio of op amp |
| $I_{b1}$ | Input bias current of $A_1$ |
| $I_{b2}$ | Input bias current of $A_2$ |
| $R_1$ to $R_3$ | Resistors which set gain of input stages $A_1$ and $A_2$ |
| $R_4$ to $R_7$ | Resistors which set gain of output stage $A_3$ |
| $v_1$ | Input voltage to noninverting terminal of $A_1$ |
| $v_2$ | Input voltage to noninverting terminal of $A_2$ |
| $v_3$ | Output voltage of $A_1$ |
| $v_4$ | Output voltage of $A_2$ |
| $v_{ic}$ | Common-mode input voltage to circuit |
| $V_{io}$ | Input offset voltages of $A_1$ and $A_2$ |
| $v_o$ | Output voltage of circuit |
| $\Delta V_o$ | Output offset voltage of circuit |

## DESIGN EQUATIONS

| Eq. No. | Description | Equation |
|---|---|---|
| 1 | Differential gain of circuit if $R_6/R_4 = R_7/R_5$ | $A_d = \dfrac{v_o}{v_2 - v_1} = \dfrac{R_6(R_1 + R_2 + R_3)}{R_1 R_4}$ |
| 2 | Differential gain of circuit if $R_2 = R_3$ and $R_4 = R_5 = R_6 = R_7$ | $A_d = 1 + \dfrac{2R_2}{R_1}$ |
| 3 | Common-mode gain if op amp CMRR = $\infty$ and $R_6/R_4 \neq R_7/R_5$ | $A_{cc} = \dfrac{v_o}{v_{ic}} = \dfrac{R_7 R_4 - R_5 R_6}{R_4(R_5 + R_7)}$ |
| 4 | Common-mode gain if $A_3$ has finite CMRR and $R_6/R_4 = R_7/R_5$ | $A_{co} = \dfrac{v_o}{v_{ic}} = \dfrac{R_6^2}{R_4(R_4 + R_6)\text{CMRR}}$ |
| 5 | Output offset voltage if input offset voltages of $A_1$ and $A_2$ are in opposite directions and each with a magnitude of $V_{io}$ | $\Delta V_o = \pm \dfrac{2R_6 V_{io}(R_1 + R_2 + R_3)}{R_1 R_4}$ |
| 6 | Output offset voltage if input bias currents of $A_1$ and $A_2$ are $I_{b1}$ and $I_{b2}$ | $v_o = \dfrac{R_6}{R_4}(R_3 I_{b1} - R_2 I_{b2})$ |

## REFERENCE

1. Tobey, G. E., J. G. Graeme, and Huelsman, L. P.: "Operational Amplifiers—Design and Applications," p. 206, McGraw-Hill Book Company, New York, 1971.

Chapter **10**

# Low-Pass Filters

## INTRODUCTION

There is no lack of active low-pass filter circuits in the literature. The circuit designer's task is to determine which of these numerous circuits:
1. Allow a low-pass filter to be designed in a short period of time without going through many difficult equations.
2. Provide a filter with good feedback stability.
3. Will not require a large circuit.

In this chapter we will present two of the most commonly used single op amp low-pass-filter circuits. These are simple circuits utilizing single feedback. The first is a second-order filter which will be described in detail including a design procedure and example. A third-order filter will then be described in approximately the same level of detail. By cascading second- and third-order filters, one may produce filters of almost any order. Procedures for cascaded filter design are given in the references at the end of the chapter.

### 10.1 SECOND-ORDER LOW-PASS FILTER

**ALTERNATE NAMES** Unity-gain low-pass filter, active $RC$ low-pass filter, active inductorless low-pass filter.

**EXPLANATION OF OPERATION** This circuit provides two complex poles with adjustable damping. By proper choice of $R_1$, $R_2$, $C_1$, and $C_2$ of Fig. 10.1, the

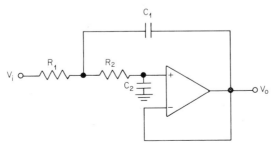

**Fig. 10.1** Single-feedback second-order low-pass filter using a unity-gain amplifier.

## 10-2 LOW-PASS FILTERS

transfer function can be made to exhibit the range of characteristics shown in Fig. 10.2. The curves of Fig. 10.2 were obtained with an actual circuit using a 741 op amp and a pole frequency $f_{cp}$ of 1,000 Hz. Using the design steps for this circuit (which will be listed later), the following component values were calculated:

Bessel:           $R_1 = R_2 = 10{,}800 \; \Omega$
                   $C_1 = 0.0133 \; \mu\text{F}$
                   $C_2 = 0.01 \; \mu\text{F}$
Butterworth:    $R_1 = R_2 = 10{,}800 \; \Omega$
                   $C_1 = 0.02 \; \mu\text{F}$
                   $C_2 = 0.01 \; \mu\text{F}$
3-dB Chebyshev: $R_1 = R_2 = 49{,}400 \; \Omega$
                   $C_1 = 0.01 \; \mu\text{F}$
                   $C_2 = 1{,}470 \; \text{pF}$

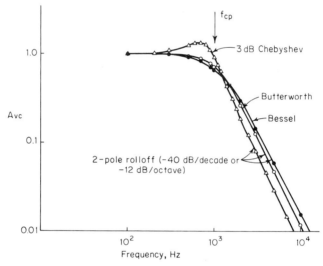

**Fig. 10.2** Test data of three typical two-pole low-pass filters using the circuit of Fig. 10.1.

The transfer function of Fig. 10.1 is

$$A_{vc} = \frac{V_o}{V_i} = \frac{1}{s^2(C_1 C_2 R_1 R_2) + s[C_2(R_1 + R_2)] + 1}$$

The locations of the two complex poles are

$$s_1, s_2 = \frac{-C_2(R_1 + R_2) \pm [C_2^2(R_1 + R_2)^2 - 4C_1 C_2 R_1 R_2]^{1/2}}{2 C_1 C_2 R_1 R_2}$$

Figure 10.3 shows the pole locations in the $s$ domain for all filters discussed in this section. In this figure the corner frequency has been normalized to $\omega = 1$ rad/s.

The damping factor $\zeta$ determines the shape of $A_{vc}$ in the frequency region near $f_{cp}$. Low values of $\zeta$ cause the frequency-response curve to have more peaking near the pole frequency. This term is related to the familiar circuit $Q$ by

$$\zeta = \frac{1}{2Q}$$

The circuit transfer function can be put in the classical form to help us find $\zeta$:

$$A_{vc} = \frac{1}{s^2 + 2\zeta\omega_n s + \omega_n^2}$$

where $\omega_n = 2\pi f_{cp}$ is the natural resonant radian frequency of the circuit. We now can determine $\zeta$ to be

$$\zeta = \frac{R_1 + R_2}{2}\left(\frac{C_2}{R_1 R_2 C_1}\right)^{1/2}$$

The formula for $\zeta$ gives a value of 0.383 for the Chebyshev filter plotted in Fig. 10.2. Similarly, values of $\zeta = 0.866$ and 0.707 are obtained for these particular Bessel and Butterworth filters, respectively.

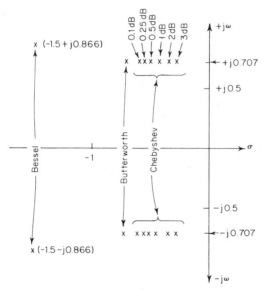

**Fig. 10.3** Pole locations in $s$ domain of the transfer function of Fig. 10.1.

**TABLE 10.1** Unscaled Capacitor Values for Fig. 10.1

| Type of two-pole low-pass filter | $\zeta$ | $C_1''$, F | $C_2''$, F |
|---|---|---|---|
| Bessel | 0.8659 | 0.9066 | 0.6799 |
| Butterworth | 0.7072 | 1.414 | 0.7071 |
| Chebyshev (0.1-dB peak) | 0.6516 | 1.638 | 0.6955 |
| Chebyshev (0.25-dB peak) | 0.6179 | 1.778 | 0.6789 |
| Chebyshev (0.5-dB peak) | 0.5789 | 1.949 | 0.6533 |
| Chebyshev (1-dB peak) | 0.5228 | 2.218 | 0.6061 |
| Chebyshev (2-dB peak) | 0.4431 | 2.672 | 0.5246 |
| Chebyshev (3-dB peak) | 0.3833 | 3.103 | 0.4558 |

## 10-4 LOW-PASS FILTERS

### DESIGN PARAMETERS

| Parameter | Description |
|---|---|
| $A_v$ | Op amp voltage gain as a function of frequency |
| $A_{vc}$ | Voltage gain of circuit as a function of frequency |
| $C_1, C_2$ | Final values for capacitors after both impedance and frequency scaling. Determines $f_{cp}$ and $\zeta$ |
| $C_1', C_2'$ | Intermediate values for $C_1$ and $C_2$ after frequency scaling |
| $C_1'', C_2''$ | Unscaled capacitor values from Table 10.1 |
| $f_{cp}$ | Pole (or corner) frequency of circuit |
| $Q$ | Determines height of peak in frequency response |
| $R$ | Common value of $R_1$ and $R_2$ (both $= R$) |
| $R_1, R_2$ | Determines $f_{cp}$ and $\zeta$ |
| $S_A^B$ | Sensitivity of $B$ to variations in $A$ (applies to all sensitivity functions listed in Design Equations) |
| $V_i$ | Circuit input voltage |
| $V_o$ | Circuit output voltage |
| $\omega_n$ | Natural radian frequency of poles of circuit |
| $\zeta$ | Damping factor |

### DESIGN EQUATIONS

| Eq. No. | Description | Equation |
|---|---|---|
| 1 | Transfer function (voltage gain) of circuit | $A_{vc} = \dfrac{V_o}{V_i} = \dfrac{1}{s^2(C_1 C_2 R_1 R_2) + s[C_2(R_1 + R_2)] + 1}$<br><br>or $A_{vc} = \dfrac{1}{s^2 + 2\zeta\omega_n s + \omega_n^2}$ |
| 2 | Location of two complex poles of Eq. 1 in $s$ domain | $s_1, s_2 = \dfrac{-C_2(R_1 + R_2) \pm [C_2^2(R_1 + R_2)^2 - 4C_1 C_2 R_1 R_2]^{1/2}}{2 C_1 C_2 R_1 R_2}$ |
| 3 | Relationship between initial and final capacitor values | $C_1 = \dfrac{C_1''}{2\pi f_{cp} R}$<br><br>$C_2 = \dfrac{C_2''}{2\pi f_{cp} R}$ |
| 4 | Damping factor of circuit | $\zeta = \dfrac{R_1 + R_2}{2}\left(\dfrac{C_2}{R_1 R_2 C_1}\right)^{1/2}$<br><br>$= \left(\dfrac{C_2}{C_1}\right)^{1/2}$ if $R_1 = R_2$ |
| 5 | Relationship between circuit $Q$ and damping factor | $Q = \dfrac{1}{2\zeta}$ |
| 6 | Relationship between pole frequency and natural radian frequency | $\omega_n = 2\pi f_{cp}$ |
| 7 | Sensitivity of $f_{cp}$ to variations in $R_1, R_2, C_1,$ or $C_2$ | $S_{R_1}^{f_{cp}} = S_{R_2}^{f_{cp}} = S_{C_1}^{f_{cp}} = S_{C_2}^{f_{cp}} = -\dfrac{1}{2}$ |

SECOND-ORDER LOW-PASS FILTER  10-5

| Eq. No. | Description | Equation |
|---|---|---|
| | | NOTES: 1. $S_{R_1}^{f_{cp}} = -\frac{1}{2}$ means that if $R_1$ increases in value by 1%, $f_{cp}$ will decrease in frequency by $\frac{1}{2}\%$ |
| | | 2. After all sensitivities of a given parameter are computed, they are algebraically added to determine the total result |
| 8 | Sensitivity of $\zeta$ to variations in $R_1$, $R_2$, $C_1$, or $C_2$ | $S_{R_1}^{\zeta} = \frac{1}{2} - \frac{1}{4\pi\zeta f_{cp} R_1 C_1}$ |
| | | $S_{R_2}^{\zeta} = \frac{1}{2} - \frac{1}{4\pi\zeta f_{cp} R_1 C_1}$ |
| | | $S_{C_1}^{\zeta} = \frac{1}{2} - \left(\frac{1}{R_1} + \frac{1}{R_2}\right) \frac{1}{4\pi\zeta f_{cp} C_1}$ |
| | | $S_{C_2}^{\zeta} = \frac{1}{2}$ |
| 9 | Required op amp open-loop gain to assure accuracy | $A_v \gg \frac{C_1}{2C_2}$ at $f_{cp}$ and lower frequencies |

## DESIGN PROCEDURE

Several approaches are given in the literature for designing this circuit. The one we have chosen to present here (Ref. 1) requires only very simple calculations. Its only disadvantage is that the capacitor values are different, whereas some design approaches result in $C_1 = C_2$.

## DESIGN STEPS

*Step 1.* Choose $C_1''$ and $C_2''$ from Table 10.1 according to the type of filter required.

*Step 2.* Using the required corner frequency $f_{cp}$, perform the following frequency scaling:

$$C_1' = \frac{C_1''}{2\pi f_{cp}} \qquad C_2' = \frac{C_2''}{2\pi f_{cp}}$$

*Step 3.* Choose a value $R = R_1 = R_2$ which will produce practical sizes for $C_1$ and $C_2$ according to

$$C_1 = \frac{C_1'}{R} \qquad C_2 = \frac{C_2'}{R}$$

This procedure is called impedance scaling.

NOTE: The remaining steps are not required unless the designer has time for gaining further insight into error sources, etc.

*Step 4.* Compute the damping factor $\zeta$ using Eq. 4. Compare the result with data in Table 10.1 to verify that the correct filter has been designed.

*Step 5.* If required, use Eq. 7 to compute the sensitivity of $f_{cp}$ to variations in $R_1$, $R_2$, $C_1$, and $C_2$. Likewise, Eq. 8 may be used to determine the sensitivity of $\zeta$ to changes in $R_1$, $R_2$, $C_1$, and $C_2$.

## 10-6 LOW-PASS FILTERS

*Step 6.* From the op amp data sheet determine $A_v$ at $f_{cp}$. This value of $A_v$ must satisfy Eq. 9 by at least a factor of 100 in order to keep the actual frequency response (Fig. 10.2) less than 0.2 dB from the ideal frequency response.

**EXAMPLE OF SECOND-ORDER LOW-PASS FILTER DESIGN** The six design steps will be numerically illustrated through an example. The results of tests on a circuit designed using these steps were previously shown in Fig. 10.2 (3-dB Chebyshev).

*Design Requirements*

$f_{cp} = 1{,}000$ Hz
Peaking $\approx$ 3 dB (Chebyshev)
Maximum capacitor size $\approx 0.01\ \mu\text{F}$

*Device Data*

$A_v(1{,}000\ \text{Hz}) = 1{,}000$
$\left.\begin{array}{l}\Delta R_1 = 0.018\ R_1\\ \Delta R_2 = 0.018\ R_2\\ \Delta C_1 = 0.01\ C_1\\ \Delta C_2 = 0.01\ C_2\end{array}\right\}$ ($-55$ to $+125^\circ$C)

*Step 1.* From Table 10.1 we obtain $C_1'' = 3.103$ F and $C_2'' = 0.4558$ F.
*Step 2.* Frequency scaling:

$$C_1' = \frac{C_1''}{2\pi f_{cp}}$$
$$= \frac{3.103}{2\pi \times 1{,}000} = 4.94 \times 10^{-4}$$

$$C_2' = \frac{C_2''}{2\pi f_{cp}}$$
$$= \frac{0.4558}{2\pi \times 1{,}000} = 7.25 \times 10^{-5}$$

*Step 3.* Since $C_1$ is always the largest capacitor in this design approach, we scale $R$ so that $C_1 = 0.01\ \mu\text{F}$.

$$R = \frac{C_1'}{C_1} = \frac{4.94 \times 10^{-4}}{10^{-8}} = 49{,}400\ \Omega$$

Also,
$$C_2 = \frac{C_2'}{R} = \frac{7.25 \times 10^{-5}}{49{,}400} = 1{,}470\ \text{pF}$$

*Step 4*

$$\zeta = \left(\frac{C_2}{C_1}\right)^{1/2} = \left(\frac{1.47 \times 10^{-9}}{10^{-8}}\right)^{1/2} = 0.383$$

*Step 5.* The sensitivity functions are:

$$S_{R_1}^{f_{cp}} = S_{R_2}^{f_{cp}} = S_{C_1}^{f_{cp}} = S_{C_2}^{f_{cp}} = -\frac{1}{2}$$

The fractional variations of $R_1$ and $R_2$ over temperature are

$$\frac{\Delta R_1}{R_1} = \frac{\Delta R_2}{R_2} = 0.018\ (-55\ \text{to}\ +125^\circ\text{C})$$

Thus if these resistances increase as temperature increases, $f_{cp}$ will decrease by $1/2(0.018) f_{cp} = 9$ Hz as the temperature increases from $-55$ to $+125°C$. The fractional variations of $C_1$ and $C_2$ over temperature are

$$\frac{\Delta C_1}{C_1} = \frac{\Delta C_2}{C_2} = 0.01 \; (-55 \text{ to } +125°C)$$

If the capacitances increase as temperature increases, $f_{cp}$ will decrease by $1/2(0.01)1,000 = 5$ Hz as the temperature increases from $-55$ to $+125°C$. Changes to $\zeta$ as temperature varies are

$$S_{R_1}^{\zeta} = \frac{1}{2} - \frac{1}{4\pi \zeta f_{cp} R_1 C_1}$$

$$= \frac{1}{2} - \frac{1}{4\pi(0.383)1,000(49,400)10^{-8}} = 0.0794$$

If $\Delta R_1/R_1$ increases by $0.018$ as the temperature increases from $-55$ to $+125°C$, then $\zeta$ will correspondingly increase by $0.0794(0.383)0.018 = 5.5 \times 10^{-4}$. It should be realized that the generator driving $R_1$ has a finite resistance. This resistance (and its variations) must be incorporated into all $R_1$ calculations.

Continuing the calculations:

$$S_{R_2}^{\zeta} = \frac{1}{2} - \frac{1}{4\pi \zeta f_{cp} R_2 C_1}$$

$$= \frac{1}{2} - \frac{1}{4\pi(0.383)1,000(49,400)10^{-8}} = 0.0794$$

Also, $\quad S_{C_1}^{\zeta} = \frac{1}{2} - \left(\frac{1}{R_1} + \frac{1}{R_2}\right) \frac{1}{4\pi \zeta f_{cp} C_1}$

$$= \frac{1}{2} - \left(\frac{1}{49,400} + \frac{1}{49,400}\right) \frac{1}{4\pi(0.383)1,000(10^{-8})} = -0.341$$

$$S_{C_2}^{\zeta} = \frac{1}{2}$$

*Step 6.* We must satisfy

$$A_v \gg \frac{C_1}{2C_2} = \frac{10^{-8}}{2(1.47 \times 10^{-9})} = 3.4$$

at 1,000 Hz. This is assured by more than 100, since $A_v(1,000 \text{ Hz}) \geq 1,000$ in most monolithic op amps.


## REFERENCES

1. Shepard, R. R.: Active Filters: Part 12–Short Cuts to Network Design, *Electronics*, Aug. 18, 1969, p. 82.
2. Al-Nasser, F.: Tables Shorten Design Time for Active Filters, *Electronics*, Oct. 23, 1972, p. 113.
3. Tobey, G. E., J. G. Graeme, and L. P. Huelsman: "Operational Amplifiers–Design and Applications," p. 296, McGraw-Hill Book Company, New York, 1971.

## 10.2 THIRD-ORDER LOW-PASS FILTER

**ALTERNATE NAMES** Unity-gain low-pass filter, active low-pass filter, active inductorless low-pass filter, active $RC$ filter.

**EXPLANATION OF OPERATION** The circuit shown in Fig. 10.4 is nearly identical to the circuit shown in Fig. 10.1 except for the additional $RC$ input stage. These two additional passive parts add a pole on the negative real axis. Compare Fig. 10.3 with Fig. 10.6. The resulting three-pole (third-order) filter rolls off at $-60$ dB/decade or $-18$ dB/octave at frequencies above $f_{cp}$.

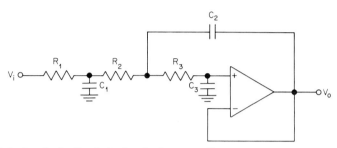

**Fig. 10.4** Single-feedback third-order low-pass filter using a unity-gain amplifier.

This is shown in Fig. 10.5, where the actual response curves of three circuits with different peaking characteristics are plotted. A 741 op amp was used with a selected pole frequency of 1,000 Hz.

The transfer function of Fig. 10.4 is

$$A_{vc} = \frac{V_o}{V_i} = \frac{1}{s^3 A + s^2 B + sC + 1}$$

where $A = C_1 C_2 C_3 R_1 R_2 R_3$
$B = C_1 C_2 C_3 (R_1 + R_2) + C_1 C_3 R_1 (R_2 + R_3)$
$C = C_1 R_1 + C_3 (R_1 + R_2 + R_3)$

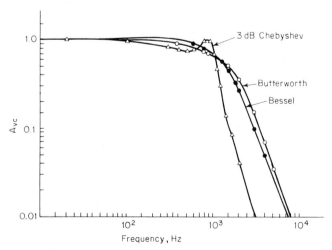

**Fig. 10.5** Measured frequency response of three typical third-order filters using the circuit of Fig. 10.4.

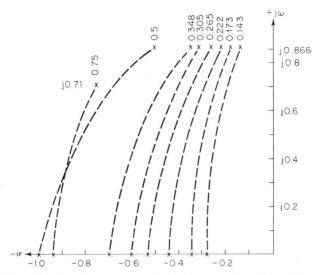

**Fig. 10.6** Pole locations in $s$ domain of the transfer function of Fig. 10.4.

The locations of the three poles in the $s$ domain are not easily expressed in terms of circuit components. Table 10.2 lists the pole locations for a normalized cutoff frequency of 1 rad/s.

The damping factor $\zeta$ is not defined for third-order systems. However, the peaking of the frequency plot can be controlled by choosing appropriate capacitance values in Table 10.3.

**TABLE 10.2  Pole Locations in $s$ Domain for the Single-Feedback Third-Order Low-Pass Filter**

| Type of three-pole low-pass filter | Location of real pole | Location of two complex poles |
|---|---|---|
| Bessel | $-0.942$ | $-0.746 \pm j\, 0.711$ |
| Butterworth | $-1.000$ | $-0.500 \pm j\, 0.866$ |
| Chebyshev (0.1-dB peak) | $-0.696$ | $-0.348 \pm j\, 0.866$ |
| Chebyshev (0.25-dB peak) | $-0.609$ | $-0.305 \pm j\, 0.866$ |
| Chebyshev (0.5-dB peak) | $-0.530$ | $-0.265 \pm j\, 0.866$ |
| Chebyshev (1-dB peak) | $-0.444$ | $-0.222 \pm j\, 0.866$ |
| Chebyshev (2-dB peak) | $-0.345$ | $-0.173 \pm j\, 0.866$ |
| Chebyshev (3-dB peak) | $-0.286$ | $-0.143 \pm j\, 0.866$ |

**TABLE 10.3  Unscaled Capacitor Values for Fig. 10.4**

| Type of three-pole low-pass filter | $C_1''$ | $C_2''$ | $C_3''$ |
|---|---|---|---|
| Bessel | 0.9880 | 1.423 | 0.2538 |
| Butterworth | 1.392 | 3.546 | 0.2024 |
| Chebyshev (0.1-dB peak) | 1.825 | 6.653 | 0.1345 |
| Chebyshev (0.25-dB peak) | 2.018 | 8.551 | 0.1109 |
| Chebyshev (0.5-dB peak) | 2.250 | 11.23 | 0.08950 |
| Chebyshev (1-dB peak) | 2.567 | 16.18 | 0.06428 |
| Chebyshev (2-dB peak) | 3.113 | 27.82 | 0.03892 |
| Chebyshev (3-dB peak) | 3.629 | 43.42 | 0.02533 |

## 10-10 LOW-PASS FILTERS

### DESIGN PARAMETERS

| Parameter | Description |
|---|---|
| $A_v$ | Op amp voltage gain as a function of frequency |
| $A_{vc}$ | Voltage gain of circuit as a function of frequency |
| $C_1, C_2, C_3$ | Final values for capacitors after both impedance and frequency scaling. These capacitors determine $f_{cp}$ and the magnitude of peaking near $f_{cp}$ |
| $C_1', C_2', C_3'$ | Intermediate values for $C_1$, $C_2$, and $C_3$ after frequency scaling |
| $C_1'', C_2'', C_3''$ | Unscaled capacitor values from Table 10.3 |
| $f_{cp}$ | Pole or corner frequency of circuit |
| $R$ | Common value of $R_1$, $R_2$, and $R_3$ |
| $R_1, R_2, R_3$ | Determines $f_{cp}$ along with $C_1$, $C_2$, and $C_3$ |
| $V_i$ | Input voltage to circuit |
| $V_o$ | Output voltage of circuit |

### DESIGN EQUATIONS

| Eq. No. | Description | Equation |
|---|---|---|
| 1 | Transfer function (voltage gain) of circuit | $A_{vc} = \dfrac{V_o}{V_i} = \dfrac{1}{s^3 A + s^2 B + sC + 1}$ where $A = C_1 C_2 C_3 R_1 R_2 R_3$ $B = C_1 C_2 C_3 (R_1 + R_2) + C_1 C_3 R_1 (R_2 + R_3)$ $C = C_1 R_1 + C_3 (R_1 + R_2 + R_3)$ |
| 2 | Relationship between initial and final capacitor values | $C_1 = \dfrac{C_1''}{2\pi f_{cp} R}$ $C_2 = \dfrac{C_2''}{2\pi f_{cp} R}$ $C_3 = \dfrac{C_3''}{2\pi f_{cp} R}$ |
| 3 | Recommended minimum $A_v$ at $f_{cp}$ | $A_v(f_{cp}) \geq 100$ |

### DESIGN PROCEDURE

At least two simplified design approaches are possible for this circuit. One could assume $R = R_1 = R_2 = R_3$ and solve for $R$ and the capacitor values. Conversely, we could let $C = C_1 = C_2 = C_3$ and solve for $C$ and the resistor values. The design steps will utilize the first method in conjunction with Table 10.3.

### DESIGN STEPS

*Step 1.* Choose $C_1''$, $C_2''$, and $C_3''$ from Table 10.3 according to the type of filter required.

*Step 2.* Using the required corner frequency $f_{cp}$, perform the following frequency scaling:

$$C_1' = \frac{C_1''}{2\pi f_{cp}}$$

$$C_2' = \frac{C_2''}{2\pi f_{cp}}$$

$$C_3' = \frac{C_3''}{2\pi f_{cp}}$$

*Step 3.* Choose a value $R = R_1 = R_2 = R_3$ which will produce convenient sizes for $C_1$, $C_2$, and $C_3$ according to

$$C_1 = \frac{C_1'}{R}$$

$$C_2 = \frac{C_2'}{R}$$

$$C_3 = \frac{C_3'}{R}$$

This procedure is called impedance scaling.

*Step 4.* To minimize errors due to the op amp, verify that the following is satisfied:

$$A_v(\text{at } f_{cp}) \geq 100$$

**EXAMPLE OF THIRD-ORDER LOW-PASS-FILTER DESIGN** The four design steps will be used to design a third-order Bessel filter.

*Design Requirements*
$f_{cp} = 1,000$ Hz
Maximum capacitor size 0.01 $\mu$F

*Device Data*
$A_v(1,000 \text{ Hz}) = 1,000$

*Step 1.* From Table 10.3 we get

$$C_1'' = 0.9880$$
$$C_2'' = 1.423$$
$$C_3'' = 0.2538$$

*Step 2.* Frequency scaling is performed as follows:

$$C_1' = \frac{C_1''}{2\pi f_{cp}} = \frac{0.9880}{2,000\pi} = 1.572 \times 10^{-4}$$

$$C_2' = \frac{C_2''}{2\pi f_{cp}} = \frac{1.423}{2,000\pi} = 2.265 \times 10^{-4}$$

$$C_3' = \frac{C_3''}{2\pi f_{cp}} = \frac{0.2538}{2,000\pi} = 4.039 \times 10^{-5}$$

*Step 3.* $C_2$ is always the largest capacitor, so we let

$$C_2 = 0.01 \ \mu\text{F}$$

and solve for $R$, $C_1$, and $C_3$ as follows:

$$R = \frac{C_2'}{C_2} = \frac{2.265 \times 10^{-4}}{10^{-8}} = 22,650 \ \Omega$$

$$C_1 = \frac{C_1'}{R} = \frac{1.572 \times 10^{-4}}{22,650} = 6,940 \ \text{pF}$$

$$C_3 = \frac{C_3'}{R} = \frac{4.039 \times 10^{-5}}{22,650} = 1,783 \ \text{pF}$$

*Step 4.* Verification of Eq. 3:

$$A_v(1,000 \text{ Hz}) = 1,000 > 100$$

## REFERENCES

1. Al-Nasser, F.: Tables Shorten Design Time for Active Filters, *Electronics*, Oct. 23, 1972, p. 113.
2. Shepard, R. R.: Active Filters: Part 12 – Short Cuts to Network Design, *Electronics*, Aug. 18, 1969, p. 82.

Chapter **11**

# High-Pass Filters

## INTRODUCTION

This chapter is similar in many respects to Chap. 10. We will present the same two circuits discussed in that chapter. However, the $R$'s and $C$'s are exchanged. This changes the circuits from low-pass to high-pass filters. As before, the resulting circuits are
1. Easily designed with very simple calculations.
2. Stable against oscillation and parameter drift.
3. Implemented with only one op amp.

## 11.1 SECOND-ORDER HIGH-PASS FILTER

**ALTERNATE NAMES** Unity-gain high-pass filter, active high-pass filter, active inductorless high-pass filter, ac-coupled voltage follower, active $RC$ filter.

**EXPLANATION OF OPERATION** The circuit shown in Fig. 11.1 provides zero response at dc and unity gain from $f_{cp}$ up to the frequency where the op amp gain crosses unity. By proper choice of $R_1$, $R_2$, $C_1$, and $C_2$ the transfer function can be made to exhibit the range of characteristics shown in Fig. 11.2. These curves were obtained from an actual circuit using a 741 op amp. The $f_{cp}$ was chosen to be 100 Hz. It will be noted that this circuit is a high-pass filter only for those frequencies between $f_{cp}$ and $f_u$ of the op amp. Since $f_u$

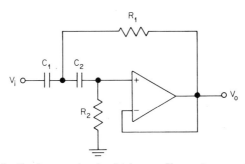

**Fig. 11.1** Single-feedback second-order high-pass filter using a unity-gain amplifier.

## 11-2 HIGH-PASS FILTERS

varies with temperature, the filter will have progressively more error as $f_u$ is approached.

The transfer function of Fig. 11.1 is

$$A_{vc} = \frac{V_o}{V_i} = \frac{s^2}{s^2 + s(1/R_2C_1 + 1/R_2C_2) + 1/R_1R_2C_1C_2}$$

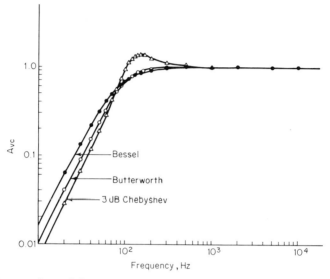

**Fig. 11.2** Test data of three typical two-pole high-pass filters using the circuit of Fig. 11.1.

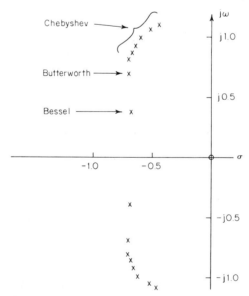

**Fig. 11.3** Pole locations in $s$ domain of the transfer function of Fig. 11.1.

The locations of the two complex poles are

$$s_1, s_2 = -\frac{1}{2R_2}\frac{C_1+C_2}{C_1C_2} \pm \left[\left(\frac{C_1+C_2}{2R_2C_1C_2}\right)^2 - \frac{1}{R_1R_2C_1C_2}\right]^{1/2}$$

Figure 11.3 shows the pole locations in the $s$ domain for all filters discussed in this section. Each pole also has a zero at $s = 0 + j0$ due to the blocking capacitor $C_1$. In Fig. 11.3 the corner frequency has been normalized to 1 rad/s.

The damping factor $\zeta$ determines the shape of $A_{vc}$ in the frequency region near $f_{cp}$. Low values of $\zeta$ cause the frequency-response curve to have more peaking near the pole (corner) frequency. This term is related to the familiar circuit $Q$ by

$$\zeta = \frac{1}{2Q}$$

## DESIGN PARAMETERS

| Parameter | Description |
|---|---|
| $A_v$ | Op amp voltage gain as a function of frequency |
| $A_{vc}$ | Voltage gain of circuit as a function of frequency |
| $C, C_1, C_2$ | Common value of $C_1$ and $C_2$. Determines both $f_{cp}$ and $\zeta$ |
| $f_{cp}$ | Poles, or corner frequency, of circuit |
| $f_u$ | Op amp unity-gain crossover frequency |
| $Q$ | Determines height of peak in frequency response |
| $R_1, R_2$ | Final values for resistors after both impedance and frequency scaling |
| $R_1', R_2'$ | Intermediate values for $R_1$ and $R_2$ after frequency scaling |
| $R_1'', R_2''$ | Unscaled resistor values from Table 11.1 |
| $s_1, s_2$ | Locations of complex poles in $s$ domain |
| $S_A^B$ | Sensitivity of parameter $B$ to a change in parameter $A$ |
| $V_i$ | Circuit input voltage |
| $V_o$ | Circuit output voltage |
| $\omega_n$ | Natural radian frequency of poles of circuit |
| $\zeta$ | Damping factor |

**TABLE 11.1** Unscaled Resistor Values for Fig. 11.1

| Type of two-pole high-pass filter | $\zeta$ | $R_1'$ | $R_2'$ |
|---|---|---|---|
| Bessel | 0.8659 | 1.103 | 1.471 |
| Butterworth | 0.7072 | 0.7072 | 1.414 |
| Chebyshev (0.1-dB peak) | 0.6516 | 0.6105 | 1.438 |
| Chebyshev (0.25-dB peak) | 0.6179 | 0.5624 | 1.473 |
| Chebyshev (0.5-dB peak) | 0.5789 | 0.5131 | 1.531 |
| Chebyshev (1-dB peak) | 0.5228 | 0.4509 | 1.650 |
| Chebyshev (2-dB peak) | 0.4431 | 0.3743 | 1.906 |
| Chebyshev (3-dB peak) | 0.3833 | 0.3223 | 2.194 |

## DESIGN EQUATIONS

| Eq. No. | Description | Equation |
|---|---|---|
| 1 | Transfer function (voltage gain) of circuit | $A_{vc} = \dfrac{V_o}{V_i} = \dfrac{s^2}{s^2 + s[(1/R_2C_1) + (1/R_2C_2)] + (1/R_1R_2C_1C_2)}$ |

## 11-4 HIGH-PASS FILTERS

| Eq. No. | Description | Equation |
|---|---|---|
| 2 | Location of two complex poles of Eq. 1 | $s_1, s_2 = -\dfrac{1}{2R_2}\dfrac{C_1+C_2}{C_1 C_2} \pm \left[\left(\dfrac{C_1+C_2}{2R_2 C_1 C_2}\right)^2 - \dfrac{1}{R_1 R_2 C_1 C_2}\right]^{1/2}$ |
| 3 | Damping factor of circuit | $\zeta = \dfrac{1}{2}\left(\dfrac{R_1 C_1}{R_2 C_2}\right)^{1/2} + \dfrac{1}{2}\left(\dfrac{R_1 C_2}{R_2 C_1}\right)^{1/2}$ |
| 4 | Sensitivity of $f_{cp}$ to variations in $R_1$, $R_2$, $C_1$, or $C_2$ | $S^{f_{cp}}_{R_1} = S^{f_{cp}}_{R_2} = S^{f_{cp}}_{C_1} = S^{f_{cp}}_{C_2} = -\dfrac{1}{2}$ <br> NOTES: 1. $S^{f_{cp}}_{R_1} = -\dfrac{1}{2}$ means that if $R_1$ increases in value by 1%, $f_{cp}$ will decrease in frequency by ½%. <br> 2. After all sensitivities of a given parameter are computed, they are algebraically added to determine the total result |
| 5 | Sensitivity of $\zeta$ to variations in $R_1$, $R_2$, $C_1$, or $C_2$ | $S^{\zeta}_{R_1} = \dfrac{1}{2}$ <br> $S^{\zeta}_{R_2} = \dfrac{1}{2} - \dfrac{1}{2\zeta\omega_n R_2}\left(\dfrac{1}{C_1} + \dfrac{1}{C_2}\right)$ <br> $S^{\zeta}_{C_1} = \dfrac{1}{2} - \dfrac{1}{2\zeta\omega_n R_2 C_1}$ <br> $S^{\zeta}_{C_2} = \dfrac{1}{2} - \dfrac{1}{2\zeta\omega_n R_2 C_2}$ |
| 6 | Relationship between initial and final resistor values | $R_1 = KR'_1$ <br> $R_2 = KR'_2$ <br> where $K = \dfrac{1}{2\pi f_{cp} C_1} = \dfrac{1}{2\pi f_{cp} C_2}$ |
| 7 | Required op amp open-loop gain to assure accuracy | $A_v \geq 100$ at all frequencies where high-pass operation is required |

### DESIGN PROCEDURE

The design approach to be presented requires very simple calculations even though Eqs. 1 to 7 may appear quite difficult. The end result is capacitors of equal value but resistors of different values. Other approaches may give equally sized resistors and differently sized capacitors. Setting two variables equal at the outset is the key to simplified design.

### DESIGN STEPS

*Step 1.* Choose $R'_1$ and $R'_2$ from Table 11.1 according to the type of filter required.

## SECOND-ORDER HIGH-PASS FILTER 11-5

*Step 2.* Using the chosen corner frequency $f_{cp}$, perform the following frequency scaling:

$$C = \frac{1}{2\pi f_{cp}}$$

*Step 3.* Choose a constant $K$ which will provide practical capacitor sizes for $C_1$ and $C_2$ according to

$$C_1 = C_2 = \frac{C}{K}$$

*Step 4.* Calculate values for resistors with

$$R_1 = KR'_1 \qquad R_2 = KR'_2$$

These last two steps are called impedance scaling.

NOTE: Steps 1 to 4 cover the basic design of the filter. The remaining steps are provided only for the designer who wishes to obtain more insight into the filter operation, error sources, etc.

*Step 5.* Compute the damping factor $\zeta$ using Eq. 3. Compare the result with data in Table 11.1 to verify that the correct filter has been designed.

*Step 6.* If required, use Eq. 4 to compute the sensitivity of $f_{cp}$ to variations in $R_1$, $R_2$, $C_1$, and $C_2$. Likewise, Eq. 5 can be used to determine the sensitivity of $\zeta$ to changes in $R_1$, $R_2$, $C_1$, and $C_2$.

*Step 7.* Determine the range of frequencies where $A_v \geq 100$ from the op amp data sheet. This value of $A_v$ must be maintained in order to keep the actual frequency response within 0.1 dB of the theoretical response. This is even more important as temperature changes, since $A_v$ has a strong dependence on temperature in most op amps.

**EXAMPLE OF SECOND-ORDER HIGH-PASS FILTER** A Butterworth high-pass filter will be designed to illustrate the seven design steps. The response of an actual filter built with this procedure was shown in Fig. 11.2 along with Chebyshev and Bessel filters.

*Design Requirements*
$f_{cp} = 100$ Hz
Response = Butterworth ($\zeta = 0.707$)
Maximum capacitor size = 0.1 $\mu$F

*Device Data*
$A_v \geq 100$ up to 10 kHz

*Step 1.* From Table 11.1 we get

$$R'_1 = 0.7072 \qquad R'_2 = 1.414$$

*Step 2.* Frequency scaling is performed.

$$C = \frac{1}{2\pi f_{cp}} = \frac{1}{2\pi \times 100} = 1.592 \times 10^{-3}$$

*Step 3.* If we want the capacitor sizes of $C_1$ and $C_2$ to be 0.1 $\mu$F,

$$K = \frac{C}{C_1} = \frac{C}{C_2} = \frac{1.592 \times 10^{-3}}{10^{-7}} = 1.592 \times 10^4$$

**11-6 HIGH-PASS FILTERS**

*Step 4.* The final resistor values become
$$R_1 = KR_1' = 1.592 \times 10^4(0.7072) = 11,255 \text{ }\Omega$$
$$R_2 = KR_2' = 1.592 \times 10^4(1.414) = 22,505 \text{ }\Omega$$

*Step 5.* The damping factor is checked at this point:
$$\zeta = \frac{1}{2}\left(\frac{R_1 C_1}{R_2 C_2}\right)^{1/2} + \frac{1}{2}\left(\frac{R_1 C_2}{R_2 C_1}\right)^{1/2}$$
$$= \frac{1}{2}\left(\frac{R_1}{R_2}\right)^{1/2} + \frac{1}{2}\left(\frac{R_1}{R_2}\right)^{1/2} = \left(\frac{R_1}{R_2}\right)^{1/2}$$

since $C_1 = C_2$. Thus,
$$\zeta = \left(\frac{11,255}{22,505}\right)^{1/2} = 0.7072$$

This checks out with Table 11.1.

*Step 6.* The sensitivity of $f_{cp}$ to component variations is as follows:
$$S_{R_1}^{fcp} = S_{R_2}^{fcp} = S_{C_1}^{fcp} = S_{C_2}^{fcp} = -\frac{1}{2}$$

Therefore, if any of these passive components increase in value by 1 percent, $f_{cp}$ will decrease in frequency by $\frac{1}{2}$ percent.

The sensitivity of $\zeta$ to component variations is

$$S_{R_1}^{\zeta} = \frac{1}{2}$$

$$S_{R_2}^{\zeta} = \frac{1}{2} - \frac{1}{2\zeta\omega_n R_2}\left(\frac{1}{C_1} + \frac{1}{C_2}\right)$$
$$= \frac{1}{2} - \frac{1/10^{-7} + 1/10^{-7}}{2(0.7072)2\pi(100)22,505} = -0.5000$$

$$S_{C_1}^{\zeta} = \frac{1}{2} - \frac{1}{2\zeta\omega_n R_2 C_1}$$
$$= \frac{1}{2} - \frac{1}{2(0.7072)2\pi(100)22,505(10^{-7})} = 1.310 \times 10^{-6}$$

$$S_{C_2}^{\zeta} = \frac{1}{2} - \frac{1}{2\zeta\omega_n R_2 C_2}$$
$$= \frac{1}{2} - \frac{1}{2(0.7072)2\pi(100)22,505(10^{-7})} = 1.310 \times 10^{-6}$$

*Step 7.* Since $A_v > 100$ from dc to 10 kHz, the response curve should be stable from $f_{cp}$ to 10 kHz. A 100 percent variation of $A_v$ should therefore cause less than a 1 percent change (<0.1 dB) in $A_{vc}$ within this frequency range.

## REFERENCES

1. Tobey, G. E., J. G. Graeme, and L. P. Huelsman: "Operational Amplifiers—Design and Applications," p. 298, McGraw-Hill Book Company, New York, 1971.
2. Shepard, R. R.: Active Filters: Part 12, Short Cuts to Network Design, *Electronics*, Aug. 18, 1969, p. 82.

## 11.2 THIRD-ORDER HIGH-PASS FILTER

**ALTERNATE NAMES** Unity-gain high-pass filter, active high-pass filter, active inductorless high-pass filter, ac-coupled voltage follower, active $RC$ filter.

**EXPLANATION OF OPERATION** The circuit shown in Fig. 11.4 provides zero response at dc and unity gain from $f_{cp}$ up to the frequency where the op amp gain crosses unity. By proper choice of the six passive components the transfer function can be made to exhibit the range of characteristics shown in Fig. 11.5. These curves were obtained from an actual circuit utilizing a 741 op amp. The $f_{cp}$ was chosen to be 100 Hz. It should be noted that this circuit is a high-pass filter only for those frequencies between $f_{cp}$ and $f_u$ of the op amp. Experience tells us that $f_u$ varies with temperature. The filter will accordingly have more error drift in the region near $f_u$.

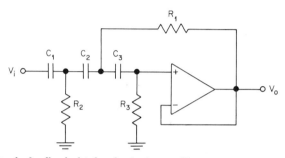

**Fig. 11.4** Single-feedback third-order high-pass filter using a unity-gain amplifier.

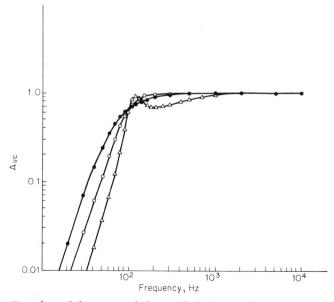

**Fig. 11.5** Test data of three typical three-pole high-pass filters using the circuit of Fig. 11.4.

## 11-8 HIGH-PASS FILTERS

The transfer function of Fig. 11.4 is

$$A_{vc} = \frac{V_o}{V_i} = \frac{s^3}{s^3 + Ds^2 + Es + F}$$

where
$$D = \frac{1}{R_3}\left(\frac{1}{C_1} + \frac{1}{C_2} + \frac{1}{C_3}\right) + \frac{1}{R_2 C_1}$$

$$E = \frac{1}{R_3}\left(\frac{1}{R_1 C_2 C_3} + \frac{1}{R_1 C_1 C_3} + \frac{1}{R_2 C_1 C_3} + \frac{1}{R_2 C_1 C_2}\right)$$

$$F = \frac{1}{R_1 R_2 R_3 C_1 C_2 C_3}$$

### DESIGN PARAMETERS

| Parameter | Description |
| --- | --- |
| $A_v$ | Op amp voltage gain as a function of frequency |
| $A_{vc}$ | Voltage gain of circuit as a function of frequency |
| $C$ | Initial capacitor size |
| $C_1, C_2, C_3$ | Final capacitor sizes |
| $D, E, F$ | Variables used in Eq. 1 |
| $f_{cp}$ | Corner frequency of filter |
| $K$ | Constant used in filter design |
| $R_1, R_2, R_3$ | Final resistor sizes |
| $R_1', R_2', R_3'$ | Initial unscaled resistor sizes |
| $V_i$ | Input voltage to circuit |
| $V_o$ | Output voltage from circuit |

### DESIGN EQUATIONS

| Eq. No. | Description | Equation |
| --- | --- | --- |
| 1 | Transfer function (voltage gain) of circuit | $A_{vc} = \dfrac{V_o}{V_i} = \dfrac{s^3}{s^3 + Ds^2 + Es + F}$ <br> where $D = \dfrac{1}{R_3}\left(\dfrac{1}{C_1} + \dfrac{1}{C_2} + \dfrac{1}{C_3}\right) + \dfrac{1}{R_2 C_1}$ <br> $E = \dfrac{1}{R_3}\left(\dfrac{1}{R_1 C_2 C_3} + \dfrac{1}{R_1 C_1 C_3} + \dfrac{1}{R_2 C_1 C_3} + \dfrac{1}{R_2 C_1 C_2}\right)$ <br> $F = \dfrac{1}{R_1 R_2 R_3 C_1 C_2 C_3}$ |
| 2 | Relationship between initial and final resistor values | $R_1 = KR_1'$ <br> $R_2 = KR_2'$ <br> $R_3 = KR_3'$ <br> where $K = \dfrac{1}{2\pi f_{cp} C_1} = \dfrac{1}{2\pi f_{cp} C_2} = \dfrac{1}{2\pi f_{cp} C_3}$ |
| 3 | Required op amp open-loop gain to assure accuracy | $A_v \geq 100$ at all frequencies where high-pass operation is required |

### DESIGN PROCEDURE

The following design steps require only simple calculations even though the characteristic equation (Eq. 1) of a third-order filter is quite cumbersome. A good feature of the procedure is that it gives equally sized capacitors. This is usually convenient, since it is easier to have a wide range of resistors available instead of capacitors.

## DESIGN STEPS

*Step 1.* Choose $R'_1$, $R'_2$, and $R'_3$ from Table 11.2 according to the type of filter required.

*Step 2.* Using the chosen corner frequency $f_{cp}$, perform the following frequency scaling:

$$C = \frac{1}{2\pi f_{cp}}$$

*Step 3.* Choose a constant $K$ which will provide practical capacitor sizes for $C_1$, $C_2$, and $C_3$ according to

$$C_1 = C_2 = C_3 = \frac{C}{K}$$

*Step 4.* Calculate values for resistors with

$$R_1 = KR'_1 \qquad R_2 = KR'_2 \qquad R_3 = KR'_3$$

These last two steps are called impedance scaling.

*Step 5.* Determine the range of frequencies where $A_v \geq 100$ using the op amp data sheet. This value of $A_v$ must be maintained in order to keep the actual frequency response within 0.1 dB of the theoretical response. This is even more important as temperature changes, since $A_v$ has a strong dependence on temperature in most op amps.

**TABLE 11.2** Unscaled Resistor Values for Fig. 11.4

| Type of three-pole high-pass filter | $R'_1$ | $R'_2$ | $R'_3$ |
|---|---|---|---|
| Bessel | 0.7027 | 1.012 | 3.940 |
| Butterworth | 0.2820 | 0.7184 | 4.941 |
| Chebyshev (0.1-dB peak) | 0.1503 | 0.5479 | 7.435 |
| Chebyshev (0.25-dB peak) | 0.1169 | 0.4955 | 9.017 |
| Chebyshev (0.5-dB peak) | 0.08905 | 0.4444 | 11.17 |
| Chebyshev (1-dB peak) | 0.06180 | 0.3896 | 15.56 |
| Chebyshev (2-dB peak) | 0.03595 | 0.3212 | 25.69 |
| Chebyshev (3-dB peak) | 0.02303 | 0.2756 | 39.48 |

**EXAMPLE OF THIRD-ORDER HIGH-PASS FILTER** A Chebyshev filter with 3-dB peaking will be designed to illustrate the five design steps. The response of an actual filter built with this procedure was shown in Fig. 11.5.

*Design Requirements*
$f_{cp} = 100$ Hz
Response $= 3$-dB Chebyshev
Maximum capacitor size $= 0.1$ $\mu$F

*Device Data*
$A_v \geq 100$ up to 10 kHz

*Step 1.* From Table 11.2 we get

$$R'_1 = 0.02303 \qquad R'_2 = 0.2756 \qquad R'_3 = 39.48$$

**11-10 HIGH-PASS FILTERS**

*Step 2.* Frequency scaling is performed.

$$C = \frac{1}{2\pi f_{cp}} = \frac{1}{2\pi \times 100} = 1.592 \times 10^{-3}$$

*Step 3.* If we want the capacitor sizes to be 0.1 $\mu$F,

$$K = \frac{C}{C_1} = \frac{C}{C_2} = \frac{C}{C_3} = \frac{1.592 \times 10^{-3}}{10^{-7}} = 1.592 \times 10^4$$

*Step 4.* The final resistor values become

$R_1 = KR_1' = 1.592 \times 10^4 \, (0.02303) = 366.6 \, \Omega$
$R_2 = KR_2' = 1.592 \times 10^4 \, (0.2756) = 4{,}388 \, \Omega$
$R_3 = KR_3' = 1.592 \times 10^4 \, (39.48) = 628{,}500 \, \Omega$

*Step 5.* Since $A_v \geq 100$ from dc to 10 kHz, the filter response should be stable from $f_{cp}$ to 10 kHz. A 100 percent variation of $A_v$ should therefore cause less than a 1 percent change ($<0.2$ dB) in $A_{vc}$ within this frequency range.

## REFERENCES

1. Al-Nasser, F.: Tables Shorten Design Time for Active Filters, *Electronics*, Oct. 23, 1972, p. 113.
2. Shepard, R. R.: Active Filters: Part 12 – Short Cuts to Network Design, *Electronics*, Aug. 18, 1969, p. 82.

Chapter **12**

# Bandpass Filters

## INTRODUCTION

Information regarding many types of bandpass amplifiers using op amps is readily available in the literature. In this chapter we will provide logically organized design information on two of the most popular bandpass circuits. Both these filter types, multiple-feedback and biquadratic, will be described in detail. The discussions will include step-by-step design procedures and fully worked out examples.

## 12.1 MULTIPLE-FEEDBACK BANDPASS FILTER

**ALTERNATE NAMES** Dual-feedback bandpass filter, active resonator, active filter, active bandpass amplifier, active $RC$ filter.

**EXPLANATION OF OPERATION** This circuit, shown in Fig. 12.1, is useful for several reasons:
 1. It requires only one op amp.
 2. Adjustment of the resonant frequency $f_o$ can be performed with one resistor $R_2$.
 3. If $Q$ is less than 10, the sensitivity of $Q$ and $f_o$ to component variations is not large.

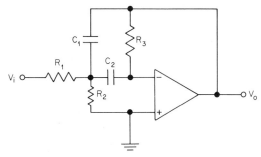

**Fig. 12.1** Bandpass filter using multiple feedback.

## 12-2 BANDPASS FILTERS

4. If $Q$ is less than 10, a large spread in calculated component values will not occur.

5. One resistor, $R_3$, may be used to adjust both $Q$ and the midband gain $H$. This resistor also affects $f_o$, so $Q$ and $H$ should always be adjusted before $f_o$.

This is an inverting circuit with a transfer function expressed as

$$A_{vc} = \frac{V_o}{V_i} = \frac{-As}{s^2 + Bs + C}$$

where

$$A = \frac{1}{R_1 C_1}$$

$$B = \frac{1/C_1 + 1/C_2}{R_3}$$

$$C = \frac{1/R_1 + 1/R_2}{R_3 C_1 C_2}$$

Capacitors are more difficult to trim, so the design of this circuit often begins by assuming $C = C_1 = C_2$, where $C$ is some practical value. We can now state the effect of the three resistors on $f_o$, $H$, and $\Delta f$ ($\Delta f = f_o/Q$).

$$R_1 = \frac{1}{2\pi \Delta f H C}$$

$$R_2 = \left[ 2\pi C \left( \frac{2 f_o^2}{\Delta f} - \Delta f H \right) \right]^{-1/2}$$

$$R_3 = \frac{1}{\pi \Delta f C}$$

We note from the above that
1. $R_1$ affects both $\Delta f$ and $H$.
2. $R_2$ affects $f_o$, $\Delta f$, and $H$; however, the effect on $\Delta f$ and $H$ is small.
3. $R_3$ affects only $\Delta f$.

We can also invert these three equations to get

$$f_o = \frac{1}{2\pi} \left[ \frac{1}{R_3 C_1 C_2} \left( \frac{1}{R_1} + \frac{1}{R_2} \right) \right]^{1/2}$$

$$Q = \frac{f_o}{\Delta f} = \frac{[R_3(1/R_1 + 1/R_2)]^{1/2}}{(C_2/C_1)^{1/2} + (C_1/C_2)^{1/2}}$$

$$H = \frac{R_3 C_2}{R_1(C_1 + C_2)}$$

### DESIGN PARAMETERS

| Parameter | Description |
|---|---|
| $A_v(f_o)$ | Open-loop voltage gain of op amp at frequency $f_o$ |
| $A_{vc}$ | Closed-loop voltage gain of circuit as a function of frequency |
| $C, C_1, C_2$ | Common value for capacitors in circuit |
| $\Delta C/C$ | Fractional change in capacitance of a capacitor $C$ over a specified temperature range |
| $f_o$ | Resonant frequency of circuit |
| $\Delta f$ | Frequency difference between $-3$-dB points on $A_{vc}$ response curve (bandwidth) |

## MULTIPLE-FEEDBACK BANDPASS FILTER

| Parameter | Description |
|---|---|
| $H$ | Voltage gain of circuit at $f_o$ |
| $I_b$ | Input bias current of op amp |
| $Q$ | Quality factor of circuit |
| $R_1$ | Resistor which controls input resistance of circuit |
| $R_2$ | Resistor which principally controls resonant frequency of circuit |
| $R_3$ | Resistor which affects only $Q$ of circuit |
| $\Delta R/R$ | Fractional change in the resistance of a resistor $R$ over a specified temperature range |
| $S_A^B$ | Sensitivity of parameter $B$ to changes in parameter $A$ |
| $V_i$ | Input voltage to circuit |
| $V_o$ | Output voltage from circuit |
| $V_{oo}$ | Circuit output offset voltage |

## DESIGN EQUATIONS

| Eq. No. | Description | Equation |
|---|---|---|
| 1 | Voltage gain of circuit | $A_{vc} = \dfrac{V_o}{V_i} = \dfrac{-As}{s^2 + Bs + C}$ <br> where $A = \dfrac{1}{R_1 C_1}$ <br> $B = \dfrac{1/C_1 + 1/C_2}{R_3}$ <br> $C = \dfrac{1/R_1 + 1/R_2}{R_3 C_1 C_2}$ <br> $s = 2\pi f$ |
| 2 | Voltage gain of circuit at $f_o$ | $H = \dfrac{R_3 C_2}{R_1 (C_1 + C_2)}$ |
| 3 | Bandpass-filter center frequency | $f_o = \dfrac{1}{2\pi} \left( \dfrac{1/R_1 + 1/R_2}{R_3 C_1 C_2} \right)^{1/2}$ |
| 4 | $Q$ of circuit | $Q = \dfrac{[R_3(1/R_1 + 1/R_2)]^{1/2}}{(C_2/C_1)^{1/2} + (C_1/C_2)^{1/2}}$ |
| 5 | Bandwidth (3 dB down from gain at $f_o$) | $\Delta f = \dfrac{f_o}{Q} = \dfrac{1/C_1 + 1/C_2}{2R_3}$ |
| 6 | $R_1$ in terms of other parameters | $R_1 = \dfrac{1}{2\pi \Delta f H C}$ <br> where $C = C_1 = C_2$ |
| 7 | $R_2$ in terms of other parameters | $R_2 = \left[ 2\pi C \left( \dfrac{2f_o^2}{\Delta f} - \Delta f H \right) \right]^{-1/2}$ |
| 8 | $R_3$ in terms of other parameters | $R_3 = \dfrac{1}{\pi \Delta f C}$ |
| 9 | Sensitivity of $f_o$ to component parameter changes | $S_{R_3}^{f_o} = S_{C_1}^{f_o} = S_{C_2}^{f_o} = -\dfrac{1}{2}$ |
| 10 | | $S_{R_1}^{f_o} = \dfrac{-1}{8\pi^2 f_o^2 R_1 R_3 C_1 C_2}$ |

## 12-4 BANDPASS FILTERS

| Eq. No. | Description | Equation |
|---|---|---|
| 11 | | $S^{f_o}_{R_2} = \dfrac{-1}{8\pi^2 f_o^2 R_2 R_3 C_1 C_2}$ |
| 12 | Sensitivity of $Q$ to component parameter changes | $S^{Q}_{R_1} = \dfrac{R_1}{2(R_1 + R_2)} - \dfrac{1}{2}$ |
| 13 | | $S^{Q}_{R_3} = \dfrac{1}{2}$ |
| 14 | | $S^{Q}_{C_1} = \dfrac{Q}{2\pi f_o R_3 C_1} - \dfrac{1}{2}$ |
| 15 | | $S^{Q}_{C_2} = \dfrac{Q}{2\pi f_o R_3 C_2} - \dfrac{1}{2}$ |

### DESIGN PROCEDURE

This list of design steps assumes capacitor values are more difficult to trim than are resistor values. It also assumes both capacitor values are identical. The algebra is greatly simplified, since several variables disappear from the equations.

### DESIGN STEPS

*Step 1.* Choose values for $f_o$, $H$, and $Q$. As mentioned previously, this circuit design should not be attempted if $Q > 10$ is required. The chosen op amp places some restrictions on the choices for $f_o$ and $H$. Using the op amp open-loop-frequency plot, make sure that $H < 0.01 A_v$ at $f_o$. This will guarantee that a 100 percent change in $A_v$ will have much less than a 1 percent effect on $f_o$ and $H$.

*Step 2.* Let $C = C_1 = C_2$ be some practical value. Compute the following:

$$R_3 = \dfrac{2Q}{2\pi f_o C}$$

If $R_3$ is too large, then $I_b$, the op amp input bias current, will cause a dc offset at $V_o$ with a magnitude of

$$V_{oo} = I_b R_3$$

If this offset is larger than allowed for the application, choose a higher $C$. Recompute $R_3$ and $V_{oo}$.

*Step 3.* Find $R_1$ from

$$R_1 = \dfrac{Q}{2\pi f_o C H}$$

*Step 4.* Compute $R_2$ from

$$R_2 = \dfrac{Q}{(2\pi f_o C)(2Q^2 - H)}$$

*Step 5.* Determine the sensitivity of $f_o$ to component parameter variations using Eqs. 9 to 11.

*Step 6.* Determine the sensitivity of $Q$ to component parameter variations using Eqs. 12 to 15.

**MULTIPLE-FEEDBACK BANDPASS FILTER**    12-5

*Step 7.* Verify that $H$ is correct using Eq. 2.
*Step 8.* Verify that $f_o$ is correct using Eq. 3.
*Step 9.* Verify that $Q$ is correct using Eq. 4.
*Step 10.* Verify that $\Delta f$ is correct using Eq. 5.

**EXAMPLE OF BANDPASS-FILTER DESIGN** A bandpass filter with a center frequency of 1,000 Hz will now be designed using the 10 design steps.

*Design Requirements*
$f_o = 1{,}000$ Hz
$H = 10$
$Q = 5$
$\Delta f = f_o/Q = 200$ Hz
$V_{oo}(\max) = \pm 1$ V

*Device Data*
$A_v(1{,}000 \text{ Hz}) = 1{,}000$
$I_b = 10^{-8}$ A
$\dfrac{\Delta R_1}{R_1} = \dfrac{\Delta R_2}{R_2} = \dfrac{\Delta R_3}{R_3} = +0.02 \; (-55 \text{ to } +125°C)$

$\dfrac{\Delta C_1}{C_1} = \dfrac{\Delta C_2}{C_2} = -0.03 \; (-55 \text{ to } +125°C)$

*Step 1.* The required parameters of the circuit are $f_o = 1{,}000$ Hz, $H = 10$, and $Q = 5$. We next verify that

$$H \leq 0.01 \, A_v \, (f_o)$$
$$10 \leq 0.01 \, (1{,}000)$$
$$10 \leq 10$$

This is obviously satisfied.

*Step 2.* Our initial choice for $C$ will be 0.01 μF. We next compute

$$R_3 = \frac{2Q}{2\pi f_o C} = \frac{2(5)}{2\pi(1{,}000)10^{-8}} = 159 \text{ k}\Omega$$

The output offset is now checked:

$$V_{oo} = I_b R_3 = (10^{-8}) \, 1.6 \times 10^5 = 1.6 \text{ mV}$$

*Step 3.* $R_1$ is computed:

$$R_1 = \frac{Q}{2\pi f_o CH} = \frac{5}{2\pi(1{,}000)10^{-8}(10)} = 7{,}960 \; \Omega$$

*Step 4.* $R_2$ is determined:

$$R_2 = \frac{Q}{(2\pi f_o C)(2Q^2 - H)} = \frac{5}{(2\pi \times 1{,}000 \times 10^{-8})(2 \times 5^2 - 10)} = 1{,}990 \; \Omega$$

*Step 5.* The sensitivity of $f_o$ to various parameter variations is computed:

$R_3$: $\qquad\qquad\qquad\qquad\qquad S_{R_3}^{f_o} = -\dfrac{1}{2}$

Since $\Delta R_3/R_3$ increases by 0.02 as the temperature increases from $-55$ to $+125°C$, $f_o$ will change by $-\frac{1}{2} \times 0.02 \times 1{,}000$ Hz $= -10$ Hz.

**12-6 BANDPASS FILTERS**

$C_1$ and $C_2$:
$$S^{f_o}_{C_1} = S^{f_o}_{C_2} = -\frac{1}{2}$$

We have indicated that $\Delta C_1/C_1$ and $\Delta C_2/C_2$ both change by $-0.03$ as the temperature changes from $-55$ to $+125°C$. This will make $f_o$ increase by $(-\frac{1}{2}) \times (-0.03) \times 1{,}000 = +15$ Hz for each capacitor.

$R_1$:
$$S^{f_o}_{R_1} = \frac{-1}{8\pi^2 f_o^2 R_1 R_3 C_1 C_2} = \frac{-1}{8\pi^2 (10^6) 7{,}960 (159{,}000) 10^{-8} 10^{-8}} = -0.1$$

If $\Delta R_1/R_1 = +0.02$ as the temperature varies from $-55$ to $+125°C$, $f_o$ will change $-0.1 \times 0.02 \times 1{,}000 = -10$ Hz.

$R_2$:
$$S^{f_o}_{R_2} = \frac{-1}{8\pi^2 f_o^2 R_2 R_3 C_1 C_2}$$
$$= \frac{-1}{8\pi^2 (10^6) 1{,}990 (159{,}000) 10^{-8} 10^{-8}} = -0.4$$

This will cause a $-8$-Hz shift in $f_o$.

The total shift in frequency caused by variations in $R_1$, $R_2$, $R_3$, $C_1$, and $C_2$ is $-10 + 15 + 15 - 10 - 8 = 2$ Hz.

*Step 6.* The sensitivity of $Q$ to various parameter variations is computed:

$R_1$:
$$S^{Q}_{R_1} = \frac{R_1}{2(R_1 + R_2)} - \frac{1}{2} = \frac{7{,}960}{2(7{,}960 + 1{,}990)} - \frac{1}{2} = -0.1$$

The $\Delta R_1/R_1 = +0.02$ change will cause $Q$ to change by $(-0.1) \times (0.02) \times 5 = -0.01$ as the temperature increases from $-55$ to $+125°C$.

$R_3$:
$$S^{Q}_{R_3} = \frac{1}{2}$$

This will cause $Q$ to change by $\frac{1}{2} \times 0.02 \times 5 = +0.05$ as the temperature increases from $-55$ to $+125°C$.

$C_1$ or $C_2$:
$$\left. \begin{array}{l} S^{Q}_{C_1} = \dfrac{Q}{2\pi f_o R_3 C_1} - \dfrac{1}{2} \\[6pt] S^{Q}_{C_2} = \dfrac{Q}{2\pi f_o R_3 C_2} - \dfrac{1}{2} \end{array} \right\} \text{identical since } C_1 = C_2$$

$$= \frac{5}{2\pi(1{,}000)159{,}000(10^{-8})} - \frac{1}{2}$$
$$= 4.86 \times 10^{-4}$$

If $C_1$ and $C_2$ each decrease by $-0.03$ as the temperature changes from $-55$ to $+125°C$, $Q$ will change by $4.86 \times 10^{-4} (-0.03) 5 = -7.2 \times 10^{-6}$.

The total shift in $Q$ caused by variations in $R_1$, $R_3$, $C_1$, and $C_2$ is $-0.01 + 0.05 - 2 \times 7.2 \times 10^{-6} = +0.04$.

*Step 7.* Equation 2 is used to verify $H$:
$$H = \frac{R_3 C_2}{R_1(C_1 + C_2)} = \frac{(159{,}000)10^{-8}}{7{,}960(10^{-8} + 10^{-8})} = 9.987$$

*Step 8.* Equation 3 is used to verify $f_o$:
$$f_o = \frac{1}{2\pi}\left(\frac{1/R_1 + 1/R_2}{R_3 C_1 C_2}\right)^{1/2} = \frac{1}{2\pi}\left(\frac{1/7{,}960 + 1/1{,}990}{159{,}000 \times 10^{-8} \times 10^{-8}}\right)^{1/2} = 1{,}000.29 \text{ Hz}$$

*Step 9.* Equation 4 is used to verify $Q$:

$$Q = \frac{[R_3(1/R_1 + 1/R_2)]^{1/2}}{(C_2/C_1)^{1/2} + (C_1/C_2)^{1/2}} = \frac{[159{,}000(1/7{,}960 + 1/1{,}990)]^{1/2}}{(10^{-8}/10^{-8})^{1/2} + (10^{-8}/10^{-8})^{1/2}}$$
$$= 4.997$$

*Step 10.* Equation 5 is used to verify $\Delta f$:

$$\Delta f = \frac{1/C_1 + 1/C_2}{2\pi R_3} = \frac{1/10^{-8} + 1/10^{-8}}{2\pi \times 159{,}000} = 200.19$$

## REFERENCES

1. Geffe, P. R.: Designers Guide to Active Bandpass Filters, Parts 1 to 5, *EDN*, Feb. 5, 1974, p. 68; Mar. 5, 1974, p. 40; Apr. 5, 1974, p. 46; May 5, 1974, p. 63; June 5, 1974, p. 65.
2. Robinson, L.: Active Bandpass Filter with Adjustable Center Frequency and Constant Bandwidth, *EEE*, February 1968, p. 124.
3. Tobey, G. E., J. G. Graeme, and L. P. Huelsman: "Operational Amplifiers – Design and Applications," p. 291, McGraw-Hill Book Company, New York, 1971.

## 12.2 STATE-VARIABLE BANDPASS FILTER

**ALTERNATE NAMES** Biquadratic filter, biquad filter, active resonator, active filter, active *RC* filter, active bandpass amplifier.

**EXPLANATION OF OPERATION** Analog-computer technology provides a very stable bandpass filter shown in Fig. 12.2. This circuit requires three op amps

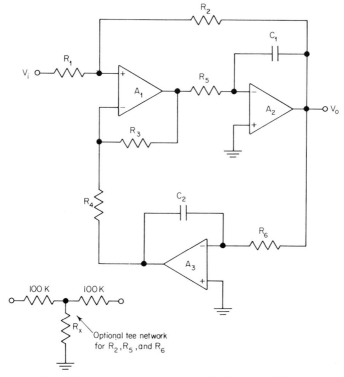

**Fig. 12.2** An inverting state-variable bandpass filter.

## 12-8 BANDPASS FILTERS

which, preferably, should be in one package for thermal tracking. This circuit has several advantages over single op amp bandpass filters. First, if components are properly selected, the passband center frequency $f_o$ can be made independent of circuit $Q$. Second, the sensitivity of $f_o$ and $Q$ to parameter variations is very low in the state-variable filter. Third, high circuit $Q$'s are possible ($Q \gg 5$).

The filter is made up of two integrators ($A_2$ and $A_3$) and a summing amplifier. The passband center frequency is

$$f_o = \frac{1}{2\pi}\left(\frac{R_3}{R_4 R_5 R_6 C_1 C_2}\right)^{1/2}$$

The circuit $Q$ is

$$Q = \frac{1 + R_2/R_1}{1 + R_3/R_4}\left(\frac{R_3 R_5 C_1}{R_4 R_6 C_2}\right)^{1/2}$$

If we initially set $R_3 = R_4$ (fixed resistors) and $C = C_1 = C_2$ (fixed capacitors), these equations reduce to

$$f_o = \frac{1}{2\pi(R_5 R_6 C^2)^{1/2}}$$

and

$$Q = \frac{(1 + R_2/R_1)(R_5/R_6)^{1/2}}{2}$$

Suppose we let $R_5$ and $R_6$ be ganged pots with identical resistances. In this case $R_5/R_6$ always equals unity and $Q$ depends only on $R_1$ and $R_2$. If the common value of $R_5$ and $R_6$ is $R$, the equations reduce to

$$f_o = \frac{1}{2\pi RC}$$

$$Q = \frac{R_1 + R_2}{2R_1}$$

The ganged pots $R_5$ and $R_6$ are used to set $f_o$ while $R_2$ is used for $Q$ adjustment.

The transfer function for the circuit shown in Fig. 12.2 is

$$A_{vc} = \frac{V_o}{V_i} = \frac{-sA}{s^2 + sB + C}$$

where $A = \dfrac{1}{R_5 C_1}\dfrac{1 + R_3/R_4}{1 + R_1/R_2}$

$B = \dfrac{1}{R_5 C_1}\dfrac{1 + R_3/R_4}{1 + R_2/R_1}$

$C = \dfrac{R_3}{R_4}\dfrac{1}{R_5 R_6 C_1 C_2}$

### DESIGN PARAMETERS

| Parameter | Description |
|---|---|
| $A_{vc}$ | Closed-loop voltage gain of circuit as a function of frequency |
| $C_1, C_2$ | Capacitors which help determine bandpass center frequency |
| $f_o$ | Bandpass center frequency |
| $H$ | Voltage gain of circuit at resonance |

## STATE-VARIABLE BANDPASS FILTER

| Parameter | Description |
|---|---|
| $I_b$ | Op amp input bias current |
| $Q$ | Quality factor of circuit |
| $R_1, R_2$ | Resistors which determine $Q$ of circuit |
| $R_3, R_4$ | Resistors which set gain of $A_1$ |
| $R_5, R_6$ | Resistors which control $f_o$ of circuit |
| $R_x$ | Resistor used in tee networks to make large simulated resistances in place of $R_5$ and $R_6$ |
| $S_A^B$ | Sensitivity of parameter $B$ to variations in parameter $A$ |
| $\omega_o$ | Radian frequency of passband ($\omega_o = 2\pi f_o$) |

## DESIGN EQUATIONS

| Eq. No. | Description | Equation |
|---|---|---|
| 1 | Passband center frequency | $f_o = \dfrac{1}{2\pi}\left(\dfrac{R_3}{R_4 R_5 R_6 C_1 C_2}\right)^{1/2}$ |
| 2 | Circuit $Q$ | $Q = \dfrac{1 + R_2/R_1}{1 + R_3/R_4}\left(\dfrac{R_3 R_5 C_1}{R_4 R_6 C_2}\right)^{1/2}$ |
| 3 | Circuit gain at $f_o$ | $H = R_2/R_1$ |
| 4 | Voltage gain of circuit (in classical form) | $A_{vc} = \dfrac{V_o}{V_i} = \dfrac{sH\omega_o/Q}{s^2 + s(\omega_o/Q) + \omega_o^2}$ where $\omega_o = 2\pi f_o$ $s = j2\pi f$ |
| 5 | Sensitivity of $f_o$ to component parameter variations | $S_{R_3}^{f_o} = \dfrac{1}{2}$ $S_{R_4}^{f_o} = S_{R_5}^{f_o} = S_{R_6}^{f_o} = -\dfrac{1}{2}$ $S_{C_1}^{f_o} = S_{C_2}^{f_o} = -\dfrac{1}{2}$ |
| 6 | Sensitivity of $Q$ to component parameter variations | $S_{R_1}^Q = S_{R_2}^Q = \dfrac{R_2}{R_1 + R_2}$ $S_{R_3}^Q = S_{R_4}^Q = \dfrac{1}{2} - \dfrac{R_3}{R_3 + R_4}$ $S_{R_5}^Q = S_{R_6}^Q = \dfrac{1}{2}$ $S_{C_1}^Q = S_{C_2}^Q = \dfrac{1}{2}$ |
| 7 | Voltage gain of circuit | $A_{vc} = \dfrac{V_o}{V_i} = \dfrac{-sA}{s^2 + sB + C}$ where $A = \dfrac{1}{R_5 C_1}\dfrac{1 + R_3/R_4}{1 + R_1/R_2}$ $B = \dfrac{1}{R_5 C_1}\dfrac{1 + R_3/R_4}{1 + R_2/R_1}$ $C = \dfrac{R_3}{R_4}\dfrac{1}{R_5 R_6 C_1 C_2}$ |

## 12-10  BANDPASS FILTERS

### DESIGN PROCEDURE

This circuit has two design requirements: (1) the passband center frequency $f_o$ and (2) the circuit $Q$. The gain $H$ at $f_o$ is fixed once $Q$ is chosen. The design steps to follow will show an optimum way to calculate circuit-component values using the required $f_o$ and $Q$. The sensitivity of $f_o$ and $Q$ to component parameter variations is dependent only on the quality of passive components used.

### DESIGN STEPS

*Step 1.* Compute nominal values for $R_3$ and $R_4$ using

$$R_3 = R_4 = \frac{10^8}{f_o}$$

These values are not critical and may be selected from one-half to twice the computed $R_3$. However, $R_3 = R_4$ must be maintained. If $A_1$ is a conventional bipolar monolithic op amp, do not allow $R_3 = R_4$ to go above 1 MΩ. If $A_1$ has a low $I_b$, $R_3$ and $R_4$ may be 10 MΩ or more.

*Step 2.* Select a common value for $C_1$ and $C_2$ in the vicinity of

$$C_1 = C_2 = \frac{10^{-7}}{f_o}$$

Again, as in step 1, a value for $C_1 = C_2$ from one-half to twice the computed value may be used.

*Step 3.* Compute the required common values for $R_5$ and $R_6$ from

$$R_5 = R_6 = \frac{1}{2\pi f_o C_1}$$

As in step 1, these resistors may cause offset problems if they become much larger than 1 MΩ (or 10 MΩ if $A_2$ and $A_3$ have low input bias currents). If this is a problem, $C_1$ and $C_2$ can be adjusted upward and $R_5 = R_6$ recalculated. If $C_1$ and $C_2$ are already too large, $R_5$ and $R_6$ can each be replaced with the tee network shown in Fig. 12.2. The value for $R_x$ in each tee network is

$$R_x = \frac{10^{10}}{R_5 - 2 \times 10^5}$$

Offsets at the output of $A_2$ can be further reduced by returning the noninverting input of $A_2$ to ground through a resistor with the same resistance as $R_5$. Likewise, $A_3$ offsets can be reduced by returning the noninverting input of $A_3$ to ground through a resistor equal to $R_6$.

*Step 4.* Set $R_1 = R_3$. Compute $R_2$ from

$$R_2 = R_1(2Q - 1)$$

*Step 5.* Use Eq. 3 to compute the circuit gain $H$ at resonance.

*Step 6.* Compute the sensitivity of $f_o$ to component parameter variations using Eq. 5.

*Step 7.* Compute the sensitivity of $Q$ to component parameter variations using Eq. 6.

**EXAMPLE OF BANDPASS-FILTER DESIGN** A filter with a center frequency of 100 Hz will be designed using the seven steps. This example will illustrate the ease with which this circuit can be designed even though the transfer function is quite complex.

*Design Requirements*
$f_o = 100$ Hz
$Q = 50$

*Device Data* ($\Delta T = -55$ to $+125°C$)

$\dfrac{\Delta R}{R} = +0.018$ (all resistors)

$\dfrac{\Delta C}{C} = -0.027$ (all capacitors)

Op amp type: quad 741
$I_b = 0.5 \ \mu A$ (max)

*Step 1.* Nominal values for $R_3$ and $R_4$ are $10^8/f_o = 10^8/100 = 1$ M$\Omega$.
*Step 2.* Nominal values for $C_1$ and $C_2$ are $10^{-7}/f_o = 10^{-7}/100 = 1,000$ pF.
*Step 3.* Required values for $R_5$ and $R_6$ are

$$R_5 = R_6 = \frac{1}{2\pi f_o C_1} = \frac{1}{2\pi (100) 10^{-9}} = 1.59 \text{ M}\Omega$$

These resistors will cause an output offset voltage in $A_2$ and $A_3$ of $I_b R_5 = 0.5 \times 10^{-6} \times 1.59 \times 10^6 = 0.8$ V. To prevent this offset at the output of $A_2$ and $A_3$, we will use two of the tee circuits shown in Fig. 12.2.

$$R_x = \frac{10^{10}}{R_5 - 2 \times 10^5} = \frac{10^{10}}{1.59 \times 10^6 - 2 \times 10^5} = 7,194 \ \Omega$$

As an additional guard against offsets we can return the noninverting inputs of $A_2$ and $A_3$ to ground through 100 k$\Omega$ + 7.194 = 107-k$\Omega$ resistors.

*Step 4.* We set $R_1 = R_3 = 1$ M$\Omega$. $R_2$ is found from

$$R_2 = R_1(2Q - 1) = 10^6(2 \times 50 - 1) = 99 \text{ M}\Omega$$

This value of $R_2$ appears to be impractical. We can use the same tee recommended in step 3 to solve this problem. The required value of $R_x$ is

$$R_x = \frac{10^{10}}{R_2 - 2 \times 10^5} = \frac{10^{10}}{9.9 \times 10^7 - 2 \times 10^5} = 101 \ \Omega$$

*Step 5.* The circuit gain $H$ at resonance is

$$H = \frac{R_2}{R_1} = \frac{9.9 \times 10^7}{10^6} = 99$$

*Step 6.* Sensitivity-function computations for $f_o$ variations are as follows:
RESISTORS

$$S^{f_o}_{R_3} = \frac{1}{2}$$

$$S^{f_o}_{R_4} = S^{f_o}_{R_5} = S^{f_o}_{R_6} = -\frac{1}{2}$$

The $\Delta R/R = +0.018$ ($-55$ to $+125°C$) specified for all resistors will cause $f_o$ to increase by $\frac{1}{2}(0.018)100$ Hz = 0.9 Hz owing to $R_3$. Resistors $R_4$, $R_5$, and $R_6$ will each cause $f_o$ to decrease by the same amount.

CAPACITORS

$$S^{f_o}_{C_1} = S^{f_o}_{C_2} = -\frac{1}{2}$$

The $\Delta C/C = -0.027$ specified for each capacitor will cause $f_o$ to increase by $-\frac{1}{2}(-0.027)100 \text{ Hz} = 1.35 \text{ Hz}$.

The total shift in $f_o$ due to parameter variations of the above six components will be $+0.9 -0.9 -0.9 -0.9 +1.35 +1.35 = +0.9$ Hz. This 0.9 percent positive frequency shift will occur as the temperature increases from $-55$ to $+125°C$.

*Step 7.* Sensitivity-function computations for $Q$ variations are as follows:

RESISTORS $R_1$ AND $R_2$

$$S_{R_1}^Q = S_{R_2}^Q = \frac{R_2}{R_1 + R_2} = \frac{9.9 \times 10^7}{10^6 + 9.9 \times 10^7} = 0.99$$

This will cause $Q$ to vary by $0.99(0.018)50 = 0.891$ as the temperature varies from $-55$ to $+125°C$.

RESISTORS $R_3$ AND $R_4$

$$S_{R_3}^Q = S_{R_4}^Q = \frac{1}{2} - \frac{R_3}{R_3 + R_4} = \frac{1}{2} - \frac{10^6}{10^6 + 10^6} = 0$$

RESISTORS $R_5$ AND $R_6$

$$S_{R_5}^Q = S_{R_6}^Q = \frac{1}{2}$$

The change in $Q$ due to this sensitivity function is $\frac{1}{2}(0.018) \times 50 = 0.45$.

CAPACITORS

$$S_{C_1}^Q = S_{C_2}^Q = \frac{1}{2}$$

The change in $Q$ due to changes in each capacitance is $\frac{1}{2}(-0.027)50 = -0.675$.

The total change in $Q$ as the temperature varies from $-55$ to $+125°C$ will be

$$0.891 + 0.891 + 0.45 + 0.45 - 0.675 - 0.675 = 1.332$$

This is a 2.7 percent change in $Q$.

## REFERENCES

1. Kerwin, W. J., L. P. Huelsman, and R. W. Newcomb: State Variable Synthesis for Insensitive Integrated Circuit Transfer Functions, *IEEE J. Solid-State Circuits*, vol. SC-2, pp. 87–92, September 1967.
2. Tobey, G. E., J. G. Graeme, and L. P. Huelsman: "Operational Amplifiers—Design and Applications," p. 307, McGraw-Hill Book Company, New York, 1971.

Chapter **13**

# Bandstop Filters

## INTRODUCTION

These circuits are especially useful in systems containing unwanted signals of fixed frequencies. Often the only way to eliminate the unwanted signals is to pass the main signal through several bandstop filters. Each filter will reduce one unwanted frequency to a tolerable level. Since no filter is ideal, frequencies on either side of the bandstop frequency $f_o$ will also be slightly affected. If the response curve of the filter is very sharp, its effect on the system may be minimal. However, a highly stable, very sharp bandstop (or notch) filter is practical only if the unwanted signal frequency is also very stable.

## 13.1 ACTIVE INDUCTOR BANDSTOP FILTER

**ALTERNATE NAMES** Notch filter, active bandstop filter, active $RC$ notch filter, parasitic suppressor, hum-reduction circuit.

**EXPLANATION OF OPERATION** The circuit shown in Fig. 13.1 provides unity gain for all frequencies from dc to $f_u$ except at $f_o$, the notch frequency. The voltage gain of the circuit at $f_o$ may be 50 or 60 dB below unity with a careful selection of components. The notch frequency is tuned using $C_1$ or $C_2$. These components actually affect both notch frequency and notch depth. The final notch depth (sharpness) is controlled using $R_1$. $C_1$ or $C_2$ may become fairly large with a low notch frequency. In this case $C_1$ or $C_2$ may use a large fixed capacitor in parallel with a trimmer capacitor.

The transfer function of the circuit is

$$A_{vc} = \frac{V_o}{V_i} = \frac{j2\pi C_2 R^2(f^2 - f_o^2)}{f(R + R_3) + j2\pi C_2 R^2(f^2 - f_o^2)}$$

where $R = R_4 = R_5$ and $f_o$, the notch frequency, is

$$f_o = \frac{1}{2\pi R(C_1 C_2)^{1/2}}$$

When $f = f_o$, the numerator goes to zero and the gain $A_{vc}$ ideally should also equal zero. In practice, the gain of this circuit will be 50 or 60 dB ($\frac{1}{316}$ to $\frac{1}{1,000}$) below unity if the following relationship is exactly satisfied:

## 13-2 BANDSTOP FILTERS

$$\frac{R_1}{R_2} = \frac{R_3}{R_4 + R_5} = \frac{R_3}{2R}$$

The notch-sharpness adjustment $R_1$ is used to satisfy this equation.

The bandstop filter shown in Fig. 13.1 is a variation of the basic differential amplifier shown in Fig. 9.1. There are two differences between the circuits:

1. The resistor going to ground ($R_4$ in Fig. 9.1) has been replaced with $C_1$, $C_2$, $R_4$, $R_5$, and $A_2$. This network simulates a series $RLC$ network. At its resonant frequency $f_o$ it becomes a pure resistance $R_4 + R_5$.

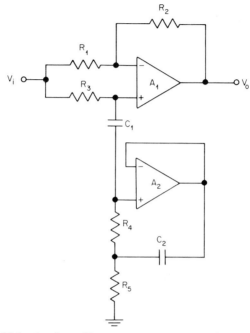

**Fig. 13.1** Bandstop filter which utilizes an active inductor.

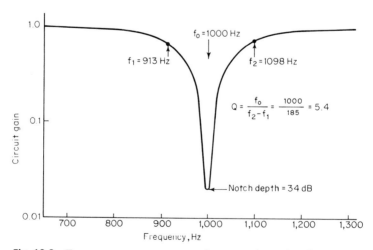

**Fig. 13.2** Frequency-response curve of active-inductor bandstop filter.

## ACTIVE INDUCTOR BANDSTOP FILTER 13-3

2. Both differential inputs in Fig. 13.1 are tied together. Therefore, if this were a balanced differential amplifier (as in Fig. 9.1), the output would be zero. For this circuit (Fig. 13.1), zero output occurs only at the resonant frequency where $R_1/R_2 = R_3/(R_4 + R_5)$. At all other frequencies the differential amplifier is out of balance and the circuit gain is +1.

## DESIGN PARAMETERS

| Parameter | Description |
|---|---|
| $A_1$ | Op amp for differential amplifier |
| $A_2$ | Op amp for simulated inductance |
| $A_{vc}$ | Voltage gain of circuit |
| $C_1$ | Capacitor which is C portion of effective series $RLC$ circuit. This part affects $f_o$ only |
| $C_2$ | Part of simulated inductance in $RLC$ circuit. This part affects both $f_o$ and $Q$ |
| $f_o$ | Notch center frequency |
| $f_u$ | Unity-gain crossover frequency of $A_1$ |
| $f_1, f_2$ | Frequencies where circuit response is 3 dB below the frequencies far removed from $f_o$ |
| $Q$ | Quality factor of circuit $Q = f_o/(f_2 - f_1)$ |
| $R_1, R_2, R_3$ | Gain-determining resistors |
| $R_4, R_5$ | Resistive portion of effective $RLC$ circuit |
| $R$ | Common value for $R_4$ and $R_5$ ($R = R_4 = R_5$) |
| $R_{in}$ | Input resistance of circuit |
| $V_i$ | Input voltage to circuit |
| $V_o$ | Output voltage from circuit |

## DESIGN EQUATIONS

| Eq. No. | Description | Equation |
|---|---|---|
| 1 | Voltage gain of circuit | $A_{vc} = \dfrac{V_o}{V_i} = \dfrac{j2\pi C_2 R^2 (f^2 - f_o^2)}{f(R + R_3) + j2\pi C_2 R^2 (f^2 - f_o^2)}$ where $R = R_4 = R_5$ |
| 2 | Notch frequency | $f_o = \dfrac{1}{2\pi R(C_1 C_2)^{1/2}}$ |
| 3 | Resistor ratios required for proper operation | $\dfrac{R_1}{R_2} = \dfrac{R_3}{R_4 + R_5} = \dfrac{R_3}{2R}$ |
| 4 | $Q$ of circuit | $Q = \dfrac{\pi f_o C_2 R}{2}$ |
| 5 | Input resistance of circuit at $f_o$ | $R_{in} = \dfrac{R_1(R_3 + 2R)}{R_1 + R_3 + 2R}$ |
| 6 | Input resistance of circuit at all frequencies not near $f_o$ | $R_{in} = R_1$ |

## DESIGN PROCEDURE

In order to simplify the calculations for this bandstop filter, we begin by fixing the nominal value of all resistors. The two capacitors then depend only on the resistor choices, the notch frequency, and the notch sharpness.

## 13-4 BANDSTOP FILTERS

### DESIGN STEPS

*Step 1.* Set $R_1 = R_2 = R_3 = 2R_{in}$ where $R_{in}$ is equal to or above the minimum required input resistance.
*Step 2.* Set $R = R_4 = R_5 = R_1/2$.
*Step 3.* Compute $C_2 = 2Q/\pi f_o R$.
*Step 4.* Compute $C_1 = 1/[(2\pi f_o R)^2 C_2]$. The smaller of $C_1$ or $C_2$ may be used to tune $f_o$.
*Step 5.* Compute Eq. 2 to verify that the correct notch frequency has been implemented.

**EXAMPLE OF BANDSTOP-FILTER DESIGN** We will design a medium-$Q$ bandstop filter for 1,000 Hz to illustrate the five design steps numerically.

*Design Requirements*
$f_o = 1,000$ Hz
$Q = 5$
$R_{in}(\min) = 10,000$ Ω

*Step 1*
$$R_1 = R_2 = R_3 = 2(R_{in}) = 2(10,000\ \Omega) = 20,000\ \Omega$$

*Step 2*
$$R = R_4 = R_5 = \frac{R_1}{2} = \frac{20,000}{2} = 10,000\ \Omega$$

*Step 3*
$$C_2 = \frac{2Q}{\pi f_o R} = \frac{10}{\pi 1,000 \times 10^4} = 0.318\ \mu F$$

*Step 4*
$$C_1 = \frac{1}{(2\pi f_o R)^2 C_2} = \frac{1}{(2\pi 10^3 \times 10^4)^2\ 3.18 \times 10^{-7}} = 795\ \text{pF}$$

*Step 5*
$$f_o = \frac{1}{2\pi R (C_1 C_2)^{1/2}} = \frac{1}{2\pi (10^4)(0.159 \times 10^{-6} \times 1.59 \times 10^{-9})^{1/2}} = 1,000\ \text{Hz}$$

### REFERENCE

1. Harris, R. J.: The Design of an Operational Amplifier Notch Filter, *Proc. IEEE*, October 1968, p. 1722.

### 13.2 TWIN-TEE BANDSTOP FILTER

**ALTERNATE NAMES** Notch filter, active bandpass filter, active $RC$ notch filter, parasitic suppressor, hum-reduction circuit.

**EXPLANATION OF OPERATION** The circuit shown in Fig. 13.3 provides a means to adjust circuit $Q$ without affecting notch frequency. The circuit $Q$ is adjustable from approximately 0.3 to 50 using $R_4$. The minimum $Q$ is obtained when the $R_4$ wiper is at ground potential. Notch depth and frequency are controlled with the six components in the twin tee. The basic six-component twin tee ($C_1$, $C_2$, $C_3$, $R_1$, $R_2$, and $R_3$) typically provides a maximum $Q$ of 0.3, approximately. $A_1$ and $A_2$ provide bootstrapping back to the twin-tee ground

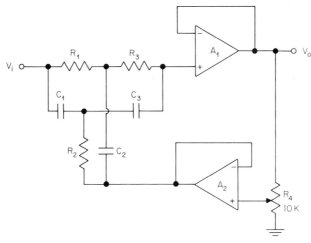

**Fig. 13.3** Twin-tee bandstop filter with adjustable $Q$.

point, thus making a maximum $Q$ of 50 possible. Figure 13.4 shows the range of adjustment $R_4$ provides.

The transfer function of the circuit is

$$A_{vc} = -\frac{V_o}{V_i} = \frac{s^3 + As^2 + Bs + C}{s^3 + Ds^2 + Es + C}$$

where $A = \dfrac{R_2(R_1 + R_3)C_1C_3}{\Delta}$

$B = \dfrac{R_2(C_1 + C_3)}{\Delta}$

$C = \dfrac{1}{\Delta}$

$D = \dfrac{R_2(R_1 + R_3)C_1C_3 + R_1R_3C_2C_3 + R_1R_2C_2(C_1 + C_3)}{\Delta}$

$E = \dfrac{R_2(C_1 + C_3) + R_1C_2 + (R_1 + R_3)C_3}{\Delta}$

$\Delta = R_1R_2R_3C_1C_2C_3$

The notch frequency is

$$f_o = \frac{1}{2\pi}\left(\frac{C_1 + C_3}{C_1C_2C_3R_1R_3}\right)^{1/2}$$

The design of this circuit is simplified if the following relations are used:

$R_1 = R_3 = 2R_2$

$C_1 = C_3 = \dfrac{C_2}{2}$

$R_1C_1 = R_2C_2 = R_3C_3$

## 13-6 BANDSTOP FILTERS

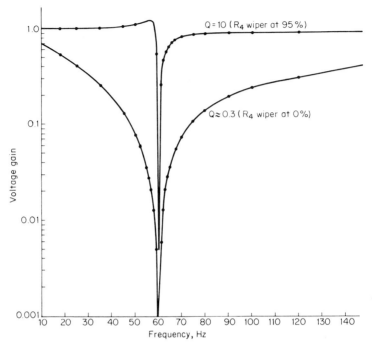

**Fig. 13.4** Gain as a function of frequency for the twin-tee bandstop filter.

### DESIGN PARAMETERS

| Parameter | Description |
|---|---|
| $A_1$ | Buffer amplifier with high input resistance which does not load twin-tee network |
| $A_2$ | Low-output-resistance buffer which bootstraps ground return of twin tee to circuit output voltage |
| $A_{vc}$ | Voltage gain of circuit |
| $C_1, C_2, C_3$ | Capacitors which determine notch frequency |
| $f_o$ | Notch frequency |
| $Q$ | Quality factor of circuit |
| $R_1, R_2, R_3$ | Resistors which determine notch frequency |
| $R_4$ | Potentiometer used to adjust the $Q$ |
| $R_g$ | Generator output resistance |
| $V_i$ | Circuit input voltage |
| $V_o$ | Circuit output voltage |

### DESIGN EQUATIONS

| Eq. No. | Description | Equation |
|---|---|---|
| 1 | Voltage gain of circuit | $A_{vc} = \dfrac{V_o}{V_i} = \dfrac{s^3 + Ds^2 + Es + C}{s^3 + Ds^2 + Es + C}$ <br> where $A = \dfrac{R_2(R_1 + R_3)C_1C_3}{\Delta}$ |

| Eq. No. | Description | Equation |
|---|---|---|
| | | $B = \dfrac{R_2(C_1 + C_3)}{\Delta}$ |
| | | $C = \dfrac{1}{\Delta}$ |
| | | $D = \dfrac{R_2(R_1 + R_3)C_1C_3 + R_1R_3C_2C_3 + R_1R_2C_2(C_1 + C_3)}{\Delta}$ |
| | | $E = \dfrac{R_2(C_1 + C_3) + R_1C_2 + (R_1 + R_3)C_3}{\Delta}$ |
| | | $\Delta = R_1R_2R_3C_1C_2C_3$ |
| 2 | Notch frequency | $f_o = \dfrac{1}{2\pi}\left(\dfrac{C_1 + C_3}{C_1C_2C_3R_1R_3}\right)^{1/2}$ |
| 3 | Recommended relationship between resistors | $R_1 = R_3 = 2R_2 \geq 100\, R_g$ |
| 4 | Recommended relationship between capacitors | $C_1 = C_3 = \dfrac{C_2}{2}$ |
| 5 | Recommended relationships between resistors and capacitors | $R_1C_1 = R_2C_2 = R_3C_3$ |

## DESIGN PROCEDURE

Since the equations describing this circuit are so complex, a simplified approach is necessary. The steps can be used for a first-cut design, with refinements made afterward.

### DESIGN STEPS

*Step 1.* Choose $R_1 = R_3$ equal to a practical value greater than $100\, R_g$. Set $R_2 = R_1/2$.

*Step 2.* $C_1$ and $C_3$ are found by combining Eqs. 2 through 5:

$$C_1 = C_3 = \frac{1}{4\pi f_o R_2}$$

*Step 3.* Using Eq. 4, we now determine $C_2$ and $C_3$:

$$C_2 = 2C_1 \qquad C_3 = C_1$$

*Step 4.* Verify that $f_o$ is correct using Eq. 2:

$$f_o = \frac{1}{2\pi}\left(\frac{C_1 + C_3}{C_1C_2C_3R_1R_3}\right)^{1/2}$$

**EXAMPLE OF BANDSTOP-FILTER DESIGN** The design of a 60-Hz bandstop filter using the four design steps will be presented.

*Design Requirements*
 $f_o = 60$ Hz
 Largest resistor = 2 MΩ

## 13-8 BANDSTOP FILTERS

*Design Parameters*
$R_g = 600\ \Omega$

**Step 1.** We first compute $100\ R_g = 100(600) = 600\ \text{k}\Omega$. Set $R_1 = R_3 = 2\ \text{M}\Omega$ to satisfy Eq. 3. We then set $R_2 = R_1/2 = 1\ \text{M}\Omega$.

**Step 2.** We determine $C_1$ and $C_3$ as follows:

$$C_1 = C_3 = \frac{1}{4\pi f_o R_2} = \frac{1}{4\pi(60)10^6} = 1{,}320\ \text{pF}$$

**Step 3.** $C_2$ is simply $C_2 = 2C_1 = 2(1{,}320) = 2{,}640\ \text{pF}$.

**Step 4.** The resonant frequency is double-checked:

$$f_o = \frac{1}{2\pi}\left(\frac{C_1 + C_3}{C_1 C_2 C_3 R_1 R_3}\right)^{1/2}$$

$$= \frac{1}{2\pi}\left(\frac{1.32 \times 10^{-9} + 1.32 \times 10^{-9}}{1.32 \times 10^{-9} \times 2.64 \times 10^{-9} \times 1.32 \times 10^{-9} \times 2 \times 10^6 \times 2 \times 10^6}\right)^{1/2}$$

$$= 60.2\ \text{Hz}$$

## REFERENCES

1. Dobkin, B.: "High Q Notch Filter," National Semiconductor Corp. Linear Brief LB-5, 1969.
2. Ramey, R. L., and E. J. White: "Matrices and Computers in Electronic Circuit Analysis," p. 36, McGraw-Hill Book Company, New York, 1971.

Chapter **14**

# Frequency Control

## INTRODUCTION

In this chapter we present design information on several circuits which manipulate frequency or perform a function related to frequency. The first circuit presents a method for doubling the input frequency. The second circuit provides the designer with an analog voltage proportional to the difference between two input frequencies. Other chapters in this handbook contain more frequency-related circuits. A frequency-to-voltage converter, or FM detector, is described in Chap. 7. A frequency divider, or bistable multivibrator, is shown in Chap. 20. Chapter 21 describes a voltage-to-frequency converter (voltage-controlled oscillator).

## 14.1 FREQUENCY DOUBLER

**ALTERNATE NAMES** Frequency multiplier, harmonic generator.

**EXPLANATION OF OPERATION** The circuit shown in Fig. 14.1 is a practical frequency doubler since it requires only one +5-V supply. $A_1$ and $A_3$ are comparators which operate from a wide range of supply voltages. $A_2$ is an integrator with dc negative feedback which forces $v_2$ to stay out of saturation. The output IC is an exclusive OR gate.

The timing diagram in Fig. 14.2 is useful to visualize how the circuit operates. Both $A_1$ and $A_3$ are operated as level-detecting comparators. $A_1$ is set to trip at $V_C$ and $A_3$ is set to trip at $V_B$. Resistors $R_1$ to $R_2$ provide hysteresis (positive feedback) in the $A_1$ circuit. This forces the $A_1$ output voltage to make clean transitions between states each time $v_i$ passes through the trip point. In many applications we can let $V_C$ = zero volts. The trip voltage $V_C$ for a raising waveform [zero to $v_i$(peak)] is slightly higher than $V_A$ because of the hysteresis:

$$V_C = V_A + \frac{5R_1}{R_1 + R_2 + R_3}$$

For a falling waveform [$v_i$(peak) to zero] the trip voltage is $V_A$, since $v_1 = 0$ and no hysteresis voltage is fed back.

In $A_3$, the output voltage $v_3$ changes states each time $v_2$ passes through $V_B$.

14-1

## 14-2 FREQUENCY CONTROL

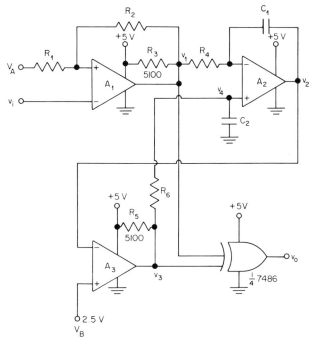

**Fig. 14.1** A frequency doubler with a TTL-compatible output.

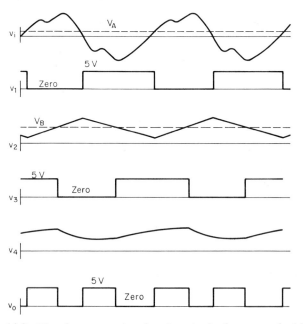

**Fig. 14.2** Waveforms at various locations in the frequency doubler.

# FREQUENCY DOUBLER

The outputs from $A_1$ and $A_3$ are rectangular waveforms with duty cycles which depend on the relationship of $V_A$ to $v_i$ and $V_B$ to $v_2$.

Since $A_2$ is an integrator, $v_2$ is a triangular waveform. Voltage $v_3$ will be a symmetrical square wave (50 percent duty cycle) only if the average voltage of $v_2$ equals $V_B$. The input voltage $v_i$ may not be symmetrical, so $V_A$ may be set to obtain the required symmetry.

The dc feedback voltage $v_4$ is equal to the average value of $v_3$. This averaging is done with a brute-force filter composed of $R_6$ and $C_2$. The dc feedback voltage forces the average value of $v_3$ to equal the average value of $v_1$. Both $v_1$ and $v_3$ are rectangular waveforms with an ON voltage of +5 V and an OFF voltage of zero. Therefore, $v_1$ and $v_3$ will have identical ON and OFF times if the $v_4$ feedback is utilized. With the integrator in the loop, however, $v_3$ will be delayed by one-half of the $v_1$ ON time.

The circuit output voltage $v_o$ is high when $v_3$ or $v_1$ is high, but not when both are high. Since $v_i$ may not be a 50 percent duty cycle waveform, $v_o$ will not be a 50 percent waveform. If a 50 percent duty cycle is required, $V_A$ can be appropriately adjusted.

## DESIGN PARAMETERS

| Parameter | Description |
|---|---|
| $A_1, A_3$ | Comparators or op amps used as comparators |
| $A_2$ | Dual-input integrator |
| $A_{v3}$ | Open-loop dc voltage gain of $A_3$ |
| $C_1, C_2$ | Integrating capacitors |
| $f_i(\min)$ | Limiting lower frequency where integrator output voltage reaches $\pm$ saturation |
| $f_i(\max)$ | Limiting upper frequency where output voltage changes of $A_2$ are not sufficient to switch $A_3$ on and off |
| $I_{b1}$ | Input bias current of $A_1$ |
| $I_{b2}$ | Input bias current of $A_2$ |
| $R_1, R_2$ | Resistors which determine hysteresis of $A_1$ |
| $R_3, R_5$ | Output pull-up resistors of $A_1$ and $A_3$ (required only if $A_1$ and $A_3$ have open collector outputs) |
| $R_4, R_6$ | Integrating resistors |
| $V_A, V_B$ | Bias voltages |
| $V_C$ | Bias voltage plus hysteresis for $A_1$ |
| $v_i$ | Circuit input voltage |
| $v_o$ | Circuit output voltage |
| $v_1$ to $v_4$ | Voltage waveforms at various locations in circuit |
| $v_1(\text{on})$ | Peak voltage of $v_1$ |
| $v_1(\text{off})$ | Minimum voltage of $v_1$ |

## DESIGN EQUATIONS

| Eq. No. | Description | Equation |
|---|---|---|
| 1 | Trip voltage of $A_1$ for rising $v_i$ waveform | $V_C = V_A + \dfrac{5R_1}{R_1 + R_2 + R_3}$ |
| 2 | Integrator output voltage | $v_2 = \dfrac{\int (v_4 - v_1)dt}{(R_3 + R_4)C_2} + v_4$ |
| 3 | Nominal resistor values | $R_1 \ll \dfrac{V_A}{I_{b1}}$ <br> The minimum $R_1$ should be $\approx 1\ \text{k}\Omega$ |

## 14-4 FREQUENCY CONTROL

| Eq. No. | Description | Equation |
|---|---|---|
| 4 | | $R_2 = \dfrac{5R_1 - (R_1 + R_3)(V_C - V_A)}{V_C - V_A}$ |
| 5 | | $R_3 = 1$ to $10$ k$\Omega$ |
| 6 | | $R_4 \ll \dfrac{5}{I_{b2}} - R_3$ |
| 7 | | $R_5 = 1$ to $10$ k$\Omega$ |
| 8 | | $R_6 \ll \dfrac{5}{I_{b2}} - R_5$ |
| 9 | Nominal capacitor values | $C_1 \approx \dfrac{1}{(R_3 + R_4)f_i(\min)}$ |
| 10 | | $C_2 \approx \dfrac{5}{(R_5 + R_6)f_i(\min)}$ |
| 11 | Minimum frequency of operation | $f_i(\min) \approx \dfrac{1}{(R_3 + R_4)C_1}$ |
| 12 | Maximum frequency of operation | $f_i(\max) \approx \dfrac{v_i(\text{on})A_{v3}}{2R_4 C_1 v_3(\text{on})}$ |

### DESIGN PROCEDURE

If this circuit is designed around one supply voltage (+5 V), several parts can be saved. Design equations 1 through 5 have assumed this simplification. We will also assume a symmetrical sine wave is driving the circuit and that a TTL-compatible output is required.

### DESIGN STEPS

*Step 1.* Choose a $V_C$ which is 0.1 to 10 percent larger than $V_A$. A large $V_C$ (1 to 10 percent) is required only if $v_i$ contains excessive noise. Select $R_1$ such that it satisfies Eq. 3:

$$R_1 \approx \frac{10^{-3} V_A}{I_{b1}}$$

*Step 2.* Assuming 111-type comparators are to be used, we can let $R_3 = R_5 = 5{,}100 \ \Omega$. Compute the required $R_2$ using Eq. 4:

$$R_2 = \frac{5R_1 - (R_1 + R_3)(V_C - V_A)}{V_C - V_A}$$

*Step 3.* Compute $R_4$ and $R_6$ from Eqs. 6 and 8:

$$R_4 = \frac{10^{-3}}{I_{b2}} - R_3$$

$$R_6 = \frac{10^{-3}}{I_{b2}} - R_5$$

*Step 4.* Capacitor values depend on the minimum input frequency $f_i(\min)$. Recommended capacitor values come from application of Eqs. 9 and 10:

$$C_1 = \frac{1}{(R_3 + R_4)f_i(\min)}$$

$$C_2 = \frac{5}{(R_5 + R_6)f_i(\min)}$$

**EXAMPLE OF FREQUENCY-DOUBLER DESIGN** Suppose we want a frequency doubler for an audio application. Electronic music synthesizers require frequency doublers (or multipliers) to generate "rich" sounds. The input-output voltages and frequency range must be compatible with standard studio practice.

*Design Requirements*
$f_i(\min) = 20$ Hz
$v_o$ = TTL compatible
$V_i(\text{noise}) = 0.1$-V peak
$v_i$ = sine wave varying from zero to +4 V (NOTE: Many comparators do not allow the input voltage to exceed the limits of the two supply voltages. If our supply voltages are +5 V and ground, $v_i$ must be within the 0- to +5-V range at all times)

*Device Data*
$I_{b1} = 10^{-7}$ A (111 comparator)
$I_{b2} = 2 \times 10^{-9}$ A (108 op amp)

Step 1. Our choice for $V_A$ will be ground potential with $V_C = 0.2$ V, since the peak noise in $v_i$ is 0.1 V. This small offset will not make $v_1$ very nonsymmetrical. $R_1$ is computed from

$$R_1 \approx \frac{10^{-3} V_A}{I_{b1}} = 0$$

We will set $R_1 = 1{,}000 \; \Omega$.

Step 2. $R_2$ is determined from

$$R_2 = \frac{5R_1 - (R_1 + R_3)(V_C - V_A)}{V_C - V_A}$$

$$= \frac{5(1{,}000) - (1{,}000 + 5{,}100)(0.2 - 0)}{0.2 - 0} = 18.9 \text{ k}\Omega$$

We will let $R_1 = 18$ k$\Omega$.

Step 3. $R_4$ is determined from Eq. 6:

$$R_4 = \frac{10^{-3}}{I_{b2}} - R_3 = \frac{10^{-3}}{2 \times 10^{-9}} - 5{,}100$$

$$= 495 \text{ k}\Omega$$

We will use $R_4 = 470$ k$\Omega$. The calculation for $R_6$ is identical, so we also set $R_6 = 470$ k$\Omega$.

Step 4. Capacitor values are as follows:

$$C_1 = \frac{1}{(R_3 + R_4)\,f_i(\min)} = \frac{1}{(5{,}100 + 470{,}000)\,20} = 0.1 \; \mu\text{F}$$

$$C_2 = \frac{5}{(R_5 + R_6)\,f_i(\min)} = \frac{5}{(5{,}100 + 470{,}000)\,20} = 0.5 \; \mu\text{F}$$

## 14-6 FREQUENCY CONTROL

**REFERENCE**

1. "Linear Applications," p. AN41-4, National Semiconductor Corp., January 1972.

## 14.2 FREQUENCY-DIFFERENCE DETECTOR

**ALTERNATE NAMES** Frequency comparator, bipolar frequency-difference detector, heterodyne circuit, frequency discriminator.

**EXPLANATION OF OPERATION** While this circuit is less accurate than a digital frequency-difference detector, it has several advantages:
1. It can interface directly with analog circuits on the input and output.
2. It provides both sign and magnitude of the difference frequency.
3. It is simple to understand and requires only a few parts.

As shown in Fig. 14.3, each input line triggers a single shot once per cycle. These output pulses are all of identical amplitude and duration. One pulse train drives the noninverting input of a low-pass filter and the other pulse train drives the inverting input of the same low-pass filter. The dc output of the low-pass filter will therefore be proportional to the difference of the average values of these two pulse trains. Since the average value of each pulse train is proportional to that particular input frequency, $v_o$ will be proportional to the difference in the two input frequencies.

If a 74123 dual single shot is utilized, triggering can be performed on

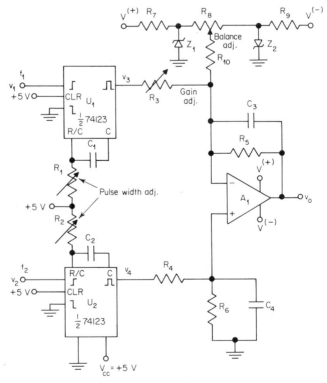

**Fig. 14.3** A frequency-difference detector.

either a rising or falling waveform. Figure 14.4 assumes triggering on the rising waveform. For a rising waveform the output pulse will begin (at the latest) when the input rises to 2 V. If the trailing-edge input is used, triggering occurs (at the latest) when the input signal falls to 0.8 V.

The output pulses from the two single shots should be identical in all respects if balanced operation is desired. Since the dual single shot is operating from a +5-V supply, both pulse trains will have identical amplitudes. The OFF amplitudes of $v_3$ and $v_4$ are also required to be identical, since the low-pass filter responds to the average value of each pulse train. The average value of each pulse train is

$$v_3(\text{av}) = v_3(\text{on}) \, T_1 f_1 + v_3(\text{off})$$

and

$$v_4(\text{av}) = v_4(\text{on}) \, T_2 f_2 + v_4(\text{off})$$

The dc output voltage from the circuit is

$$v_o(\text{dc}) = \frac{R_6}{R_4} [v_4(\text{on}) T_2 f_2 + v_4(\text{off})] - \frac{R_5}{R_3} [v_3(\text{on}) T_1 f_1 + v_3(\text{off})] + V_{oo}$$

where $R_6/R_4$ is the op amp noninverting gain, $-R_5 R_3$ is the op amp inverting gain, and $V_{oo}$ is the op amp output offset voltage due to $V_{io}$ and $I_{io}$. In practice we attempt to set $R_6/R_4 = R_5/R_3$, $v_3(\text{on}) = v_4(\text{on})$, $T_1 = T_2$, and $v_3(\text{off}) = v_4(\text{off})$. If we also adjust $R_8$ to cancel out the effects of $V_{oo}$, the dc output voltage reduces to

$$v_o(\text{dc}) = \frac{R_6 v_4(\text{on}) T_2 (f_2 - f_1)}{R_4}$$

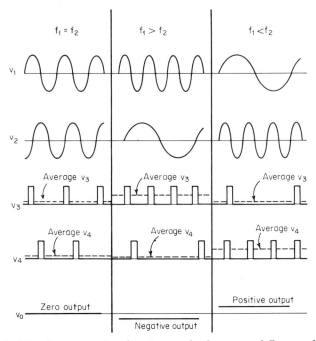

**Fig. 14.4** Waveforms at various locations in the frequency-difference detector.

## 14-8 FREQUENCY CONTROL

These conditions can always be met if $R_1$, $R_3$, and $R_8$ are made adjustable. The pulse widths are typically set to be 0.1 to 0.2 of the smaller of $1/f_1$ or $1/f_2$. The pulse widths as given on the 74123 data sheet are:

$$T_1 \approx 0.25\, R_1 C_1 \text{ s}$$
$$T_2 \approx 0.25\, R_2 C_2 \text{ s}$$

### DESIGN PARAMETERS

| Parameter | Description |
|---|---|
| $A_1$ | Dual-input low-pass filter |
| $C_1, C_2$ | Determines pulse widths of $T_1$ and $T_2$ |
| $C_3, C_4$ | Determines low-pass properties of $A_1$ circuit |
| $f_1$ | Reference frequency |
| $f_2$ | Unknown frequency |
| $f_{max}$ | Maximum of $f_1$ or $f_2$ |
| $I_b$ | $A_1$ input bias current |
| $I_{io}$ | $A_1$ input offset current |
| $I_{Z1}, I_{Z2}$ | Nominal recommended currents for $Z_1$ and $Z_2$ |
| $R_1, R_2$ | Determines pulse widths of $T_1$ and $T_2$ |
| $R_3$ to $R_6$ | Determines gain and low-pass properties of $A_1$ circuit |
| $R_7$ to $R_{10}$ | Offset adjustment circuit to set $v_o(\text{dc}) = 0$ when $f_1 = f_2$ |
| $T_1, T_2$ | Widths of pulses in $v_3$ and $v_4$ pulse trains |
| $U_1, U_2$ | Single-shot multivibrators |
| $v_o(\text{dc})$ | Circuit output dc voltage |
| $v_1$ to $v_4$ | Voltage waveforms (see Fig. 14.4) |
| $V_{io}$ | Input offset voltage of $A_1$ |
| $V_{oo}$ | Output offset voltage of $A_1$ |

### DESIGN EQUATIONS

| Eq. No. | Description | Equation |
|---|---|---|
| 1 | Circuit output voltage assuming all error sources and offsets are trimmed out with $R_1$, $R_3$, and $R_8$ | $v_o(\text{dc}) = \dfrac{R_6 v_4(\text{on}) T_2 (f_2 - f_1)}{R_4}$ |
| 2 | Pulse width of $T_1$ (ON time) | $T_1 \approx 0.25 R_1 C_1$ s |
| 3 | Pulse width of $T_2$ (ON time) | $T_2 \approx 0.25 R_2 C_2$ s |
| 4 | Maximum $T_1$ and $T_2$ | $T_1 = T_2 \leq \dfrac{0.2}{f_{max}}$ |
| 5 | Average value of $v_3$ | $v_3(\text{av}) = v_3(\text{on}) T_1 f_1 + v_3(\text{off})$ |
| 6 | Average value of $v_4$ | $v_4(\text{av}) = v_4(\text{on}) T_2 f_2 + v_4(\text{off})$ |
| 7 | Required gain-setting resistor ratios | $\dfrac{R_5}{R_3} = \dfrac{R_6}{R_4} = \dfrac{v_o(\text{max})}{v_4(\text{on}) T_2 \lvert f_2 - f_1 \rvert_{max}}$ |
| 8 | Offset-adjustment network resistor relationships | $R_7 = R_9$ <br> $R_8 \gg R_7$ <br> $R_{10} \approx R_8$ <br> $R_{10} \gg R_3$ |

## FREQUENCY-DIFFERENCE DETECTOR  14-9

| Eq. No. | Description | Equation |
|---|---|---|
| 9 | Recommended size of $R_3$ | Minimum load resistor allowed for single shot $< R_3 \ll \dfrac{v_3(\text{on})}{I_b}$ |
| 10 | Recommended sizes for $C_3$ and $C_4$ | $C_3 = C_4 \gg \dfrac{1}{2\pi f_{\max} R_5}$ |
| 11 | Recommended size for $R_1$ and $R_2$ | $5\text{ k}\Omega < R_1 = R_2 < 50\text{ k}\Omega$ (if using a 74123) |
| 12 | Output offset voltage of $A_1$ | $V_{oo} = V_{io}\left(1 + \dfrac{R_5}{R_3}\right) + I_{io} R_5$ |

### DESIGN PROCEDURE

A dual single shot from the medium-power TTL family is assumed in this procedure. If op amp single shots are utilized, Chap. 20 should be consulted.

### DESIGN STEPS

*Step 1.* Choose pulse widths $T_1$ and $T_2$ which satisfy Eq. 4:

$$T_1 = T_2 \leq \frac{0.2}{f_{\max}}$$

*Step 2.* Select $R_1$ and $R_2$ in the 5- to 50-k$\Omega$ range. The lower values of $R_1$ and $R_2$ allow $T_1$ and $T_2$ to extend down to 40 ns. Calculate values for $C_1$ and $C_2$ from

$$C_1 = C_1 \approx \frac{4T_1}{R_1}$$

The closest practical capacitor size to those calculated above can be used. The resistors $R_1$ and $R_2$ can be readjusted accordingly.

*Step 3.* Calculate a value for $R_3$ using Eq. 9:

$$R_3 \ll \frac{v_3(\text{on})}{I_b}$$

A minimum value for $R_3$ is 470 $\Omega$ if the 74123 single shot is used. If $R_3$ is adjustable, a 470-$\Omega$ resistor should be placed in series with it.

*Step 4.* The value of $R_5$ is found using Eq. 7:

$$R_5 = \frac{R_3\, v_o(\max)}{v_4(\text{on}) T_2 |f_2 - f_1|_{\max}}$$

Let $R_6 = R_5$ and $R_4 = R_3$. Check to see that $A_v(f_{\max})$ is at least five or ten times larger than $R_5/R_3$. If this is not true, the gain of the integrator stage must be lowered by reducing $R_6$ and $R_5$. An additional dc-coupled buffer amplifier may be used if a larger $v_o(\max)$ is needed.

*Step 5.* $C_3$ and $C_4$ are determined using Eq. 10:

$$C_3 = C_4 \gg \frac{1}{2\pi f_{\max} R_5}$$

These capacitors control the amount of ripple noise in $v_o$. An inequality by a factor of 100 to 1,000 may be required for Eq. 10.

*Step 6.* The balancing-network resistors are approximated using Eqs. 8:

$$R_8 = R_{10} \approx 100 \, R_3$$

$$R_7 = R_9 = \frac{V_{Z1}}{I_{Z1}}$$

Check to make sure that $R_8 \gg R_7$. If this is not so, some compromise in the above values may be required.

**EXAMPLE OF FREQUENCY-DIFFERENCE-DETECTOR DESIGN** Suppose we need a circuit which must monitor the stability of a 10-kHz signal relative to a highly stable 10-kHz clock. We want a positive dc output if the unknown frequency is higher than the reference. This requires that we make $f_2$ the unknown and $f_1$ the reference. A $\pm 1$-kHz deviation is expected for $f_2$.

*Design Requirements*

$f_1 = $ 10-kHz reference sine wave
$v_1 = $ 8 V peak-to-peak (zero to +8 V)
$f_2 = 10 \pm 1$ kHz (unknown)
$v_2 = $ 6 V peak-to-peak (zero to +6 V)
$f_{max} = $ highest $f_2 = 11$ kHz
$v_o(\text{dc, max}) = \pm 1$ V for a $f_1 - f_2 = \pm 1$ kHz
$V^{(\pm)} = \pm 15$ V

*Device Data*

$v_3(\text{on}) = v_4(\text{on}) = 3.9$ V (assuming $V_{cc} = 5.0$ V)
$V_3(\text{off}) = v_4(\text{off}) = 0.2$ V
$V_{io} = 2$ mV
$I_b = 300$ nA
$I_{io} = 30$ nA
$v_0(\text{sat}) = \pm 13$ V
$V_{Z1} = V_{Z2} = 6.4$ V at $I_{Z1} = I_{Z2} = 0.5$ mA

*Step 1.* $T_1 = T_2$ is calculated from

$$T_1 = T_2 = \frac{0.2}{f_{max}} = \frac{0.2}{11,000} = 18.1 \, \mu s$$

*Step 2.* Let the first-cut $R_1$ and $R_2$ equal 10 k$\Omega$. $C_1$ and $C_2$ are computed as follows:

$$C_1 = C_2 = \frac{4T_1}{R_1} = \frac{4(18.1 \times 10^{-6})}{10^4} = 7,240 \text{ pF}$$

If we use standard 6,800-pF capacitors for $C_1$ and $C_2$,

$$R_1 = R_2 = \frac{4T_1}{C_1} = \frac{4(18.1 \times 10^{-6})}{6,800 \times 10^{-12}} = 10,600 \, \Omega$$

*Step 3.* $R_3$ must be much less than

$$R_3 \ll \frac{v_3(\text{on})}{I_b} = \frac{3.9}{300 \times 10^{-9}} = 13 \text{ M}\Omega$$

This will be easy to satisfy. We will let $R_3$ be a 2-k$\Omega$ potentiometer in series with a 1-k$\Omega$ resistor. The median $R_3$ will be 2,500 $\Omega$.

*Step 4.* $R_5$ is determined by using Eq. 7:

$$R_5 = \frac{R_3 v_o(\text{max})}{v_4(\text{on}) T_2 |f_2 - f_1|_{max}} = \frac{2,500 \, (1)}{3.9(18.1 \times 10^{-6})1,000} = 35.4 \text{ k}\Omega$$

We also assume
$$R_6 = R_5 = 35.4 \text{ k}\Omega$$
$$R_4 = R_3 = 2,500 \text{ }\Omega$$

The gain of most monolithic op amps is $\approx 100$ at 10 kHz. Since $R_5/R_3 = 14.16$, little gain error will occur.

*Step 5.* $C_3$ and $C_4$ are found from

$$C_3 = C_4 \gg \frac{1}{2\pi f_{\max} R_5} = \frac{1,000}{2\pi(1,000)354,000} = 0.45 \text{ }\mu\text{F}$$

We will let $C_3 = C_4 = 0.47 \text{ }\mu\text{F}$.

*Step 6.* The balancing-network calculations follow:

$$R_8 = R_{10} \approx 100 R_3 = 100(2,500) = 250 \text{ k}\Omega$$

$$R_7 = R_9 \approx \frac{V^{(+)} - V_{Z1}}{I_{Z1}} = \frac{15 - 6.4}{0.5 \times 10^{-3}} = 17,200 \text{ }\Omega$$

This is sufficiently lower than $R_8$ so that $R_8$ will not load $R_7$ and $R_9$.

# REFERENCE

1. Campbell, J. D.: A Simple Frequency Difference Detector, *EEE*, November 1970, p. 80.

Chapter **15**

# Integrators and Differentiators

## INTRODUCTION

An integrator is a low-pass filter, and a differentiator is a high-pass filter. Each of these two circuits therefore performs a mathematical function which is the inverse of the other. Even though these circuits are related mathematically, the practical problems associated with each are completely different. The prime differentiator problems are noise and instability. The integrator is prone to dc drift and offset.

## 15.1 DIFFERENTIATOR

**ALTERNATE NAMES** First-derivative circuit, differentiating amplifier, high-pass filter.

**EXPLANATION OF OPERATION** An ideal differentiator produces an instantaneous output voltage which is precisely proportional to the instantaneous derivative of the input voltage. The basic circuit which will perform this function requires the op amp, $C_1$, and $R_f$ in Fig. 15.1. The other parts ($R_1$, $C_f$, $R_p$, and $C_p$) have been added for reasons to be explained below. $R_1$ and $C_f$, if properly sized, will stabilize the feedback loop, which is inherently unstable in the basic differentiator circuit. Referring to Fig. 15.2, if $R_1$ is not present, the open- and closed-loop frequency responses of the circuit intersect at 40 dB/decade (12 dB/octave). As shown in Chap. 3, the intersection should be 20 dB/decade (6 dB/octave) or less to provide absolute stability. Most op amps have another open-loop pole near $f_u$ which leads to more instability. $C_f$ does not allow the closed-loop gain curve to intersect the open-loop curve until the unity-gain crossover frequency has been exceeded. This provides an additional measure of stability.

$R_p$ prevents $I_b$ from producing a dc offset at the op amp output. $R_p$ can be made adjustable from $<R_f$ to $>R_f$ to cancel out the effects of both $I_b$ and $I_{io}$.

Capacitor $C_p$ is required to bypass the thermal noise of $R_p$ to ground and

**15-2 INTEGRATORS AND DIFFERENTIATORS**

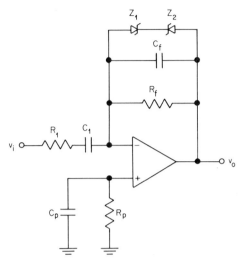

**Fig. 15.1** A low-noise differentiator with overload protection and good feedback stability.

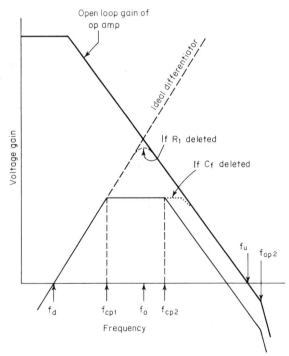

**Fig. 15.2** Differentiator frequency-response curves for three circuit configurations compared with the op amp open-loop response.

to maintain loop stability. It is added only if $R_p$ is greater than 5 or 10 k$\Omega$. $C_p$ is chosen so that its reactance is less than one-tenth $R_p$ for all frequencies down to $f_d$, if possible. See Chap. 3 for a discussion of the feedback-stability aspects of $C_p$ and $R_p$.

Since this circuit is sensitive to the slope of the input signal, the designer must be aware of the relation between maximum input slope and maximum output voltage:

$$v_o(\text{max}) = R_f C_1 \frac{dv_i}{dt}(\text{max})$$

If the possibility exists that input slopes steeper than the above maximum slope may be present, the zener-diode clamp is recommended. These zeners will prevent $C_1$ from acquiring a charge when unipolar noise bursts occur. This avoids a temporary paralysis of the input which could possibly last for a period of time after the burst.

True differentiation will occur for frequencies below $f_{cp1}$. Beyond this frequency the circuit behaves as a voltage amplifier with progressively lower gain at higher frequencies.

## DESIGN PARAMETERS

| Parameter | Description |
|---|---|
| $A_{vc}$ | Closed-loop gain of differentiator as a function of frequency |
| $C_1$ | Input capacitor required in basic differentiator |
| $C_f$ | Feedback capacitor utilized for stability |
| $C_p$ | Bypass capacitor utilized to hold op amp noninverting input at ac ground |
| $f_o$ | Geometric center frequency of passband for closed-loop circuit |
| $f_{op1}$ | First-pole frequency of op amp |
| $f_{op2}$ | Second-pole frequency of op amp |
| $f_{cp1}$ | First-pole frequency of closed-loop circuit |
| $f_{cp2}$ | Second-pole frequency of closed-loop circuit |
| $f_d$ | Characteristic frequency of differentiator, i.e., the low-frequency unity-gain crossover frequency of closed-loop circuit |
| $f_u$ | The unity-gain crossover frequency of the op amp |
| $I_{io}$ | Op amp input offset current |
| $I_n$ | Op amp equivalent input noise current |
| $R_1$ | Input resistor utilized for stability |
| $R_f$ | Feedback resistor required in basic differentiator |
| $R_{id}$ | Op amp differential input resistance |
| $R_p$ | Resistor used to cancel effects of the op amp input bias current |
| $v_i$ | Circuit input voltage |
| $V_{io}$ | Op amp input offset voltage |
| $V_n$ | Op amp equivalent input noise voltage |
| $v_o$ | Circuit output voltage |
| $V_{on}$ | Total circuit output noise voltage |
| $V_{oni}$ | Circuit output noise voltage due to op amp equivalent input noise current |
| $V_{onv}$ | Circuit output noise voltage due to op amp equivalent input noise voltage |

## DESIGN EQUATIONS

| Eq. No. | Description | Equation |
|---|---|---|
| 1 | Ideal output voltage in time domain for $f < f_{cp1}$ | $v_o = -R_f C_1 \dfrac{dv_i}{dt}$ |
| 2 | Ideal circuit gain for all frequencies | $A_{vc} = \dfrac{v_o}{v_i} = \dfrac{-jf/2\pi R_1 C_f}{(jf + 1/2\pi R_1 C_1)(jf + 1/2\pi R_f C_f)}$ $= \dfrac{-sR_f C_1}{(1 + sR_1 C_1)(1 + sC_f R_f)}$ |
| 3 | Low-frequency unity-gain crossover frequency (i.e., the differentiator characteristic frequency) | $f_d = \dfrac{1}{2\pi R_f C_1}$ |
| 4 | First-pole frequency of circuit (due to input circuit) | $f_{cp1} = \dfrac{1}{2\pi R_1 C_1}$ |
| 5 | Second-pole frequency of circuit (due to feedback circuit) | $f_{cp2} = \dfrac{1}{2\pi R_f C_f}$ |
| 6 | Midband frequency $f_o$ for best performance, noise and stability trade-off | $f_o \le (f_d f_u)^{1/2}$ <br> NOTE: <br> $f_o = (f_{cp1} f_{cp2})^{1/2}$ |
| 7 | Relationship between $f_d$, the maximum expected input-voltage rate of change and the maximum desired output voltage | $\left.\dfrac{dv_i}{dt}\right|_{\max} = \dfrac{1}{2\pi v_o(\max)} = \dfrac{1}{2\pi R_f C_1}$ |
| 8 | Differentiation error caused by nonideal op amp input characteristics | $\left.\dfrac{dv_i}{dt}\right|_{\text{error}} = \dfrac{1}{C_1}\left[\left(\dfrac{1}{R_1} + \dfrac{1}{R_{id}}\right)V_{io} + I_{io}\right]$ |
| 9 | Differentiator output rms noise due to equivalent input-voltage noise of op amp | $V_{onr} = \dfrac{V_n \{s^2 + s[(R_1 C_1 + R_f C_1 + R_f C_f)/R_1 R_f C_1 C_f] + (1/R_1 R_f C_1 C_f)\}}{(s + 1/R_1 C_1)(s + 1/R_f C_f)}$ |

10 Differentiator output rms noise due to equivalent input-current noise of op amp

$$V_{oni} = \frac{I_{ns}/R_1 C_f}{(s + 1/R_1 C_1)(s + 1/R_f C_f)}$$

11 Total output rms noise due to op amp noise

$$V_{on} = (V_{oni}^2 + V_{onv}^2)^{1/2}$$

12 Optimum $R_p$

$$R_p = R_f$$

13 Optimum $C_p$

$$C_p = \frac{10}{2\pi f_d R_p}$$

NOTE: This should be a low-leakage capacitor so it will not degrade the effect of $R_p$.

14 Optimum $C_1$

$C_1 < 1$ $\mu$F (if possible or $<10$ $\mu$F (with caution)

NOTE: $C_1$ should be a low-leakage capacitor such as polystyrene or polycarbonate

15 Optimum $R_f$

Choose $R_f$ such that the peak feedback current $v_o(\max)/R_f$ is $\approx 500$ $\mu$A.

## 15-6 INTEGRATORS AND DIFFERENTIATORS

### DESIGN PROCEDURE

This circuit should present no design difficulties as long as the preceding noise and stability equations are followed. $R_1$ is the critical part which reduces both noise and instability. $C_f$ provides additional protection against instability for op amps having excessive high-frequency phase lag. The zener diodes are required only if large unipolar noise bursts are expected and if a temporary paralysis cannot be tolerated.

### DESIGN STEPS

*Step 1.* Choose the $\pm$ maximum limits for $v_o$. Let the $\pm$ power-supply voltages be at least 3 V greater than $\pm v_o(\text{max})$.

*Step 2.* Choose $R_f$ so that 500 $\mu$A flows through it in the presence of $\pm v_o(\text{max})$.

$$R_f = \frac{v_o(\text{max})}{5 \times 10^{-4}}$$

*Step 3.* Assuming the maximum input slope is known, determine the size of $C_1$ using Eq. 7:

$$C_1 = \frac{v_o(\text{max})}{R_f \left.\dfrac{dv_i}{dt}\right|_{\text{max}}}$$

NOTE: If $C_1$ is larger than 0.1 $\mu$F, make sure its leakage resistance is at least 100 times larger than $R_f$. Otherwise this leakage resistance $(R_x)$ will cause the differentiator to act as an inverting amplifier having a gain of $-R_f/R_x$. The output signal resulting from this will be added to that caused by differentiation, thus causing a distorted output waveform.

*Step 4.* Compute the differentiator characteristic frequency $f_d$ from

$$f_d = \frac{1}{2\pi R_f C_1}$$

*Step 5.* Let $R_p = R_f$. Compute the optimum $C_p$ using Eq. 13:

$$C_p = \frac{10}{2\pi f_d R_p}$$

NOTE: The leakage resistance of $C_p$ should be at least 100 times larger than $R_p$.

*Step 6.* If the op amp is guaranteed by the manufacturer to be stable with 100 percent feedback (unity-gain amplifier), let $f_o = (f_d f_u)^{1/2}$. If the op amp second pole $(f_{op2})$ occurs at a lower frequency than $f_u$, say at a frequency of $Af_u$, let

$$f_o = (Af_d f_u)^{1/2}$$

(i.e., if $f_{op2} = 500$ kHz and $f_u = 1$ MHz, $A = 0.5$).

*Step 7.* Let the two closed-loop poles be placed at

$$f_{cp1} = \frac{f_o}{2} \quad \text{and} \quad f_{cp2} = 2 f_o$$

*Step 8.* Determine $R_1$ from Eq. 4:

$$R_1 = \frac{1}{2\pi f_{cp1} C_1}$$

**DIFFERENTIATOR  15-7**

*Step 9.* Determine $C_f$ from Eq. 5:

$$C_f = \frac{1}{2\pi f_{cp2} R_f}$$

*Step 10.* Determine the differentiation error caused by nonideal op amp input parameters using Eq. 8.

*Step 11.* Compute the output noise as a function of frequency using Eqs. 9, 10, and 11.

**DESIGN EXAMPLE** As a practical example suppose we need to generate a rectangular waveform from a sawtooth waveform. The maximum input slope is given to be $\pm 0.1$ V/$\mu$s. The resulting maximum output voltage should be $\pm 10$ V.

*Device Data* (108 op amp)
$V_{io} = 2$ mV max at $+25°$C
$I_{io} = 0.2$ nA max at $+25°$C
$V_n \approx 50$ nV/Hz$^{1/2}$
$I_n$ (not specified for 108)
$f_{op1} = 20$ Hz (minimum compensation)
$f_{op2} = 2$ MHz (minimum compensation)
$f_u = 3$ MHz (minimum compensation)
$R_{id} = 3 \times 10^7$ $\Omega$

*Step 1.* If we choose the power-supply voltages to be $\pm 15$ V, no possibility of nonlinear behavior due to op amp saturation will exist.

*Step 2.*

$$R_f = \frac{v_o(\max)}{5 \times 10^{-4}} = \frac{10}{5 \times 10^{-4}} = 20 \text{ k}\Omega$$

*Step 3.*

$$C_1 = \frac{v_o(\max)}{R_f \left.\dfrac{dv_i}{dt}\right|_{\max}} = \frac{10}{(20{,}000)\, 0.1/10^{-6}} = 0.005 \text{ }\mu\text{F}$$

*Step 4.*

$$f_d = \frac{1}{2\pi R_f C_1} = \frac{1}{2\pi \times 20{,}000 \times 5 \times 10^{-9}} = 1{,}592 \text{ Hz}$$

*Step 5.* Let $R_p = R_f = 20$ k$\Omega$ also.
We compute $C_p$ from

$$C_p = \frac{10}{2\pi f_d R_p} = \frac{10}{2\pi \times 1{,}592 \times 20{,}000} = 0.05 \text{ }\mu\text{F}$$

*Step 6.* Since the second pole of the op amp $f_{op2}$ occurs at a frequency slightly below the unity-gain crossover frequency $f_u$, the factor A must be used.

$$A = \frac{f_{cp2}}{f_u} = \frac{2}{3}$$

thus
$$f_o = (A f_d f_u)^{1/2}$$
$$= (0.667 \times 1{,}592 \times 3 \times 10^6)^{1/2} = 56{,}441 \text{ Hz}$$

## 15-8 INTEGRATORS AND DIFFERENTIATORS

*Step 7.*

$$f_{cp1} = \frac{f_o}{2} = 28.2 \text{ kHz}$$

$$f_{cp2} = 2f_o = 113 \text{ kHz}$$

*Step 8.*

$$R_1 = \frac{1}{2\pi f_{cp1} C_1} = \frac{1}{2\pi \times 28,200 \times 5 \times 10^{-9}} = 1,129 \text{ }\Omega$$

*Step 9.*

$$C_f = \frac{1}{2\pi f_{cp2} R_f} = \frac{1}{2\pi \times 113,000 \times 20,000} = 70 \text{ pF}$$

*Step 10.* The differentiation error caused by the nonideal op amp input parameters is

$$\left.\frac{dv_i}{dt}\right|_{\text{error}} = \frac{1}{C_1}\left[\left(\frac{1}{R_1} + \frac{1}{R_{id}}\right)V_{io} + I_{io}\right]$$

$$= \frac{(1/1,384 + 1/3 \times 10^7) 2 \times 10^{-3} + 2 \times 10^{-10}}{5 \times 10^{-9}}$$

$$= 289 \text{ V/s error or}$$
$$289 \text{ } \mu\text{V}/\mu\text{s error}$$

We can determine the circuit output-voltage error using Eq. 1:

$$v_o(\text{error}) = -R_f C_1 \left.\frac{dv_i}{dt}\right|_{\text{error}} = -20,000 \times 5 \times 10^{-9} \times 289 = -0.0289 \text{ V}$$

*Step 11.* Since the 108 data sheets do not specify equivalent input-current noise, the differentiator output noise is simply

$$V_{on} = \frac{V_n\{s^2 + s[(R_1C_1 + R_fC_1 + R_fC_f)/R_1R_fC_1C_f] + (1/R_1R_fC_1C_f)\}}{(s + 1/R_1C_1)(s + 1/R_fC_f)}$$

$$= \frac{50[s^2 + (2.4 \times 10^5)s + 8.96 \times 10^8] \text{ nV}}{2 \times 10^7 (s + 1.45 \times 10^5)(s + 5.81 \times 10^5)}$$

$$= \frac{50(s + 3,793)(s + 2.36 \times 10^5)}{(s + 1.45 \times 10^5)(s + 5.81 \times 10^5)} \text{ nV}$$

$$= \frac{50(jf + 604)(jf + 37,560)}{(jf + 23,077)(jf + 92,469)} \text{ nV}$$

The noise ranges from 50 nV down to <1 nV over the differentiator useful frequency range ($f_d$ to $f_{cp1}$).

## REFERENCES

1. Best, R. E.: Differentiator Noise Is No Problem, *Electron. Des.*, June 21, 1966, p. 92.
2. Tobey, G. E., J. G. Graeme, and L. P. Huelsman: "Operational Amplifiers—Design and Applications," p. 218, McGraw-Hill Book Company, New York, 1971.
3. Philbrick, G. A.: "Applications Manual for Computing Amplifiers," p. 48, Nimrod Press, Boston, Mass., 1966.

## 15.2 INTEGRATOR

**ALTERNATE NAMES** Integrating amplifier, integral amplifier, definite-integral circuit, analog integrator, low-pass filter.

**EXPLANATION OF OPERATION** An ideal integrator produces an output voltage which is proportional to the integral of the input voltage. In other words, the output is proportional to the product of the amplitude and duration of the input. The integrator performs this mathematical operation on an instantaneous basis, producing an output proportional to the sum of the products of instantaneous voltages and vanishingly small increments of time. The result is an output exactly proportional to the area under a waveform.

The circuit shown in Fig. 15.3 performs integration by using an op amp to force the same current through both $R_1$ and $C_f$. The voltage across the feedback capacitor is related to capacitor current by

$$v_c = -\frac{1}{C_f}\int i_f dt$$

Since the circuit causes $i_f$ to equal the input current ($i_i = v_i/R_1$),

$$v_c = v_o = -\frac{1}{R_1 C_f}\int v_i dt$$

The gain of the circuit is given by $-1/R_1 C_f$. Thus the output voltage will change by $-1/R_1 C_f$ V/s for each volt of input. Numerically, the circuit performs integration in the following manner. For a start, assume $v_i = v_o = 0$, $R_1 = 10$ k$\Omega$, and $C_f = 1$ $\mu$F. Under these conditions no current will flow through $R_1$ or $C_f$. If a $-1$-V dc level is suddenly applied to $v_i$, a current of $v_i/R_1 = -1/10^4 = -100$ $\mu$A will immediately flow in $R_1$. If we assume the op amp draws no current, 100 $\mu$A must also immediately flow through $C_f$. To obtain a dc current of 100 $\mu$A through $C_f$, we require a linear positive ramp

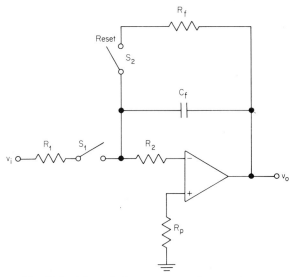

**Fig. 15.3** A low-drift integrator with manual reset.

## 15-10 INTEGRATORS AND DIFFERENTIATORS

at $v_o$. The current through the capacitor must satisfy $i_f = C_f(dv_c/dt)$. The ramp at $v_o$ will therefore have a slope of

$$\frac{dv_o}{dt} = \frac{i_f}{C_f} = \frac{-v_i}{R_1 C_f} = \frac{-(-1)}{10^4 \times 10^{-6}} = 100 \text{ V/s}$$

This is equivalent to saying

$$v_o = -\frac{1}{R_1 C_f} \int v_i \, dt = 100t$$

Since the input waveform can be sinusoidal or nonlinear, we have used the lowercase generalized nomenclature. The uppercase nomenclature is only for sinusoidal ac operations or dc parameters.

The op amp input offset voltage $V_{io}$ and input offset current $I_{io}$ add errors to the above equations. If $R_p$ is not included in the circuit, the error caused by $I_{io}$ is replaced by a larger error due to input bias current $I_b$. The output voltage with these errors included is

$$v_o = -\frac{1}{R_1 C_f}\left(\int v_i \, dt \pm \int V_{io} \, dt\right) \pm \frac{1}{C} \int I_{io} \, dt \pm V_{io}$$

Note that $V_{io}$ causes a small step voltage $\pm V_{io}$ and a $\pm$ ramp with a gain of $1/R_1 C_f$. However, $I_{io}$ causes a $\pm$ ramp with a gain of $1/C_f$. This latter error source may be reduced if the designer has the option available to increase $C_f$ and lower $R_1$. If this is done, one must realize that the input resistance is lowered and the leakage component of $C_f$ will probably be increased. The leakage current through $C_f$ must be less than $I_b$.

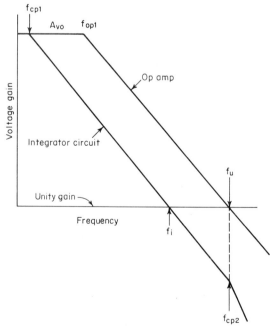

**Fig. 15.4** Frequency characteristics of a typical integrator compared with a typical op amp.

If the finite gain of the op amp $A_{vo}$ is considered, the transfer function of an integrator is

$$A_{vc} = \frac{v_o}{v_i} = \frac{-A_{vo}f_{op1}f_i}{(jf + A_{vo}f_{op1})(jf + f_i/A_{vo})}$$

where $A_{vo}, f_{op1}$, and $f_i$ are defined in Fig. 15.4.

The response of a practical integrator to a step function of amplitude $-V$ is

$$v_o = A_{vo}V\left[1 - \frac{\exp(-t/A_{vo}R_1C_f)}{1 - f_i/A_{vo}f_{op1}} - \frac{\exp(-2\pi f_{op1}t)}{1 - f_{op1}A_{vo}/f_i}\right]$$

## DESIGN PARAMETERS

| Parameter | Description |
|---|---|
| $A_{vo}$ | Open-loop voltage gain of op amp at dc |
| $C_f$ | Feedback capacitor |
| $f_{cp1}$ | First-pole frequency of circuit |
| $f_{cp2}$ | Second-pole frequency of circuit |
| $f_i$ | Characteristic frequency of integrator |
| $f_{op1}$ | First-pole frequency of op amp |
| $f_u$ | Unity-gain crossover frequency of op amp |
| $i_f$ | Feedback capacitor current |
| $I_b$ | Op amp input bias current |
| $I_{io}$ | Op amp input offset current |
| $R_1$ | Input resistor to circuit |
| $R_2$ | Resistor used to protect op amp inverting input from large transients through $C_f$ when power supplies are turned off (required only if $C_f \geq 0.05\ \mu F$) |
| $R_f$ | Resistor used to reset integrator by discharging $C_f$ |
| $R_p$ | Resistor used to nullify effects of $I_b$ |
| $T_o$ | Equivalent time constant of op amp dominant pole |
| $T_s$ | Time to reset integrator to zero |
| $v_c$ | Capacitor voltage |
| $v_i$ | Input voltage to circuit |
| $V_{io}$ | Op amp input offset voltage |
| $v_o$ | Output voltage of circuit |

## DESIGN EQUATIONS

| Eq. No. | Description | Equation |
|---|---|---|
| 1 | Ideal output voltage | $v_o = -\dfrac{1}{R_1C_f}\int v_i\,dt$ |
| 2 | Output voltage considering nonideal op amp input current and voltage and assuming $R_1 = R_p$. If $R_p$ is not used, $+I_b$ must replace $\pm I_{io}$ | $v_o = -\dfrac{1}{R_1C_f}\left(\int v_i\,dt \pm \int V_{io}\,dt\right) \pm \dfrac{1}{C}\int I_{io}\,dt \pm V_{io}$ |
| 3 | Transfer function of circuit considering finite op amp gain | $A_{vc} = \dfrac{v_o}{v_i} = \dfrac{-A_{vo}f_{op1}f_i}{(jf + A_{vo}f_{op1})(jf + f_i/A_{vo})}$ |
| 4 | Response of circuit to a step-voltage input of $-V$ | $v_o = A_{vo}V\left[1 - \dfrac{\exp(-t/A_{vo}R_1C_f)}{1 - f_i/A_{vo}f_{op1}} - \dfrac{\exp(-2\pi f_{op1}t)}{1 - f_{op1}A_{vo}/f_i}\right]$ |

## 15-12 INTEGRATORS AND DIFFERENTIATORS

| Eq. No. | Description | Equation |
|---|---|---|
| 5 | Response of circuit for small values of time for a step-voltage input of $-V$ | $v_o = \dfrac{V}{R_1 C_f}[t + t_o + T_o\exp(-t/T_o)]$ where $T_o = \dfrac{1}{2\pi f_{op1}}$ |
| 6 | Response of circuit for large values of time for a step-voltage input of $-V$ | $v_o = A_{vo}V[1 - \exp(-t/A_{vo}R_1 C_f)]$ |
| 7 | Optimum size for $R_1$ to minimize effects of $I_{io}$ and $V_{io}$ if $R_1 = R_p$ | $R_1 = \dfrac{V_{io}}{I_{io}}$ |
| 8 | Optimum size for $R_1$ to minimize effects of $I_b$ and $V_{io}$ if $R_p = 0$ | $R_1 = \dfrac{V_{io}}{I_b}$ |
| 9 | Optimum size for reset resistor $R_f$ if 0.1% accuracy in starting point is required | $R_f \leq \dfrac{T_s}{7C_f}$ |
| | | NOTE: $S_1$ must be open and $S_2$ closed during reset |
| 10 | First pole frequency of integrator | $f_{cp1} = \dfrac{f_i}{A_{vo}}$ |
| 11 | Integrator crossover frequency (characteristic frequency of integrator) | $f_i = \dfrac{1}{2\pi R_1 C_f}$ |

## REFERENCE

1. Tobey, G. E., J. G. Graeme, and L. P. Huelsman: "Operational Amplifiers—Design and Applications," p. 213, McGraw-Hill Book Company, New York, 1971.

# Chapter 16
# Limiters and Rectifiers

## INTRODUCTION

This chapter discusses circuits which modify signals only if their amplitude possesses certain magnitude or polarity characteristics.

The limiter does not modify the input signal until it rises to a given amplitude. Beyond this amplitude the signal is abruptly clipped. We will present design information on the conventional feedback circuit which performs limiting by clipping all waveforms above a given magnitude.

Precision rectifiers alter the input signal depending on its instantaneous polarity. A half-wave rectifier can be made to remove either the positive or negative portions of a waveform. Likewise, the output signal can be chosen to display the selected polarity in either a positive or negative format. Full-wave rectifiers accept both input polarities. However, one of the input polarities becomes inverted so that the output is unipolar. Either type of rectifier can be operated with or without a filter capacitor in the feedback loop. Without this capacitor the output waveforms are of the same shape as the input, although their polarity may be inverted or amplified. The filter capacitor transforms the circuit into a precision ac-dc converter such that the output dc level is exactly proportional to the average value of the rectified input signal.

## 16.1 AMPLITUDE LIMITER

**ALTERNATE NAMES** Limited amplifier, volume compressor, amplitude leveler, feedback limiter, precision limiter, limiting amplifier.

**EXPLANATION OF OPERATION** Amplitude limiters are required in many systems where the amplitude of a signal cannot be allowed to exceed given positive or negative limits. This function is often done utilizing resistor/zener networks. A zener in the feedback network of an op amp will also accomplish the same function without excessive loading of the input signals. Extensive coverage of these circuits is given in the literature. These circuits all suffer from one common disadvantage, namely, any time or temperature variation of the zener breakdown voltage creates a circuit error of corresponding size. This may be acceptable for applications where amplitude limiting is performed only for protection or noise reduction. However, in

## 16-2 LIMITERS AND RECTIFIERS

that class of circuits where the limiting voltage is an important system parameter, a precision limiter, as described below, is required.

In this section we will cover, in detail, the design of the precision bridge-type amplitude limiter shown in Fig. 16.1. The diode bridge is placed in the forward loop of the feedback network. This means that the effects of all nonlinear, forward resistance, and temperature characteristics of the diodes will be divided by a factor of $1/\beta A_v$ in the closed-loop circuit. The transfer characteristics of the circuit, shown in Fig. 16.2, are therefore dependent only on resistor values and the two reference voltages $V_{R1}$ and $V_{R2}$. The slope of

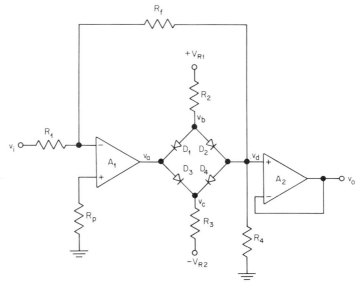

**Fig. 16.1** A precision bridge-type amplitude-limiter circuit.

the linear region is identical to that of an inverting amplifier, namely, $-R_f/R_1$. The positive limiting output voltage is given by

$$V_o(\text{sat,pos}) = \frac{V_{R1} R_f R_4}{R_f R_4 + R_f R_2 + R_4 R_2}$$

The negative limiting output voltage is

$$V_o(\text{sat,neg}) = \frac{V_{R2} R_f R_4}{R_f R_4 + R_f R_3 + R_4 R_3}$$

In the linear region of circuit operation the four diodes are all forward-biased. The voltage $v_a$ closely follows $v_d$ and the circuit gain is

$$A_{vc} = -\frac{R_f}{R_1}$$

Note that no power can flow directly from $v_a$ to $v_d$. All power for $v_d$ comes through $R_2$ or $R_3$. However, $v_b$ and $v_c$ are controlled by $v_a$, since $v_a$ is a stiff source and draws current through $D_1$ and $D_3$. If $D_1$ is conducting, $v_b$ is one diode drop above $v_a$, and likewise $v_d$ is one diode drop below $v_b$. If $D_3$ is conducting, $v_c$ is one diode drop below $v_a$ and $v_d$ is one diode drop above $v_c$. In either case, $v_a \approx v_d$. Utilizing back-to-back diodes in series with the signal

## AMPLITUDE LIMITER

flow allows cancellation of most of the temperature-induced changes of diode characteristics. Any residual error caused by a mismatch of diodes will be reduced by a factor of $1/\beta A_v$, since the diodes are within the forward loop of the feedback circuit.

If $v_a$ was not present, the maximum positive $v_d$ is set by the voltage divider composed of $V_{R1}$, $R_2$, $R_4$, and $R_f$. Likewise the most negative $v_d$ is set by the voltage divider composed of $V_{R2}$, $R_3$, $R_4$, and $R_f$. $v_a$ cannot swing $v_d$ beyond these limits. If $v_a$ swings positive beyond the upper limit, $D_1$ becomes reverse-biased and $v_d$ remains at the upper limit shown in Fig. 16.2. If $v_a$ swings negative, it reverse-biases $D_3$ at the limit shown in Fig. 16.2. Thus $D_1$ and $D_3$ are switches which cause the abrupt change in circuit characteristics when the limiting voltages are reached.

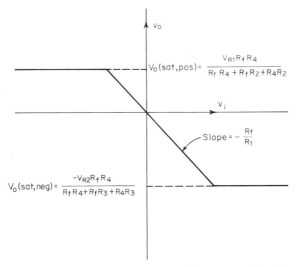

**Fig. 16.2** Transfer function of precision bridge-type amplitude limiter.

## DESIGN PARAMETERS

| Parameter | Description |
| --- | --- |
| $A_v$ | Op amp open-loop gain as a function of frequency |
| $A_{vc}$ | Closed-loop gain of circuit in linear region of operation |
| $\beta$ | Feedback ratio due to $R_f$ and $R_1$ |
| $D_1$ to $D_4$ | Diode bridge used to switch circuit from linear to limited |
| $R_1$ | Input resistor which establishes input resistance of circuit |
| $R_2$ to $R_3$ | Resistors which provide output current for diode bridge |
| $R_4$ | Resistor used to refer output of diode bridge to ground |
| $R_f$ | Feedback resistor |
| $R_p$ | Resistor used to minimize effects of op amp input bias current |
| $v_a$ to $v_d$ | Diode bridge voltages |
| $v_i$ | Input voltage to circuit |
| $v_o$ | Output voltage of circuit |
| $V_o(\text{sat,pos})$ | Positive limited output voltage |
| $V_o(\text{sat,neg})$ | Negative limited output voltage |
| $V_{R1}$ | Magnitude of positive reference voltage |
| $V_{R2}$ | Magnitude of negative reference voltage |

**16-4 LIMITERS AND RECTIFIERS**

## DESIGN EQUATIONS

| Eq. No. | Description | Equation |
|---|---|---|
| 1 | Voltage gain of circuit in linear region assuming ideal op amp and diode parameters | $A_{vc} = \dfrac{v_o}{v_i} = -\dfrac{R_f}{R_1}$ |
| 2 | Voltage gain of circuit in linear region assuming finite op amp gain | $A_{vc} = \dfrac{-R_f/R_1}{1 + 1/\beta A_v}$ |
| 3 | Positive limiting output voltage | $V_o(\text{sat,pos}) = \dfrac{V_{R1}R_f R_4}{R_f R_4 + R_f R_2 + R_2 R_4}$ |
| 4 | Negative limiting output voltage | $V_o(\text{sat,neg}) = \dfrac{-V_{R2}R_f R_4}{R_f R_4 + R_f R_3 + R_3 R_4}$ |

## REFERENCE

1. Tobey, G. E., J. G. Graeme, and L. P. Huelsman: "Operational Amplifiers—Design and Applications," p. 247, McGraw-Hill Book Company, New York, 1971.

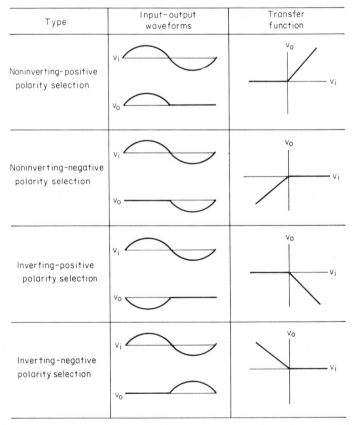

**Fig. 16.3** Input-output waveforms and transfer functions of four basic precision half-wave rectifiers.

## 16.2 PRECISION HALF-WAVE RECTIFIER

**ALTERNATE NAMES** Polarity selector, ideal half-wave rectifier, ideal diode, zero-bound circuit, precision AM detector, precision ac-dc converter, low-level ac-dc converter.

**EXPLANATION OF OPERATION** This circuit comes in four basic configurations. Figure 16.3 shows pictorially the input-output relationship of these four basic circuits along with a plot of each transfer function. A precision half-wave rectifier performs very closely to the expected response of an ideal diode. Figure 16.4B shows the response expected if one could produce an ideal diode. The ideal diode possesses several advantages relative to the silicon diode. First, the ideal diode can rectify signals down to zero volts amplitude. Second, the forward-conduction region of an ideal diode is linear.

In the inverting half-wave precision rectifier circuit shown in Fig. 16.5 the two ideal properties discussed above can be approached with nearly zero error. The circuit will rectify low-level signals with peak voltages of only $0.7/A_v$. If $A_v = 1,000$, precision linear rectification of a 0.7-mV signal is possible.

The circuit operates by providing two gains. For one polarity of input, $D_1$ is reverse-biased and $D_2$ is forward-biased. Under these conditions the gain of the circuit is $\pm R_f/R_1$ (+ for Fig. 16.6 and − for Fig. 16.5). If the opposite-polarity input is applied, $D_1$ is forward-biased and $D_2$ is reverse-biased. The gain of the circuit then becomes zero. The slope of the linear gain and the breakpoint are insensitive to temperature owing to the $1/A_v$ factor.

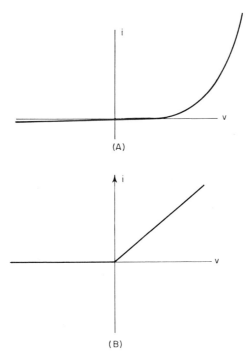

**Fig. 16.4** Typical V-I curve of a silicon diode (A) and an ideal diode (B).

## 16-6 LIMITERS AND RECTIFIERS

Figure 16.6 shows the circuit configuration for the noninverting precision half-wave rectifier.

Several sources of error are possible in the circuits of Figs. 16.5 and 16.6. If the op amp output offset voltage approaches 0.7 V, $D_1$ or $D_2$ may begin to conduct. This will add a dc component to $v_o$ which may be falsely interpreted as a rectified ac signal. The portion of this error voltage due to the op amp input offset voltage may be eliminated by adding a coupling capacitor in series with $R_1$. This will cause the dc gain of the circuit to equal unity.

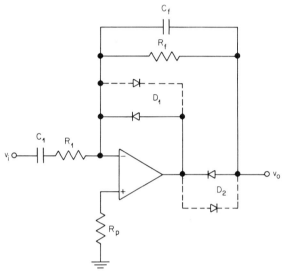

**Fig. 16.5** Inverting precision half-wave rectifier (solid lines for positive-polarity selection and dashed lines for negative-polarity selection).

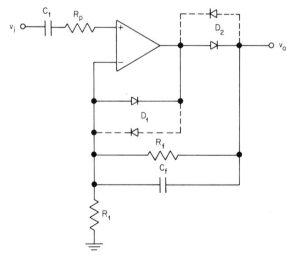

**Fig. 16.6** Noninverting precision half-wave rectifier (solid lines for positive-polarity selection and dashed lines for negative-polarity selection).

Even though mid-band ac signals will be amplified by $R_f/R_1$, the input offset voltage will be multiplied by 1. If $R_p$ is made equal to $R_f$, a further reduction in offset is possible by cancellation of the effects of each input bias current.

Capacitor $C_f$ is added to the circuit if a dc output voltage proportional to the peak input voltage is required. The magnitude of this dc voltage is given in Eq. 5.

## DESIGN PARAMETERS

| Parameter | Description |
|---|---|
| $A_v$ | Op amp open-loop gain (varies with frequency) |
| $A_{vo}$ | Op amp dc open-loop gain |
| $A_{vc}$ | Closed-loop gain of circuit during linear-gain portion of cycle |
| $B$ | Variable used in computations |
| $C_1$ | Input isolation capacitor to reduce errors due to input offset voltage |
| $C_f$ | Feedback capacitor used to provide dc output instead of half-wave rectified ac |
| $D_1$ | Provides an effective feedback resistance of zero ohms during portion of waveform when no output voltage is required |
| $D_2$ | Passes op amp output on to $V_o$ terminal during portion of waveform when undistorted output is required |
| $f_m$ | Modulating frequency of input carrier frequency $f_c$ |
| $f_c$ | Carrier frequency of input waveform |
| $f_{max}$ | Maximum frequency at which high-accuracy performance can be achieved |
| $f_u$ | Unity-gain frequency of the op amp (gain crossover frequency) |
| $I_b$ | Input bias current of op amp |
| $I_{io}$ | Input offset current of op amp |
| $R_1$ | Controls gain and input resistance of circuit |
| $R_f$ | Controls gain of circuit. $R_f$ also controls degree of filtering provided by $C$ |
| $R_{id}$ | Op amp differential input resistance |
| $R_o$ | Op amp output resistance |
| $R_p$ | Resistor used to minimize offset due to input bias current |
| $S$ | Slew rate of op amp |
| $t_{rr}$ | Reverse recovery time of diodes |
| $\dfrac{\Delta V_i}{\Delta t}$ | Fastest slew rate of input waveform |
| $v_i$ | Input voltage |
| $V_{io}$ | Input offset voltage of op amp |
| $v_o$ | Output voltage |

## DESIGN EQUATIONS

| Eq. No. | Description | Equation |
|---|---|---|
| 1 | Voltage gain of circuit if solid diode connections are used ($C_f = 0$) | INVERTING<br>$A_{vc} = \dfrac{v_o}{v_i} = -\dfrac{R_f}{R_1}$ if $v_i > 0$<br>$A_{vc} = 0$ if $v_i < 0$<br><br>NONINVERTING<br>$A_{vc} = \dfrac{v_o}{v_i} = 1 + \dfrac{R_f}{R_1}$ if $v_i > 0$<br>$A_{vc} = 0$ if $v_i < 0$ |

## 16-8 LIMITERS AND RECTIFIERS

| Eq. No. | Description | Equation |
|---|---|---|
| 2 | Voltage gain of circuit if dashed diode connections are used ($C_f = 0$) | INVERTING<br>$A_{vc} = \dfrac{v_o}{v_i} = -\dfrac{R_f}{R_1}$ if $v_i < 0$<br>$A_{vc} = 0$ if $v_i > 0$<br><br>NONINVERTING<br>$A_{vc} = \dfrac{v_o}{v_i} = 1 + \dfrac{R_f}{R_1}$ if $v_i < 0$<br>$A_{vc} = 0$ if $v_i > 0$ |
| 3 | Input resistance | INVERTING<br>$R_{in} = R_1$<br><br>NONINVERTING<br>$R_{in} = A_v R_{id}$ |
| 4 | Size of filter capacitor $C_f$ | $\dfrac{1}{2\pi f_c R_f} \ll C_f \ll \dfrac{1}{2\pi f_m R_f}$ |
| 5 | Magnitude of dc output voltage if $C_f$ utilized | INVERTING<br>$V_o(\text{dc}) = -\dfrac{0.45\, v_i(\text{rms}) R_f}{R_1}$<br><br>NONINVERTING<br>$V_o(\text{dc}) = 0.45\, v_i(\text{rms})\left(1 + \dfrac{R_f}{R_1}\right)$ |
| 6 | Maximum high-accuracy frequency of circuit (error $< 1\%$) | $f_{\max} = \dfrac{f_u}{100|A_{vc}|}$ |
| 7 | Required slew rate of op amp for a nonsinusoidal input waveform | $S > |A_{vc}|\dfrac{\Delta v_i}{\Delta t}$ |
| 8 | Required slew rate of op amp for sinusoidal input | $S > 2\pi f_c |A_{vc}| v_i(\text{peak})$ |
| 9 | Maximum dc offset voltage at op amp output | $V_o(\text{off,max}) = V_{io}\left(1 + \dfrac{R_f}{R_1}\right) + \dfrac{2I_b + I_{io}}{2R_f}$ |
| 10 | Voltage gain of circuit if op amp gain is considered | In Eqs. 1 and 2 replace $\dfrac{R_f}{R_1}$ with $\dfrac{R_f A_v}{R_1 A_v + R_f}$ |
| 11 | Optimum value of $R_p$ | IF $C_1$ USED<br>$R_p = R_f$<br><br>IF $C_1$ NOT USED<br>$R_p = \dfrac{R_1 R_f}{R_1 + R_f}$ |
| 12 | Maximum diode reverse recovery time so that output waveform will not be distorted | $t_{rr} < \dfrac{0.01}{f_c}$ |

## PRECISION HALF-WAVE RECTIFIER 16-9

| Eq. No. | Description | Equation |
|---|---|---|
| 13 | Effective forward voltage drop of precision rectifier (note that this increases with frequency as $A_v$ drops) | $V_f = \dfrac{0.7}{A_v}$ (silicon diodes assumed) |

## DESIGN PROCEDURE

Precision rectifiers are often utilized to extract a modulation frequency $f_m$ from a carrier frequency $f_c$. This is called AM demodulation. In the following design procedure we will assume such an application. Other applications, such as AGC detectors, require a consideration of the topics to be discussed below. The inverting circuit (Fig. 16.5) will be assumed.

## DESIGN STEPS

*Step 1.* Choose $R_1$ so that it equals the required input resistance of the circuit.

*Step 2.* Compute the maximum frequency $f_{max}$ at which high-accuracy performance can be expected:

$$f_{max} = \frac{f_u}{100|A_{vc}|}$$

*Step 3.* Compute the required slew rate of the op amp

$$v_i = \text{sinusoid}: S > 2\pi f_c |A_{vc}| v_i(\text{peak})$$

$$v_i = \text{nonsinusoid}: S > A_{vc} \frac{\Delta v_i}{\Delta t}$$

*Step 4.* Compute $R_f = A_{vc} R_1$.

*Step 5.* If filtering of the rectified output is not required, skip steps 5 through 7 and do not install $C_f$. Compute

$$B = \left(\frac{f_c}{f_m}\right)^{1/2}$$

*Step 6.* Compute

$$C_f = \frac{B}{2\pi f_c R_f}$$

In this equation, $B$ sets the ripple level of $f_c$ in $v_o$.

*Step 7.* Verify

$$C_f = \frac{1}{2\pi f_m R_f B}$$

In this equation, $B$ sets the degree of attenuation of $f_m$ in $v_o$.

NOTE: $C_f$ was found by a compromise above. It must be large enough to keep the $f_c$ ripple low. However, it must be low enough so that the $f_m$ modulation does not vanish. The above computation results in a geometric mean value for $C_f$.

**Step 8.** Compute the size of $R_p$ required to minimize the effects of bias-current offset

$$R_p = \frac{R_1 R_f}{R_1 + R_f}$$

**Step 9.** Compute the dc output level for the median $v_i$(rms) input.

$$v_o(\text{dc}) = \frac{\pm 0.45 v_i(\text{rms}) R_f}{R_1}$$

Polarity depends on the direction of $D_1$ and $D_2$.

**Step 10.** If a negative $v_o$ (filtered or unfiltered) is required, the solid diode connections shown in Fig. 16.5 are required. A positive output requires the dashed diode connections. The reverse recovery time $t_{rr}$ of the diodes must be less than $0.01/f_c$ or the output waveform will be distorted.

**Step 11.** Compute the dc output error caused by $V_{io}$, $I_b$, and $I_{io}$. This will vary with temperature.

$$V_o(\text{offset,max}) = (A_{vc} + 1) V_{io}(\text{max}) + \frac{2I_b + I_{io}}{2R_f}$$

**Step 12.** Examine the feedback stability of the circuit using the seven causes of instability outlined in Chap. 3.

**Step 13.** If voltage-gain accuracy over temperature and power-supply variations is important, determine the magnitude of closed-loop gain for different values of open-loop gain using

$$A_{vc} = \frac{-R_f A_v}{R_1 A_v + R_f}$$

**DESIGN EXAMPLE** Assume a demodulator design is required which will efficiently extract a low-frequency modulation from a 7-kHz carrier frequency. The upper frequency component of the modulation is 2 Hz. A positive output waveform is required, but the negative portion of the input waveform contains the required information. Therefore, an inverting circuit is used.

*Tentative Circuit-Performance Requirements*
$A_{vc} = -3$
$R_{in} = 2,000 \; \Omega$
$f_m = $ 2-Hz sine wave
$f_c = $ 7-kHz sine wave
$v_i$(peak-to-peak) = 1 V
$v_o$(peak-to-peak) = 3 V
Positive output with filtering
Maximum offset without external offset adjustment = 0.5 V

*Op Amp Parameters* (741)
$f_u = $ 800 kHz
Maximum 3 V (peak-to-peak) frequency = 100 kHz
$S(\text{min}) = 0.5 \; \text{V}/\mu\text{s}$
$V_{io}(\text{max}) = 6 \; \text{mV}$
$I_{io}(\text{max}) = 0.5 \; \mu\text{A}$
$I_b(\text{max}) = 1.5 \; \mu\text{A}$
$\phi_m$(open-loop) = 80°
$A_v = 25{,}000$ if $V^{(\pm)} = \pm 2$ V to 250 k$\Omega$ if $V^{(\pm)} = \pm 20$ V

## PRECISION HALF-WAVE RECTIFIER

Closed-loop bandwidth (normalized) = 1 at 25°C, 1.12 at −55°C, and 0.8 at 125°C

## DESIGN STEPS

**Step 1**  $R_1 = R_{in} = 2{,}000 \ \Omega$.

**Step 2**
$$f_{max} = \frac{f_u}{100|A_{vc}|} = \frac{8 \times 10^5}{100 \times 3} = 2667 \text{ Hz}.$$

**Step 3**
$$S > 2\pi f_c |A_{vc}| v_i(\text{peak}) = 6.28 \times 7 \times 10^3 \times 3 \times 0.5 = 0.066 \text{ V}/\mu\text{s}$$

**Step 4**
$$R_2 = |A_{vc}| R_1 = 3 \times 2{,}000 = 6{,}000 \ \Omega$$

**Step 5**
$$B = \left(\frac{f_c}{f_m}\right)^{1/2} = \left(\frac{7{,}000}{2}\right)^{1/2} = 59.16$$

**Step 6**
$$C_1 = \frac{B}{2\pi f_c R_2} = \frac{59.16}{6.28 \times 7{,}000 \times 6{,}000} = 0.22 \ \mu\text{F}$$

**Step 7**
$$C_f = \frac{1}{2\pi f_m R_f B} = \frac{1}{6.28 \times 2 \times 6{,}000 \times 59.16} = 0.22 \ \mu\text{F}$$

**Step 8**
$$R_p = \frac{R_1 R_f}{R_1 + R_f} = \frac{2{,}000 \times 6{,}000}{2{,}000 + 6{,}000} = 1{,}500 \ \Omega$$

**Step 9**
$$V_o(\text{dc}) = \frac{+0.45 \ v_i(\text{rms}) \ R_f}{R_1} = \frac{0.45 \times 0.35 \times 6{,}000}{2{,}000} = +0.473 \text{ V}$$

**Step 10** Appropriate $D_1$ and $D_2$ direction chosen for a positive output. The diode chosen is the 1N191, which has a $t_{rr}$ of 0.5 $\mu$s. This satisfies the requirement $t_{rr} < 0.01/7{,}000 = 1.4 \ \mu\text{s}$. Laboratory tests showed no distortion.

**Step 11**  The maximum output offset is
$$\Delta V_o = (A_{vc} + 1) V_{io}(\text{max}) + \frac{2I_b + I_{io}}{2R_f}$$
$$= 4 \times 6 \times 10^{-3} + \frac{2 \times 1.5 \times 10^{-6} + 0.5 \times 10^{-6}}{2 \times 6{,}000} = 24 \text{ mV}$$

**Step 12** The 741 is internally compensated and has an open-loop phase margin of 80°. It is therefore unlikely that instability causes 1 or 2 in Chap. 3 will be applicable. The other potential instability causes are now considered:

3. The 741 has a maximum output resistance $R_o$ of 300 $\Omega$ at 1 MHz. The capacitive load, in conjunction with $R_o$, must not create a pole near gain

crossover—otherwise the 80° phase margin will be reduced. If the pole is set at least ten times higher than the gain crossover frequency, the phase margin will not be reduced more than 6°. The load capacitance which will reduce the phase margin from 80 to 74° is

$$C_L = \frac{1}{2\pi(10f_u)R_c} = \frac{1}{(6.28)10^7(300)} = 53 \text{ pF}$$

More load capacitance than 53 pF could be allowed, since a phase margin of 74° is still quite high.

4. $R_f$ is so low that little phase lag of the feedback network is likely. In fact, with $C_f$ installed, the feedback network causes a lead in the loop gain.

5. The resistance between the positive input and ground is 1500 Ω.

6. Careful board layout is required—depending on the op amp pin locations.

7. Ceramic 0.1-μF capacitors between each power-supply terminal to ground are usually sufficient. Some op amps require much less than this.

*Step 13.* The changes of dc closed-loop gain resulting from changes of dc open-loop gain are computed as follows:

$$A_{vc}(A_{vo} = 250 \text{ k}, \text{dc}) = \frac{-R_f A_v}{R_1 A_v + R_f} = \frac{-6{,}000 \times 250{,}000}{2{,}000 \times 250{,}000 + 6{,}000} = 2.99996$$

$$A_{vc}(A_{vo} = 25 \text{ k}, \text{dc}) = \frac{-6{,}000 \times 25{,}000}{2{,}000 \times 25{,}000 + 6{,}000} = 2.9996$$

The closed-loop gain changes only 0.01 percent from a 90 percent reduction of open-loop gain.

## REFERENCES

1. Tobey, G. E., J. G. Graeme, and L. P. Huelsman: "Operational Amplifiers—Design and Applications," p. 245, McGraw-Hill Book Company, New York, 1971.
2. Millman, J., and C. C. Halkias: "Integrated Electronics: Analog Digital Circuits and Systems," p. 572, McGraw-Hill Book Company, New York, 1972.
3. Smith, J. I.: "Modern Operational Circuit Design," p. 36, John Wiley & Sons, Inc., New York, 1971.
4. Kreeger, R.: Ac-to-dc Converters for Low Level Input Signals, *EDN*, Apr. 5, 1973, p. 60.

Chapter **17**

# Logarithmic Circuits

## INTRODUCTION

Logarithmic and antilogarithmic amplifiers are basic building blocks for many nonlinear circuits. They are also intrinsically useful by themselves as analog compressors and expanders. In this chapter we will first discuss, in detail, a high-precision–low-drift log amplifier. Second, design details will be systematically presented for an antilog amplifier.

Log and antilog amplifiers are mathematical inverses of each other. Their principal usefulness is in applications such as multipliers, dividers, and square-root circuits. We will discuss some of these more complex applications in Chap. 19.

## 17.1 DIFFERENTIAL LOGARITHMIC AMPLIFIER

**ALTERNATE NAMES** Log ratio circuit, log amplifier, log converter, data compressor, log subtracting circuit.

**EXPLANATION OF OPERATION** The differential log amplifier shown in Fig. 17.1 provides several useful types of transfer functions. Since it has differential inputs and adjustable gain, transfer functions of the following forms are possible:

$$v_o = K \log_a v_1$$

$$v_o = K \log_a \left(\frac{v_1}{v_2}\right) = K \log_a v_1 - K \log_a v_2$$

$$v_o = K \log_a \left(\frac{1}{v_2}\right) = -K \log_a v_2$$

where $K$ = gain of circuit
$a$ = base of logarithm, which can be set at $e$, 10, or any other useful number

There is really only one constant in the above equations, since logarithms of different bases are related to each other by a constant. For example, the relationship between logarithms to the bases 10 and $e$ is

$$\frac{\log_{10} x}{\log_e x} = 0.4343$$

17-1

## 17-2 LOGARITHMIC CIRCUITS

To simplify the following discussion, we will therefore work exclusively with natural logarithms (base $e = 2.71828$).

In Fig. 17.1 if we assume $R_1 = R_2$, $R_3 = R_4$, $R_5 = R_6$, $R_9 = R_{10}$, $R_{11} = R_{12}$, and $R_{13} = R_{14}$, the transfer function of the circuit is

$$v_o = \frac{kTR_{15}R_{13}}{q(R_{16} + R_T)R_{11}} \ln \left(\frac{v_1/R_7}{v_2/R_8}\right)$$

Resistor $R_T$ is a device having a positive linear temperature coefficient (a silicon resistor). It is used to cancel the $T$ in the transfer function. Other-

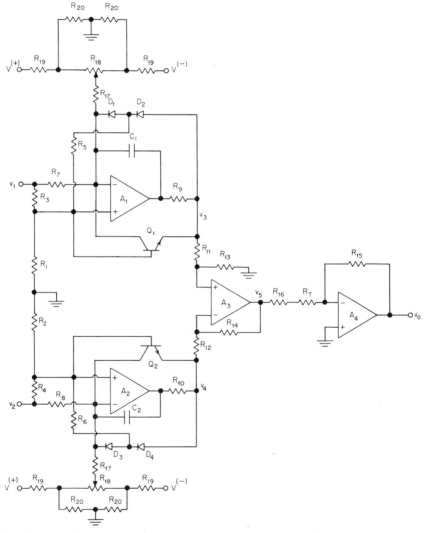

**Fig. 17.1** A differential-input logarithmic amplifier with adjustable gain and logarithm base.

wise $v_o$ would vary linearly with temperature. $T$ is the temperature in degrees Kelvin (273 K = 0°C).

Several other design tricks will keep this circuit from drifting with temperature. Transistors $Q_1$ and $Q_2$ should be a matched pair of devices on one chip. Ideally they should be gain-regulated such as the $\mu A726$ temperature-controlled differential pair (Fairchild). This device has active temperature-regulating circuitry on the same chip as the matched pair so that external temperature sources have no effect on transistor parameters. $A_1$ and $A_2$ should also be a matched pair of op amps. Perhaps $A_1$ to $A_4$ could be a high-quality quad set of op amps on one chip.

This circuit is designed only for positive input voltages. Diodes $D_1$ to $D_4$ clamp the outputs of $A_1$ and $A_2$ to zero if negative input voltages are accidentally applied. The output $v_o$, however, can swing positive or negative as shown in the plotted transfer function (Fig. 17.2).

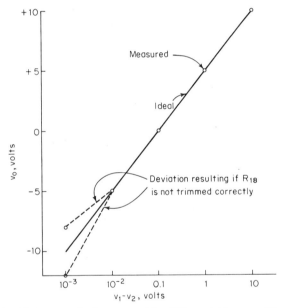

**Fig. 17.2** Transfer function of the differential log amplifier.

Feedback stability of $A_1$ and $A_2$ is controlled with $R_9$, $R_{10}$, $C_1$, and $C_2$. Selection of these parts is quite difficult, since stability depends on the feedback factor, and the feedback factor depends on the input voltages $v_1$ and $v_2$. Experience has shown that the following equations provide reasonable values for these four components:

$$R_9 = R_{10} = \frac{v_3(\max) - 0.7}{v_1(\max)/R_7 + v_3(\max)/R_{11}}$$

$$C_1 = C_2 = \frac{1}{\pi f_u R_9}$$

A tolerance of ±20 percent is sufficient for these parts.

## 17-4 LOGARITHMIC CIRCUITS

### DESIGN PARAMETERS

| Parameter | Description |
|---|---|
| $A$ | Apparent resistance of $R_T$ at 0 K ($-273°C$) |
| $A_1, A_2$ | Differential-input log amplifiers |
| $A_3$ | Summing differential amplifier |
| $A_4$ | Temperature-compensating circuit |
| $B$ | Slope of $R_T$ as a function of temperature |
| $\beta$ | Current gain of $Q_1$ and $Q_2$ |
| $C_1$ to $C_2$ | Feedback-stabilizing capacitors |
| $D_1$ to $D_4$ | Diodes used to clamp $A_1$ and $A_2$ if a negative input voltage is accidentally applied to $v_1$ or $v_2$ |
| $E$ | Fractional error in transfer function |
| $f_u$ | Frequency of open-loop unity gain ($A_1$ and $A_2$) |
| $I_{b1}$ to $I_{b2}$ | Input bias current of $A_1$ and $A_2$ |
| $\Delta I_{b1}$ to $\Delta I_{b2}$ | Change of input bias current of $A_1$ and $A_2$ |
| $k$ | Boltzmann's constant = $1.380 \times 10^{-23}$ J/K |
| $q$ | Electronic charge = $1.60 \times 10^{-19}$ C |
| $Q_1$ to $Q_2$ | Transistors which provide logarithmic characteristics to circuit |
| $R_1$ to $R_4$ | Resistors to compensate for bulk-resistance effects of $Q_1$ and $Q_2$ at high levels |
| $R_5$ to $R_6$ | Part of diode clamp circuits |
| $R_7$ to $R_8$ | Gain-determining resistors |
| $R_9$ to $R_{10}$ | Feedback-compensating resistors |
| $R_{11}$ to $R_{14}$ | Gain-determining resistors of differential summing amplifier $A_3$ |
| $R_{15}$ to $R_{16}$ | Determines gain of temperature-compensating circuit |
| $R_{17}$ to $R_{20}$ | $A_1$ and $A_2$ input offset adjustment to trim circuit at $v_o = 0$ |
| $R_{Q1}$ to $R_{Q2}$ | Effective collector-emitter resistance of $Q_1$ and $Q_2$ |
| $R_T$ | Positive-temperature-coefficient resistor |
| $T$ | Temperature in kelvins (0 K = $-273°C$) |
| $V^{(\pm)}$ | Power-supply voltages |
| $v_o$ | Circuit output voltage |
| $v_1$ to $v_2$ | Circuit input voltages |
| $\Delta v_1$ to $\Delta v_2$ | Measurement errors in $v_1$ or $v_2$ |
| $V_{io1}$ to $V_{io2}$ | Input offset voltages of $A_1$ and $A_2$ |
| $\Delta V_{io1}$ to $\Delta V_{io2}$ | Change in input offset voltages of $A_1$ and $A_2$ |

### DESIGN EQUATIONS

| Eq. No. | Description | Equation |
|---|---|---|
| 1 | Output-circuit voltage assuming $R_1 = R_2$, $R_3 = R_4$, $R_5 = R_6$, $R_9 = R_{10}$, $R_{11} = R_{12}$, and $R_{13} = R_{14}$ | $v_o = \dfrac{kTR_{15}R_{13}}{q(R_{16} + R_T)R_{11}} \ln\left(\dfrac{v_1/R_7}{v_2/R_8}\right)$ |
| 2 | Error in measurement of $v_1$ due to $A_1$ input errors (important at low $v_1$ levels) | $\Delta v_1 = \pm V_{io1} + R_7 I_{b1}$ |
| 3 | Error in measurement of $v_2$ due to $A_2$ input errors (important at low $v_2$ levels) | $\Delta v_2 = \pm V_{io2} + R_8 I_{b2}$ |

NOTE: The above two errors can be mostly canceled out at one temperature with the $R_{18}$ potentiometers. If $A_1$ and $A_2$ have identical parameter drifts with temperature, $\Delta v_1$ and $\Delta v_2$ will cancel over temperature

## DIFFERENTIAL LOGARITHMIC AMPLIFIER

| Eq. No. | Description | Equation |
|---|---|---|
| 4 | Required relationship among $R_1$, $R_3$, and $R_7$ to cancel the effects of bulk resistance in $Q_1$ (important at high $v_1$ levels) | $R_{Q1} = \dfrac{R_1 R_7}{R_1 + R_3}$ |
| 5 | Required relationship among $R_2$, $R_4$, and $R_8$ to cancel the effects of bulk resistance in $Q_2$ (important at high $v_2$ levels) | $R_{Q2} = \dfrac{R_2 R_8}{R_2 + R_4}$ |
| 6 | Approximate dynamic range possible for $v_1$ input (assuming no cancellation of errors in $A_3$) | $\dfrac{v_1(\max)}{v_1(\min)} \approx \dfrac{kTE^2 R_7 / q R_{Q1}(\min)}{\Delta V_{io1}(\max) + R_7 \Delta I_{b1}(\max)}$ |
| 7 | Approximate dynamic range possible for $v_2$ input (assuming no cancellation of errors in $A_3$) | $\dfrac{v_2(\max)}{v_2(\min)} \approx \dfrac{kTE^2 R_8 / q R_{Q2}(\min)}{\Delta V_{io2}(\max) + R_8 \Delta I_{b2}(\max)}$ |
| 8 | Resistor values $R_1$, $R_2$ | $R_1 = R_2 = 10 \ \Omega$ |
| 9 | $R_3$ | $R_3 = \dfrac{R_1 R_7}{R_{Q1}(\min)} - R_1$ |
| 10 | $R_4$ | $R_4 = \dfrac{R_2 R_8}{R_{Q2}(\min)} - R_2$ |
| 11 | $R_5$, $R_6$ | $R_5 = R_6 = 10 \ k\Omega$ |
| 12 | $R_7$, $R_8$ | $R_7 = R_8 \approx \dfrac{\Delta V_{io1}/\Delta T}{\Delta I_{b1}/\Delta T}$ |
| 13 | $R_9$, $R_{10}$ | $R_9 = R_{10} = \dfrac{(kT/q)\ln(v_1/v_2)(\max)}{[v_1(\max)/R_7] - [kT \ln(v_1/v_2)(\max)/qR_{11}]}$ (assume $T = 300$ K) |
| 14 | $R_{11}$, $R_{12}$ | $R_{11} = R_{12} = 10 \ k\Omega$ |
| 15 | $R_{13}$, $R_{14}$ | $R_{13} = R_{14} = R_{11} \left[ \dfrac{qv_o(\max)}{kT \ln(v_1/v_2)(\max)} \right]^{1/2}$ (assume $T = 300$ K) |
| 16 | $R_{15}$ | $R_{15} = BT \left[ \dfrac{qv_o(\max)}{kT \ln(v_1/v_2)(\max)} \right]^{1/2}$ (assume $T = 300$ K) |
| 17 | $R_{16}$ | $R_{16} = -A$ |
| 18 | $R_{17}$ | $R_{17} \geqq 10 R_7$ |
| 19 | $R_{18}$ | $R_{18} \geqq \dfrac{R_{17}}{10}$ |
| 20 | $R_{19}$ | $R_{19} \approx \dfrac{V^{(+)} R_{20}}{100 \ V_{io1}(\max)}$ |
| 21 | $R_{20}$ | $R_{20} \leqq \dfrac{R_{18}}{100}$ |

## 17-6 LOGARITHMIC CIRCUITS

| Eq. No. | Description | Equation |
|---|---|---|
| 22 | Resistance of positive-temperature-coefficient device | $R_T \approx A + BT$ |
| 23 | Capacitor values: $C_1$, $C_2$ | $C_1 = C_2 \approx \dfrac{1}{\pi f_u R_9}$ |
| 24 | Bulk resistance of $Q_1$ | $R_{Q1} = \dfrac{kTR_7}{qv_1\beta}$ |
| 25 | Bulk resistance of $Q_2$ | $R_{Q2} = \dfrac{kTR_8}{qv_2\beta}$ |

### DESIGN PROCEDURE

Many complicating factors must be considered to design a high-quality log amplifier. We will assume that linearity is the most important parameter and let the other parameters be controlled by physical limitations imposed by the design equations.

### DESIGN STEPS

*Step 1.* If $A_1$ and $A_2$ are identical op amps, $R_7$ and $R_8$ are determined from Eq. 12.

*Step 2.* Calculate the minimum expected bulk resistances of $Q_1$ and $Q_2$ using Eqs. 24 and 25.

*Step 3.* Compute values for $R_3$ and $R_4$ using Eqs. 8, 9, and 10 and the results of steps 1 and 2.

*Step 4.* Let $R_{11} = R_{12} = 10$ k$\Omega$. Determine values for $R_9$ and $R_{10}$ using Eq. 13.

*Step 5.* Compute values for $R_{13}$ and $R_{14}$ from Eq. 15.

*Step 6.* The resistance of most positive-coefficient temperature-sensitive resistors (silicon resistors) can be described by the form shown in Eq. 22. If the temperature could be extended down to 0 K ($-273°$C), its resistance would theoretically be $A$. The slope of resistance as a function of temperature is $B$. Use these numbers to compute $R_{15}$ and $R_{16}$ using Eqs. 16 and 17.

*Step 7.* Calculate values for $C_1$ and $C_2$ using Eq. 23.

*Step 8.* Calculate approximate values for $R_{17}$ to $R_{20}$ using Eqs. 18 to 21.

**EXAMPLE OF LOG-AMPLIFIER DESIGN**  Suppose we wish to precondition analog data before they are applied to an analog-to-digital converter. Our goal is to keep the A/D error small by compressing the input voltage (which encompasses four orders of magnitude) into $\pm 10$ V (slightly more than 1 order of magnitude). The A/D converter accepts voltages from $-10$ to $+10$ V. The input voltage to the log amplifier ranges from 0.001 to 10 V. The output voltage is to be zero when the input voltage is 0.1 V. The required transfer function, along with test data from a circuit designed with the following steps, is shown in Fig. 17.2.

*Design Requirements*

$v_o(\text{max}) = \pm 10$ V
$v_1(\text{min}) = 10^{-3}$ V
$v_1(\text{max}) = 10$ V

$v_2 = 0.1$ V (fixed)
$V^{(\pm)} = \pm 15$ V

*Device Data*

$\Delta V_{io1} = \Delta V_{io2} = \pm 0.3$ mV (0 to $+75°$C)
$V_{io1}(\text{max}) = 1$ mV
$\Delta I_{b1} = \Delta I_{b2} = 2$ nA
$f_u = 10^6$ Hz
$R_T = 500 \, \Omega$ at 300 K
$A = -730 \, \Omega$
$B = 4.1 \, \Omega/\text{K}$
$\beta = 150(\text{max})$, 100 at $I_c(\text{max})$

*Step 1.* $R_7$ and $R_8$ must be approximately

$$R_7 = R_8 \approx \frac{\Delta V_{io1}/\Delta T}{\Delta I_{b1}/\Delta T} \approx \frac{3 \times 10^{-4}}{2 \times 10^{-9}} \approx 150 \text{ k}\Omega$$

*Step 2.* The minimum bulk resistances of $Q_1$ and $Q_2$ are

$$R_{Q1}(\text{min}) = R_{Q2}(\text{min}) = \frac{kTR_7}{qv_1(\text{max}) \beta [\text{at } I_c(\text{max})]}$$

$$= \frac{(1.38 \times 10^{-23} \text{ J/K})(273 \text{ K})1.5 \times 10^5 \, \Omega}{(1.6 \times 10^{-19} \text{ C})(10 \text{ V})(100)}$$

$$= 3.53 \, \Omega$$

*Step 3.* From Eq. 8: $R_1 = R_2 = 10 \, \Omega$. Equations 9 and 10 provide

$$R_3 = R_4 = \frac{R_1 R_7}{R_{Q1}(\text{min})} - R_1 = \frac{10(1.5 \times 10^5)}{3.53} - 10$$

$$= 425 \text{ k}\Omega$$

*Step 4.* We let $R_{11} = R_{12} = 10$ k$\Omega$ as recommended. $R_9$ and $R_{10}$ are found from

$$R_9 = R_{10} = \frac{(kT/q)\ln(v_1/v_2)(\text{max})}{[v_1(\text{max})/R_7] - [kT \ln(v_1/v_2)(\text{max})/qR_{11}]}$$

$$= \frac{[(1.38 \times 10^{-23})(300)/1.6 \times 10^{-19}]\ln(10/0.1)}{(10/1.5 \times 10^5) - [(1.38 \times 10^{-23})(300)\ln(10/0.1)/(1.6 \times 10^{-19})10^4]}$$

$$= 2{,}176 \, \Omega$$

*Step 5.* $R_{13}$ and $R_{14}$ are now found:

$$R_{13} = R_{14} = R_{11} \left[ \frac{qv_o(\text{max})}{kT \ln(v_1/v_2)(\text{max})} \right]^{1/2}$$

$$= 10^4 \left[ \frac{1.6 \times 10^{-19} \times 10}{1.38 \times 10^{-23}(300)\ln(10/0.1)} \right]^{1/2} = 91.6 \text{ k}\Omega$$

*Step 6.* Equation 16 gives us $R_{15}$:

$$R_{15} = BT \left[ \frac{qv_o(\text{max})}{kT \ln(v_1/v_2)(\text{max})} \right]^{1/2}$$

$$= (4.1)(300) \left[ \frac{1.6 \times 10^{-19} \times 10}{1.38 \times 10^{-23}(300)\ln(10/0.1)} \right]^{1/2}$$

$$= 11.3 \text{ k}\Omega \qquad \text{(during test this resistor is trimmed so that } v_o = 10 \text{ when } v_1 = 10)$$

**17-8  LOGARITHMIC CIRCUITS**

Equation 17 provides: $R_{16} = -A = -(-730) = 730 \ \Omega$

*Step 7.* Capacitor values are

$$C_1 = C_2 \approx \frac{1}{\pi f_u R_9} \approx \frac{1}{\pi 10^6 \times 2{,}176} \approx 146 \text{ pF}$$

*Step 8.* $R_{17}$ to $R_{20}$ are determined using Eqs. 18 to 21:

$$R_{17} \geq 10 \ R_7 = 10 \ (150 \text{ k}\Omega) = 1.5 \text{ M}\Omega$$

$$R_{19} \leq \frac{R_{17}}{10} = \frac{1.5 \times 10^6}{10} = 150 \text{ k}\Omega$$

$$R_{20} \leq \frac{R_{18}}{10} = \frac{150{,}000}{10} = 1{,}500 \ \Omega$$

$$R_{19} \approx V^{(\pm)} \frac{R_{20}}{100 V_{io1}(\text{max})}$$

$$\approx \frac{15(1{,}500)}{100(10^{-3})} \approx 225 \text{ k}\Omega$$

## REFERENCES

1. Morgan, D. R.: Get the Most out of Log Amplifiers by Understanding the Error Sources, *EDN*, Jan. 10, 1973, p. 52.
2. Sheingold, D., and F. Pouliot: The Hows and Whys of Log Amps, *Electron. Des.*, Feb. 1, 1974, p. 52.

## 17.2 ANTILOGARITHMIC AMPLIFIER

**ALTERNATE NAMES**  Antilog amplifier, inverse-log amplifier, antilog converter, data expander, exponential amplifier.

**EXPLANATION OF OPERATION**  This circuit is merely a modification of the log circuit described in the last section. The antilog function is implemented in Fig. 17.3 by changing the connections to $Q_1$ and $Q_2$ in the log circuit of Fig. 17.1.

The relationship between input and output voltages in this circuit is

$$v_o = D \exp(-E v_1)$$

where $D = \dfrac{R_5 V_R (R_7 + R_8)}{R_1 R_7}$

$$E = -\frac{q(R_T + R_3)}{kT(R_T + R_2 + R_3)}$$

The $D$ term is equal to the required output voltage when the input voltage is zero.

The $R_T$ and $R_2$ terms in $E$ establish the dynamic range of the antilog amplifier. For a given $\pm v_1(\text{max})$ input range, the total number of decades excursion of $v_o$ increases as $E$ increases in magnitude. The maximum input voltage range $\pm v_1(\text{max})$ is controlled mainly by the $R_T/(R_T + R_2)$ ratio. In some applications this ratio can be made quite large, so $\pm v_1(\text{max})$ may be many tens or hundreds of volts, if required. The resistor $R_T$ must have a positive temperature coefficient to cancel the $T$ term in Eq. 1 (see Design Equations).

The emitter saturation currents $I_{E1}$ and $I_{E2}$ will have no effect on $v_o$ if they track over temperature. If they are not equal, they merely affect $D$ (see

ANTILOGARITHMIC AMPLIFIER 17-9

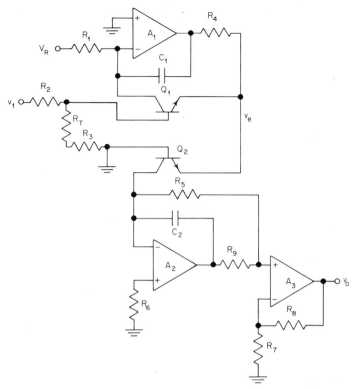

**Fig. 17.3** Antilog amplifier formed by changing the input and feedback circuits of Fig. 17.1.

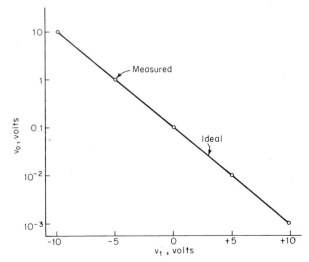

**Fig. 17.4** Response of an antilog amplifier using component values computed in the example.

## 17-10  LOGARITHMIC CIRCUITS

Eqs. 1 and 2); however, they still should be selected so their ratio remains constant over temperature.

The dynamic range of $v_o$ is limited by $V_n$, $I_n$, $V_{io}$, $I_b$, and $I_{io}$ of both op amps. In practice, only three or four decades of dynamic range is possible without going to expensive op amps and complex temperature-compensating circuits.

### DESIGN PARAMETERS

| Parameter | Description |
|---|---|
| $A$ | Apparent resistance of $R_T$ at 0 K ($-273°C$) |
| $A_1$ to $A_2$ | Nonlinear amplifiers |
| $A_3$ | Buffer to extend output-voltage range |
| $B$ | Slope of $R_T$ as a function of temperature |
| $C_1$ to $C_2$ | Feedback-stabilizing capacitors |
| $D$ | Output voltage when $v_i = 0$ |
| $E$ | Determines dynamic range of circuit |
| $I_b$ | Op amp input bias current ($A_1$ to $A_2$) |
| $I_{io}$ | Op amp input offset current ($A_1$ to $A_2$) |
| $I_n$ | Equivalent rms input noise current of an op amp ($A_1$ to $A_2$) |
| $k$ | Boltzmann's constant = $1.380 \times 10^{-23}$ J/K |
| $N$ | Number of decades response of circuit on either side of $v_o = D$ |
| $q$ | Electronic charge = $1.60 \times 10^{-19}$ C |
| $Q_1$ to $Q_2$ | Transistors used to provide nonlinear transfer function |
| $R_1$ | Sets reference current into $A_1$ |
| $R_2$ to $R_3$ | Part of temperature-compensation circuit |
| $R_4$, $R_9$ | Provides feedback stability |
| $R_5$ | Establishes gain of $A_2$ circuit |
| $R_6$ | Cancels effect of $I_b$ in $A_2$ |
| $R_7$ to $R_8$ | Establishes gain of $A_3$ buffer stage |
| $R_T$ | Provides temperature compensation |
| $T$ | Temperature in kelvins (0°C = 273 K) |
| $v_o$ | Circuit output voltage |
| $v_1$ | Circuit input voltage |
| $v_e$ | Voltage at emitters of $Q_1$ to $Q_2$ |
| $V_{io}$ | Input offset voltage of an op amp |
| $V_n$ | Equivalent rms input voltage noise of an op amp |
| $V_R$ | Reference voltage |

### DESIGN EQUATIONS

| Eq. No. | Description | Equation |
|---|---|---|
| 1 | Circuit output voltage | $v_o = D \exp(-E v_1)$ where $D = \dfrac{R_5 V_R (R_7 + R_8)}{R_1 R_7}$ $E = -\dfrac{q(R_T + R_3)}{kT(R_T + R_2 + R_3)}$ |
| 2 | Voltage at junction of transistor emitters | $v_e = \dfrac{v_1(R_T + R_3)}{R_T + R_2 + R_3} - \dfrac{kT}{q} \ln \left( \dfrac{V_R I_{E2} R_5}{v_o I_{E1} R_1} \right)$ |
| 3 | $R_T$, $R_2$, and $R_3$ required to provide $\pm N$ decades response of $v_o$ about $D$ (nominal $v_o$) | $\dfrac{R_T + R_3}{R_T + R_2 + R_3} = \dfrac{kTN \ln 10}{q|v_1(\max)|}$ |

## ANTILOGARITHMIC AMPLIFIER

| Eq. No. | Description | Equation |
|---|---|---|
| 4 | Resistor values, $R_1$ | $R_1 = 10^6$ (chosen such that it does not load down $V_R$) |
| 5 | $R_2$ | $R_2 = (R_T + R_3)\left[\dfrac{q|v_1(\text{max})|}{kTN \ln 10} - 1\right]$ <br> ($T = 300$ K) |
| 6 | $R_3$ | $R_3 = -A$ |
| 7 | $R_4$ | $R_4 = 10$ k$\Omega$ |
| 8 | $R_5$, $R_6$ | $R_5 = R_6 = \dfrac{DR_1}{10 V_R}$ |
| 9 | Capacitor $C_1$ | $C_1 = 100$ pF |
| 10 | Resistance of $R_T$ as a function of temperature | $R_T = A + BT$ |

### DESIGN PROCEDURE

This antilog amplifier will provide the best performance over temperature if the transistors are a single-chip matched set. The op amps should be high-quality devices with low $V_n$, $I_n$, $V_{io}$, $I_b$, and $I_{io}$. Offset-adjustment terminals would also be a desirable feature. Resistor $R_T$ provides most of the temperature compensation if $R_2 \gg R_T$. The circuit is trimmed for $v_o = D$ when $v_1 = 0$ using $R_5$. Likewise, $v_o(\text{max})$ is trimmed when $v_1$ is fully negative using $R_2$.

### DESIGN STEPS

*Step 1.* The temperature characteristics of $R_T$ must first be obtained. Most positive temperature-sensitive resistors (silicon resistors) can be described by the relationship shown in Eq. 10. The apparent resistance of $R_T$ at 0 K is $A$. The slope of resistance as a function of temperature is $B$. Use $A$ to determine $R_3$ using Eq. 6. (Note that $A$ is a negative number, so $R_3$ will be positive.)

*Step 2.* Calculate a value for $R_2$ using Eq. 5.

*Step 3.* Calculate values for $R_5$ and $R_6$ using Eq. 8.

**EXAMPLE OF AN ANTILOG-AMPLIFIER DESIGN** The three design steps will be numerically illustrated through the design of an antilog amplifier which is the inverse of the log amplifier designed in Sec. 17.1. This circuit will produce $v_o = +10^{-3}$ V if $v_1 = +10$ V and $v_o = +10$ V if $v_1 = -10$ V. If $v_1 =$ zero, $v_o$ should be 0.1 V, which is the geometric center of $10^{-3}$ V and 10 V.

*Design Requirements*

$D = 0.1$ V (output when $v_1 = 0$)
$N = 2$ ($\pm$ decades response on either side of $D$)
$\pm v_1(\text{max}) = \pm 10$ V
$V_R = +5.00$ V

*Device Data*

$A = -730$ $\Omega$
$B = 4.1$ $\Omega$/K
$R_T = 500$ $\Omega$ at 300 K

*Step 1.* Resistor $R_3$ is simply
$$R_3 = -A = -(-730) = 730 \ \Omega$$
*Step 2.* $R_2$ is calculated from
$$R_2 = (R_T + R_3)\left[\frac{q|v_1(\text{max})|}{kTN \ln 10} - 1\right]$$
$$= (500 + 730)\left[\frac{1.6 \times 10^{-19}(10)}{1.38 \times 10^{-23}(300)2 \ln 10} - 1\right] = 102 \text{ k}\Omega$$
*Step 3.* $R_5$ and $R_6$ are found from
$$R_5 = R_6 = \frac{DR_1}{10V_R} = \frac{0.1(10^6)}{10\ (5)} = 2{,}000 \ \Omega$$

## REFERENCES

1. Sheingold, D., and F. Pouliot: The Hows and Whys of Log Amps, *Electron. Des.*, Feb. 1, 1974, p. 52.
2. "Logarithmic Converters," National Semiconductor Corp., Application Note AN-30, November 1969.

Chapter **18**

# Modulators

## INTRODUCTION

We will present systematized design information on three types of modulators in this chapter. Each of these modulators requires a corresponding demodulator, detector, discriminator, or decoder at the other end of the system to restore the information to its original form. These other circuits are discussed in Chaps. 7, 8, 14, and 16.

The first circuit we present will be the amplitude modulator — sometimes called the linear modulator. This circuit is similar in function to that used by AM broadcast stations to superimpose an audio signal on a high-frequency carrier signal. Multipliers, discussed in Chap. 19, can also be used for AM modulation.

Frequency modulation, or voltage-to-frequency conversion, is covered in Chap. 21.

The second circuit to be presented is the pulse-amplitude modulator. This is similar to the amplitude modulator except that the circuit is optimized for pulse-handling efficiency. The last circuit will be a pulse-width modulator. This is a good modulator to use in digital systems which operate at fixed pulse voltages. It is also quite useful in high-power systems such as motor drives and switching power supplies.

## 18.1 AMPLITUDE MODULATOR

**ALTERNATE NAMES** AM modulator, linear modulator, linear amplitude modulator.

**EXPLANATION OF OPERATION** The carrier frequency is applied to the $v_c$ terminal as shown in Fig. 18.1. The modulation input, at a lower frequency than $v_c$, is applied to $v_m$. An inverted replica of $v_m$ appears at $v_1$. Its amplitude and offset are controlled by $R_2$ and $R_4$. The offset adjustment $R_4$ controls the modulation depth. The amplitude adjustment $R_2$ determines the final peak-to-peak amplitude of $v_o$.

On positive $v_c$ cycles, switches $S_1$ and $S_4$ are turned on. In this state they look like a resistor ($R_{on}$) of 50 to 500 Ω, depending on the type of CMOS switch chosen. During this same time, switches $S_2$ and $S_3$ are off. In the off

## 18-2 MODULATORS

state these devices look like a resistor ($R_{off}$) of many megohms. On negative $v_c$ cycles just the opposite occurs: $S_1$ and $S_4$ are off and $S_2$ and $S_3$ are on.

Chopped versions of $v_1$ appear at $v_2$ and $v_3$. Figure 18.2 shows that $v_2$ and $v_3$ are chopped on opposite half cycles of $v_c$. The amplitudes of $v_2$ and $v_3$ are smaller than $v_1$ depending on the size of the $R_8/(R_6 + R_8)$ and $R_{11}/(R_7 + R_{11})$ ratios. Voltage waveform $v_2$ is then inverted to create $v_4$. Waveforms $v_3$ and $v_4$ are combined in the summing amplifier $A_4$. Figure 18.2 shows the final $v_o$ result. The peak-to-peak magnitude of $v_o$ can again be adjusted using $R_{12}$, if required.

The only significant error source in this circuit is the magnitudes of $R_{on}$ and $R_{off}$ relative to $R_6$, $R_7$, $R_8$, and $R_{11}$. For example, when $S_1$ is off and $S_2$ is on, the voltage at $v_2$ should be zero. Instead, it is approximately $v_2(\text{off}) \approx v_1 R_{on}/R_{off}$. Conversely, when $S_1$ is on and $S_2$ is off, the voltage at $v_2$ should be exactly $R_8/(R_6 + R_8)$. In this case switch errors give us $v_2(\text{on}) \approx v_1 R_8/(R_{on} + R_6 + R_8)$. If $R_8$ is small, the error is significant. The errors at $v_3$ are similar in nature and are listed in the design equations.

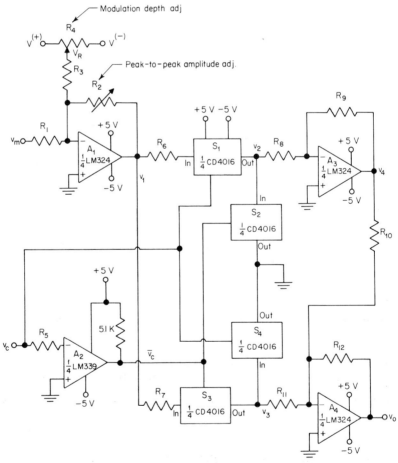

**Fig. 18.1** A precision amplitude modulator which uses CMOS switches.

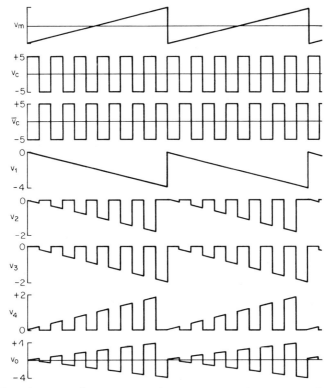

**Fig. 18.2** Voltage waveforms at various locations in Fig. 18.1. A sawtooth modulation waveform is assumed.

## DESIGN PARAMETERS

| Parameter | Description |
|---|---|
| $A_1$ | Buffer amplifier which sets modulation depth and system gain |
| $A_2$ | Comparator used to invert $v_c$ |
| $A_3$ | Amplifier used to invert $v_2$ |
| $A_4$ | Summing amplifier used to combine $v_3$ and $v_4$ |
| $I_{b4}$ | Input bias current of $A_4$ |
| $R$ | Common value for $R_6$ to $R_{11}$ |
| $R_1$ to $R_2$ | Determines system gain for $v_m$ |
| $R_3$ to $R_4$ | Sets modulation depth |
| $R_5$ | Limits current into $A_2$ |
| $R_6$, $R_8$ to $R_{10}$ | Establishes magnitude of negative portion of output waveform |
| $R_7$, $R_{11}$ | Establishes magnitude of positive portion of output waveform |
| $R_{12}$ | Can be used in conjunction with $R_2$ to adjust system gain |
| $R_{on}$, $R_{off}$ | $S_1$ to $S_4$ on and off resistances |
| $R_{in}$ | Input resistance seen by $v_m$ input |
| $S_1$ to $S_4$ | CMOS switches |
| $v_o$ | Circuit output-voltage waveform |
| $\Delta v_o$ | DC offset in $v_o$ caused by $A_4$ |
| $v_1$ to $v_4$ | Voltage waveforms as shown in Figs. 18.1 and 18.2 |
| $v_c, \bar{v}_c$ | Carrier input voltage and its inverse |

## 18-4 MODULATORS

| Parameter | Description |
|---|---|
| $v_m$ | Modulation input voltage |
| $V_N$ | Most negative value of $V_R$ |
| $V_P$ | Most positive value of $V_R$ |
| $V_R$ | Reference voltage used to adjust modulation depth |
| $V^{(\pm)}$ | Power-supply voltages |

### DESIGN EQUATIONS

| Eq. No. | Description | Equation |
|---|---|---|
| 1 | Output voltage $v_o$ when $S_1$ is on and $S_2$ is off (neglecting switch errors and op amp errors) | $v_o(-) = \dfrac{-R_2 R_9 R_{12}}{R_{10}(R_6 + R_8)} \left(\dfrac{v_m}{R_1} + \dfrac{V_R}{R_3}\right)$ |
| 2 | Output voltage $v_o$ when $S_3$ is on and $S_4$ is off (neglecting switch errors and op amp errors) | $v_o(+) = \dfrac{R_2 R_{12}}{(R_7 + R_{11})} \left(\dfrac{v_m}{R_1} + \dfrac{V_R}{R_3}\right)$ |
| 3 | Peak-to-peak output voltage $v_o$ when switch errors are included | $v_o(\pm) \approx \dfrac{\pm R_2 R_{12} R_{\text{off}}}{(R + R_{\text{on}})(R + R_{\text{off}}) + R\, R_{\text{off}}} \left(\dfrac{v_m}{R_1} + \dfrac{V_R}{R_3}\right)$ |
| 4 | Voltage $v_1$ as a function of $v_m$ and $V_R$ | $v_1 = -R_2 \left(\dfrac{v_m}{R_1} + \dfrac{V_R}{R_3}\right)$ |
| 5 | Voltage $v_2$ while $S_1$ is on and $S_2$ is off | $v_2 = \dfrac{R_8 v_1}{R_6 + R_8}$ |
| 6 | Voltage $v_3$ while $S_3$ is on and $S_4$ is off | $v_3 = \dfrac{R_{11} v_1}{R_7 + R_{11}}$ |
| 7 | Voltage $v_4$ | $v_4 = -\dfrac{R_9 v_2}{R_8}$ |
| 8 | Voltage $v_o$ | $v_o = -R_{12}\left(\dfrac{v_4}{R_{10}} + \dfrac{v_3}{R_{11}}\right)$ |
| 9 | Resistor values $R_1$ | $R_1 \gtrsim R_{\text{in}}$ required |
| 10 | $R_2$ | $R_2 < \dfrac{R_1(|V^{(\pm)}|-2)}{|v_m|\,\text{max}}$ or $< \dfrac{R_3(|V^{(\pm)}|-2)}{|V_R|\,\text{max}}$ whichever is smaller |
| 11 | $R_3, R_4$ | $R_3 = R_4 = R_1$ |
| 12 | $R_5$ | $R_5$ = zero to 100 kΩ (use manufacturer's recommendation for $A_2$) |
| 13 | $R_6$ to $R_{11}$ | $R = R_6 = R_7 = R_8 = R_9 = R_{10} = R_{11} = (R_{\text{on}} R_{\text{off}})^{1/2}$ |
| 14 | $R_{12}$ | $R_{12} = 2 R_{11}$ |
| 15 | Output offset due to $I_{b4}$ | $\Delta v_o = I_{b4} R_{12}$ |

### DESIGN PROCEDURE

We assume all op amps are operating from the same $V^{(\pm)}$ supplies. If a larger output is required, all stages or perhaps only $A_4$ can be connected to higher-voltage supplies. Fewer power supplies are required if the comparator ($A_2$), the four switches, and the three op amps can use the same voltages.

## DESIGN STEPS

*Step 1.* Choose a comparator, switches, and op amps which utilize the same supply voltages. The $\pm$ supply voltages must be at least 2 V higher than the $\pm$ peak values expected at $v_o$.

*Step 2.* Choose $R_1$ to be greater than or equal to the minimum allowed input resistance at $v_m$.

*Step 3.* Choose a range of values for $V_R$ which is slightly larger than the expected positive and negative peaks of $v_m$. Set $V_P$ and $V_N$ equal to the two limits of this range.

*Step 4.* Use Eqs. 11 and 12 to obtain values for $R_3$, $R_4$, and $R_5$. Resistor $R_5$ is usually between 1 and 5 k$\Omega$, but the comparator ($A_2$) specification sheet should be consulted.

*Step 5.* Solve both portions of Eq. 10 to determine an optimum value for $R_2$. Let $R_2$ be less than the smaller value calculated.

*Step 6.* Use the CMOS switch specification to estimate $R_{on}$ and $R_{off}$ for these particular values of $V^{(\pm)}$, $v_1(\text{max})$, and $\pm v_c$. Calculate values for $R_6$ through $R_{11}$ using Eq. 13.

*Step 7.* Rearrange Eq. 1 as follows to compute a value for $R_{12}$ (let $R = R_6 = R_8 = R_9 = R_{10}$):

$$R_{12} = \frac{4\ R\ R_1\ v_o(\text{peak})}{3\ R_2\ v_m(\text{peak})}$$

To rearrange this equation, we assumed the nominal value for $V_R$ would be $v_m/2$.

If $R_{12}$ is too large, the bias current of $A_4$ will cause an output offset. Calculate this offset using $\Delta v_o = I_{b4}\ R_{12}$. If the offset is more than can be tolerated, decrease $R_{12}$ until a satisfactory offset is achieved. This change will lower the overall circuit gain, so an opposite change to the circuit transfer function must be made elsewhere. The best option is to lower all the $R$ resistors by the same magnitude that $R_{12}$ was lowered. This will slightly increase switching errors, but this is usually of less concern than a dc offset in $v_o$.

*Step 8.* Solve Eqs. 1 and 2 to make sure the correct positive and negative peak output voltages will be achieved.

*Step 9.* Solve Eq. 3 to determine the worst-case expected errors in $v_o$.

**EXAMPLE OF AN AM MODULATOR DESIGN** Suppose we wish to modulate a 2-kHz carrier with transponder information between 3 and 100 Hz. The modulation signal has a maximum peak-to-peak amplitude of 4 V. The modulated carrier is required to have a peak-to-peak amplitude of 6 V.

*Design Requirements*

$v_m = \pm 2$ V centered on zero
$v_o = \pm 3$ V centered on zero
$V^{(\pm)} = \pm 5$ V
$R_{in} \geq 5{,}000\ \Omega$
$\Delta v_o(\text{max}) = \pm 0.1$ V

*Device Data* (+25°C)

$R_{on} = 580\ \Omega$
$R_{off} = 4 \times 10^8\ \Omega$ (125-nA leakage with 5 V)
$I_{b4} = 500$ nA (out of device, since this op amp has a PNP input stage)

## 18-6 MODULATORS

*Step 1.* We choose one-fourth of an LM339 quad comparator for $A_2$. This device works quite well from supply voltages of $\pm 5$ V. If a $\pm 5$-V $v_c$ is not available, the other three sections of the comparator can be utilized to square up and level shift the carrier input to the proper waveform. The 4016 quad CMOS switch is used for $S_1$ through $S_4$. Three sections of the LM324 quad op amp are utilized for $A_1$, $A_3$, and $A_4$. Both these devices also operate fairly efficiently from $\pm 5$ V. Slightly better performance can be achieved in switch performance using $\pm 10$ V, however.

*Step 2.* We will let $R_{in} = R_1 = 10$ k$\Omega$. This satisfies the 5-k$\Omega$ minimum input resistance.

*Step 3.* We will let $V_P = +5$ V and $V_N = -5$ V so that the same supply voltages can be used. This means, of course, that the $\pm 5$ V must now be regulated. Otherwise, the depth of modulation (percent modulation) will vary with supply voltage.

*Step 4.* Equation 11 provides us with $R_3 = R_4 = R_1 = 10$ k$\Omega$. For the LM339 comparator we can let $R_5 =$ zero.

*Step 5.* The two portions of Eq. 10 are computed as follows:

$$R_2 < \frac{R_1[|V^{(\pm)}| - 2]}{|v_m| \text{ max}}$$
$$< \frac{10^4 (5 - 2)}{2}$$
$$< 15{,}000 \ \Omega$$

Also,
$$R_2 < \frac{R_3[|V^{(\pm)}| - 2]}{|V_R| \text{ max}}$$
$$< \frac{10^4 (5 - 2)}{5}$$
$$< 6{,}000 \ \Omega$$

We therefore let $R_2 = 5{,}000 \ \Omega$.

*Step 6.* The common value for $R_6$ through $R_{11}$ is $R = (R_{on}R_{off})^{1/2} = 482$ k$\Omega$.

*Step 7.* A value for $R_{12}$ is found from

$$R_{12} = \frac{4RR_1 \ v_o(\text{peak})}{3R_2 \ v_m(\text{peak})} = \frac{4(482{,}000)10^4(3)}{3(5{,}000) \ 2} = 1.928 \ \text{M}\Omega$$

This resistance seems quite high, so we now calculate the dc offset in $v_o$ due to $A_4$ input bias current:

$$v_o = I_{b4}R_{12} = (-5 \times 10^{-7}) \ 1.928 \times 10^6 = -0.965 \ \text{V}$$

This is about ten times too large, so we make the following resistor changes (these changes will lower $\Delta v_o$ to 0.0964 V):

$$R_{12} = 193 \ \text{k}\Omega$$
$$R = R_6 = R_7 = R_8 = R_9 = R_{10} = R_{11} = 48 \ \text{k}\Omega$$

*Step 8.* As a double check on our calculations above, we put values into Eqs. 1 and 2 to find the peak-to-peak $v_o$ (assume $V_R = v_m(\text{peak})/2$)

$$v_o(-) = \frac{-R_2 R_9 R_{12}}{R_{10}(R_6 + R_8)} \left[ \frac{v_m(\text{peak})}{R_1} + \frac{v_m(\text{peak})}{2 R_3} \right]$$
$$= \frac{-5{,}000(48{,}000)193{,}000}{48{,}000(48{,}000 + 48{,}000)} \left[ \frac{2}{10^4} + \frac{2}{2(10^4)} \right] = -3.015625 \ \text{V}$$

$$v_o(+) = \frac{R_2 R_{12}}{R_7 + R_{11}} \left[ \frac{v_m(\text{peak})}{R_1} + \frac{v_m(\text{peak})}{2 R_3} \right]$$

$$= \frac{5{,}000(193{,}000)}{48{,}000 + 48{,}000} \left[ \frac{2}{10^4} + \frac{2}{2(10^4)} \right] = +3.015625 \text{ V}$$

These peak voltages can be trimmed if $R_2$ or $R_{12}$ is made adjustable.

*Step 9.* The peak-to-peak output voltage with switch errors accounted for is now calculated using Eq. 3:

$$v_o(\pm) \approx \frac{\pm R_2 R_{12} R_{\text{off}}}{(R + R_{\text{on}})(R + R_{\text{off}}) + RR_{\text{off}}} \left[ \frac{v_m(\text{peak})}{R_1} + \frac{v_m(\text{peak})}{2 R_3} \right]$$

$$\approx \frac{\pm 5{,}000(193{,}000) \, 4 \times 10^8}{(48{,}000 + 580)(48{,}000 + 4 \times 10^8) + 48{,}000(4 \times 10^8)} \left[ \frac{2}{10^4} + \frac{2}{2(10^4)} \right]$$

$$\approx \pm 2.997334$$

Even though the $R$ resistors had to be lowered by a factor of 10, the switch errors cause only an 18.2-mV error (3.015625 − 2.997334).

## REFERENCES

1. Kelly, R. G.: Linear Modulator Has Excellent Temperature Stability, *EEE*, July 1968, p. 102.
2. Althouse, J.: Linear Amplifier Circuit Eliminates Transformers, *Electronics*, Mar. 21, 1966, p. 99.

## 18.2 PULSE-AMPLITUDE MODULATOR

**ALTERNATE NAMES** Pulse-height modulator, PAM, analog gate, gated amplifier, single-channel multiplexer, sampling gate.

**EXPLANATION OF OPERATION** This circuit is quite similar to the amplitude modulator in the previous section. In this case, however, a unipolar output is required; so the circuit is about 50 percent the size of Fig. 18.1. The pulses are applied to $v_c$ and the modulation waveform drives the circuit at $v_m$. For the circuit shown in Fig. 18.3, $v_c$ must cross through zero. If a logic circuit such as a TTL device is driving $v_c$, the + input of $A_2$ should be biased to +1.4 V. The object, of course, is to switch $A_2$ on and off so that $\bar{v}_c$ (an inverted $v_c$) appears at its output.

The modulation input $v_m$ can be centered about zero, or it can be exclusively positive or negative. The modulation depth adjustment $R_4$ is used to place $v_1$ at the correct bias point.

Each time $v_c$ goes high, $S_1$ turns on and $S_2$ turns off. During this time the voltage at $v_2$ is $v_1 R_6/(R_5 + R_6)$. In between pulses, when $v_c$ is low, $S_1$ is off and $S_2$ is on. During this time the voltage at $v_2$ is very small ($\approx v_1 R_{\text{on}}/R_{\text{off}}$). The voltage at $v_o$ is inverted with a magnitude of $v_o = -v_2 R_7/R_6$.

### DESIGN PARAMETERS

| Parameter | Description |
|---|---|
| $A_1$ | Amplifier which controls the modulation gain and the depth of modulation |
| $A_2$ | Comparator used to invert $v_c$ |
| $A_3$ | Buffer amplifier used to provide a low circuit output resistance and a constant load resistance to the switch circuit |
| $I_{b3}$ | Input bias current of $A_3$ |

## 18-8 MODULATORS

| Parameter | Description |
|---|---|
| $R_1$ to $R_2$ | Controls modulation gain |
| $R_3$ to $R_4$ | Controls modulation range |
| $R_5$ | This resistor, along with $R_6$, must be much larger than $R_{on}$ so that switch errors are minimized |
| $R_6$ to $R_7$ | Along with $R_5$, these resistors control the gain of $A_3$ |
| $R_{on}$, $R_{off}$ | On and off resistances of $S_1$ and $S_2$ |
| $S_1$, $S_2$ | CMOS switches |
| $v_c$, $\bar{v}_c$ | Input carrier voltage (pulse train to be modulated) |
| $v_m$ | Modulation waveform used to control height of output pulses |
| $V_N$ | Negative reference voltage |
| $V_P$ | Positive reference voltage |
| $V_R$ | Reference voltage at wiper of $R_4$ |
| $v_1$ | Output voltage from input buffer |
| $v_2$ | Output voltage from switches |
| $v_o$ | Circuit output pulse train |
| $\Delta v$ | Offset in output voltage due to input bias current of $A_3$ |

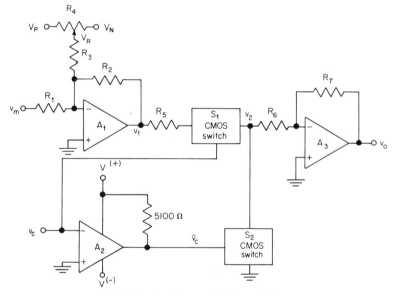

**Fig. 18.3** A pulse-amplitude modulator.

### DESIGN EQUATIONS

| Eq. No. | Description | Equation |
|---|---|---|
| 1 | Output voltage $v_o$ when $S_1$ is on and $S_2$ is off (neglecting switch errors and op amp errors) | $v_o(\text{on}) = \dfrac{R_2 R_7}{R_5 + R_6}\left(\dfrac{v_m}{R_1} + \dfrac{V_R}{R_3}\right)$ |
| 2 | Output voltage $v_o$ when $S_1$ is on and $S_2$ is off (switch errors included) | $v_o(\text{on}) = \dfrac{R_2 R_7 R_{off}\,[(v_m/R_1) + (V_R/R_3)]}{(R_5 + R_{on})(R_6 + R_{off}) + R_6 R_{off}}$ |

| Eq. No. | Description | Equation |
|---|---|---|
| 3 | Output voltage $v_o$ when $S_1$ is off and $S_2$ is on (due to switch errors only) | $v_o(\text{off}) = \dfrac{R_2 R_7 R_{on} [(v_m/R_1) + (V_R/R_3)]}{(R_5 + R_{on})(R_6 + R_{off}) + R_6 R_{on}}$ |
| 4 | Output offset due to $I_{b3}$ | $\Delta v_o = I_{b3} R_7$ |
| 5 | Voltage $v_1$ | $v_1 = -R_2 \left( \dfrac{v_m}{R_1} + \dfrac{V_R}{R_3} \right)$ |
| 6 | Voltage $v_2$ while $S_1$ is on and $S_2$ is off (switch errors included) | $v_2 = \dfrac{R_6 R_{off} v_1}{(R_5 + R_{on})(R_6 + R_{off}) + R_6 R_{off}}$ |

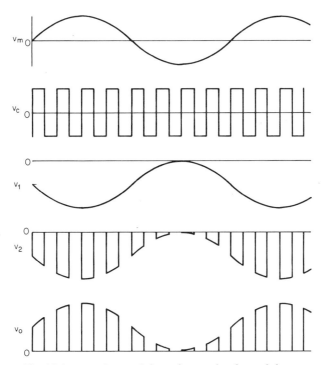

**Fig. 18.4** Waveforms of the pulse-amplitude modulator.

## REFERENCES

1. Stojanovic, B.: Pulse Height Modulator Multiplies Voltage by Frequency, *EEE*, July 1968, p. 99.
2. Tobey, G. E., J. G. Graeme, and L. P. Huelsman: "Operational Amplifiers – Design and Applications," p. 398, McGraw-Hill Book Company, New York, 1971.

## 18.3 PULSE-WIDTH MODULATOR

**ALTERNATE NAMES** Voltage-to-pulse-width converter, two-state amplifier, switching-mode amplifier, pulse-duration modulator, PWM circuit.

## 18-10 MODULATORS

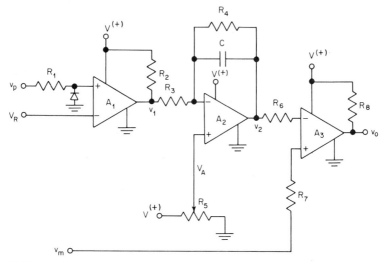

**Fig. 18.5** A pulse-width modulator utilizing two comparators ($A_1$ and $A_3$) and an integrator ($A_2$).

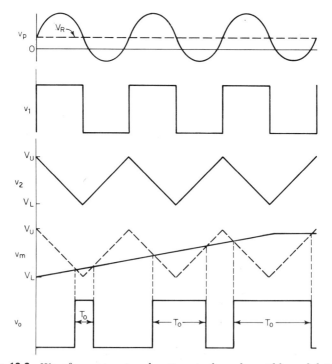

**Fig. 18.6** Waveforms at various locations in the pulse-width modulator.

**EXPLANATION OF OPERATION** Pulse-width modulation is used in a variety of applications where highly accurate and efficient control of power is required. PWM techniques are widely used in power supplies, motor controls, and even some hi-fi amplifiers. The circuit shown in Fig. 18.5 performs pulse-width modulation with two comparators ($A_1$ and $A_3$) and an integrator ($A_2$). A sine wave or a rectangular waveform with 30 to 70 percent duty cycle is applied to $v_p$. $A_1$ changes the $v_p$ input to a 50 percent duty cycle square wave using its high open-loop gain along with an appropriate choice for $V_R$. Rectangular-input waveforms must have approximately 50 percent duty factor. Voltage waveform $v_1$ is integrated with $A_2$ to form a precise triangular waveform at $v_2$ (see Fig. 18.6).

The modulation input $v_m$ is compared with the triangular waveform $v_2$ in the $A_3$ comparator. Whenever the triangular waveform is larger than $v_m$, $v_o$ is low. If $v_m$ is larger, $v_o$ is high. The output-voltage pulse width is precisely proportional to $v_m$ from pulse widths of 0 to 100 percent of the period of $v_p$.

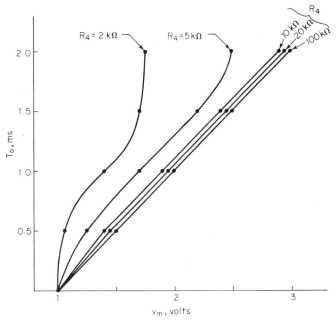

**Fig. 18.7** Response of a typical PWM with various values of $R_4$. It is assumed that $R_2 = 1,000\Omega$, $R_3 = 10$ k$\Omega$, $C = 0.1$ $\mu$F, and $f = 500$ Hz.

High accuracy and low drift require a good integrator ($A_2$). The comparators contribute little to the system errors.

For the sake of economy and compactness this design assumes operation from one positive power supply. The comparators can be in a common package or even one-half of a quad comparator chip. The op amp, however, must not drift, so it should be of good quality.

The feedback resistor $R_4$ is required to minimize $A_2$ drift. It must be small enough to prevent drift yet large enough so that it does not destroy the linearity of the modulator. Figure 18.7 shows the effect of this resistor on linearity.

## DESIGN PARAMETERS

| Parameter | Description |
|---|---|
| $A_1, A_3$ | Standard open-collector comparators |
| $A_2$ | Op amp with high-quality input characteristics |
| $C$ | Integration capacitor |
| $f_p$ | Input carrier frequency |
| PWM | Pulse-width modulation (modulator) |
| $R_1, R_6$ to $R_7$ | Provides comparator input protection if needed |
| $R_2, R_8$ | Pull-up resistors for open-collector outputs of comparators |
| $R_3$ | Integration resistor |
| $R_4$ | Provides sufficient dc feedback to stabilize $A_2$ gain to keep integrator in linear region |
| $R_5$ | Biases $A_2$ with a voltage equal to the average value of $v_1$. This part is used to place $v_2$ in the linear region of $A_2$ |
| $T_o$ | Output pulse width |
| $v_o$ | Output-voltage waveform |
| $v_1$ | Integrator input-voltage waveform |
| $v_2$ | Integrator output-voltage waveform |
| $V_A$ | Bias voltage for $A_2$ |
| $V_L$ | Lower limit of integrator output waveform (this is set equal to the lowest expected $v_m$ input) |
| $v_m$ | Modulation input voltage—ranges from $V_L$ to $V_U$ |
| $v_p$ | Pulse or carrier input |
| $V_R$ | Switching point for $A_1$ (typically +1.4 V) |
| $V_{sat}$ | Voltage $v_1$ during saturation of $A_1$ output transistor |
| $V_U$ | Upper limit of integrator output waveform (this is set equal to the highest expected $v_m$ input) |
| $V^{(\pm)}$ | Power-supply voltages |

## DESIGN EQUATIONS

| Eq. No. | Description | Equation |
|---|---|---|
| 1 | Output pulse width as a function of $v_m$ | $T_o \approx \dfrac{v_m - V_L}{f_p(V_U - V_L)}$ (where $V_L$ and $V_U$ are adjusted to put entire $v_2$ in linear range of $A_2$ using $R_5$) |
| 2 | Integrator output $v_2$ during $v_1$ on time assuming $R_4C > (R_2 + R_3)C$ | $v_2 \approx V_U - \dfrac{t[V^{(+)} - V_A]}{C(R_2 + R_3)} + V_A$ |
| 3 | Integrator output $v_2$ during $v_1$ off time assuming $R_4C > R_3C$ | $v_2 \approx V_L - \dfrac{t(V_{sat} - V_A)}{CR_3} + V_A$ |
| 4 | Required nominal $V_A$ | $V_A \approx v_1$ (average) if $R_2 \ll R_3$ |
| 5 | Peak-to-peak triangle magnitude at integrator output | $v_2(\text{peak-to-peak}) \approx \dfrac{V^{(+)}}{4f_p CR_3} = v_m(\max) - v_m(\min)$ |
| 6 | Resistor values, $R_1$ | $R_1 = 0$ to 100 k$\Omega$ (depending on maximum allowable input currents and voltages for $A_1$) |
| 7 | $R_2, R_8$ | $R_2 = R_8 = 500$ to 5,000 $\Omega$ [should be $\ll R_3$ yet not so small that $A_1$ pulls excessive $V^{(+)}$ current] |

| Eq. No. | Description | Equation |
|---|---|---|
| 8 | $R_3$ | $R_3 = \dfrac{V^{(+)}}{4f_p C [v_m(\max) - v_m(\min)]}$ |
| 9 | $R_4$ | $R_4 = 2(R_2 + R_3)$ |
| 10 | $R_5$ | $R_5 = 1$ to $10$ k$\Omega$ |
| 11 | $R_6$ to $R_7$ | $R_6 = R_7 = R_1$ |
| 12 | Feedback capacitor size | $C = 0.01$ to $1$ $\mu$F (see step 3 of Design Procedure) |

## DESIGN PROCEDURE

To simplify the calculations for this circuit, we have abbreviated many of the design equations. This procedure will provide a first-cut set of resistor values. Resistor $R_5$ should be adjustable as shown in Fig. 18.5 to trim the integrator range of operation. Otherwise the circuit has few problems and is easily designed.

## DESIGN STEPS

*Step 1.* Select a value for $V^{(+)}$ which is equal to the required $v_o$ pulse height. This supply voltage should be at least 3 V higher than the expected $V_U$.

*Step 2.* Select values for $R_1$, $R_2$, $R_6$, $R_7$, and $R_8$ according to the types of comparators picked for $A_1$ and $A_3$. If the data sheet provides no recommendation, let all these resistors equal 2,000 $\Omega$.

*Step 3.* Select a capacitor $C$ using the following "rough" selection guide:

| $f_p$, Hz | $C$, $\mu$F |
|---|---|
| 1–10 | 10 |
| 10–100 | 1 |
| 100–1,000 | 0.1 |
| 1,000–10,000 | 0.01 |

*Step 4.* Calculate a value for $R_3$ using Eq. 8. If $R_3$ is not at least ten times larger than $R_2$, either of the following can be done:
 1. Lower $R_2$ until $R_3 = 10R_2$. Be careful that the new $R_2$ does not cause excessive output current to flow in $A_1$ during the $v_1$ low state.
 2. Raise $R_3$ until $R_3 = 10R_2$. Capacitor $C$ must be lowered so that Eq. 8 is still satisfied.

*Step 5.* Find values for $R_4$ and $R_5$ using Eqs. 9 and 10.

**EXAMPLE OF A PULSE-WIDTH-MODULATOR DESIGN** As a numerical illustration, we will design a pulse-width modulator which operates at 2 kHz. The 10-V output pulse should vary from zero width to 0.5-ms width as the modulation signal varies from 3 to 6 V.

*Design Requirements*

$T_o = 0$, corresponding to $v_m = 3$ V
$T_o = 500$ $\mu$s, corresponding to $v_m = 6$ V
$f_p = 2,000$ Hz
$v_o(\text{peak}) = 15$ V

*Device Data*

$V_{sat}(A_1) = 0.3$ V (measured)

*Step 1.* With a $v_o(\text{peak}) = 15$-V requirement, we choose $V^{(+)} = 15$ V.

*Step 2.* The LM339 quad comparator is chosen for $A_1$ and $A_3$. This device can sink a maximum of 20 mA into its output terminal through $R_2$ or $R_8$. If we want to keep power consumption down, suppose we let the current through $R_2$ and $R_8$ be 5 mA. This fixes the resistor values at $R_2 = R_8 = 15$ V/5 mA $= 3,000$ Ω. Application notes for this device indicate that no input resistor is required. However, to prevent loading of $A_2$, $v_p$, and $v_m$, we will let $R_1 = R_6 = R_7 = 2,000$ Ω.

*Step 3.* In accordance with the proposed "rough" selection guide, we will let $C = 0.01$ μF.

*Step 4.* We now use Eq. 8 to find $R_3$:

$$R_3 \approx \frac{V^{(+)}}{4 f_p C [v_m(\text{max}) - v_m(\text{min})]}$$

$$\approx \frac{15}{4(2,000)10^{-8}(6-3)} = 62.5 \text{ k}\Omega$$

This is much larger than ten times $R_2$, so no recalculation is necessary.

*Step 5.* Resistor $R_4$ is found from

$$R_4 = 2(R_2 + R_3) = 2(3,000 + 62,500) = 65.5 \text{ k}\Omega$$

We will choose a value of 10 kΩ for $R_5$.

## REFERENCE

1. Schmid, H.: Digital Meters for Under $100, *Electronics*, Nov. 28, 1966, p. 88.

Chapter **19**

# Multipliers and Dividers

## INTRODUCTION

Multiplication and division of two or more analog quantities can be implemented with a variety of methods. Traditionally, the most popular approach has been to use a log amplifier in conjunction with an antilog amplifier. These two circuits can also be used to generate nearly any fractional exponent either below or above 1. At the cost of four op amps and two differential amplifiers both multiplication and division can be simultaneously provided with the log-antilog circuit. We will discuss the log-antilog multiplier/divider in Sec. 19.2.

Section 19.1 will provide design information on the voltage-controlled FET multiplier. This is probably the simplest and most easily implemented multiplier in the literature. Satisfactory performance can be achieved with this circuit only after the FET transfer characteristics are made linear. This is done with proper biasing and local feedback around the FET.

## 19.1 FET-CONTROLLED MULTIPLIER

**ALTERNATE NAMES** Voltage-controlled amplifier, analog multiplier, linear multiplier.

**EXPLANATION OF OPERATION** Field-effect transistors make nearly ideal voltage-controlled resistors. Their range of operation, however, is limited by several constraints which must be understood before their utility can be fully exploited. Field-effect transistors have a voltage-controlled drain-to-source resistance of $R_{DS} = V_p^2/I_{DSS}|v_c - 2V_p|$. We will let $R_{DS}$ be the drain-to-source resistance of both $Q_1$ and $Q_2$, since they are identical devices at the same temperature and use the same $v_c$.

The output of the $A_1$ stage in Fig. 19.1 is

$$v_o = -\frac{v_1 R_2 R_{10}}{(R_1 + R_2)R_{DS}}$$

But $R_{DS}$ depends on the control voltage $v_c$. This control voltage depends on $v_2$ and $V_R$. For proper circuit operation $V_R$ must be positive and $v_2$ can range from zero to a specified negative limit. These polarities must be observed. The current through $R_{DS}$ is identical to the current through $R_9$ since $A_2$ draws

## 19-2 MULTIPLIERS AND DIVIDERS

insignificant current. The control voltage $v_c$ will force $R_{DS}$ to the correct resistance so that these currents are equal. Since the inverting input of $A_2$ tries to remain at ground potential (because of feedback), we can draw the following conclusion:

$$I_{Q2} = \frac{v_4 - 0}{R_{DS}} = I_{R9} = \frac{0 - v_2}{R_9}$$

The voltage divider $R_3$ to $R_4$ provides us with

$$v_4 = \frac{V_R R_4}{R_3 + R_4}$$

Combining these two equations gives

$$R_{DS} = \frac{R_4 R_9 V_R}{|v_2|(R_3 + R_4)}$$

$Q_1$ and $Q_2$ must be a matched pair on a single chip such as the 2N5196. The control voltage $v_c$ drives the gates of both $Q_1$ and $Q_2$. This guarantees that $R_{DS1} = R_{DS2}$. The final result for $v_o$ is found by substituting the $R_{DS}$ expression into the $v_o$ expression:

$$v_o = -\frac{v_1|v_2|R_2 R_{10}(R_3 + R_4)}{V_R R_4 R_9(R_1 + R_2)}$$

If we allow $R_1 = R_3$, $R_2 = R_4$, and $R_9 = R_{10}$, the above equation reduces to

$$v_o = -\frac{v_1|v_2|}{V_R}$$

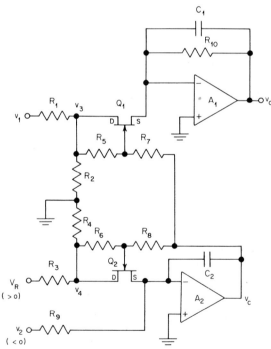

**Fig. 19.1** A linear multiplier which utilizes the voltage-controlled resistance property of field-effect transistors.

This circuit can even be used as a divider by allowing $V_R$ to be an input variable. However, the available range for $V_R$ is only several volts, which makes this option inadvisable. Also, if one attempts to drive the $R_3$ to $R_{DS}$ to $R_9$ circuit backward by letting $V_R$ be negative and $v_2$ positive, a lock-up condition will occur. The $A_2$ circuit has positive feedback in this situation. Suppose the inverting input of $A_2$ is a few microvolts more positive than ground. This will drive $v_c$ negative, which causes $R_{DS}$ to increase. This causes the $A_2$ inverting input to go more positive. In several microseconds (or less) $v_c$ is driven to the negative saturation voltage of $A_2$. This causes $R_{DS2}$ and $R_{DS1}$ to lock up at maximum resistance. The same thing occurs if the $A_2$ inverting terminal starts out several microvolts negative. $A_2$ quickly goes to positive saturation. We conclude that the polarities noted in Fig. 19.1 must be observed.

This circuit will exhibit less than 5 percent distortion only if the following are considered:

1. $R_9$ must be larger than $R_{DS}(\min)$.
2. The voltages at $v_3$ and $v_4$ must always be less than $\pm 1$ V.
3. The control voltage $v_c$ must operate only in the range from zero down to $2V_p$.
4. FETs with large $V_p$ are used. Note, however, that $A_2$ must be able to drive $v_c$ down to $2V_p$.
5. $R_{DS}(\min)$ must be at least 100 times larger than $R_2$ or $R_4$ so that $R_{DS}$ does not load the $R_1$ to $R_2$ and $R_3$ to $R_4$ dividers. This will cause an error which varies according to the magnitude of $v_2$.

## DESIGN PARAMETERS

| Parameter | Description |
|---|---|
| $C_1, C_2$ | Capacitors required if $A_1$ or $A_2$ tend to be unstable ($\approx$ 100 to 500 pF) |
| FET | Field-effect transistor |
| $I_{DSS}$ | The drain-to-source current of an FET if $V_{GS}=0$ and $V_{DS}=5$ or 10 V |
| $I_{Q1}, I_{Q2}$ | Drain-to-source current through $Q_1$ or $Q_2$ |
| $I_{R9}$ | Current through $R_9$ |
| $Q_1, Q_2$ | Field-effect transistors |
| $R_1$ to $R_4$ | Input attenuators |
| $R_5$ to $R_8$ | FET biasing and feedback resistors |
| $R_9$ to $R_{10}$ | Determines overall gain of circuit |
| $R_{DS}$ | Drain-to-source resistance of an FET at a specified gate-to-source voltage |
| $R_{G1}, R_{GR}$ | Generator resistances for $V_1$ and $V_R$ |
| $v_o$ to $v_4$, $v_c$ | Voltages as noted in Fig. 19.1 |
| $V_p$ | FET pinch-off voltage (where $R_{DS}$ approaches infinity) |
| $V_R$ | Reference input voltage |
| $V^{(\pm)}$ | Positive and negative supply voltages |

## DESIGN EQUATIONS

| Eq. No. | Description | Equation |
|---|---|---|
| 1 | Output voltage of circuit | $v_o = -\dfrac{v_1|v_2|R_2R_{10}(R_3+R_4)}{V_R R_4 R_9 (R_1+R_2)}$ |
| 2 | Output voltage of circuit if $R_1=R_3$, $R_2=R_4$, $R_9=R_{10}$ | $v_o = -\dfrac{v_1|v_2|}{V_R}$ |

## 19-4 MULTIPLIERS AND DIVIDERS

| Eq. No. | Description | Equation |
|---|---|---|
| 3 | $R_{DS}$ as a function of control voltage $v_c$ | $R_{DS} = \dfrac{V_p^2}{I_{DSS}|v_c - 2V_p|}$ |
| 4 | Minimum value for $R_{DS}$ if non-linearity is to be minimized | $R_{DS}(\min) = \dfrac{|V_p|}{2I_{DSS}}$ |
| 5 | Maximum recommended value for $R_{DS}$ | $R_{DS}(\max) = 10 R_{DS}(\min)$ |
| 6 | Minimum and maximum recommended $|v_2|$ | $|v_2| = \dfrac{V_R R_4 R_9}{R_{DS}(R_3 + R_4)}$ <br> NOTE: Use the minimum and maximum $R_{DS}$ from Eqs. 4 and 5 |
| 7 | Optimum range for $v_c$ | $v_c$ range $= 2V_p$ to zero |
| 8 | Resistor values $R_1, R_3$ | $R_1 = R_3 > 100\, R_{G1}$ or $100\, R_{GR}$ (whichever is larger) |
| 9 | $R_2, R_4$ | $R_2 = R_4 < \dfrac{R_{DS}(\min)}{100}$ |
| 10 | $R_5, R_6, R_7, R_8$ | $R_5 = R_6 = R_7 = R_8 > 1{,}000\, R_2$ |
| 11 | $R_9, R_{10}$ | $R_9 = R_{10} = \dfrac{|v_o|_{\max}(R_1 + R_2)|V_p|}{2|v_1|_{\max} R_2 I_{DSS}}$ |

### DESIGN PROCEDURE

The design of this circuit begins by choosing a good-quality matched FET pair with a high $V_p$. The op amp $A_2$ must be able to swing to a negative output voltage twice the value of $V_p$.

### DESIGN STEPS

*Step 1.* Choose a single-chip set for $Q_1$ and $Q_2$ which track $I_{DSS}$ and $V_p$ over temperature. A high value for $V_p$ is also desirable. Choose good-quality op amps for $A_1$ and $A_2$ which will drive $v_c$ more negative than $2V_p$ with several volts margin. Let the $\pm$ power supplies be compatible with $\pm v_o(\max)$ and $2V_p$ with several volts margin.

*Step 2.* Compute $R_{DS}(\min)$ from Eq. 4. Choose a common value for $R_2$ and $R_4$ which is less than 1 percent of $R_{DS}(\min)$. Choose a common value for $R_1$ and $R_3$ which is at least 100 times larger than $R_{G1}$ or $R_{GR}$.

*Step 3.* Calculate a nominal value for $R_5$ through $R_8$ using Eq. 10.

*Step 4.* Compute values for $R_9$ and $R_{10}$ using Eq. 11.

*Step 5.* Determine the allowable range of $v_2$ from Eqs. 4, 5, and 6.

**EXAMPLE OF MULTIPLIER DESIGN** This circuit is a simple multiplier, and as would be expected, it has several limitations which should be recognized. Both $v_1$ and $v_2$ have definite limits over which linearity of $v_o = -v_1 v_2 / V_R$ can be expected. We will assume for this example that $Q_1$ and $Q_2$ are a 2N5196 dual-FET device and the op amp outputs can drive $\pm 15$ V.

*Design Requirements*

$v_o = -v_1|v_2|/5$
$\pm v_o(\text{max}) = \pm 5$ V
$V^{(\pm)} = \pm 15$ V
$v_1$ range $= \pm 5$ V
$V_R = +5.00$ V regulated

*Device Data*

$V_p\ (Q_1,Q_2) = -3$ V $\left.\begin{array}{l}\end{array}\right\}$ measured on a 2N5196
$I_{DSS}\ (Q_1,Q_2) = 0.85$ mA

*Step 1.* Dual-FET specification sheets do not usually specify the degree of matching and tracking over temperature of $I_{DSS}$ and $V_p$. Fairly good results can be achieved if the dual device is on a single chip and has guaranteed tracking of 5 or 10 $\mu$V/°C for differential gate-source voltage. This is often the only parameter specified over temperature. For this example we choose the 2N5196 dual FET which has a specified pinch-off voltage of $-0.7$ to $-4$ V. The device tested in this example measured $-3$ V for both $Q_1$ and $Q_2$. The maximum required drive from $A_2$ is $2V_p = 2(-3) = -6$ V. We will use the 747 op amp for this application. The power supplies will be $\pm 15$ V so that the maximum $v_o = \pm 5$ V can be achieved.

*Step 2.* The minimum $R_{DS}$ is

$$R_{DS}(\min) = \frac{|V_p|}{2I_{DSS}} = \frac{3}{2 \times 0.85 \times 10^{-3}} = 1{,}765\ \Omega$$

The common value for $R_2$ and $R_4$ must be

$$R_2 = R_4 < \frac{1{,}765}{100} = 17.65\ \Omega$$

We will use 17.4-$\Omega$ precision resistors.

Assume the source resistances for $v_1$ and $v_2$ are 50 $\Omega$. We should make $R_1 = R_3 \geq 100(50) = 5{,}000\ \Omega$ at least. We will use 5,110-$\Omega$ precision resistors.

*Step 3.* With Eq. 10 we get

$$R_5 = R_6 = R_7 = R_8 \geq 1{,}000\ R_2 = 1{,}000(17.4) = 17.4\ \text{k}\Omega$$

We will use 100-k$\Omega$ resistors.

*Step 4.* $R_9$ and $R_{10}$ are found with Eq. 11:

$$R_9 = R_{10} = \frac{|v_o|_{\max}(R_1 + R_2)|V_p|}{2|v_1|_{\max}R_2 I_{DSS}}$$

$$= \frac{5(5{,}110 + 17.4)3}{2(5)17.4(8.5 \times 10^{-4})} = 520\ \text{k}\Omega$$

*Step 5.* We first compute the minimum and maximum allowable $R_{DS}$ from Eqs. 4 and 5:

$$R_{DS}(\min) = \frac{|V_p|}{2I_{DSS}} = \frac{3}{2(0.85 \times 10^{-3})} = 1{,}765\ \Omega$$

$$R_{DS}(\max) = 10 R_{DS}(\min) = 10(1{,}765) = 17{,}650\ \Omega$$

The range of allowable $v_2$ values is now determined with the help of Eq. 6:

## 19-6 MULTIPLIERS AND DIVIDERS

$$|v_2|_{min} = \frac{V_R R_4 R_9}{R_{DS}(max)(R_3 + R_4)} = \frac{5(17.4)520{,}000}{17{,}650(5{,}110 + 17.4)} = 0.50 \text{ V}$$

$$v_2(max) = \frac{V_R R_4 R_9}{R_{DS}(min)(R_3 + R_4)}$$

$$= \frac{5(17.4)520{,}000}{1{,}765(5{,}110 + 17.4)} = 5 \text{ V}$$

Figure 19.2 shows the response of a circuit built according to the above calculations. The worst-case linearity error was 4.4 percent of full-scale output.

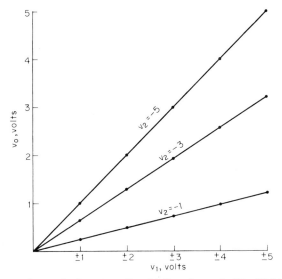

**Fig. 19.2** Measured transfer function of multiplier shown in Fig. 19.1 built according to the design steps.

### REFERENCES

1. Christie, W. C.: Multiply and Divide with a Dual Photo Resistor, *Electron. Des.*, vol. 21, p. 108, Oct. 10, 1968.
2. Mollinga, T.: The FET as a Voltage Controlled Resistor, *EEE*, January 1970, p. 58.
3. Graeme, J. G.: "Applications of Operational Amplifiers—Third-Generation Techniques," p. 97, McGraw-Hill Book Company, New York, 1973.

### 19.2 LOG-ANTILOG MULTIPLIER/DIVIDER

**ALTERNATE NAMES** Analog multiplier, analog divider, analog multiplier/divider, one-quadrant multiplier/divider.

**EXPLANATION OF OPERATION** Although operation is restricted to the first quadrant (for $v_1$, $v_2$, $v_3$, and $v_o$), this circuit is extremely useful in a variety of applications. Linearity errors of less than 1 percent can be achieved if input offsets of $A_1$, $A_2$, and $A_3$ are properly handled. This minimal error is possible over two to three decades of input voltages.

The circuits of $A_1$, $A_2$, and $A_3$ are single-ended log amplifiers. Two of these circuits are utilized in the log ratio circuit shown in Fig. 17.1. Note that $Q_1$ and $Q_2$ are effectively connected in series. This results in addition of logarithms or, in other words, multiplication of $v_1$ and $v_2$. The output from $A_1$ and $A_2$ drives the $A_4$ antilog circuit and thereby produces the product of $v_1$ and $v_2$ at $v_o$.

The output from $A_3$ is treated differently. This voltage subtracts current from $Q_4$. Subtraction of logarithmic quantities results in division. The final result is

$$v_o = \frac{v_1 v_2}{v_3}$$

The foregoing remarks will now be treated in mathematical terminology. Each output voltage from the three log amplifiers is equal to the emitter-base voltage of the transistor in its feedback loop. The following can be stated:

$$v_4 = -\frac{kt}{q} \ln \frac{v_1}{R_1 I_{ES1}} = -v_{BE1}$$

$$v_5 = -\frac{kt}{q} \ln \frac{v_2}{R_2 I_{ES2}} = -v_{BE2}$$

$$v_6 = -\frac{kt}{q} \ln \frac{v_3}{R_3 I_{ES3}} = -v_{BE3}$$

The bases of $Q_1$ and $Q_3$ are at ground potential (through 10 $\Omega$). Starting at the base of $Q_1$, we trace the following voltage loop:

$$v_{BE1} + v_{BE2} - v_{BE4} - v_{BE3} = 0$$

Transistor $Q_4$ has the following voltage-current relationship:

$$v_{BE4} = -\frac{kt}{q} \ln \frac{I_{C4}}{\alpha_4 I_{ES4}}$$

If this equation and the other equations for emitter-base voltages are substituted into the voltage-loop equation, we get

$$-\frac{kt}{q} \ln \frac{v_1}{R_1 I_{ES1}} - \frac{kt}{q} \ln \frac{v_2}{R_2 I_{ES2}} + \frac{kt}{q} \ln \frac{v_3}{R_3 I_{ES3}} + \frac{kt}{q} \ln \frac{I_{C4}}{\alpha_4 I_{ES4}} = 0$$

If $I_{ES1} = I_{ES2} = I_{ES3} = I_{ES4}$ and $\alpha_4 = 1$, this reduces to

$$-\ln \frac{v_1}{R_1} - \ln \frac{v_2}{R_2} + \ln \frac{v_3}{R_3} = -\ln I_{C4}$$

or

$$\ln I_{C4} = \ln \frac{v_1 v_2 R_3}{R_1 R_2 v_3}$$

Taking the antilog of each side,

$$I_{C4} = \frac{v_1 v_2}{v_3} \frac{R_3}{R_1 R_2}$$

the output voltage $v_o$ is related to $I_{C4}$ by $v_o = I_{C4} R_{10}$. The output voltage now becomes

$$v_o = \frac{v_1 v_2}{v_3} \frac{R_3 R_{10}}{R_1 R_2}$$

If we let $R_1 = R_2 = R_3 = R_{10}$,

$$v_o = \frac{v_1 v_2}{v_3}$$

## 19-8 MULTIPLIERS AND DIVIDERS

The circuit of Fig. 19.3 can be easily converted into a square-root circuit. If $v_o$ is connected to $v_3$ and $v_2 = 1$ V,

$$v_o = \frac{v_1(1)}{v_o} \quad \text{or} \quad v_o = v_1^{1/2}$$

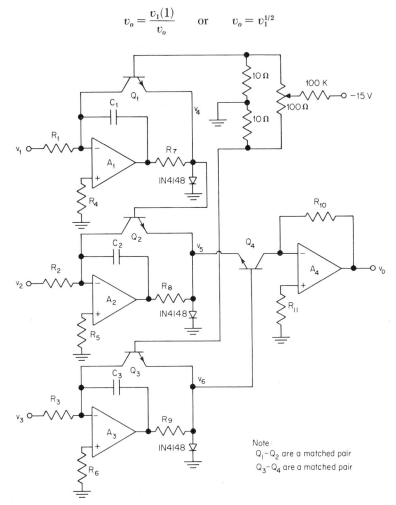

**Fig. 19.3** A multiplier/divider which uses three log amplifiers and an antilog amplifier.

### DESIGN PARAMETERS

| Parameter | Description |
|---|---|
| $A_1$ to $A_3$ | Log-amplifier-circuit op amps |
| $A_4$ | Antilog-amplifier-circuit op amp |
| $\alpha_4$ | Common-base current gain of $Q_4$ |
| $\beta_1$ to $\beta_4$ | Common-emitter current gains of $Q_1$ to $Q_4$ |
| $C_1$ to $C_4$ | Capacitors required for feedback stability |
| $f_{u1}$ to $f_{u3}$ | Unity-gain crossover frequency of $A_1$ to $A_3$ |
| $I_{C4}$ | Collector current of $Q_4$ |
| $I_{ES1}$ to $I_{ES4}$ | Emitter saturation currents of $Q_1$ to $Q_4$ |

## LOG-ANTILOG MULTIPLIER/DIVIDER

| Parameter | Description |
|---|---|
| $I_{io}$ | Input offset current of $A_1$, $A_2$, or $A_3$ |
| $k$ | Boltzmann's constant $= 1.38 \times 10^{-23}$ J/K |
| $q$ | Electronic charge $= 1.6 \times 10^{-19}$ C |
| $Q_1$ to $Q_3$ | Transistors used in log amplifiers |
| $Q_4$ | Transistor used in antilog amplifier |
| $R_1$ to $R_3$ | Resistors used to determine gain of input stages and to provide a high input resistance |
| $R_4$ to $R_6$, $R_{11}$ | Compensates for input bias currents of $A_1$ to $A_4$ |
| $R_7$ to $R_9$ | Utilized to provide feedback stability in $A_1$ to $A_3$ circuits |
| $R_{10}$ | Determines overall gain of circuit along with $R_1$ to $R_3$ |
| $R_{in}$ | Input resistance of any input |
| $R_{S1}$ to $R_{S3}$ | Source resistances of $v_1$ to $v_3$ sources |
| $T$ | Temperature in kelvins (273 K $= 0°$C) |
| $v_o$ to $v_6$ | Voltages as noted in Fig. 19.3 |
| $V_{BE1}$ to $V_{BE4}$ | Base-to-emitter voltages of $Q_1$ to $Q_4$ |
| $V_{io}$ | Input offset voltage of $A_1$, $A_2$, or $A_3$ |

### DESIGN EQUATIONS

| Eq. No. | Description | Equation |
|---|---|---|
| 1 | Output voltage of circuit as a function of the three input voltages | $v_o = \dfrac{v_1 v_2}{v_3} \dfrac{R_3 R_{10}}{R_1 R_2}$ |
| 2 | Output voltages of three log amplifiers | $v_4 = -\dfrac{kt}{q} \ln \dfrac{v_1}{R_1 I_{ES1}}$ |
|   |   | $v_5 = -\dfrac{kt}{q} \ln \dfrac{v_2}{R_2 I_{ES2}}$ |
|   |   | $v_6 = -\dfrac{kt}{q} \ln \dfrac{v_3}{R_3 I_{ES3}}$ |
| 3 | Output voltage of antilog amplifier as a function of its input current | $v_o = I_{C4} R_{10}$ where $I_{C4} = \dfrac{v_1 v_2}{v_3} \dfrac{R_3}{R_1 R_2}$ |
| 4 | Resistor values $R_1$, $R_2$, $R_3$ | $R_1 = R_2 = R_3 = R_{in}$ required for each input<br>NOTE: The source resistance must be at least 1,000 times smaller than each input resistor |
| 5 | $R_4$ | $R_4 = R_1$ |
| 6 | $R_5$ | $R_5 = R_2$ |
| 7 | $R_6$ | $R_6 = R_3$ |
| 8 | $R_7$ | $R_7 \approx \dfrac{(kt/q) \ln v_1(\text{max})}{[v_1(\text{max})/R_1] - [v_2(\text{max})/R_2 \beta_2]}$ |
| 9 | $R_8$ | $R_8 \approx \dfrac{(kt/q) \ln v_2(\text{max})}{[v_2(\text{max})/R_2] + [v_o(\text{max})/R_{10}]}$ |
| 10 | $R_9$ | $R_9 \approx \dfrac{(kt/q) \ln v_3(\text{max})}{[v_3(\text{max})/R_3] - [v_o(\text{max})/R_{10} \beta_4]}$ |
| 11 | $R_{10}$ | $R_{10} = \dfrac{R_1 R_2 v_o(\text{max})}{R_3 (v_1 v_2 / v_3)(\text{max})}$ |

| Eq. No. | Description | Equation |
|---|---|---|
| 12 | $R_{11}$ | $R_{11} = R_{10}$ |
| 13 | Capacitor values $C_1$ | $C_1 \approx \dfrac{1}{\pi f_{u1} R_7}$ |
| 14 | $C_2$ | $C_2 \approx \dfrac{1}{\pi f_{u2} R_8}$ |
| 15 | $C_3$ | $C_3 \approx \dfrac{1}{\pi f_{u3} R_9}$ |

## DESIGN PROCEDURE

We assume limits have already been placed on the magnitudes of $v_1$, $v_2$, $v_3$, and $v_o$. Next we choose good-quality op amps which have low input offsets and stability in the unity-gain configuration. The matched transistor sets $Q_1, Q_2$ and $Q_3, Q_4$ should have no more than 10 $\mu$V/°C differential $v_{BE}$ temperature coefficient.

## DESIGN STEPS

*Step 1.* Choose values for $R_1$, $R_2$, and $R_3$ which are at least 100 or 1,000 times larger than the source resistances of $v_1$, $v_2$, and $v_3$. Also let $R_4 = R_1$, $R_5 = R_2$, and $R_6 = R_3$.

*Step 2.* Choose good-quality op amps for $A_1$ through $A_4$. The input offset currents should be less than $v_1(\min)/R_1$, $v_2(\min)/R_2$, or $v_3(\min)/R_3$. Likewise, the input offset voltages should be smaller than $v_1(\min)$, $v_2(\min)$, or $v_3(\min)$.

*Step 3.* Use Eq. 11 to determine a value for $R_{10}$. Set $R_{11} = R_{10}$.

*Step 4.* Nominal values for $R_7$, $R_8$, and $R_9$ are found using Eqs. 8, 9, and 10. Any resistor within ±20 percent of the computed value will be sufficient.

*Step 5.* Compute nominal values for $C_1$, $C_2$, and $C_3$ using Eqs. 13, 14, and 15. Again, ±20 percent tolerances are satisfactory.

**EXAMPLE OF MULTIPLIER/DIVIDER DESIGN** Let us assume a compact multiplier/divider is required using a quad op amp such as the 324. The inputs must range from 10 mV to 5 V, and $v_o$ must not exceed 10 V.

*Design Requirements*

$v_o$ = 10 mV to 10 V
$v_1$ = 10 mV to 5 V
$v_2$ = 10 mV to 5 V
$v_3$ = 10 mV to 5 V

*Device Data* (Room Temperature)

$V_{io}$ = 7 mV maximum
$I_{io}$ = 50 nA maximum
$R_{S1} = R_{S2} = R_{S3} < 100$ Ω
$\beta_1 = \beta_2 = \beta_3 = \beta_4 = 100$
$f_{u1} = f_{u2} = f_{u3} = 10^6$ Hz

*Step 1.* Given the source resistances of 100 Ω, we choose $R_1 = R_2 = R_3 = 1,000\, R_{S1} = 10^3(100) = 100$ kΩ. We also let $R_4 = R_5 = R_6 = 100$ kΩ.

Step 2. All three inputs are identical, so we compute

$$\frac{v_1(\min)}{R_1} = \frac{0.01}{10^5} = 100 \text{ nA}$$

This current is only twice the value of $I_{io}(\max)$, so $v_1(\min) = 10$ mV is truly a minimum. The maximum input offset voltage is 7 mV. This is also large enough so that $v_1$, $v_2$, or $v_3$ will not produce accurate results below 10 mV.

Step 3. $R_{10}$ and $R_{11}$ are computed from

$$R_{10} = R_{11} = \frac{R_1 R_2 v_o(\max)}{R_3(v_1 v_2/v_3)(\max)}$$

$$= \frac{10^5 \times 10^5 \times 10}{10^5(10)} = 100 \text{ k}\Omega$$

Step 4. Nominal values for $R_7$, $R_8$, and $R_9$ are

$$R_7 \approx \frac{(kt/q) \ln v_1(\max)}{[v_1(\max)/R_1] - [v_2(\max)/R_2\beta_2]} = \frac{[1.38 \times 10^{-23}(300)/1.6 \times 10^{-19}] \ln 5}{(5/10^5) - (5/10^5 \times 100)}$$
$$\approx 833 \text{ }\Omega$$

We will use $R_7 = 1$ k$\Omega$.

$$R_8 \approx \frac{(kt/q) \ln v_2(\max)}{[v_2(\max)/R_2] + [v_o(\max)/R_{10}]}$$

$$\approx \frac{[1.38 \times 10^{-23}(300)/1.6 \times 10^{-19}] \ln 5}{(5/10^5) + (10/10^5)} = 278 \text{ }\Omega$$

We will use $R_8 = 300$ $\Omega$.

$$R_9 \approx \frac{(kt/q) \ln v_3(\max)}{[v_3(\max)/R_3] - [v_o(\max)/R_{10}\beta_4]}$$

$$\approx \frac{[1.38 \times 10^{-23}(300)/1.6 \times 10^{-19}] \ln 5}{(5/10^5) - (10/10^5 \times 100)} = 850 \text{ }\Omega$$

We will use $R_9 = 1$ k$\Omega$.

Step 5. The feedback-stability capacitors $C_1$, $C_2$, and $C_3$ are now computed:

$$C_1 \approx \frac{1}{\pi f_{u1} R_7} = \frac{1}{\pi (10^6) 10^3} = 318 \text{ pF}$$

We will use $C_1 = 330$ pF.

$$C_2 \approx \frac{1}{\pi f_{u2} R_8} = \frac{1}{\pi (10^6) 300} = 1{,}061 \text{ pF}$$

We will use $C_2 = 1{,}000$ pF.

$$C_3 \approx \frac{1}{\pi f_{u3} R_9} = \frac{1}{\pi (10^6) 10^3} = 318 \text{ pF}$$

We will use $C_3 = 330$ pF.

## REFERENCES

1. Counts, L., and D. Sheingold: Analog Dividers: What Choice Do You Have?, *EDN*, May 5, 1974, p. 55.
2. National Semiconductor Corp. Application Note AN-4.
3. Dobkin, R. C.: Logarithmic Converters, *IEEE Spectrum*, November 1969, p. 69.

Chapter **20**

# Multivibrators

## INTRODUCTION

Many circuit functions which are ordinarily performed with digital microcircuits can also be performed with op amp circuits. Three of these digital-type functions will be discussed in this chapter. The astable multivibrator, or free-running rectangular-waveform generator, will be discussed first. This will be a generalized circuit which produces nonsymmetrical waveforms where both amplitude and time symmetry can be individually selected. The bistable multivibrator, commonly known as the flip-flop, will also be described in detail. Lastly, the monostable multivibrator, often called the one-shot or single-shot, will be presented. This last circuit will also contain a detailed example showing numerical results of the design steps.

These circuits are implemented much more easily (and require less space) with digital microcircuits. However, op amps provide a much wider selectable range of circuit parameters. The upper and lower amplitudes in digital microcircuits are usually fixed at +5 and ground or −12 and ground, etc. Op amp circuits provide the possibility of using any upper and lower amplitude which falls within the range of ±20 V. Longer time constants are also possible using op amps with high input impedances.

## 20.1 ASTABLE MULTIVIBRATOR

**ALTERNATE NAMES** Free-running square-wave generator, rectangular-waveform generator, square-wave generator.

**EXPLANATION OF OPERATION** The circuit shown in Fig. 20.1 generates a rectangular waveform with selectable positive and negative pulse widths. There are effectively three circuits in Fig. 20.1: (1) an integrator composed of $C$ and $R_1$ (or $R_2$), (2) a comparator, performed by the op amp, and (3) a latch composed of the positive-feedback network $R_4$ and $R_5$ in conjunction with the op amp.

Circuit operation is as follows: Assume $v_o$ has just switched to $+V_{Z2}$. At this instant $v_1$ is at a voltage of $-\beta V_{Z1}$ and $v_2$ has just switched to $+\beta V_{Z2}$. The positive feedback through $R_4$ causes $v_2$ to be more positive than $v_1$, so the op amp remains locked in this state. Meanwhile, the positive $v_o$ causes a current

## 20-2 MULTIVIBRATORS

to flow through $D_1$ and $R_1$ which charges $C$. The voltage across the capacitor increases according to

$$v_1 = (\beta V_{Z1} + V_{Z2} - V_D)\left[1 - \exp\left(-\frac{t}{R_1 C}\right)\right] - \beta V_{Z1}$$

When this voltage rises to just slightly above $+\beta V_{Z2}$, the op amp switches states and $v_o$ drops to $-V_{Z1}$. Simultaneously, $v_2$ drops to $-\beta V_{Z1}$. The negative

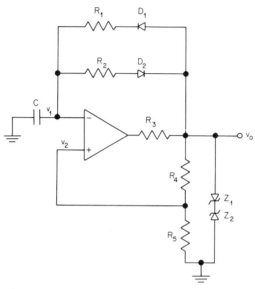

**Fig. 20.1** Nonsymmetrical astable multivibrator.

output voltage begins to discharge $C$ through $R_2$ and $D_2$. During this period the capacitor voltage follows

$$v_1 = (V_{Z1} + \beta V_{Z2} - V_D)\exp\left(-\frac{t}{R_2 C}\right) - (V_{Z1} - V_D)$$

When $v_1$ decreases to slightly below $-\beta V_{Z1}$, the op amp switches states and the cycle repeats itself. Diodes $D_1$ and $D_2$ therefore allow totally different positive and negative pulse durations.

### DESIGN PARAMETERS

| Parameter | Description |
|---|---|
| $\beta$ | Voltage feedback ratio due to $R_4$ and $R_5$ |
| $C$ | Capacitor used to determine durations of $T_1$ and $T_2$ |
| $D_1$ | Diode which conducts during $T_1$ |
| $D_2$ | Diode which conducts during $T_2$ |
| $I_b$ | Op amp input bias current |
| $R_1$ | Resistor used to determine duration of $T_1$ |
| $R_2$ | Resistor used to determine duration of $T_2$ |
| $R_3$ | Resistor used to control zener current |
| $R_4, R_5$ | Network for positive feedback |
| $T_1$ | Positive pulse duration |

# ASTABLE MULTIVIBRATOR

| Parameter | Description |
|---|---|
| $T_2$ | Negative pulse duration |
| $v_1$ | Voltage across $C$ |
| $v_2$ | Voltage at op amp noninverting terminal |
| $v_o$ | Output voltage of circuit |
| $V_{Z1}$ | Breakdown voltage of $Z_1$ plus the forward breakdown voltage of $Z_2$ |
| $V_{Z2}$ | Breakdown voltage of $Z_2$ plus the forward breakdown voltage of $Z_1$ during time when $v_o$ is positive |
| $Z_1, Z_2$ | Zener diodes used to establish positive and negative limits of output waveform |

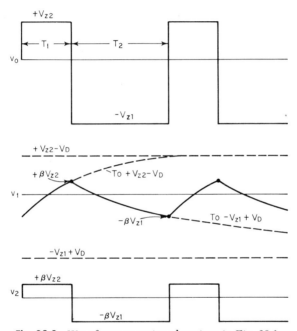

**Fig. 20.2** Waveforms at various locations in Fig. 20.1.

## DESIGN EQUATIONS

| Eq. No. | Description | Equation |
|---|---|---|
| 1 | Duration of $T_1$ (in seconds) | $T_1 = R_1 C \ln \left[ \dfrac{V_{Z2} + \beta V_{Z1} - V_D}{V_{Z2}(1 - \beta) - V_D} \right]$ where $\beta = \dfrac{R_5}{R_4 + R_5}$ |
| 2 | Duration of $T_2$ (in seconds) | $T_2 = R_2 C \ln \left[ \dfrac{V_{Z1} + \beta V_{Z2} - V_D}{V_{Z1}(1 - \beta) - V_D} \right]$ |
| 3 | Frequency of oscillation | $f = \dfrac{1}{T_1 + T_2}$ |

**20-4 MULTIVIBRATORS**

| Eq. No. | Description | Equation |
|---|---|---|
| 4 | Duty cycle | $D = \dfrac{R_1}{R_1 + R_2} = \dfrac{T_1}{T_1 + T_2}$ |
| 5 | Maximum size for $R_1$ | $R_1 \ll \dfrac{V_{Z2}}{I_b}$ |
| 6 | Maximum size for $R_2$ | $R_2 \ll \dfrac{V_{Z1}}{I_b}$ |

## REFERENCE

1. Graeme, J. G.: "Applications of Operational Amplifiers—Third-Generation Techniques," p. 164, McGraw-Hill Book Company, New York, 1973.

## 20.2 BISTABLE MULTIVIBRATOR

**ALTERNATE NAMES** Flip-flop, bipolar flip-flop, bistable flip-flop. one-bit memory, binary counter.

**EXPLANATION OF OPERATION** The flip-flop possesses two completely stable output states. A transition from one state to the other, or vice versa, occurs only upon the arrival of a trigger pulse. In the circuit of Fig. 20.3, two

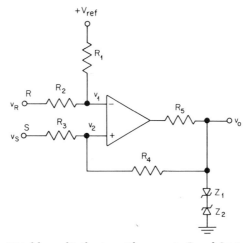

**Fig. 20.3** Bistable multivibrator with separate $R$ and $S$ trigger inputs.

separate trigger-pulse inputs are utilized. This type of flip-flop is commonly called the $R$-$S$ flip-flop. A trigger pulse at the $S$ input sets the output to a high state, and a trigger pulse at the $R$ input resets the output to a low state. These input triggers do not need to be pulses, since dc coupling is used throughout the circuit. In effect, the circuit is a dual-input level detector with hysteresis (see Chap. 5).

Operation of the circuit is as follows: Assume that the output voltage $v_o$

is in the low state, $-V_{Z1}$. Also assume the $S$ and $R$ inputs are zero. The voltage at the op amp noninverting terminal $v_2$ is equal to $\beta v_o$ where $\beta = R_3/(R_3 + R_4)$. Thus

$$v_2(\text{low}) = \frac{-R_3 V_{Z1}}{R_3 + R_4}$$

If the $R$ input is zero, the voltage at $v_1$ is

$$V_1 = \frac{V_{\text{ref}} R_2}{R_1 + R_2}$$

If $v_1 > v_2$, the op amp output will remain locked in the low state.

If a pulse (or any other waveform) is impressed on $S$ with an amplitude sufficient to drive $v_2 > v_1$, the op amp output will switch to the high state of

$$v_o = V_{Z2}$$

The voltage at $v_2$ will likewise change to a steady-state value of $v_2(\text{high}) = R_3 V_{Z2}/(R_3 + R_4)$. During the time that $v_s$ is present, a fraction of $v_s$ equal to $v_s R_4/(R_3 + R_4)$ will be superimposed on $v_2$. This is shown in Fig. 20.4.

If a signal is impressed on $R$ having a sufficient amplitude such that $v_1 > v_2$, the op amp output will switch (or reset) to the low state. As shown in Fig. 20.4, if the set pulse is of a longer duration than the reset pulse, $v_o$ may immediately switch back to the high state upon removal of the reset pulse. This will occur if $v_2 > v_1$ after the termination of the reset pulse.

This circuit can also operate with only one trigger input. Either the $S$ or $R$ input can be used as the single-input trigger terminal. For example, if

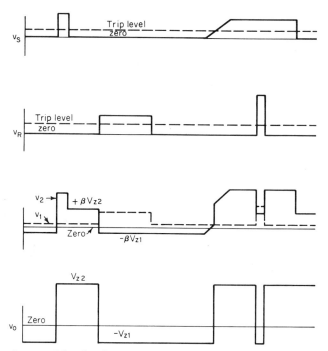

**Fig. 20.4** Output of flip-flop for various waveforms at the $R$ and $S$ input terminals.

## 20-6 MULTIVIBRATORS

the $S$ input is to be used, the $R$ input would be grounded. A set trigger would need to have characteristics as described above. A reset trigger pulse, however, would be a negative pulse on the $S$ input. The reset pulse must be more negative than

$$v_N = \frac{V_{\text{ref}} R_2 (R_3 + R_4)}{R_4 (R_1 + R_2)} - \frac{R_3 V_{Z2}}{R_4}$$

It is worth noting that if the $R$ input is grounded, the circuit operation is identical to the noninverting level detector with hysteresis (see Sec. 5.4). If the $S$ input is grounded, operation is identical to the inverting level detector with hysteresis.

### DESIGN PARAMETERS

| Parameter | Description |
| --- | --- |
| $\beta$ | Voltage feedback ratio of $R_3$ and $R_4$ |
| $R$ | Reset input terminal |
| $R_1, R_2$ | Voltage divider used to establish trip level |
| $R_3, R_4$ | Voltage divider used to establish hysteresis (positive feedback) |
| $R_5$ | Resistor used to control zener-diode current |
| $S$ | Set input terminal |
| $v_o$ | Output voltage |
| $v_1$ | Voltage at op amp inverting terminal |
| $v_2$ | Voltage at op amp noninverting terminal |
| $V_{\text{ref}}$ | Reference-voltage source used to establish trip level of circuit |
| $V_{Z1}$ | Breakdown voltage of $Z_1$ plus forward breakdown voltage of $Z_2$ |
| $V_{Z2}$ | Breakdown voltage of $Z_2$ plus forward breakdown voltage of $Z_1$ |
| $Z_1, Z_2$ | Zener diodes which determine magnitudes of output high and low states |

### DESIGN EQUATIONS

| Eq. No. | Description | Equation |
| --- | --- | --- |
| 1 | Magnitude of high-state output voltage | $v_o(\text{high}) = V_{Z2}$ |
| 2 | Magnitude of low-state output voltage | $v_o(\text{low}) = -V_{Z1}$ |
| 3 | Magnitude of $v_2$ during high state (assuming $S$ input is present) | $v_2(\text{high}) = \dfrac{R_3 V_{Z2} + R_4 v_S}{R_3 + R_4}$ |
| 4 | Magnitude of $v_2$ during low state (assuming $S$ input is present) | $v_2(\text{low}) = \dfrac{-R_3 V_{Z1} + R_4 v_S}{R_3 + R_4}$ |
| 5 | Magnitude of $v_1$ | $v_1 = \dfrac{R_2 V_{\text{ref}} + R_1 v_R}{R_1 + R_2}$ |
| 6 | Size of negative pulse required to reset circuit using $S$ input | $v_R(\text{set terminal}) = \dfrac{V_{\text{ref}} R_2 (R_3 + R_4)}{R_4 (R_1 + R_2)} - \dfrac{R_3 V_{Z2}}{R_4}$ |

## 20.3 MONOSTABLE MULTIVIBRATOR

**ALTERNATE NAMES**  One-shot, single-shot, triggered pulse generator.

**EXPLANATION OF OPERATION**  In the standby condition, the circuit output voltage is positive with a magnitude of $V_{Z2}$. $v_2$ is held at a fraction $\beta$ of this positive voltage owing to the $R_5$ to $R_6$ voltage divider. Likewise, $v_3$ is held at $V_D$ ($\approx 0.6$) V by the forward breakdown voltage of $D_1$. By making $\beta V_{Z2} > V_D$, the circuit is stable with the op amp positive input slightly more positive than the op amp negative input.

When a negative trigger pulse appears at $v_2$, it is amplified by the op amp. Positive feedback through $R_5$ and $R_6$ makes $v_2$ even more negative. The op amp then locks up with the output equal to $-V_{Z1}$. $v_2$ is locked up at $-\beta V_{Z1}$. $R_3$ begins to charge $C_1$ with a negative voltage. $D_1$ no longer conducts, as it is back-biased. When $C_1$ is charged down to a voltage slightly more negative than $-\beta V_{Z1}$, the op amp switches states again. The output returns to $V_{Z2}$. $v_2$ returns to $\beta V_{Z2}$ and $C_1$ starts charging positively through $R_2$ and $R_3$ until $D_1$ conducts. When $D_1$ starts conducting, the circuit is reset and ready for a new trigger pulse.

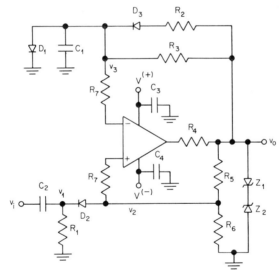

**Fig. 20.5**  Basic monostable multivibrator.

### DESIGN PARAMETERS

| Parameter | Description |
|---|---|
| $\beta$ | Feedback factor for positive input terminal of the op amp |
| $C_1$ | Determines pulse width and reset time |
| $C_2$ | Part of input differentiator |
| $C_3, C_4$ | Bypass capacitors—typically 0.01- to 0.1-$\mu$F ceramic for most op amps |
| $D_1$ | Clamps negative input of op amp at +0.6 V during standby time |
| $D_2$ | Selects negative trigger pulses from differentiator circuit |
| $D_3$ | Allows reset time to be shorter than pulse width |
| $I_4$ | Standby current through $R_4$ |

## 20-8 MULTIVIBRATORS

| Parameter | Description |
|---|---|
| $I_b$ | Bias current into the op amp negative input terminal |
| $I_{io}$ | Input offset current of op amp |
| $R_1$ | Part of input differentiator |
| $R_2$ | Determines reset time of circuit |
| $R_3$ | Determines pulse width |
| $R_4$ | Current limiter for $Z_1$ and $Z_2$ |
| $R_5, R_6$ | Voltage divider for positive feedback (determines feedback factor $\beta$) |
| $R_p$ | Parallel resistance of $R_1$ and $R_6$ |
| $R_{out}(+)$ | Output resistance of circuit during standby (positive output) |
| $R_{out}(-)$ | Output resistance of circuit during pulse (negative output) |
| $R_{Z1}$ | Dynamic zener impedance (resistance) of $Z_1$ at chosen current level |
| $R_{Z2}$ | Dynamic zener impedance (resistance) of $Z_2$ at chosen current level |
| $S$ | Slew rate of op amp |
| $TC_{Z1}$ | Temperature coefficient of $Z_1$ breakdown voltage (in %/°C) |
| $TC_{Z2}$ | Temperature coefficient of $Z_2$ breakdown voltage (in %/°C) |
| $T_p$ | Pulse width |
| $T_r$ | Reset time—determines when circuit is ready for next trigger pulse |
| $v_i$ | Input trigger waveform |
| $v_1$ | Differentiated input waveform |
| $v_2$ | Sum of 2 waveforms: (1) Negative trigger pulse formed by $C_2$, $R_1$, and $D_2$. (2) A fraction $\beta$ of the output waveform $V_o$ |
| $v_3$ | Slowly varying waveform (due to $C_1$) which determines pulse width and reset time |
| $v_o$ | Output pulse |
| $V_{Z1}$ | Breakdown voltage of $Z_1$ plus forward breakdown voltage of $Z_2$ |
| $V_{Z2}$ | Breakdown voltage of $Z_2$ plus forward breakdown voltage of $Z_1$ |
| $V^{(+)}$ | Positive supply voltage |
| $V^{(-)}$ | Negative supply voltage |
| $V(+,sat)$ | Maximum positive output saturated voltage of op amp |
| $V(-,sat)$ | Maximum negative (most negative) output saturated voltage of op amp |
| $X_{C2}$ | Reactance of $C_2$ to input waveform |
| $Z_1$ | Determines negative peak voltage of output pulse |
| $Z_2$ | Determines magnitude of positive standby output voltage |
| $Z_{in}$ | Input impedance of circuit to trigger waveform. For fast-trigger waveforms this is essentially equal to $R_p$ |

### DESIGN EQUATIONS

| Eq. No. | Description | Equation |
|---|---|---|
| 1 | Pulse width | $T_p = R_3 C_1 \ln\left(\dfrac{1 + 0.6/V_{Z2}}{1 - \beta}\right)$ |
| 2 | Pulse width if $R_5 = R_6$ and $V_{Z2} > 3$ V | $T_p = 0.8\, R_3 C_1$ |
| 3 | Reset time | $T_r = \dfrac{R_2 R_3 C_1}{R_2 + R_3} \ln\left(\dfrac{V_{Z2} + \beta V_{Z1}}{V_{Z2} - 0.6}\right)$ |
| 4 | Reset time if $R_5 = R_6$ and $V_{Z1} = V_{Z2} > 3$ V | $T_r = \dfrac{0.5\, R_2 R_3 C_1}{R_2 + R_3}$ |
| 5 | $v_3$ during pulse | $v_3(t) = (V_{Z2} + 0.6) \exp\left(\dfrac{-t}{R_3 C_1}\right) - V_{Z2}$ |
| 6 | $v_3$ during reset | $v_3(t) = (V_{Z2} + \beta V_{Z1})\left\{1 - \exp\left[\dfrac{-t(R_2 + R_3)}{R_2 R_3 C_1}\right]\right\} - \beta V_{Z1}$ |

## MONOSTABLE MULTIVIBRATOR

| Eq. No. | Description | Equation |
|---|---|---|
| 7 | $v_2$ standby required | $v_2(\text{standby}) \geq 0.8$ V |
| 8 | Input voltage required | $v_i(\text{peak-to-peak, required}) \geq 1.5\,\beta V_{Z2}$ |
| 9 | Differentiator time constant required | $\dfrac{V_{Z1} + V_{Z2}}{2S} < R_1 C_2 < \dfrac{T_p}{2}$ |
| 10 | Loading of $R_1$ | $R_6 \ll R_1$ |
| 11 | Standby output resistance | $R_{\text{out}}(+) = \dfrac{R_{Z2} R_4}{R_{Z2} + R_4}$ |
| 12 | Output resistance during pulse | $R_{\text{out}}(-) = \dfrac{R_{Z1} R_4}{R_{Z1} + R_4}$ |
| 13 | Feedback factor | $\beta = \dfrac{R_6}{R_5 + R_6}$ |
| 14 | Input impedance | $Z_{\text{in}} \approx X_{C2} + \dfrac{R_1 R_6}{R_1 + R_6}$ |
| 15 | Standby output current | $I_4(\text{standby}) = \dfrac{V(+,\text{sat}) - V_{Z2}}{R_4}$ |

### DESIGN PROCEDURE

The amplitude and pulse width are usually the most important output characteristics for this circuit. In the following design procedure we assume these two parameters are of prime importance. Input and output impedances are also specified, but some compromise is possible. The input trigger voltage required and the maximum op amp output current are also specified. If these last two requirements cannot be met, input/output buffer stages may be necessary.

### DESIGN STEPS

*Step 1.* Choose $Z_1$ and $Z_2$. $Z_2$ sets the level of positive standby output voltage and as shown in Fig. 20.6, $Z_1$ sets the negative peak voltage of the pulse. $Z_1$ and $Z_2$ may be deleted if the op amp saturation characteristics are to be used for maximum and minimum voltages.

*Step 2.* $R_4$ must be chosen using two criteria. It must be low enough so that the circuit will have sufficient driving power. It must be high enough so that standby current $I_4$ through $Z_2$ is not excessive. The minimum $R_4$ is found from

$$R_4(\text{min}) = \dfrac{V(+,\text{sat}) - V_{Z2}}{I_4(\text{max})}$$

The maximum $R_4$ is determined by the maximum allowable circuit output resistances. The circuit output resistances are $R_{\text{out}}(+)$ for standby and $R_{\text{out}}(-)$ during the pulse. $R_{\text{out}}(+)$ is the parallel combination of $R_4$ and the dynamic resistance of $Z_2$. The dynamic resistance of a zener depends on its bias point. Thus the dynamic impedance of $Z_2$ depends on the size of $R_4$. In most cases, however, $R_{Z2} \ll R_4$ and $R_4$ need not be included in the calculation for $R_{\text{out}}(+)$. The other output resistance $R_{\text{out}}(-)$ is similarly computed.

## 20-10 MULTIVIBRATORS

In most cases it is nearly equal to $R_{Z1}$. The final compromise for $R_4$ will determine the standby current in $R_4$:

$$I_4 = \frac{V(+,\text{sat}) - V_{Z2}}{R_4}$$

*Step 3.* Find the sum of $R_5$ and $R_6$ as follows:

$$R_5 + R_6 = \frac{100 V_{Z2}}{I_4}$$

NOTE: For simplified algebra, set $R_5 = R_6 = 50 V_{Z2}/I_4$ and skip steps 4 and 5.

*Step 4.* $R_6$ is then found from

$$R_6 = \frac{100}{I_4}$$

*Step 5.* Compute

$$R_5 = \frac{100(V_{Z2} - 1)}{I_4}$$

*Step 6.* Solve

$$\beta = \frac{R_6}{R_5 + R_6}$$

*Step 7.* Set $R_1 = 10\, R_6$.

*Step 8.* Compute the parallel resistance $R_p$ of $R_1$ and $R_6$:

$$R_p = \frac{R_1 R_6}{R_1 + R_6}$$

Is $R_p > Z_{\text{in}}(\text{minimum})$? If yes, Eq. 14 is satisfied. If no, the resistances of $R_1$, $R_5$, and $R_6$ must be increased.

*Step 9.* Is the slew rate of the op amp much faster than the rate required to form a good-quality pulse? Figure 20.6 shows the pulse degradation resulting when the op amp slew rate is not fast enough to form a pulse with steep sides. For example, assume $T_p$ is to be 100 $\mu$s wide and 10 V from standby to the (negative) pulse peak. If $S = 0.5$ V/$\mu$s, the rise and fall times of the pulse (0 to 100 percent) will each be 20 $\mu$s.

*Step 10.* Find

$$C_2 = \frac{T_p}{R_1}$$

Also, check to make sure that $[(V_{Z1} + V_{Z2})/2S] < T_p$. If this inequality is not true, the chosen op amp is too slow for the chosen pulse amplitude $V_{Z1} + V_{Z2}$ and pulse width $T_p$. If $R_1 C_2$ is too small, triggering will not occur. If $R_1 C_2$ is too large, multiple output pulses will occur for each input pulse.

*Step 11.* Set $R_3 = R_5 + R_6$.

*Step 12.* Compute

$$C_1 = \frac{T_p}{R_3 \ln\left[(1 + V_D/V_{Z2})/(1 - \beta)\right]} \quad \text{if } R_5 \neq R_6$$

or

$$C_1 = \frac{T_p}{0.8\, R_3} \quad \text{if } R_5 = R_6 \text{ and } V_{Z2} > 3 \text{ V}$$

*Step 13.* Another option exists for the designer at this point. If the reset time $T_r$ is not critical, it can be set equal to $T_p$. This means $R_2$ and $D_3$ can be deleted from the circuit and step 14 skipped.

*Step 14.* Resistor $R_2$ is found as follows:

$$R_2 = \frac{R_3 T_r}{R_3 C_1 \ln\{[V_{Z2}(1+\beta)]/(V_{Z2}-V_D)\} - T_r}$$

*Step 15.* Check to see if $R_2$ and $R_3$ are in the range 10 kΩ to 1 MΩ. Also, make sure $C_1$ is in the range 1,000 pF to 2 μF. Operation outside these ranges may produce oscillation problems or drift of $T_p$. Wide-bandwidth op amps (unity-gain crossover above several MHz) may perform satisfactorily with $C_1$ below 1,000 pF and $R_2$, $R_3$ below 10 kΩ.

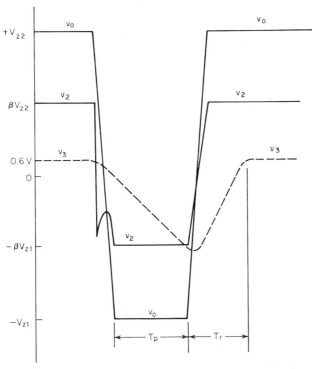

**Fig. 20.6** Monostable multivibrator waveforms $v_2$, $v_3$, and $v_o$.

*Step 16.* The next calculation determines the required size of trigger waveform:

$$v_i(\text{peak-to-peak, required}) = 1.5\, \beta V_{Z2}$$

If $v_i$(peak-to-peak) is less than 80 percent of the value computed above, triggering cannot be guaranteed.

*Step 17.* Error sources must be considered next.

1. If $R_3$ is large ($>1$ MΩ), op amp input bias $I_b$ and offset $I_{io}$ currents will affect $T_p$. These currents vary with temperature. The current through $R_3$ during the pulse ($\approx V_{Z1}/R_3$) must be large compared with the maximum current $I_b$ into the op amp negative input terminal. In summary, for a stable $T_p$,

$$I_b(\text{max}) + I_{io}(\text{max}) \ll \frac{V_{Z1}}{R_3}$$

2. The pulse voltages (standby and peak negative) will vary with temperature, since these voltages are determined by $Z_1$ and $Z_2$. Temperature-stable zeners can be used to hold these voltages constant.

### DESIGN EXAMPLE

*Tentative Circuit-Performance Requirements*

$T_p = 200$ μs
$T_r = 50$ μs
$-V_{Z1} = -5.7$ V (1N5231)
$V_{Z2} = +10.6$ V (1N5240)
$v_i$ (peak-to-peak, available) $= 4$ V (square wave)
$Z_{in} = 5,000$ Ω, minimum
$R_{out}(+) = 50$ Ω, maximum
$R_{out}(-) = 50$ Ω, maximum
$I_4 = 10$ mA, maximum

*Op Amp Parameters (μA 741)*

$S = 0.5$ V/μs
$I_{io} = 300$ nA
$I_b = 800$ nA
$V(+,\text{sat}) = +13$ V
$V(-,\text{sat}) = -13$ V

*Other Critical Parameters*

$R_{Z1} = 35$ Ω, maximum at 10 mA (1N5231)
$R_{Z2} = 35$ Ω, maximum at 10 mA (1N5240)
$TC_{Z1} = \pm 0.03\%/°C$, maximum
$TC_{Z2} = +0.075\%/°C$, maximum
$V^{(+)} = +15$ V
$V^{(-)} = -15$ V

*Step 1* $Z_1$ will be an 1N5231 zener diode which has a typical breakdown voltage of 5.1 V. Adding this 5.1 V to the 0.6-V forward breakdown voltage of $Z_2$ results in $-V_{Z1} = -5.7$ V, as required. Likewise, using a 10-V 1N5240 for $Z_2$ results in a $V_{Z2}$ of 10.6 V.

*Step 2* $$R_4(\text{min}) = \frac{V(+,\text{sat}) - V_{Z2}}{I_4(\text{max})} = \frac{13 - 10.6}{0.01} = 140 \text{ Ω}$$

$$R_{out}(+) = \frac{R_4(\text{min}) \times R_{Z2}}{R_4(\text{min}) + R_{Z2}} = \frac{240 \times 35}{240 + 35} = 30.55 \text{ Ω}$$

$$R_{out}(-) = \frac{R_4(\text{min}) \times R_{Z1}}{R_4(\text{min}) + R_{Z1}} = \frac{240 \times 35}{240 + 35} = 30.55 \text{ Ω}$$

$I_4$ is set to 10 mA as originally specified since $R_{out}(+)$ and $R_{out}(-)$ are sufficiently low.

*Step 3* $$R_5 + R_6 = \frac{100 V_{Z2}}{I_4} = \frac{100 \times 10.6}{0.01} = 106,000 \text{ Ω}$$

*Step 4* $$R_6 = \frac{100}{I_4} = \frac{100}{0.01} = 10^4 \text{ Ω}$$

*Step 5* $$R_5 = \frac{100(V_{Z2} - 1)}{I_4} = \frac{100(10.6 - 1)}{0.01} = 96,000 \text{ Ω}$$

Step 6 $\quad\beta = \dfrac{R_6}{R_5 + R_6} = \dfrac{10{,}000}{106{,}000} = 0.0943$

Step 7 $\quad R_1 = 10R_6 = 10^5 \ \Omega$

Step 8 $\quad R_p = \dfrac{R_1 R_6}{R_1 + R_6} = \dfrac{10^5 \times 10^4}{1.1 \times 10^5} = 9{,}090.9 \ \Omega$

This resistance is greater than $Z_{in}(\min)$, so we can proceed.

Step 9 $\quad S = 0.5$ V/$\mu$s, $T_p = 200$ $\mu$s, and peak-to-peak pulse output is $10.7 + 5.1 = 15.8$ V. The op amp will slew 15.8 V in $15.8/S = 15.8/0.5 = 21.6$ $\mu$s.

Step 10 $\quad C_2 = \dfrac{T_p}{R_1} = \dfrac{2 \times 10^{-4}}{10^5} = 2{,}000$ pF

$$\dfrac{V_{Z1} + V_{Z2}}{2S} = \dfrac{5.7 + 10.6}{2 \times 0.5} = 16.3 \ \mu s < T_p = 200 \ \mu s$$

The inequality is satisfied.

Step 11 $\quad R_3 = R_5 + R_6 = 106{,}000 \ \Omega$

Step 12

$$C_1 = \dfrac{T_p}{R_3 \ln[(1 + 0.6/V_{Z2})/(1-\beta)]} = \dfrac{2 \times 10^{-4}}{1.06 \times 10^5 \ln[(1 + 0.6/10.6)/(1-0.0943)]}$$
$= 0.012 \ \mu F$

Step 13 $\quad T_r < T_p$ is specified.

Step 14

$$R_2 = \dfrac{R_3 T_r}{R_3 C_1 \ln\{[V_{Z2}(1+\beta)]/(V_{Z2}-0.6)\} - T_r}$$

$$= \dfrac{(1.06 \times 10^5)(0.5 \times 10^{-4})}{(1.06 \times 10^5)(0.12 \times 10^{-6}) \ln\{[10.6(1+0.0943)]/(10.6-0.6)\} - 0.5 \times 10^{-4}}$$
$= 41{,}410 \ \Omega$

Step 15 $\quad C_1, R_2,$ and $R_3$ are all within the limits specified in this step.

Step 16

$$v_i(\text{peak-to-peak, required}) = 1.5\beta V_{Z2} = 1.5(0.0943)(10.6)$$
$$= 1.50 \text{ V (peak-to-peak)}$$

Step 17
1. $I_b(\max) + I_{io}(\max) = 800$ nA $+ 300$ nA $= 1.1 \ \mu$A.

$$\dfrac{V_{Z1}}{R_3} = \dfrac{5.6}{1.06 \times 10^5} = 52.8 \ \mu A$$

Since $52.8 \gg 1.1$, $T_p$ will be constant with temperature. This assumes $C_1$ and $R_3$ are also low-drift components.

2. $TC_{Z1} = \pm 0.03\%$/°C for the IN5231
$TC_{Z2} = +0.075\%$/°C for the IN5240

The two output-voltage levels $V_{Z1}$ and $V_{Z2}$ will have the same temperature coefficients as stated above.

## REFERENCES

1. Tobey, G. E., J. G. Graeme, and L. P. Huelsman: "Operational Amplifiers—Design and Applications," p. 392, McGraw-Hill Book Company, New York, 1971.
2. Millman, J., and C. C. Halkias: "Integrated Electronics—Analog Digital Circuits and Systems," p. 581, McGraw-Hill Book Company, New York, 1972.

# Chapter 21
# Oscillators

## INTRODUCTION

In this chapter we will provide detailed design information on two popular oscillator circuits. First to be described is the popular Wien-bridge sine-wave oscillator. The circuit presented will be a superior design which has controlled amplitude and frequency stability. The second circuit presented is a voltage-controlled square-wave generator.

Other types of oscillators are presented elsewhere in this handbook. Chapter 20 contains detailed design information on the basic square-wave generator. Several types of waveform generators are presented in Chap. 27.

## 21.1 WIEN-BRIDGE SINE-WAVE OSCILLATOR

**ALTERNATE NAMES** Phase-shift oscillator, AGC oscillator, sine-wave generator.

**EXPLANATION OF OPERATION** A Wien bridge is made up of a series $RC$ circuit in one branch of a bridge and a parallel $RC$ circuit in another branch. In the Wien-bridge oscillator shown in Fig. 21.1 these components are $R_1$, $R_2$, $C_1$, and $C_2$. The circuit will oscillate at that frequency where the phase of $V_1$ is identical to the phase of $V_o$. This frequency, in terms of circuit components, is

$$f_o = \frac{1}{2\pi(R_1 R_2 C_1 C_2)^{1/2}}$$

Oscillation cannot be sustained unless the positive feedback through $R_1$, $R_2$, $C_1$, and $C_2$ is exactly equal to the forward gain controlled by $R_3$, $R_5$, and $R_6$. The feedback factor (gain from $V_o$ to $V_1$) through the Wien bridge is (at $f_o$)

$$A_f = 1 + \frac{R_2}{R_1} + \frac{C_1}{C_2}$$

The forward gain of the amplifier is (at dc)

$$A_{vc} = 1 + \frac{R_3}{R_5} + \frac{R_3 R_4}{R_5 R_1}$$

## 21-2 OSCILLATORS

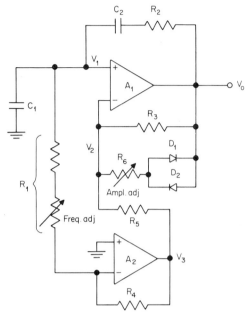

**Fig. 21.1** A Wien-bridge oscillator which requires only one component for tuning.

These gains will be equal if we satisfy the following equalities:

$R_2 = R_3 = R_4 = R_5$ (in practice $R_3$ is made 5 to 10 percent higher)
$C_1 = C_2$

In basic Wien-bridge oscillators $R_1$ is returned to ground. In the present circuit, however, it is returned to a virtual ground at the inverting input of $A_2$. This additional circuit, composed of $A_2$, $R_4$, and $R_5$, is added so that a single part $R_1$ can be used for tuning. The $A_2$ circuit forces the positive feedback to equal the negative feedback for all values of $R_1$.

The circuit composed of $D_1$, $D_2$, and $R_6$ is used to maintain $V_o$ amplitude stability. If $V_o$ tries to increase owing to load changes, diodes $D_1$ and $D_2$ conduct harder. This makes $R_3$ appear to be smaller, which lowers the gain of $A_1$ and restores $V_o$ to its correct value. The diodes keep $R_6$ out of the circuit until a firm oscillation is present. Otherwise the circuit would have too much negative feedback and would not start.

If $R_1$ is to be variable, its limits should be controlled. The minimum $R_1$ is constrained by the maximum available gain of $A_2$ at $f_o$. Equation 5 must be satisfied, so that the gain of the $A_2$ circuit $(-R_4/R_1)$ will never attempt to exceed the open-loop gain of $A_2$ at $f_o$. Conversely, very large values of $R_1$ will cause $A_1$ to have a dc output offset of $I_{b1}R_1$. The maximum $R_1$ is therefore constrained by the maximum allowable output offset.

### DESIGN PARAMETERS

| Parameter | Description |
|---|---|
| $A_1$ | Op amp which oscillates |
| $A_2$ | Op amp used to keep gain of $A_1$ constant as $R_1$ is adjusted |
| $A_f$ | Feedback factor $V_1/V_o$ of Wien bridge |

# WIEN-BRIDGE SINE-WAVE OSCILLATOR

| Parameter | Description |
|---|---|
| $A_{vc}$ | Gain of $A_1$ circuit from $V_1$ to $V_o$ assuming Wien bridge is not present |
| $C_1$ to $C_2$ | Determines frequency of oscillation along with $R_1$ and $R_2$ |
| $D_1$ to $D_2$ | Used to control gain (and output amplitude) of $A_1$ circuit |
| $f_o$ | Frequency of oscillation |
| $f_{u2}$ | Unity-gain crossover frequency of $A_2$ |
| $I_{b1}$ | Input bias current of $A_1$ |
| $R_1(\min)$ | Fixed portion of $R_1$ |
| $R_1(\max)$ | Variable portion of $R_1$ |
| $R_2$ | Controls frequency of oscillation along with $R_1$, $C_1$, and $C_2$ |
| $R_3$ | Sets basic gain of $A_1$ circuit |
| $R_4$ to $R_5$ | Controls effect of $A_2$ circuit on the gain of $A_1$ circuit |
| $R_6$ | Used to adjust stability and also output amplitude (to a lesser extent) |
| $V_1$ to $V_3$ | Voltages at various nodes in Fig. 21.1 |
| $V_o$ | Output voltage |
| $\Delta V_o$ | Output offset due to input bias current of $A_1$ |
| $V^{(\pm)}$ | Power-supply voltages |

## DESIGN EQUATIONS

| Eq. No. | Description | Equation |
|---|---|---|
| 1 | Frequency of oscillation | $f_o = \dfrac{1}{2\pi(R_1 R_2 C_1 C_2)^{1/2}}$ |
| 2 | Feedback factor of Wien bridge at $f_o$ | $A_f(f_o) = \dfrac{V_1}{V_o} - 1 + \dfrac{R_2}{R_1} + \dfrac{C_1}{C_2}$ |
| 3 | Gain of $A_1$ circuit from $V_1$ to $V_o$ (assuming Wien bridge is disconnected) | $A_{vc} = 1 + \dfrac{R_3}{R_5} + \dfrac{R_3 R_4}{R_5 R_1}$ |
| | Recommended resistor values: | |
| 4 | $R_1$ (variable portion) | $R_1(\max) \leq \dfrac{\Delta V_o(\max)}{I_{b1}}$ |
| 5 | $R_1$ (fixed portion) | $R_1(\min) \geq \left(\dfrac{R_4^2}{4\pi^2 R_2 C_1 C_2 f_{u2}^2}\right)^{1/3}$ |
| 6 | $R_2$, $R_4$, $R_5$ | $R_2 = R_4 = R_5 = [R_1(\max)]^{1/2}$ |
| 7 | $R_3$ | $R_3 \approx 1.1\, R_2$ |
| 8 | $R_6$ | $R_6 \approx 100\, R_3$ |
| 9 | Recommended capacitor values $C_1\, C_2$ | $C_1 = C_2 = \dfrac{1}{2\pi f_o R_2}$ |
| 10 | Maximum $f_o$ | $f_o(\max) \approx \left(\dfrac{f_{u2}}{4\pi^2 R_2 R_4 C_1 C_2}\right)^{1/3}$ |
| 11 | Minimum $f_o$ | $f_o(\min) \approx \dfrac{1}{2\pi}\left[\dfrac{I_{b1}}{R_2 C_1 C_2 \Delta V_o(\max)}\right]^{1/2}$ |

## DESIGN PROCEDURE

We begin by assuming the midband frequencies are most important. We then perform calculations to determine the maximum and minimum frequency limits of the oscillator.

## 21-4 OSCILLATORS

### DESIGN STEPS

*Step 1.* Compute a value for $R_1(\text{max})$ using Eq. 4.

*Step 2.* Sequentially apply Eqs. 6, 7, and 8 to determine the other resistor values.

*Step 3.* Determine nominal values for $C_1$ and $C_2$ using Eq. 9.

*Step 4.* Compute the fixed portion of $R_1$ using Eq. 5. The variable portion of $R_1$ should be a log-taper potentiometer if one wants a constant octave/degree control of frequency.

*Step 5.* Compute the approximate frequency limits expected from the oscillator using Eqs. 10 and 11.

*Step 6.* Double-check all previous calculations by computing $f_o$ with Eq. 1 at $R_1(\text{min})$, $R_1(\text{max})$, and the square root of $R_1(\text{max})$.

**WIEN-BRIDGE-OSCILLATOR DESIGN EXAMPLE** An oscillator with a mid-range frequency of 1,000 Hz will be designed. The maximum output voltage (peak-to-peak) and output offset are specified. The op amps are also predetermined. We are asked to determine the upper- and lower-frequency limits of the oscillator.

*Design Requirements*

$V_o = \pm 10$ V
$\Delta V_o(\text{max}) = 0.1$ V
$f_o(\text{midband}) = 1,000$ Hz
$A_1$ and $A_2 =$ LM 324
$V^{(\pm)} = \pm 15$ V

*Device Data*

$f_{u2} = 5 \times 10^5$ Hz
$I_{b1} = 3 \times 10^{-8}$ A

*Step 1.* The adjustable portion of $R_1$ is

$$R_1(\text{max}) = \frac{\Delta V_o(\text{max})}{I_{b1}} = \frac{0.1}{3 \times 10^{-8}} = 3.3 \text{ M}\Omega$$

*Step 2.* Equation 6 provides us with

$$R_2 = R_4 = R_5 = [R_1(\text{max})]^{1/2} = (3.3 \times 10^6)^{1/2} = 1,830 \text{ }\Omega$$

Equation 7 is approximately

$$R_3 \approx 1.1 \, R_2 = 1.1(1,830) = 2,000 \text{ }\Omega$$

$R_6$ is found from Eq. 8:

$$R_6 \approx 100 \, R_3 = 100(2,000) = 200 \text{ k}\Omega$$

*Step 3.* Equation 9 gives us nominal values for $C_1$ and $C_2$:

$$C_1 = C_2 = \frac{1}{2\pi f_o R_2} = \frac{1}{2\pi(1,000)1,830} = 0.087 \text{ }\mu\text{F}$$

The output will not be a pure sine wave unless these capacitors are closely matched.

*Step 4.* The fixed portion of $R_1$ is

$$R_1(\text{min}) = \left(\frac{R_4^2}{4\pi^2 R_2 C_1 C_2 f_{u2}^2}\right)^{1/3}$$

$$= \left[\frac{1,830^2}{4\pi^2(1,830)(0.087 \times 10^{-6})^2 (5 \times 10^5)^2}\right]^{1/3} = 29 \text{ }\Omega$$

*Step 5.* We now substitute data into Eqs. 10 and 11 to find the oscillator range:

$$f_o(\text{max}) = \left(\frac{f_{u2}}{4\pi^2 R_2 R_4 C_1 C_2}\right)^{1/3}$$

$$= \left[\frac{5 \times 10^5}{4\pi^2 (1{,}830)^2 (0.087 \times 10^{-6})^2}\right]^{1/3} = 7{,}900 \text{ Hz}$$

$$f_o(\text{min}) = \frac{1}{2\pi}\left[\frac{I_{b1}}{R_2 C_1 C_2 \Delta V_o(\text{max})}\right]^{1/2}$$

$$= \frac{1}{2\pi}\left[\frac{3 \times 10^{-8}}{(1{,}830)(0.087 \times 10^{-6})^2 0.1}\right] = 23.4 \text{ Hz}$$

*Step 6.* The oscillator frequency is computed using Eq. 1 along with the results of steps 1, 2, 3, and 4.

$$f_o(\text{min}) = \frac{1}{2\pi \, [R_1(\text{max}) R_2 C_1 C_2]^{1/2}}$$

$$= \frac{1}{2\pi \, [(3.3 \times 10^6) 1{,}830 (0.087 \times 10^{-6})^2]^{1/2}} = 23.4$$

$$f_o(\text{nom}) = \frac{1}{2\pi \, [R_1(\text{nom}) R_2 C_1 C_2]^{1/2}}$$

$$= \frac{1}{2\pi \, [1{,}830(1{,}830)(0.087 \times 10^{-6})^2]^{1/2}} = 1{,}000 \text{ Hz}$$

$$f_o(\text{max}) = \frac{1}{2\pi \, [R_1(\text{min}) R_2 C_1 C_2]^{1/2}}$$

$$= \frac{1}{2\pi \, [29(1{,}830)(0.087 \times 10^{-6})^2]^{1/2}} = 7{,}958 \text{ Hz}$$

**REFERENCES**

1. Brokaw, P.: FET Op Amp Adds New Twist to an Old Circuit, *EDN*, June 5, 1974, p. 75.
2. Coers, G.: MOSFET Network Minimizes Audio Oscillator Distortion, *Electronics*, Jan. 3, 1972, p. 85.
3. Widlar, R. J., and J. N. Giles: Avoid Over Integration, *Electron. Des.*, Feb. 1, 1966, p. 56.

## 21.2 VOLTAGE-CONTROLLED OSCILLATOR

**ALTERNATE NAMES** VCO, voltage-controlled pulse generator, voltage-to-frequency converter, V/F converter, VFC.

**EXPLANATION OF OPERATION** The op amp circuit is an integrator which is constantly attempting to drive its output terminal high. The rate at which the output slews high is

$$\frac{\Delta v_1}{\Delta t} = \frac{v_m}{R_1 C_1}$$

## 21-6 OSCILLATORS

Whenever $v_1$ rises to $V_c$, the threshold input of the timer causes the output terminal to drop to the low state. But this output terminal is attached to the gate of $Q_1$. A low voltage at this point turns the FET on, discharges $C_1$, and lowers $v_1$ to approximately zero. The rate of discharge depends on the sizes of $C_1$ and $R_{on}$ of the FET. The timer trigger voltage, attached to the threshold terminal through $R_4$, also slews downward. When the trigger terminal drops to $V_c/2$, the timer output changes to the high state. This turns $Q_1$ off, which stops the $C_1$ discharge. The $R_4C_2$ delay network is required so that the trigger

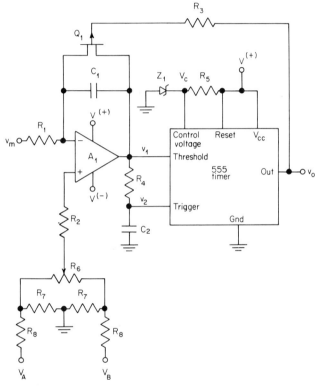

**Fig. 21.2** A voltage-controlled pulse generator which utilizes an op amp integrator in conjunction with an IC timer.

action does not occur until $C_1$ is fully discharged. This requires that $R_4C_2 > R_{on}C_1$.

The output pulse width is approximately

$$T_p \approx 0.7\, R_4 C_2$$

The width of this pulse controls the upper frequency of the VCO. If $T_o = 1\,\mu s$, the VCO has only 0.2 percent nonlinearity (of the $T_o/v_m$ transfer function) up to approximately 10 kHz. At low frequencies where the width of $T_p$ is much smaller than $T_o$,

$$f_o \approx \frac{1}{T_o} \approx \frac{2\, v_m}{V_c R_1 C_1}$$

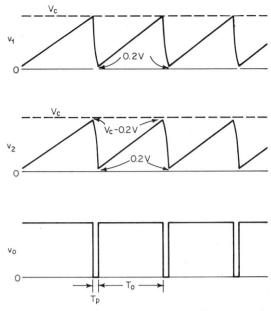

**Fig. 21.3** Waveforms at various locations in the voltage-controlled oscillator.

## DESIGN PARAMETERS

| Parameter | Description |
|---|---|
| $A_1$ | Op amp used as an integrator which is reset once per cycle |
| $C_1$ | Determines integration time along with $R_1$ |
| $C_2$ | Determines pulse width along with $R_4$ |
| $f_o$ | Output frequency |
| $f_o(\text{nom})$ | Mid-range output frequency corresponding to $v_m(\text{nom})$ |
| $I_{DSS}$ | Drain-to-source saturation current of FET (at $V_{GS}=0$, $V_{DS}=-5$ V) |
| $I_{Z1}$ | Nominal zener-diode current |
| $R_1$ | Determines input resistance of circuit and integration time of $A_1$ |
| $R_2$ | Cancels effect of input bias current of $A_1$ |
| $R_3$ | Limits drain-to-gate current in $Q_1$ |
| $R_4$ | Controls output pulse width along with $C_2$ |
| $R_5$ | Controls zener-diode current |
| $R_6$ to $R_8$ | Used to set $f_o \approx 0$ when $v_m = 0$ unless $A_1$ has offset-adjustment terminals |
| $R_{on}$ | On resistance of $Q_1$ |
| $T_o$ | Time between output pulses |
| $T_p$ | Output pulse width |
| $v_o$ | Output-voltage waveform |
| $v_1$ to $v_2$ | Voltages as shown in Figs. 21.2 and 21.3 |
| $V_A$ to $V_B$ | Reference voltages used for offset adjustment |
| $V_c$ | Control voltage established by $Z_1$ |
| $V_{io}$ | Input offset voltage of $A_1$ |
| $v_m$, $v_m(\text{nom})$ | Input modulation voltage and its nominal (mid-range) value |
| $V_p$ | Pinch-off voltage of $Q_1$ |
| $V^{(\pm)}$ | Power-supply voltages |
| $Z_1$ | Reference diode |

## 21-8 OSCILLATORS

### DESIGN EQUATIONS

| Eq. No. | Description | Equation |
|---|---|---|
| 1 | Output frequency if $T_o \gg T_p$ | $f_o \approx \dfrac{|v_m|}{V_c R_1 C_1}$ |
| 2 | Output frequency if $T_o \approx T_p$ (near upper limit of $f_o$) | $f_o \approx \dfrac{1}{0.7 R_4 C_2 + (V_c R_1 C_1/|v_m|)}$ |
| 3 | Width of negative output pulse | $T_p \approx 0.7 R_4 C_2$ |
| 4 | Width of positive output pulse | $T_o \approx \dfrac{V_c R_1 C_1}{|v_m|}$ |
|  | Resistor values: |  |
| 5 | $R_1, R_2$ | $R_1 = R_2 = \dfrac{|v_m(\text{nom})|}{V_c C_1 f_o(\text{nom})}$ |
| 6 | $R_3$ | $R_3 \approx 100 \text{ k}\Omega$ |
| 7 | $R_4$ | $R_4 \approx 10 \text{ k}\Omega$ |
| 8 | $R_5$ | $R_5 \approx \dfrac{V^{(+)} - V_c}{I_{Z1}}$ |
|  | NOTE: If offset-adjustment terminals are provided on the op amp, skip steps 9 to 11 and ground the lower end of $R_2$ | |
| 9 | $R_6$ | $R_6 \approx \dfrac{R_2}{10}$ to $\dfrac{R_2}{100}$ |
| 10 | $R_7$ | $R_7 \approx \dfrac{R_6}{100}$ to $\dfrac{R_6}{1{,}000}$ |
| 11 | $R_8$ | $R_8 \approx \dfrac{R_7 V_A}{10 V_{io}}$ (let $V_A = -V_B$) |
|  | Capacitor values: |  |
| 12 | $C_1$ | $C_1 = \dfrac{T_p}{10 R_{\text{on}}}$ |
| 13 | $C_2$ | $C_2 = \dfrac{T_p}{0.7 R_4}$ |
| 14 | FET on resistance (at $V_{GS} = 0$) | $R_{\text{on}} = \dfrac{V_p}{2 I_{DSS}}$ |

### DESIGN PROCEDURE

These design steps are fairly straightforward, owing to the simplicity of the circuit. Timer ICs in conjunction with op amps make the design of a large class of circuits greatly simplified. We begin this procedure by assuming the pulse width $T_o$ and the nominal $f_o$ [and its corresponding $v_m(\text{nom})$] are specified.

### DESIGN STEPS

*Step 1.* Choose an FET having a $V_p$ less than $V^{(+)}$. Compute $R_{\text{on}}$ using Eq. 14.

*Step 2.* Calculate a nominal value for $C_1$ using Eq. 12. Use the standard value closest to that calculated.

*Step 3.* Resistors $R_1$ and $R_2$ are now found from Eq. 5. Use a nominal $v_m$ and the corresponding nominal $f_o$.

*Step 4.* The selection of $R_3$ is not critical. For high-speed operation (>10 kHz) one should probably keep $R_3$ in the 100-k$\Omega$ range. Likewise, a nominal choice for $R_4$ is 10 k$\Omega$.

*Step 5.* Calculate approximate values for $R_5$ to $R_8$ using Eqs. 8 to 11.

*Step 6.* Compute a value for $C_2$ with Eq. 13.

**EXAMPLE OF A VCO DESIGN** Suppose we want a VCO having a range from nearly dc to 10 kHz. The (negative) pulse width is to be approximately 10 percent of the period at 10 kHz. Let the nominal $v_m$ of $-5$ V correspond to $f_o = 5$ kHz.

*Design Requirements*

$f_o(\text{nom}) = 5$ kHz
$f_o(\text{min}) = $ dc
$f_o(\text{max}) = 10$ kHz
$v_m(\text{nom}) = -5$ V
$v_m(\text{min}) = 0$ (shorted to ground)
$v_m(\text{max}) = -10$ V
$V^{(\pm)} = \pm 15$ V
$V_A = -V_B = +15$ V

*Device Data*

$V_{io}(\text{max}) = 5$ mV
$V_p = 3.2$ V (2N2608, measured)
$I_{DSS} = 3.2$ mA (2N2608, measured)
$I_{Z1} = 0.5$ mA (1N4566)
$V_c = 6.4$ V (1N4566)

*Step 1.* The 2N2608 FET has a pinch-off voltage of 1 to 4 V. We will assume 3.2 V in the following.

$$R_{on} = \frac{V_p}{2I_{DSS}} = \frac{3.2}{2(3.2 \times 10^{-3})} = 500 \ \Omega$$

*Step 2.* Equation 12 provides a nominal value for $C_1$:

$$C_1 = \frac{T_p}{10R_{on}} = \frac{1}{10f_o(\text{max})10R_{on}} = \frac{1}{10(10^4)10(500)} = 2{,}000 \text{ pF}$$

NOTE: $T_p = 1/10 \, f_o(\text{max}) = 1/10(10^4) = 10 \ \mu s$.

*Step 3.* Use Eq. 5 to find $R_1$ and $R_2$:

$$R_1 = R_2 = \frac{|v_m(\text{nom})|}{V_c C_1 f_o(\text{nom})} = \frac{|-5|}{6.4(2 \times 10^{-9})5{,}000} = 78 \text{ k}\Omega$$

*Step 4.* Let $R_3 = 100$ k$\Omega$ and $R_4 = 10$ k$\Omega$.

*Step 5.* Other resistor values are computed:

$$R_5 \approx \frac{V^{(+)} - V_c}{I_{Z1}} = \frac{15 - 6.4}{5 \times 10^{-4}} = 17{,}200 \ \Omega$$

$$R_6 \approx \frac{R_2}{10} = \frac{78{,}000}{10} = 7{,}800 \ \Omega \text{ (use 10 k}\Omega\text{)}$$

$$R_7 \approx \frac{R_6}{100} = \frac{10{,}000}{100} = 100 \ \Omega$$

$$R_8 \approx \frac{R_7 V_A}{10 V_{io}} = \frac{100(15)}{10(5 \times 10^{-3})} = 30 \text{ k}\Omega$$

*Step 6.* Capacitor $C_2$ becomes

$$C_2 = \frac{T_p}{0.7 R_4} = \frac{10^{-5}}{0.7(10^4)} = 1{,}430 \text{ pF}$$

# REFERENCE

1. Klement, C.: Voltage to Frequency Converter Constructed with Few Components is Accurate to 0.2%, *Electron. Des.*, vol. 13, p. 12, June 21, 1973.

Chapter **22**

# Parameter Enhancement and Simulation

## INTRODUCTION

The versatility of op amps allows the circuit designer many opportunities to create circuits possessing new and unique characteristics. Some of these unique circuits are those which multiply capacitance or simulate inductance. In this chapter we will present one circuit of each type. Numerous designs for parameter enhancement exist in the literature. The circuits chosen for presentation here we feel would be most useful for the application-oriented designer.

## 22.1 CAPACITANCE MULTIPLIER

**ALTERNATE NAMES**  Low-Q capacitor simulator, simulated capacitor.

**EXPLANATION OF OPERATION**  In some low-level high-impedance circuits very large capacitors are required. For example, suppose a designer needs a nonpolarized 100,000-$\mu$F capacitor for a low-voltage application. This requirement is difficult to implement with passive components. A capacitance multiplier could be used if the application does not require a high-Q capacitor.

This circuit uses the high gain of the op amp to multiply capacitance. The effective capacitance seen between point $Z_{in}$ and ground in Fig. 22.1A is

$$C_{in} = \frac{R_3 C_1}{R_1}$$

As shown in Fig. 22.1B, this capacitor has an effective series resistance which lowers its Q. This series resistance is $R_s = R_1$.

The circuit also has an effective leakage current through the capacitor. The effective leakage current does not depend on the voltage between $Z_{in}$ and ground. It has a value of

$$I_L = \frac{V_{io} + I_{io} R_3}{R_1}$$

22-1

## 22-2 PARAMETER ENHANCEMENT AND SIMULATION

This equation assumes $R_2 = R_3$. Otherwise, the input bias current of the op amp would cause an even larger $I_L$. It is obvious that a high-quality op amp is required if a low-leakage simulated capacitor is desired.

If $C_{in}$ is to be nonpolarized, $C_1$ must also be nonpolarized.

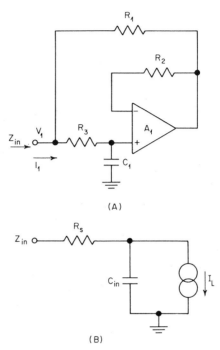

**Fig. 22.1** A single-ended capacitance multiplier (A) and its equivalent circuit (B).

### DESIGN PARAMETERS

| Parameter | Description |
| --- | --- |
| $C_1$ | Capacitor which is multiplied |
| $C_{in}$ | Effective input capacitance of circuit |
| $I_i$ | Current into circuit at $Z_{in}$ terminal |
| $I_{io}$ | Input offset current of op amp |
| $I_L$ | Effective leakage current through input capacitance |
| $Q$ | Quality factor of circuit |
| $R_1$ | Resistor which controls the effective series resistor of $Z_{in}$ |
| $R_2$ | Makes $A_1$ a noninverting unity-gain amplifier |
| $R_3$ | Controls size of effective $C_{in}$ |
| $R_s$ | Effective series resistance in $Z_{in}$ |
| $s$ | $s = j\,2\pi f$ |
| $V_1$ | Voltage at $Z_{in}$ terminal |
| $V_{io}$ | Input offset voltage of op amp |
| $Z_{in}$ | Input impedance of circuit |

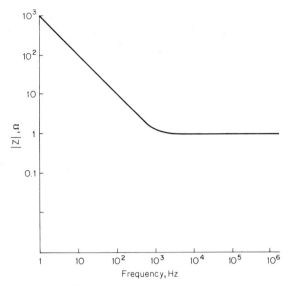

**Fig. 22.2** The input impedance of a typical capacitor multiplier as a function of frequency.

### DESIGN EQUATIONS

| Eq. No. | Description | Equation |
|---|---|---|
| 1 | Value of simulated capacitance | $C_{in} = \dfrac{R_3 C_1}{R_1}$ |
| 2 | Value of effective series resistance | $R_s = R_1$ |
| 3 | Effective leakage current of capacitor if $R_2 = R_3$ | $I_L = \dfrac{V_{io} + I_{io} R_3}{R_1}$ |
| 4 | Input impedance at $C_{in}$ | $Z_{in} = R_1 + \dfrac{R_1}{s R_3 C_1}$ |
| 5 | $Q$ of circuit | $Q = \dfrac{1}{2\pi f R_s C_{in}}$ |
| 6 | Required $R_2$ | $R_2 = R_3$ |

### REFERENCES

1. Schmutz, L. E.: Transistor Gain Boosts Capacitor Value, *Electronics*, July 25, 1974, p. 116.
2. National Semiconductor Corp. Applications Note AN-29, December 1969.
3. George A. Philbrick Researches, Inc., Applications Manual for Computing Amplifiers, 1966, p. 97, Nimrod Press, Inc., Boston, Mass.

## 22.2 INDUCTANCE SIMULATOR

**ALTERNATE NAMES** Simulated inductor, capacitance-to-inductance converter, low-$Q$ inductance simulator.

## 22-4 PARAMETER ENHANCEMENT AND SIMULATION

**EXPLANATION OF OPERATION** This circuit utilizes a triple op amp, eight resistors, and a small capacitor. Stage $A_1$ is an integrator which has a dc gain of $R_2/R_1$ and a pole at $f_{cp1} = 1/2\pi R_2 C$. The transfer function of this first stage is

$$\frac{V_2}{V_1} = \frac{R_2/R_1}{1 + sR_2C}$$

The voltage-to-current converter $A_2$ to $A_3$ has a transfer function of

$$I_o = \frac{V_2 R_4 R_7 / R_3 R_6}{R_8 + Z_L[1 - (R_4 R_7/R_5 R_6)]}$$

If we let

$$R_3 = R_4 = R_5 = R_6 = R_7$$

then $I_o$ reduces to

$$I_o = \frac{V_2}{R_8}$$

If we combine the above equation with the integrator transfer function, the result is

$$I_o = \frac{R_2 V_1}{R_1 R_8 (1 + sR_2C)}$$

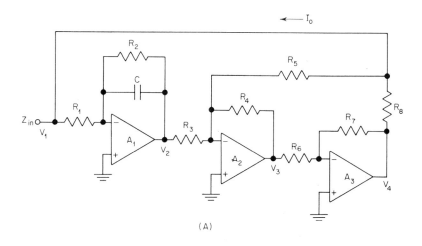

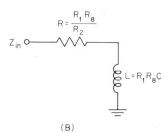

**Fig. 22.3** An inductance simulator which utilizes an integrator in conjunction with a voltage-to-current converter $(A)$. The equivalent circuit of an inductance simulator $(B)$.

This current must equal the input current, since $A_1$ draws essentially no current and $R_1$ is very large. The input impedance of this circuit is therefore

$$Z_{in} = \frac{V_1}{I_o} = \frac{R_1 R_8 (1 + sR_2 C)}{R_2}$$

The input impedance is equivalent to a series $RL$ circuit with a resistor of

$$R = \frac{R_1 R_8}{R_2}$$

and an inductance of

$$L = R_1 R_8 C$$

The $Q$ of a series $RL$ circuit is

$$Q = \frac{2\pi f L}{R}$$

The $Q$ is a direct function of frequency; so the circuit will have a $Q_{min}$ corresponding to some $f_{min}$.

The circuit also has an upper frequency limit where the closed-loop gain of the $A_1$ circuit equals the open-loop gain of $A_1$. This frequency is

$$f_{max} = \frac{f_u R_1}{R_2}$$

In Fig. 22.4 this occurs at approximately 1 kHz.

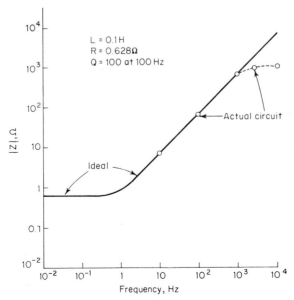

**Fig. 22.4** The impedance of a typical inductance simulator as a function of frequency.

## 22-6 PARAMETER ENHANCEMENT AND SIMULATION

### DESIGN PARAMETERS

| Parameter | Description |
|---|---|
| $C$ | Reactive element in circuit whose properties are inverted in $A_2$ and $A_3$ to create a simulated inductor |
| $f_{max}$ | Frequency where $A_1$ open-loop gain is insufficient to perform integration |
| $f_{min}$ | Minimum frequency where a specified minimum $Q$ is possible |
| $f_u$ | Unity-gain crossover frequency of $A_1$ |
| $I_o$ | Output current of $A_2$ to $A_3$ voltage-to-current converter |
| $I_b$ | Input bias current of $A_1$ |
| $L$ | Simulated inductance |
| $Q_{min}$ | Specified minimum $Q$ of circuit |
| $R$ | Effective series resistance of $L$ |
| $R_1$ to $R_2$ | Controls dc gain of integrator |
| $R_3$ to $R_8$ | Controls voltage-to-current transfer ratio of $A_2$ to $A_3$ circuit |
| $s$ | $s = j2\pi f$ |
| $V_1$ | Voltage at $Z_{in}$ terminal |
| $V_2$ | Integrator output voltage |
| $\Delta V_2$ | DC output offset voltage from integrator |
| $V^{(\pm)}$ | Power-supply voltages |
| $Z_{in}$ | Input impedance of circuit |
| $Z_L$ | Load impedance seen by voltage-to-current converter |

### DESIGN EQUATIONS

| Eq. No. | Description | Equation |
|---|---|---|
| 1 | Simulated inductance | $L = R_1 R_8 C$ |
| 2 | Series resistance of simulated inductance | $R = \dfrac{R_1 R_8}{R_2}$ |
| 3 | $Q$ of $RL$ circuit | $Q = \dfrac{2\pi f L}{R} = 2\pi f R_2 C$ |
|   | Resistor values: | |
| 4 | $R_1$ | $R_1 = \dfrac{R_2 |V_1|_{max}}{2\pi f_{min} R_2 C |V_2|_{max}}$ |
| 5 | $R_2$ | $R_2 = \left|\dfrac{\Delta V_2(max)}{10 I_b(max)}\right|$ |
| 6 | $R_3$ to $R_7$ | $R_3 = R_4 = R_5 = R_6 = R_7 = 1$ to $10$ k$\Omega$ |
| 7 | $R_8$ | $R_8 = \dfrac{L}{R_1 C}$ |
| 8 | Recommended capacitor value | $C = \dfrac{Q_{min}}{2\pi f_{min} R_2}$ |
| 9 | Maximum frequency of operation | $f_{max} = \dfrac{f_u R_1}{R_2}$ |

### DESIGN PROCEDURE

This procedure could start out assuming $L$, $R_{min}$, and $f_{min}$ are specified or $L$, $Q_{min}$, and $f_{min}$ are specified. We will assume the latter.

## DESIGN STEPS

*Step 1.* Use Eq. 3 to determine the $R_2C$ product required.

$$R_2C = \frac{Q_{min}}{2\pi f_{min}}$$

*Step 2.* Let $R_2$ be chosen so that the worst-case $V_2$ dc offset is only about 10 percent of $|V_2|_{max}$:

$$R_2 = \left|\frac{V_2(max)}{10I_b(max)}\right|$$

*Step 3.* Calculate $R_1$ from

$$R_1 = \frac{R_2|V_1|_{max}}{2\pi f_{min} R_2 C |V_2|_{max}}$$

*Step 4.* Compute a value for $C$ from

$$C = \frac{Q_{min}}{2\pi f_{min} R_2}$$

*Step 5.* The value for $R_8$ is

$$R_8 = \frac{L}{R_1 C}$$

*Step 6.* Choose a common value for $R_3$, $R_4$, $R_5$, $R_6$, and $R_7$ which is equal to or greater than the load resistance most commonly used on the $A_2$ to $A_3$ specification sheets. This value is usually between 1 and 10 k$\Omega$.

*Step 7.* If needed, compute the effective series resistance of the inductance using Eq. 2.

*Step 8.* Calculate the maximum frequency of operation using Eq. 9.

**EXAMPLE OF INDUCTANCE-SIMULATOR DESIGN** For a numerical illustration suppose we wish to create a 0.1-H choke having a minimum $Q$ of 100. The choke is to operate with a 1-V peak-to-peak sine wave which has a minimum frequency of 100 Hz. Operation below 100 Hz will be possible, but the $Q$ will decrease from 100 at 100 Hz to 1 at 1 Hz.

*Design Requirements*
$|V_1|_{max} = 1$ V
$L = 0.1$ H
$Q_{min} = 100$
$f_{min} = 100$ Hz
$V^{(\pm)} = \pm 15$ V

*Device Data*
$|V_2|_{max} = 10$ V
$I_b = 5 \times 10^{-7}$ A (maximum)
$f_u = 1$ MHz

*Step 1.* Equation 3 provides us with

$$R_2C = \frac{Q_{min}}{2\pi f_{min}} = \frac{100}{2\pi(100)} = 0.159$$

*Step 2.* The value for $R_2$ is

$$R_2 = \left|\frac{V_2(\text{max})}{10I_b(\text{max})}\right| = \frac{10}{10(5 \times 10^{-7})} = 2 \text{ M}\Omega$$

*Step 3.* We compute $R_1$ as follows:

$$R_1 = \frac{R_2|V_1|_{\text{max}}}{2\pi f_{\text{min}}R_2C|V_2|_{\text{max}}} = \frac{2 \times 10^6(1)}{2\pi(100)0.159(10)} = 2 \text{ k}\Omega$$

*Step 4.* The integration capacitor has a value of

$$C = \frac{Q_{\text{min}}}{2\pi f_{\text{min}}R_2} = \frac{100}{2\pi(100)2 \times 10^6} = 0.08 \text{ }\mu\text{F}$$

*Step 5.* We find $R_8$ from

$$R_8 = \frac{L}{R_1C} = \frac{0.1}{2{,}000(8 \times 10^{-8})} = 628 \text{ }\Omega$$

*Step 6.* We will let the common value of $R_3$ through $R_7$ be 2 k$\Omega$.

*Step 7.* The effective series resistance of the inductor is

$$R = \frac{R_1R_8}{R_2} = \frac{2{,}000(628)}{2 \times 10^6} = 0.628 \text{ }\Omega$$

*Step 8.* The maximum frequency of operation is

$$f_{\text{max}} = \frac{f_u R_1}{R_2} = \frac{10^6(2 \times 10^3)}{2 \times 10^6} = 1{,}000 \text{ Hz}$$

## REFERENCES

1. Allen, P. E., and J. A. Means: Inductor Simulation Derived from an Amplifier Roll-off Characteristic, *IEEE Trans. Circuit Theory*, July 1972, p. 395.
2. Kalinowski, J. J.: An Inductance Realization Using Two Operational Amplifiers, *IEEE Proc.*, September 1968, p. 1636.
3. Roy, S. C. D.: Inductor Realization with RC Elements, *IEEE Proc.*, September 1971, p. 1380.
4. Berndt, D. F., and S. C. D. Roy: Inductor Simulation Using a Single Unity Gain Amplifier, *IEEE J. Solid-State Circuits*, June 1969, p. 161.
5. Roy, S. C. D., and V. Nagarajan: On Inductor Simulation Using a Unity-Gain Amplifier, *IEEE J. Solid-State Circuits*, June 1970, p. 95.

# Chapter 23
# Power Circuits

## INTRODUCTION

Most op amps are low-level devices with optimized input characteristics. Although high-power op amps are available, it is sometimes advisable to combine an op amp having excellent input characteristics with a high-power output circuit. This additional circuit can be tailored to perform various types of tasks beyond the capability of the op amp. The power level may be increased, the voltage swing increased, the slew rate increased, the bandwidth increased, or the current-driving capability increased using appropriate output circuits. In this chapter we will discuss several such applications.

The first circuit to be presented increases power level, slew rate, and bandwidth. The second circuit increases both power level and peak-to-peak voltage swing.

## 23.1 OP AMP BANDWIDTH/POWER BOOSTER

**ALTERNATE NAMES** Op amp buffer, push-pull buffer amplifier, complementary transistor output circuit, slew-rate booster.

**EXPLANATION OF OPERATION** As shown in Fig. 23.1, this circuit drives the buffer transistors from the op amp power-supply terminals. The simplified circuit, in Fig. 23.2, shows that this circuit makes a feedback loop out of the buffer transistors and the op amp output transistors. This is called voltage-series feedback (Ref. 1). The resistor ratio $R_6/(R_6 + R_7)$ feeds back a portion of $V_2$ to the emitters of $Q_3$ and $Q_4$. This is degenerative feedback which lowers the output resistance, widens bandwidth, and increases slew rate.

The dc and low-frequency voltage gain of the circuit is controlled only by

$$A_{vc} = \frac{V_2}{V_1} = -\frac{R_2}{R_1}$$

The additional circuit merely reduces the portion of the work load required of the op amp so that it will operate more efficiently. The improvement in high-frequency operation is directly related to the choices for $R_6$ and $R_7$. Assume the $f_\beta$ of $Q_1$ and $Q_2$ is at least ten times greater than $f_u$ of the op amp. We can then use the following ideas to approximate the new bandwidth, full-power response, and slew rate.

## 23-2 POWER CIRCUITS

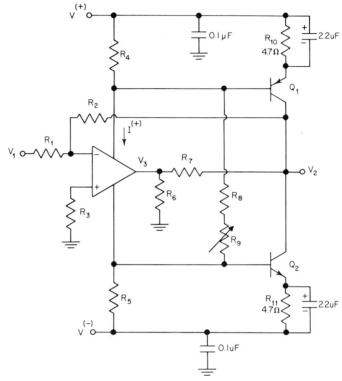

**Fig. 23.1** A complementary power booster which utilizes voltage-series feedback.

1. Let $\beta = R_6/(R_6 + R_7)$ be the feedback factor from $V_2$ back to the emitters of $Q_3$ and $Q_4$.
2. Use the open-loop gain plot and maximum output-voltage swing plot (as shown in Fig. 23.3) to estimate performance changes graphically.
3. The new closed-loop bandwidth is approximately $f'_1 = f_1/\beta$, since the op amp share of gain has been lowered from $A_{vc}$ to $\beta A_{vc}$. This is clearly illustrated in Fig. 23.3A.
4. The new full-power response frequency is likewise found by multiplying the old peak output by $\beta$ and finding the appropriate frequency. The curve of output voltage as a function of load resistance must also be factored into the above calculation. In most cases $R_6$ will lower the output-voltage curve from that shown in the data sheet.
5. As mentioned in Chap. 2, the slew rate is directly related to full-power response by

$$S_{\max} = 2\pi f_f V_{pp}$$

where $f_f$ and $V_{pp}$ are the coordinates of a point on the curve in Fig. 23.3B. Slew rate will therefore be increased in this circuit by the same factor that full-power bandwidth was increased.

The output power level is determined by the drive capability of $Q_1$ and $Q_2$. The output resistance of the buffer is approximately

$$R_o \approx \frac{[h_{ie3} + (1 + \beta_3)(R_4 + R_6)][h_{ie1} + (1 + \beta_1)R_{10}]}{\beta_1 \beta_3 R_4}$$

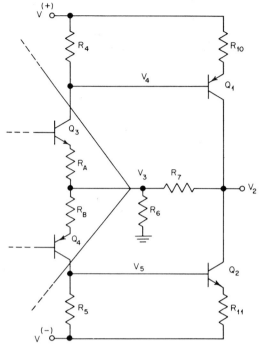

**Fig. 23.2** Simplified version of Fig. 23.1 which shows $Q_1$, $Q_3$ and $Q_2$, $Q_4$ feedback loops.

This equation assumes that the buffer is symmetrical, i.e., $h_{ie1} = h_{ie2}$ and $\beta_1 = \beta_2$. In the absence of op amp output transistor data the following estimates can be used for standard bipolar monolithic devices:

$$h_{ie3} = 3{,}000 \ \Omega$$
$$\beta_3 = 300$$

The size of $R_A$ is usually indicated on the op amp data sheet.

Resistors $R_8$ and $R_9$ are required to cancel crossover distortion. $R_9$ is adjusted at full output power and at the full-power bandwidth frequency.

## DESIGN PARAMETERS

| Parameter | Description |
|---|---|
| $A_{vc}$ | Voltage gain of complete circuit |
| $\beta$ | Feedback factor determined by $R_6$ and $R_7$ |
| $\beta_1$ | Current gain of $Q_1$ |
| $\beta_3$ | Current gain of $Q_3$ |
| $f_1$ | Small-signal bandwidth (−3 dB) of op amp at a closed-loop gain of $A_{vc}$ |
| $f'_1$ | Small-signal bandwidth (−3 dB) of entire circuit (or op amp at a closed-loop gain of $\beta A_{vc}$) |
| $f_2$ | Frequency of a given maximum op amp output (peak-to-peak) voltage |
| $f'_2$ | Frequency of a given maximum op amp output (peak-to-peak) voltage if the op amp gain is reduced by $\beta$ |
| $f_\beta$ | Small-signal bandwidth (−3 dB) of a transistor current gain |
| $f_f$ | A specified point on the op amp maximum output-voltage curve |
| $f_{op1}$ | Frequency of the first pole of the op amp |

## 23-4 POWER CIRCUITS

| Parameter | Description |
|---|---|
| $f_{op2}$ | Op amp second-pole frequency |
| $f_u$ | Unity-gain crossover frequency of op amp |
| $h_{ie1}$ | Input resistance of $Q_1$ |
| $h_{ie3}$ | Input resistance of $Q_3$ |
| $I^{(+)}$ | Current into op amp (+) power-supply terminal |
| $I_{out}$ | Output current of circuit |
| $Q_1$ to $Q_2$ | Output transistors of circuit |
| $Q_3$ to $Q_4$ | Output transistors of op amp |
| $R_1$ | Determines input resistance and gain of circuit |
| $R_2$ | Sets gain of circuit (along with $R_1$) |
| $R_3$ | Cancels effect of op amp input bias current |
| $R_4$ to $R_5$ | Determines point at which $Q_1$ and $Q_2$ are turned on |
| $R_6$ to $R_7$ | Controls feedback factor of output circuit |
| $R_8$ to $R_9$ | Used to minimize crossover distortion |
| $R_{10}$ to $R_{11}$ | Provides dc negative feedback in output transistors to prevent thermal runaway |
| $R_A$ to $R_B$ | Output current-limiting resistors in op amp |
| $R_o$ | Output resistance of circuit |
| $R_{in}$ | Input resistance of circuit |
| $S_{max}$ | Maximum slew rate of op amp |
| $V_1$ | Input voltage |
| $V_2$ | Output voltage |
| $V^{(\pm)}$ | Power-supply voltages |
| $V_{pp}$ | Peak-to-peak output voltage of op amp |

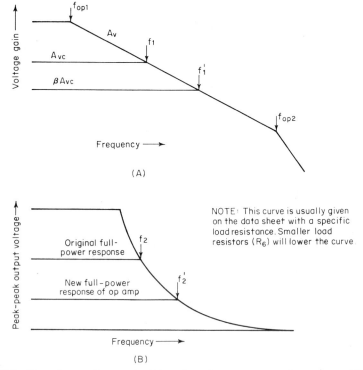

**Fig. 23.3** Open-loop voltage gain (A) and maximum output-voltage swing (B), both shown as a function of frequency for a 741 op amp.

## DESIGN EQUATIONS

| Eq. No. | Description | Equation |
|---|---|---|
| 1 | Voltage gain of circuit | $A_{vc} = \dfrac{V_2}{V_1} = -\dfrac{R_2}{R_1}$ |
| 2 | Feedback factor of output buffer stage | $\beta = \dfrac{R_6}{R_6 + R_7}$ |
| 3 | Output resistance of circuit if $h_{ie1} = h_{ie2}$, $\beta_1 = \beta_2$, and $h_{oe1} = h_{oe2}$ | $R_o \approx \dfrac{[h_{ie3} + (1+\beta_3)(R_4 + R_6)][h_{ie1} + (1+\beta_1)R_{10}]}{\beta_1 \beta_3 R_4}$ |
|  | Resistor values: | |
| 4 | $R_1$ | $R_1 = R_{in}$ required |
| 5 | $R_2$ | $R_2 = -R_1 A_{vc}$ |
| 6 | $R_3$ | $R_3 = \dfrac{R_1 R_2}{R_1 + R_2}$ |
| 7 | $R_4$ to $R_5$ | $R_4 = R_5 = \dfrac{0.7}{I^{(+)}(\text{max, peak})}$ |
| 8 | $R_6$ | $R_6 = \dfrac{\beta V_2(\text{max, peak})}{I^{(+)}(\text{max, peak})}$ |
| 9 | $R_7$ | $R_7 = \dfrac{R_6(1-\beta)}{\beta}$ |

## DESIGN PROCEDURE

This circuit can be used to extend bandwidth, slew rate, and output current by a factor of 2 to 10 over those of the basic op amp. The full-power bandwidth will simultaneously be extended by a factor of $\sqrt{2}$ to $\sqrt{10}$, approximately. Output power can be increased 20 or more using this circuit.

## DESIGN STEPS

Step 1. Let $\beta$ equal the reciprocal of the bandwidth improvement required. That is, if $f_u$ of the op amp is 1 MHz and a 10-MHz $f_u$ is required, a $\beta$ of 0.1 is necessary.

Step 2. Let $R_1$ equal the required input resistance of the circuit.

Step 3. Calculate $R_2$ from Eq. 5 using the required closed-loop voltage gain of the circuit $A_{vc}$.

Step 4. Resistor $R_3$ is found from Eq. 6, which is equivalent to the parallel resistance of $R_1$ and $R_2$.

Step 5. Determine values for $R_4$ and $R_5$ using Eq. 7. The $I^{(+)}$ (max, peak) term is the desired peak current through the op amp (+) power-supply terminal. This current must be approximately $\beta$ times the peak current allowed through $Q_1$ and $Q_2$.

Step 6. Since we want the output-voltage swing of the op amp to be $\beta$ times the swing of $V_2$, Eq. 8 gives us a reasonable choice for $R_6$.

Step 7. Calculate Eq. 9 to determine a value for $R_7$.

Step 8. Use Eq. 3 to compute the circuit output resistance.

**EXAMPLE OF POWER-BOOSTER DESIGN** A 741-type op amp is to be provided with a booster to increase output power by 9. We assume that the 2N3904–2N3906 complementary transistor pair is to be utilized.

*Design Requirements*

$V_2(\text{max, peak}) = \pm 10$ V
$V^{(\pm)} = \pm 15$ V
$I_{\text{out}}(\text{max, peak}) = 50$ mA
$f_u = 4$ MHz (total circuit)
$A_{vc} = -2$
$R_{\text{in}} = 1{,}000$ Ω

*Device Data*

$\beta_1 = 80$ at 50-mA collector current
$\beta_3 = 300$ (estimated)
$h_{ie1} = 200$
$h_{ie3} = 3{,}000$ (estimated)
$f_u = 0.4$ MHz (op amp, minimum)
$I^{(+)}(\text{max, peak}) = 5$ mA ($\pm 10$-V output into 2,000 Ω)
$R_A = 25$ Ω

*Step 1.* We want a bandwidth improvement of approximately 10, so $\beta = 1/10$.

*Step 2.* We set $R_1 = R_{\text{in}} = 1{,}000$ Ω.

*Step 3.* Equation 5 provides us with $R_2 = -R_1 A_{vc} = -1{,}000(-2) = 2{,}000$ Ω.

*Step 4.* Resistor $R_3$ must be approximately

$$R_3 = \frac{R_1 R_2}{R_1 + R_2} = \frac{1{,}000(2{,}000)}{1{,}000 + 2{,}000} = 667 \text{ Ω}$$

*Step 5.* Equation 7 provides values for

$$R_4 = R_5 = \frac{0.7}{I^{(+)}(\text{max, peak})} = \frac{0.7}{0.005} = 140 \text{ Ω}$$

*Step 6.* A reasonable choice for $R_6$ is provided if we compute Eq. 8:

$$R_6 = \frac{\beta V_2(\text{max, peak})}{I^{(+)}(\text{max, peak})} = \frac{0.1(10)}{0.005} = 200 \text{ Ω}$$

*Step 7.* $R_7$ is found using Eq. 9:

$$R_7 = \frac{R_6(1-\beta)}{\beta} = \frac{200(1-0.1)}{0.1} = 1{,}800 \text{ Ω}$$

*Step 8.* The output resistance of the circuit is approximately

$$R_o \approx \frac{[h_{ie3} + (1+\beta_3)(R_A + R_6)][h_{ie1} + (1+\beta_1)R_{10}]}{\beta_1 \beta_3 R_4}$$

$$\approx \frac{[3{,}000 + (1+300)(25+200)][200 + (1+80)4.7]}{80(300)140} \approx 12.2 \text{ Ω}$$

## REFERENCES

1. Millman, J., and C. C. Halkias: "Integrated Electronics: Analog and Digital Circuits and Systems," p. 430, McGraw-Hill Book Company, New York, 1972.
2. Gagnon, R., and R. Karwoski: Complementary Output Stage Improves Op Amp Response, *Electronics*, Sept. 25, 1972, p. 110.
3. Wooley, B. A., S. J. Wong, and D. O. Pederson: A Computer Aided Evaluation of the 741 Amplifier, *IEEE J. Solid-State Circuits*, vol. SC-6, no. 6, p. 357, December 1971.

## 23.2 OP AMP OUTPUT-VOLTAGE BOOSTER

**ALTERNATE NAMES** Op amp buffer, push-pull buffer amplifier, complementary transistor output circuit, high-voltage op amp.

**EXPLANATION OF OPERATION** This circuit operates like the circuit described in Sec. 23.1. In this voltage booster, shown in Fig. 23.4, however, one additional transistor has been included in each output loop. These new transistors $Q_1$ and $Q_2$ serve a dual function. They provide an additional stage of gain and also control the power-supply voltages applied to the op amp. The positive op amp supply voltage is

$$V^{(+)} = \frac{R_5 V_{CC}}{R_4 + R_5} - 0.7$$

Likewise, the negative op amp supply voltage is

$$V^{(-)} = \frac{R_6 V_{EE}}{R_6 + R_7} + 0.7$$

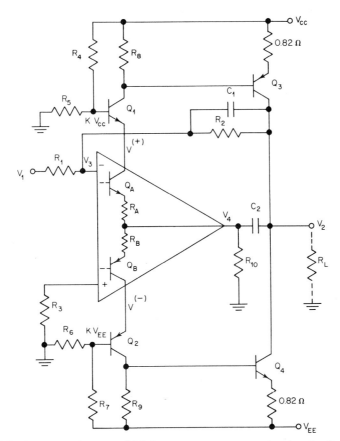

**Fig. 23.4** An op amp booster which increases output power level and voltage swing. The op amp output stage is shown to simplify the discussion.

## 23-8 POWER CIRCUITS

Resistors $R_4$ to $R_7$ are sized so that the op amp operates at its recommended supply voltages. The full supply voltage $V^{(+)} - V^{(-)}$ must be less than the $BV_{CEO}$ of either $Q_3$ or $Q_4$.

At low frequencies where $|X_{C2}| \gg R_{10}$ the voltage gain from $V_3$ to $V_4$ is $A_{vo}$ of the op amp. However, the voltage gain from $V_3$ to $V_2$ is $A_{vo}A_{VA}A_{V1}A_{V3}$, where $A_{VA}$ is the common-emitter voltage gain of $Q_4$, $A_{V1}$ is the common-base voltage gain of $Q_1$, and $A_{V3}$ is the common-emitter voltage gain of $Q_3$. The dc magnitudes of these gains are

$$A_{VA} = \frac{-\beta_A(R_4\|R_5)}{h_{ieA} + (1+\beta_A)(R_A + R_{10})}$$

$$A_{V1} = \frac{\alpha_1(R_8\|h_{ie3})}{h_{ib1}}$$

$$A_{V3} = \frac{-\beta_3(R_2\|R_L)}{h_{ie3}}$$

From $Q_A$ to $Q_3$ the current-gain increase is only that due to $\beta_3$ (of $Q_3$). Therefore, the power-gain increase of this circuit over that of the op amp is

$$A_p = \beta_3 A_{VA} A_{V1} A_{V3}$$

The voltage gain of the entire circuit is

$$A_{vc} = -\frac{R_2}{R_1}$$

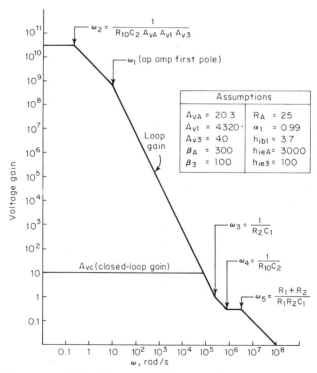

**Fig. 23.5** The open- and closed-loop gains of Fig. 23.4 shown as a function of frequency.

This gives us a total power gain of

$$A_{pc} = A_{vc}^2 \frac{R_1}{R_L} = \frac{R_2^2}{R_1 R_L}$$

This circuit can become unstable because of the additional voltage gain provided by $Q_A$, $Q_1$, and $Q_3$ (and $Q_B$, $Q_2$, and $Q_4$). The $R_2 C_1$ lead network provides phase compensation near the loop-gain unity-gain crossover frequency. The optimum value for $C_1$ is approximately

$$C_1 \approx \frac{1}{R_2 A_{VA} A_{V1} A_{V3}} \left( \frac{R_1 + R_2}{A_{vo} R_1 R_{10} C_2 \omega_1} \right)^{1/2}$$

This provides approximately 45° of phase margin. If more phase margin is required, $C_1$ could be increased by a factor of 2 or 3. For those interested in detailed loop-stability information, the loop gain is

$$\text{Loop gain} = \frac{A_{vo} \omega_1 [s + (1/R_{10} C_2)][s + (1/R_2 C_1)]}{(s + \omega_1)[s + (R_1 + R_2)/R_1 R_2 C_1][s + (1/R_{10} C_2 A_{VA} A_{V1} A_{V3})]}$$

This is plotted, along with closed-loop gain, in Fig. 23.5 for a typical circuit constructed and tested by the authors.

## DESIGN PARAMETERS

| Parameter | Description |
| --- | --- |
| $\alpha_1$ | Common-base current gain of $Q_1$ |
| $A_{pc}$ | Power gain of entire circuit |
| $A_{V1}$ | Voltage gain of $Q_1$ |
| $A_{V3}$ | Voltage gain of $Q_3$ |
| $A_{VA}$ | Voltage gain of $Q_A$ |
| $A_{vc}$ | Voltage gain of complete circuit |
| $A_{vo}$ | Voltage gain of op amp at dc |
| $\beta_A$ | Common-emitter current gain of $Q_A$ |
| $\beta_3$ | Common-emitter current gain of $Q_3$ |
| $BV_{CEO3}$ | Collector-emitter breakdown voltage of $Q_3$ |
| $BV_{CEO4}$ | Collector-emitter breakdown voltage of $Q_4$ |
| $C_1$ | Capacitor used to stabilize feedback loop of entire circuit |
| $C_2$ | Stabilizes feedback loop by placing another zero in the loop gain near the op amp second-pole frequency |
| $h_{ib1}$ | Common-base input resistance of $Q_1$ |
| $h_{ieA}$ | Common-emitter input resistance of $Q_A$ |
| $h_{ie3}$ | Common-emitter input resistance of $Q_3$ |
| $K$ | Fraction which relates size of $V^{(+)}$ to $V_{CC}$ and $V^{(-)}$ to $V_{EE}$ |
| $Q_1$ to $Q_2$ | Transistors which provide gain and also control levels of $V^{(+)}$ and $V^{(-)}$ |
| $Q_3$ to $Q_4$ | High-power complementary output transistors |
| $Q_A$ to $Q_B$ | Output transistors in op amp |
| $R_1$ | Resistor which sets circuit gain and also determines input resistance of circuit |
| $R_2$ | Controls circuit gain |
| $R_3$ | Cancels effect of op amp input bias current |
| $R_4$ to $R_5$ | Controls magnitude of $V^{(+)}$ |
| $R_6$ to $R_7$ | Controls magnitude of $V^{(-)}$ |
| $R_8$ to $R_9$ | Load resistors for $Q_1$ and $Q_2$ |
| $R_{10}$ | Provides dc load for op amp and also affects loop stability |
| $R_A$ to $R_B$ | Output current-limiting resistors in op amp |
| $R_L$ | Circuit load resistance |
| $V_1$ | Circuit input voltage |

| Parameter | Description |
|---|---|
| $V_2$ | Circuit output voltage |
| $V_3, V_4$ | Op amp input and output voltages |
| $V_{CC}$ | Positive power-supply voltage |
| $V_{EE}$ | Negative power-supply voltage |
| $V^{(+)}$ | Op amp power-supply voltages |
| $\omega_1$ | Radian frequency of op amp first pole |
| $X_{C2}$ | Reactance of $C_2$ |

## DESIGN EQUATIONS

| Eq. No. | Description | Equation |
|---|---|---|
| 1 | Voltage gain of complete circuit | $A_{vc} = -\dfrac{R_2}{R_1}$ |
| 2 | Power gain of complete circuit | $A_{pc} = A_{vc}^2 \dfrac{R_1}{R_L} = \dfrac{R_2^2}{R_1 R_L}$ |
| 3 | Voltage at op amp (+) power-supply terminal | $V^{(+)} = \dfrac{R_5 V_{CC}}{R_4 + R_5} - 0.7$ |
| 4 | Voltage at op amp (−) power-supply terminal | $V^{(-)} = \dfrac{R_6 V_{EE}}{R_6 + R_7} + 0.7$ |
| 5 | Required minimum collector-emitter breakdown voltage for $Q_3$ and $Q_4$ | $BV_{CEO3}(\min) = BV_{CEO4}(\min)$ $= 1.5(V_{CC} - V_{EE})$ |
| 6 | Voltage gain of op amp transistor $Q_A$ with connections as shown in Fig. 23.4 (common-emitter stage) | $A_{VA} = \dfrac{-\beta_A(R_4 \| R_5)}{h_{ieA} + (1 + \beta_A)(R_A + R_{10})}$ |
| 7 | Voltage gain of $Q_1$ (common-base stage) | $A_{V1} = \dfrac{\alpha_1 (R_8 \| h_{ie3})}{h_{ib1}}$ |
| 8 | Voltage gain of $Q_3$ (common-emitter stage) | $A_{V3} = \dfrac{-\beta_3(R_2 \| R_L)}{h_{ie3}}$ |
| 9 | Approximate compensation capacitor required | $C_1 \approx \dfrac{1}{R_2 A_{VA} A_{V1} A_{V3}} \left( \dfrac{R_1 + R_2}{A_{vo} R_1 R_{10} C_2 \omega_1} \right)^{1/2}$ |

## REFERENCE

1. Garza, P. P.: Getting Power and Gain out of the 741-type Op Amp, p. 99, *Electronics*, Feb. 1, 1973.

Chapter **24**

# Regulators

## VOLTAGE REGULATORS

All high-quality equipment requires voltage regulators to provide stable voltages to the various circuits. In the following pages both low- and high-voltage regulators with current limiters will be discussed. Other types of voltage regulators to be covered include a shunt regulator, precision reference, dual regulator, and switching regulator.

### 24.1  CURRENT-LIMITED VOLTAGE REGULATOR

**ALTERNATE NAMES**  Adjustable power supply, short-circuit-proof power supply, precision voltage source, buffered-zener reference supply.

**EXPLANATION OF OPERATION**  As shown in Fig. 24.1, this voltage source requires one op amp and two transistors. $Q_2$ could be eliminated if current limiting is not required. $Q_1$ could be eliminated if the op amp maximum output current is capable of driving the load.

Transistor $Q_1$ is basically an emitter follower. Its emitter is always approximately 0.6 V below its base. The base voltage is provided by the op amp, which, in turn, depends on the voltages at each of its two inputs. The op amp positive input is fixed at the zener voltage $V_R$. The negative input voltage $V_N$ depends on the output voltage according to

$$V_N = \frac{V_o R_1}{R_1 + R_f}$$

Since this is negative feedback, the output voltage will remain fixed at a voltage which makes $V_N = V_R$. Thus

$$V_o = \frac{V_N(R_1 + R_f)}{R_1} = \frac{V_R(R_1 + R_f)}{R_1}$$

The output voltage is adjusted with the rheostat which is part of $R_1$.

Regulation against load variations is accomplished as follows: Suppose $R_L$ is decreased in value such that the output current increases. Since the emitter follower has a finite output resistance, $V_o$ will drop slightly. The feedback voltage $V_N$ will also make a corresponding drop. $V_N$ will now be

## 24-2 REGULATORS

lower with respect to $V_R$, which will cause the op amp output voltage to increase. Since the emitter of $Q_1$ follows its base voltage, $V_o$ will also increase. This increase of $V_o$ will counteract most of the decrease in $V_o$ due to the increased load. The resultant change in $V_o$ will be perhaps 100 or 1,000 times less than would be expected without feedback.

Regulation against input-voltage variations $\Delta V_{cc}$ is assured for three reasons:

1. The high power-supply rejection ratio of the op amp will almost completely prevent variations in $V_{cc}$ passing through the op amp to $V_o$.

2. The emitter follower $Q_1$ has a very large power-supply rejection ratio (usually 80 to 100 dB). Since this stage has a voltage gain of only 1 (approximately), coupling of changes in $V_{cc}$ to $V_o$ is extremely small.

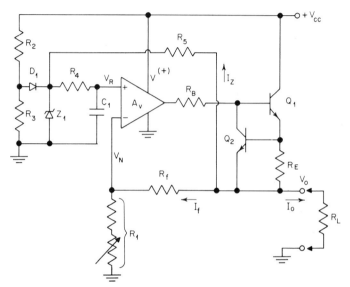

**Fig. 24.1** Positive-output current-limited voltage regulator.

3. The only other possible coupling of $V_{cc}$ to $V_o$ is through the zener reference voltage $V_R$. The relationship between changes in $V_{cc}$ and changes in $V_R$ is

$$\Delta V_R = \frac{R_Z \Delta V_{cc}}{R_x}$$

where $R_Z$ is the dynamic zener resistance and $R_x$ is the resistance between $V_{cc}$ and the zener. However, the $R_x$ for this circuit is $R_5$, which is tied to $V_o$. Therefore, $V_o$ keeps the zener-diode voltage $V_R$ constant and $V_R$ keeps $V_o$ constant. The change in $V_R$ with $V_{cc}$ will be approximately zero.

Output current limiting is provided with only two parts, $Q_2$ and $R_E$. The resistance of $R_E$ is chosen such that at a specified output current $I_o(\max)$, the voltage across $R_E$ becomes large enough to turn on $Q_2$. This pulls base current away from $Q_1$, which causes $V_o$ to drop to zero. All load resistances smaller than $R_L(\min)$ will maintain $V_o$ at zero and $I_o$ at approximately $0.6/R_E$. The pass transistor must be capable of withstanding the power generated from

$$P = \frac{V_{cc} \times 0.6}{R_E} = V_{cc} \times I_o(\max)$$

Current limiting for the op amp is provided with $R_B$. This extra resistor is required if the op amp output cannot be shorted to its negative supply terminal, which, in this case, is connected to ground.

The zener diode $Z_1$ provides the reference voltage $V_R$ to the op amp. The quality (and price) of $Z_1$ will be the major factor in determining the temperature stability of the voltage regulator. Low-priced zeners may have voltage temperature coefficients of 0.05%/°C while high-priced units go down to 0.0005%/°C. $Z_1$ is provided with a current $I_Z$ from the regulated output through $R_5$. If $R_5$ is properly sized, a temperature coefficient near zero may be possible. However, if this is to be a variable output-voltage supply, $I_Z$ will be different for each setting of $V_o$. A two-stage zener-diode circuit may be required so that the voltage driving $R_5$ is constant. This will also make $I_Z$ constant.

This circuit makes $V_o$ depend on $V_R$ and $V_R$ depend on $V_o$. Consequently, the power supply may lock up into an undesired state at the moment of turn-on. A starting circuit composed of $R_2$, $R_3$, and $D_1$ guarantees that $Z_1$ is sufficiently biased even before $Q_1$ and the op amp turn on. Once the active devices are on, the diode $D_1$ becomes reverse-biased and $R_5$ provides bias for $Z_1$.

Reduction of ripple and noise on $V_R$ is provided by $R_4$ and $C_1$. These parts are required only in voltage regulators which have very tight noise specifications.

## DESIGN PARAMETERS

| Parameter | Description |
| --- | --- |
| $A_v$ | Op amp open-loop gain as a function of frequency |
| $\beta$ | Voltage transfer ratio of feedback network $= R_1/(R_1 + R_f)$ |
| $C_1$ | Part of filter which makes $V_R$ a noise-free reference voltage |
| $D_1$ | Diode which biases $Z_1$ until op amp and $Q_1$ have turned on |
| $f_n$ | Lowest frequency of noise to be removed from $V_o$ (e.g., for power supplies operating off 60-Hz power line, make $f_n < 60$ Hz) |
| $h_{FE1}$ | $Q_1$ current gain |
| $h_{IE1}$ | $Q_1$ input resistance |
| $I_{CO1}$ | $Q_1$ collector cutoff current |
| $I_E$ | $Q_1$ emitter current |
| $I_f$ | Feedback current through $R_f$ |
| $I_{io}$ | Op amp input offset current |
| $I_n$ | Op amp equivalent input rms current noise |
| $I_o$ | Output current of circuit |
| $I_o(\max)$ | Maximum output current of circuit (where current limiting begins) |
| $I_Z$ | Zener-diode current |
| $I_{Z0}$ | Zener-diode current for zero temperature coefficient of voltage |
| $P_{oa}(\max)$ | Maximum allowable power dissipation in op amp |
| $Q_1$ | Pass transistor (for power gain) |
| $Q_2$ | Current-sensing transistor |
| $R_1$ | Part of feedback network which goes to ground |
| $R_2$ | Part of starting circuit |
| $R_3$ | Part of starting circuit |
| $R_4$ | Part of $V_R$ filter |
| $R_5$ | Feedback resistor for biasing $Z_1$ from a regulated voltage |
| $R_B$ | Resistor in base circuit of $Q_1$ for limiting op amp output current |
| $R_E$ | Resistor in emitter circuit of $Q_1$ for current limiting the circuit output |
| $R_f$ | Feedback resistor |
| $R_o$ | Output resistance of op amp |
| $R_{out}$ | Output resistance of voltage regulator |

## 24-4 REGULATORS

| Parameter | Description |
|---|---|
| $\Delta T$ | Change in temperature |
| $V^{(+)}$ | Positive power-supply voltage ($= V_{cc}$ in this circuit) |
| $V_{BE1}$ | Base-to-emitter voltage of $Q_1$ |
| $V_{cc}$ | Power-supply voltage for circuit |
| $V_{io}$ | Op amp input offset voltage |
| $V_o$ | Output voltage of circuit |
| $V_N$ | Feedback voltage to negative input of op amp |
| $V_n$ | Op amp equivalent rms input noise voltage in volts/$\sqrt{\text{Hz}}$ |
| $V_R$ | Reference voltage established by $Z_1$ |
| $Z_1$ | Zener reference diode |

### DESIGN EQUATIONS

| Eq. No. | Description | Equation |
|---|---|---|
| 1 | Resistance of $R_1 + R_f$ | $R_1 + R_f \gg R_L$ (to prevent feedback network from loading circuit) |
| 2 | Output voltage of circuit if $A_v = \infty$ | $V_o = \dfrac{V_R}{\beta} = V_R\left(1 + \dfrac{R_f}{R_1}\right)$<br>NOTE: Use both Eqs. 1 and 2 to determine $R_1$ and $R_f$ |
| 3 | Output voltage of circuit if $A_v < \infty$ | $V_o = V_R \dfrac{A_v}{1 + \beta A_v} - \dfrac{V_{BE1}}{1 + \beta A_v}$<br>where $\beta = \dfrac{R_1}{R_1 + R_f}$<br>$V_{BE1} \approx 0.6$ V |
| 4 | Minimum load current for which regulation is assured | $I_o(\text{min}) > I_{C01}(\text{max})[1 + h_{FE1}(\text{max})] - I_f - I_{Z0}$ |
| 5 | Sensitivity of $V_o$ to a temperature change $\Delta T$ (if Eq. 9 is true) | $\dfrac{\Delta V_o}{\Delta T} = \left(1 + \dfrac{R_f}{R_1}\right)\left(\pm \dfrac{\Delta V_R}{\Delta T} \pm \dfrac{\Delta V_{io}}{\Delta T}\right) \pm \dfrac{\Delta I_{io}}{\Delta T} R_f$<br>NOTE: Partial cancellation is possible if $\Delta V_R$ drifts in opposite direction of $\Delta V_{io}$ or $\Delta I_{io}$ |
| 6 | Resistor values | $R_5 = \dfrac{V_o - V_R}{I_{Z0}}$ |
| 7 | | $R_2 = 2R_5$ |
| 8 | | $R_3 = \dfrac{R_2 V_R}{2V_{cc} - V_R}$ |
| 9 | | $R_4 = \dfrac{R_1 R_f}{R_1 + R_f}$ |
| 10 | | $R_E = \dfrac{0.6}{I_o(\text{max})}$ |
| 11 | | $R_B(\text{min}) = \dfrac{V_o^2(\text{min})}{P_{oa}(\text{max})}$ |

| Eq. No. | Description | Equation |
|---|---|---|
| 12 | Closed-loop output resistance of voltage regulator | $R_{\text{out}} \approx \dfrac{R_B + h_{IE1}}{(h_{FE1} + 1)\beta A_v}$ |
| 13 | % regulation | $\dfrac{\Delta V_o}{V_o} = \dfrac{R_{\text{out}} \times 100}{R_L(\min) + R_{\text{out}}}\%$ |
| 14 | Capacitance of $C_1$ | $C_1 \gg \dfrac{1}{2\pi f_n R_4}$ |
| 15 | Noise in $V_o$ due to op amp | $V_{on} = \left[\left(1 + \dfrac{R_f}{R}\right)^2 V_n^2 + R_f^2 I_n^2\right]^{1/2}$ rms volts |
| 16 | Maximum power dissipation in $Q_1$ if output terminal is shorted to ground | $P = \dfrac{0.6 V_{cc}}{R_E}$ |

## DESIGN PROCEDURE

Depending on the design requirements, Eqs. 1 to 15 could be arranged in several logical design sequences. The following sequence of steps assumes:
1. A fixed output voltage $V_o$ is specified.
2. Feedback resistors should load circuit no more than 2 percent.
3. $A_v$ is finite, but high enough that Eq. 3 is not used.
4. Minimum $R_L$ is specified and maximum $R_L$ is to be determined.
5. Stability of $V_o$ with temperature is to be calculated.
6. Percent load regulation is to be calculated.
7. Lowest-frequency noise is specified.
8. Output noise is to be calculated.

A slightly rearranged procedure would be required if (1) $A_v$ is only 100 or 1,000, (2) the temperature stability is specified, (3) the percent regulation is specified, or (4) the output noise is specified.

## DESIGN STEPS

*Step 1.* Using the minimum $R_L$, choose $R_1 + R_f \gg R_L(\min)$. For a first cut we may try $R_1 + R_f = 100 R_L(\min)$. Equations 1 and 2 are now solved for $R_1$ and $R_f$:

$$R_f = \frac{100 R_L(\min)(V_o - V_R)}{V_o}$$

$$R_1 = 100 R_L(\min) - R_f$$

*Step 2.* $R_5$ is found as follows: Using the zener-diode data sheet, determine the zener current $I_{Z0}$ where $V_R$ exhibits the smallest change with temperature. Many high-quality zener diodes have an optimum bias point where the change in $V_R$ is less than $\pm 0.0005\%/°C$. Thus $R_5 = (V_o - V_R)/I_{Z0}$. According to Eq. 7, we also choose $R_2 = 2R_5$.

*Step 3.* Determine the maximum $R_L$ by using Eq. 4 and the fact that $I_o(\min) = V_o/R_L(\max)$. Thus,

$$R_L(\max) = \frac{V_o}{I_o(\min)} = \frac{V_o}{I_{C01}(\max)[1 + h_{FE1}(\max)] - I_f - I_{Z0}}$$

where
$$I_f = \frac{V_o}{R_f + R_1}$$

*Step 4.* If required, determine the sensitivity of $V_o$ to changes in temperature using Eq. 5. If Eq. 9 is not satisfied, i.e., $R_4 \neq R_1 R_f/(R_1 + R_f)$, then the $\pm \Delta I_{io} R_f / \Delta T$ term must be replaced with $\pm \Delta I_b R_f / \Delta T$ (worst case).

*Step 5.* Solve for $R_3$ from
$$R_3 = \frac{R_2 V_R}{2V_{cc} - V_R}$$

*Step 6.* Find $R_4$ from
$$R_4 = \frac{R_1 R_f}{R_1 + R_f}$$
and if a low-noise $V_o$ is desired find $C_1$ from
$$C_1 = \frac{10}{2\pi f_n R_4}$$

*Step 7.* If current limiting is not required, delete $Q_2$ and replace $R_E$ with a wire. If current limiting is needed, solve for
$$R_E = \frac{0.6}{I_o(\max)}$$

*Step 8.* If additional short-circuit protection is desired for the op amp, find $R_B(\min)$ from
$$R_B(\min) \approx \frac{h_{FE1}}{I_o(\max)}\left[V_{cc} - V_o - \frac{h_{FE1} P_{oa}(\max)}{I_o(\max)}\right]$$
If this number is negative, no resistor is needed.

*Step 9.* If percent regulation is to be determined, first find the closed-loop output resistance from
$$R_{\text{out}} \approx \frac{R_B + h_{IE1}(\max)}{(h_{FE1}(\min) + 1)\beta A_v}$$
then the percent regulation is
$$\% \text{ regulation} = \frac{R_{\text{out}} \times 100}{R_L(\min) + R_{\text{out}}} \%$$
This is the percent change in output voltage when the load resistance changes from $\infty$ to $R_L(\min)$. If step 3 does not allow $R_L(\max) = \infty$, the actual percent regulation will be slightly better than computed above.

*Step 10.* Noise due to the op amp can be found from
$$V_{\text{on}} = \left[\left(1 + \frac{R_f}{R_1}\right)^2 V_n^2 + R_f^2 I_n^2\right]^{1/2} \text{ rms volts}$$
Since both $V_n$ and $I_n$ vary with frequency, this equation must be solved at several frequencies of interest.

**EXAMPLE OF VOLTAGE-REGULATOR DESIGN** We will use the 10 design steps to design an actual voltage regulator. Experimental data from the actual circuit using a 741 op amp will be compared with the calculations of the 10 steps.

*Design Requirements*
$V_o = 15$ V
$V_{cc} = 20$ V minimum

$R_L = 15\ \Omega$ minimum
$V_R = 6.4$ V, 1N4566
$I_{Z0} = 0.5$ mA
$f_n = 60$ Hz
$I_o = 1.0$ A maximum

*Device Data*

$A_v > 10^4$ for all frequencies of interest
$I_{C01} = 2\ \mu$A maximum
$h_{FE1} = 50$ minimum
$\Delta V_{io} = +10\ \mu$V/°C
$\Delta I_{io} = -60$ pA/°C
$P_{oa} = 300$ mW maximum
$h_{IE1} = 2{,}000\ \Omega$ maximum
$V_n = 4 \times 10^{-8}$ V/$\sqrt{\text{Hz}}$ at 60 Hz
$I_n = 3 \times 10^{-12}$ A/$\sqrt{\text{Hz}}$ at 60 Hz
$\Delta V_R = \pm 0.001\%/°\text{C} = \pm 64\ \mu$V/°C (polarity and magnitude depend on particular diode)

*Step 1.* Since $R_L(\min) = 15\ \Omega$, $R_1 + R_f = 100\ R_L(\min) = 1{,}500\ \Omega$ shall be selected. These individual resistors become

$$R_f = \frac{100 R_L(\min)(V_o - V_R)}{V_o}$$

$$= \frac{100(15)(15 - 6.4)}{15} = 860\ \Omega$$

and
$$R_1 = 100 R_L(\min) - R_f$$
$$= 1{,}500 - 860 = 640\ \Omega$$

*Step 2.* $I_{Z0}$, the optimum zener current, is 0.5 mA. Thus,

$$R_5 = \frac{V_o - V_R}{I_{Z0}}$$

$$= \frac{15 - 6.4}{5 \times 10^{-4}} = 17.2\ \text{k}\Omega$$

and $\qquad R_2 = 34.4\ \text{k}\Omega$

*Step 3.* To determine the maximum load resistor (for which regulation can be expected) we first must find

$$I_f = \frac{V_o - V_R}{R_f}$$

$$= \frac{15 - 6.4}{860} = 10\ \text{mA}$$

We now can obtain

$$R_L(\max) = \frac{V_o}{I_{C01}(\max)[1 + h_{FE1}(\max)] - I_f - I_{Z0}}$$

$$= \frac{15}{2 \times 10^{-6}(1 + 200) - 10^{-2} - 5 \times 10^{-4}}$$

$$= \frac{15}{-0.0101} = -1{,}500\ \Omega$$

## 24-8 REGULATORS

Has something gone wrong in our calculations? No, the reason for the negative answer is simple. $I_f$ and $I_{Z0}$ are large enough that they place a constant load on the voltage regulator which is larger than the minimum load. Examination of Eq. 4 reveals that regulation is possible if $I_o(\min) + I_f + I_{Z0} > I_{co1}(\max)[1 + h_{FE1}(\max)]$. This is true for this particular regulator design because our feedback currents were chosen to be quite large ($\approx 1$ percent of $I_o$). The maximum $R_L$ we may use is therefore $\infty$.

*Step 4.* The sensitivity of $V_o$ to changes in $V_R$, $V_{io}$, and $I_{io}$ with temperature is

$$\frac{\Delta V_o}{\Delta T} = \left(1 + \frac{R_f}{R_1}\right)\left(\pm \frac{\Delta V_R}{\Delta T} \pm \frac{\Delta V_{io}}{\Delta T}\right) \pm \frac{\Delta I_{io}}{\Delta T} R_f$$

$$= \left(1 + \frac{860}{640}\right)(\pm 64 \times 10^{-6} + 10 \times 10^{-6}) - (60 \times 10^{-12})(860)$$

$$= (2.34)(+74 \text{ to } -54 \times 10^{-6}) - 0.052 \times 10^{-6}$$

$$= +173 \text{ to } -126 \ \mu\text{V}/°\text{C}$$

The direction and magnitude of this drift in $V_o$ with temperature depend on the drift of the zener voltage with respect to the op amp input offset voltage. If one has enough patience to screen (test) many zener diodes and op amps, a pair of devices which have equal and opposite drifts can probably be found.

*Step 5.* $R_3$ is found from

$$R_3 = \frac{R_2 V_R}{2V_{cc} - V_R} = \frac{2R_5 V_R}{2V_{cc} - V_R}$$

$$= \frac{2(17{,}200)(6.4)}{2(20) - 6.4} = 6{,}560 \ \Omega$$

*Step 6.* Assume we want the voltage regulator to attenuate 60-Hz ripple. If we tune the filter $R_4 C_1$ to a frequency of only 6 Hz, any 60-Hz ripple on $Z_1$ will be attenuated by a factor of 10(20 dB) before it is applied to the op amp. Thus we make

$$R_4 = \frac{R_1 R_f}{R_1 + R_f} = \frac{640 \times 860}{640 + 860} = 367 \ \Omega$$

and
$$C_1 = \frac{10}{2\pi f_n R_4} = \frac{10}{(6.28)(6)(367)} = 722 \ \mu\text{F}$$

This may seem an unreasonably large capacitor to use; however, it need only be a 10-V device. The same $f_n$ can be achieved with $R_4 = 3{,}670 \ \Omega$ and $C_1 = 72 \ \mu\text{F}$. The only compromise we make with these new values is that $R_4$ is not equal to the parallel combination of $R_1$ and $R_f$. This choice for $R_4$ was originally made so that variations of $I_b$ with temperature will not affect circuit performance. As we saw in step 4, however, $I_{io}$ does not appreciably affect the drift of $V_o$, so variations in $I_b$ will not be much worse. We can therefore make $R_4$ 10 or 100 times larger and use a more reasonable capacitance for $C_1$.

*Step 7.* An output-current limit of 1 A is specified, so $R_E$ must be

$$R_E = \frac{0.6}{I_o(\max)} = \frac{0.6}{1} = 0.6 \ \Omega$$

*Step 8.* Additional short-circuit protection to the op amp is provided by making

$$R_B = \frac{h_{FE1}}{I_o(\text{max})}\left[V_{cc} - V_o - \frac{h_{FE1}P_{oa}(\text{max})}{I_o(\text{max})}\right]$$

$$= \frac{50}{1}\left[20 - 15 - \frac{50(0.3)}{1}\right] = -100\ \Omega$$

Thus no $R_B$ is required.

*Step 9.* The closed-loop output resistance of the voltage regulator is

$$R_{\text{out}} \approx \frac{R_B + h_{IE1}(\text{max})}{[h_{FE1}(\text{min}) + 1]\beta A_v}$$

$$= \frac{750 + 2{,}000}{(50 + 1)(640/1{,}500)\ 10^4} = 0.0126\ \Omega$$

Now we can find the percent regulation as the load changes from $R_L = \infty$ to $R_L = 15\ \Omega$.

$$\%\ \text{regulation} = \frac{R_{\text{out}} \times 100\%}{R_L(\text{min}) + R_{\text{out}}}$$

$$= \frac{0.0126 \times 100\%}{15 + 0.0126} = 0.084\ \text{percent}$$

*Step 10.* Noise in $V_o$ due to the op amp will be

$$V_{\text{on}} = \left[\left(1 + \frac{R_f}{R_1}\right)^2 V_n^2 + R_f^2 I_n^2\right]^{1/2}$$

$$= \left[\left(1 + \frac{860}{640}\right)^2 (4 \times 10^{-8})^2 + (860)^2(3 \times 10^{-12})^2\right]^{1/2}$$

$$= 8.95 \times 10^{-8}\ \text{V}/\sqrt{\text{Hz}}$$

This output noise is mostly due to $V_n$.

The experimental data confirming the above calculations are as follows:

$R_1 = 634\ \Omega$   $C_1 = 77\ \mu\text{F}$
$R_2 = 34.4\ \Omega$   $V_o = 15.25\ \text{V}$
$R_3 = 6{,}630\ \Omega$   $I_o(\text{max}) = 1.2\ \text{A}$ (down to $V_o = 13\ \text{V}$)
$R_4 = 3{,}690\ \Omega$   $R_{\text{out}} < 0.05\ \Omega$
$R_5 = 17.26\ \Omega$
$R_E = 0.6\ \Omega$
$R_B = 815\ \Omega$
$R_f = 854\ \Omega$

## REFERENCES

1. Tobey, G. E., J. G. Graeme, and L. P. Huelsman: "Operational Amplifiers—Design and Applications," p. 230, McGraw-Hill Book Company, New York, 1971.
2. Millman, J., and C. C. Halkias: "Integrated Electronics—Analog Digital Circuits and Systems," p. 698, McGraw-Hill Book Company, New York, 1972.
3. Giles, J. N. (ed.): "Fairchild Semiconductor Linear Integrated Circuits Applications Handbook," p. 144, Fairchild Semiconductor, Mountain View, Calif., 1967.

## 24.2 HIGH-VOLTAGE REGULATOR

**ALTERNATE NAMES** High-voltage regulated power supply, current-limited high-voltage regulator, short-circuit-proof high-voltage supply.

## 24-10 REGULATORS

**EXPLANATION OF OPERATION**  A 30-V op amp (between + and − supply terminals) can regulate large voltages if it is biased with zener diodes as shown in Fig. 24.2. Zener diode $Z_3$ maintains 30 V across the op amp for all possible input or output voltages $V_{cc}$ or $V_o$. Zener diode $Z_1$ holds the op amp positive supply terminal 6 V above $V_o$. Zener diode $Z_2$ keeps the op amp inverting input terminal 10 V below $V_o$. $R_1$ and $R_f$ are chosen such that the op amp positive input is also held 10 V below $V_o$. Current for $Z_1$ and $Z_3$ is provided by the constant-current source made up of $Q_3$, $R_2$, $R_3$, and $Z_4$. $R_E$ and $Q_2$ provide current limiting in the event of a heavy load. The rest of the circuit

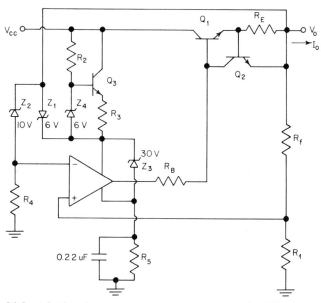

**Fig. 24.2**  A high-voltage regulator which uses a standard 30-V op amp.

is described in Sec. 24.1. A filter for the zener voltage ($R_4$ and $C_1$ of Fig. 24.1) and starting circuit ($R_2$, $R_3$, and $D_1$ of Fig. 24.1) may also be used in this circuit if necessary.

### DESIGN EQUATIONS

The design equations for this circuit are similar to those of Fig. 24.1 except for the following:

| Eq. No. | Description | Equation |
|---|---|---|
| 1 | Output voltage | $V_o = \dfrac{V_{Z2}(R_1 + R_f)}{R_f}$ |
| 2 | Resistor values | $R_1 = \dfrac{100(V_o - V_{Z2})}{I_o(\max)}$ |
| 3 |  | $R_f = \dfrac{100 \, V_o}{I_o(\max)} - R_1$ |

| Eq. No. | Description | Equation |
|---|---|---|
| 4 | | $R_3 = \dfrac{V_{Z4} - 0.6}{I_{Z1} + I_{Z3} + I_{oa} - I_{Z4}}$ <br> where $I_{Z1}$ = optimum $Z_1$ current <br> $I_{Z3}$ = optimum $Z_3$ current <br> $I_{Z4}$ = optimum $Z_4$ current <br> $I_{oa}$ = nominal op amp supply current <br> $V_{Z4}$ = nominal $Z_4$ zener voltage |
| 5 | | $R_2 = \dfrac{V_{cc} - V_{Z4} - V_{Z1} - V_o}{I_{Z4}}$ |
| 6 | | $R_4 = \dfrac{V_o - V_{Z2}}{I_{Z2}}$ <br> where $V_2$ = nominal $Z_2$ zener voltage <br> $I_{Z2}$ = optimum $Z_2$ current |
| 7 | | $R_5 = \dfrac{V_o + V_{Z1} - V_{Z3}}{I_{oa} + I_{Z3}}$ <br> where $V_{Z1}$ = nominal $Z_1$ zener voltage <br> $V_{Z3}$ = nominal $Z_3$ zener voltage |

## REFERENCE

1. English, M.: Applications for Fully Compensated Op-Amp ICs, *EEE*, January 1969, p. 62.

## 24.3 SHUNT VOLTAGE REGULATOR

**ALTERNATE NAMES** Shunt regulator, parallel regulator.

**EXPLANATION OF OPERATION** The shunt method of voltage regulation is used when high output-to-input current isolation is required. If the prime source of power must deliver a constant load current even though load resistance changes are expected, this regulator is recommended. The circuit keeps $I_{cc}$ at a constant value of $I_o + I_s$ + current into $R_Z$, $R_f$, and the op amp (see Fig. 24.3).

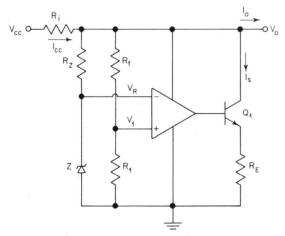

**Fig. 24.3** Shunt voltage regulator.

## 24-12 REGULATORS

If the output voltage tries to drop owing to an increased load-current demand, the following occurs: less current flows through $R_f$, which makes $V_1$ drop slightly. Since $V_1$ is attached to the positive op amp input, the op amp output will drop. This reduces the base bias of $Q_1$, which accordingly reduces $I_s$. The reduction in $I_s$ will be almost exactly equal to the increase in $I_o$. The input current $I_{cc}$ will remain essentially constant.

### DESIGN EQUATIONS

| Eq. No. | Description | Equation |
|---|---|---|
| 1 | Output voltage | $V_o = V_R \dfrac{R_1 + R_f}{R_1}$ |
| 2 | Resistor values | $R_i = \dfrac{V_{cc} - V_o}{I_i}$ where $I_{cc} \gtrsim I_o(\max)$ |
| 3 | | $R_z = \dfrac{V_o - V_z}{I_z}$ where $I_z$ = optimum zener current |
| 4 | | $R_f = \dfrac{100\, V_R}{I_{cc}}$ |
| 5 | | $R_1 = \dfrac{100\, V_o}{I_{cc}} - R_f$ |
| 6 | | $R_E = \dfrac{V_o}{I_{cc}}$ |

### REFERENCE

1. Denker, J. B., and D. A. Johnson: Hybrid Approach to Regulation Solves Power Supply Problems, *Electronics*, Aug. 2, 1973, p. 91.

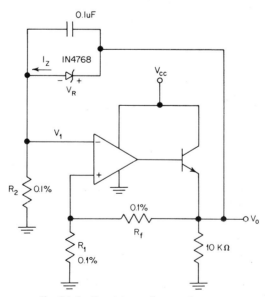

**Fig. 24.4** Precision voltage reference.

## 24.4 PRECISION VOLTAGE REFERENCE

**ALTERNATE NAMES** Reference-voltage source, precision power supply.

**EXPLANATION OF OPERATION** High-quality reference diodes used for precision sources have good voltage stability only if their current remains stable with time and temperature. The zener current is chosen such that the zener voltage has a very small temperature coefficient. In Fig. 24.4, the zener current depends on the choice of $R_1$, $R_2$, $R_f$, and $V_o$. Likewise, $V_o$ depends on $R_1$, $R_f$, and $V_R$. Since $V_R$ and $I_Z$ are given on the zener data sheet, we obtain two equations with three unknown resistances. It is therefore recommended to pick $R_f$ and solve for $R_1$ from the $V_o$ equation. The $I_Z$ equation can then be used to determine $R_2$.

### DESIGN EQUATIONS

| Description | Equation |
|---|---|
| Output voltage | $V_o = \dfrac{V_R(R_1 + R_f)}{R_f}$ |
| Zener-diode current | $I_Z = \dfrac{V_o R_1}{R_2(R_1 + R_f)}$ |
| or | $I_Z = \dfrac{V_R R_2}{R_f}$ |
| Voltage at op amp inverting input | $V_1 = \dfrac{V_o R_1}{R_1 + R_f} = I_Z R_2$ |

### REFERENCES

1. Goldfarb, W.: Single-Supply Reference Source Uses Self Regulating Zener, *Electronics*, June 7, 1973, p. 107.
2. Shah, M. J.: Stable Voltage Reference Uses Single Power Supply, *Electronics*, Mar. 13, 1972, p. 74.

## 24.5 DUAL VOLTAGE REGULATOR

**ALTERNATE NAMES** Tracking regulator, dual op amp supply, dual-polarity supply.

**EXPLANATION OF OPERATION** The stability of voltage regulators ultimately depends on the quality (and cost) of the reference diode. Many times both positive and negative supplies are required. This is often implemented by using two separate reference diodes and associated power-boosting circuitry. The circuit of Fig. 24.5 allows the use of only one reference diode for both supplies, which results in a considerable cost savings.

The + and − output voltage need not have the same absolute magnitude. Since the reference diode most likely to be chosen is the 6.4-V variety, $R_4$ and $R_5$ will each have 3.2 V applied across them. The op amps merely amplify these two voltages to produce $V_o(+)$ and $V_o(-)$. $R_f$ and $R_1$ in each op amp circuit can therefore be independently chosen to produce nearly any + or − voltage combination.

$Z_1$ is a preregulator made up of a standard low-cost zener. It helps maintain a constant current through $Z_2$, which is required for a highly stable $V_R$. $R_6$ is chosen to lessen the effects of input offset-current drift in the op amps. If $R_1(+)$ and $R_f(+)$ have resistances much different from $R_1(-)$ and $R_f(-)$, a separate $R_6$ may be needed for each op amp.

## 24-14 REGULATORS

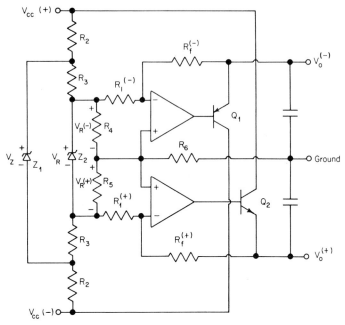

**Fig. 24.5** Dual voltage regulator.

## DESIGN EQUATIONS

| Description | Equation |
| --- | --- |
| Positive output voltage | $V_o(+) = \dfrac{R_f(+)\, V_R(+)}{R_1(+)}$ |
| Negative output voltage | $V_o(-) = \dfrac{R_f(-)\, V_R(-)}{R_1(-)}$ |
| Reference voltages | $V_R(+) = \dfrac{V_R R_5}{R_4 + R_5}$ |
|  | $V_R(-) = \dfrac{V_R R_4}{R_4 + R_5}$ |
| Offset drift-reducing resistor | $R_6$ = average of $\dfrac{R_1(-) R_f(-)}{R_1(-) + R_f(-)}$ and $\dfrac{R_1(+) R_f(+)}{R_1(+) + R_f(+)}$ |
| Resistor values | $R_3 = \dfrac{V_Z - V_R}{2 I_R}$ <br> where $I_R$ is the optimum $Z_2$ current <br><br> $R_2 = \dfrac{V_{cc}(+) - V_{cc}(-) - V_Z}{2 I_Z}$ <br> where $I_Z$ is the optimum $Z_1$ current <br><br> $R_4 = R_5 \geq \dfrac{10 V_R}{I_R}$ <br><br> $R_1(+)$ or $R_1(-) \geq \dfrac{10 V_R}{I_R}$ <br> (These two equations prevent excessive loading of $V_R$) |

# REFERENCE

1. Jones, H. T.: Build a Dual Voltage Regulator for $11, *Electron. Des.*, Dec. 23, 1971, p. 70.

## 24.6 SWITCHING VOLTAGE REGULATOR

**ALTERNATE NAMES** Switching-mode regulator, switching regulator, buck regulator.

**EXPLANATION OF OPERATION** The switching regulator is used in applications where small size and/or high efficiency is required. These advantages come about because the series pass transistor $Q_1$ switches between totally on and totally off on a periodic basis. Typical operation frequencies are 5 to 100 kHz. The switching time of $Q_1$ must be quite small compared with the on or off times if efficiency is to be kept high.

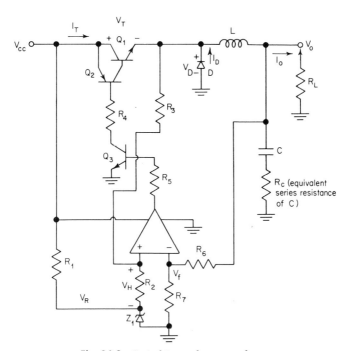

**Fig. 24.6** Switching voltage regulator.

Referring to Fig. 24.6, the circuit operates as follows: Assume, for a start, that $V_f$ is rising and suddenly $V_f > V_R + V_H$. The op amp output will accordingly be driven negative—turning $Q_3$, $Q_2$, and $Q_1$ off. The current which had been flowing in $Q_1$, $L$, and $R_L$ cannot be turned off instantaneously when $Q_1$ is turned off. The stored inductor current will therefore flow up through the diode. The current will decay, since it has no driving source, and cause $V_o$ eventually to drop. When $V_o$ drops, this causes $V_f$ to drop below $V_R$. The op amp then goes positive. $Q_3$, $Q_2$, and $Q_1$ are turned on and the inductor current begins to increase. This continues until $V_f > V_R + V_H$, which completes one cycle. Figure 24.7 shows currents and voltages in $Q_1$ and $D$ during the on and off times.

## 24-16 REGULATORS

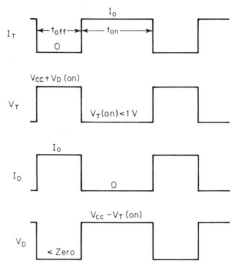

**Fig. 24.7** Idealized voltage and current waveforms of the switching regulator.

### DESIGN EQUATIONS

| Description | Equation |
|---|---|
| Approximate output voltage (see Fig. 24.7) | $V_o \approx \dfrac{t_{on} V_{cc}}{t_{on} + t_{off}}$ |
| Values for inductor and capacitor | $L = \dfrac{(V_{cc} - V_o)\, t_{on}}{2\,(I_p - I_o)}$ <br><br> $C = \dfrac{V_{cc} - V_o}{4\pi^2 f_o^2 L V_r}$ <br><br> where $I_p$ = peak-load current (usually 5 to 20% larger than $I_o$) <br> $I_o$ = rated-load current <br> $V_r$ = maximum allowed peak-to-peak ripple voltage |
| Operating frequency | $f_o = \dfrac{1}{t_{on} + t_{off}}$ <br><br> $\approx \dfrac{R_c V_R (V_{cc} - V_R)}{V_{cc} V_r L}$ <br><br> NOTE: The exact $V_o$ and $f_o$ must consider the rise-time $t_r$ and fall time $t_f$ of $Q_1$ |
| Resistor values | $R_1 = \dfrac{V_{cc} - V_R}{I_Z}$ <br><br> where $I_Z$ = optimum zener current <br><br> $R_2 = \dfrac{10\, R_1 V_r}{V_{cc} - V_R}$ <br><br> $R_3 = 10 R_1 - R_2$ <br><br> $R_4 = \dfrac{V_{cc} \beta_1 \beta_2}{I_o}$ |

| Description | Equation |
|---|---|
| where $\beta_1$ and $\beta_2$ are the minimum current gains of $Q_1$ and $Q_2$ | $R_5 = \dfrac{\beta_1 \beta_2 \beta_3 V_{cc}}{I_o}$ $R_6 = R_3$ $R_7 = \dfrac{R_6 V_R}{V_o - V_R}$ |

## REFERENCES

1. Widlar, R. J.: "Designing Switching Regulators," National Semiconductor Corp., 1969.
2. Capel, A.: New Control Technique in dc/dc Regulators for Space Applications, *IEEE Trans. Aerosp. Electron. Syst.*, vol. AES-8, no. 4, p. 472, July 1972.
3. Olla, R. S.: Switching Regulators: The Efficient Way to Power, *Electronics*, Aug. 16, 1973, p. 91.
4. Hauser, J. A.: Get with Switching Regulators, *Electron. Des.*, Apr. 25, 1968, p. 62.

## CURRENT REGULATORS

Current regulators, i.e., current sources, come in two classes. One class is those devices which provide a source of constant current. The other is often called voltage-to-current converters, wherein the output current is a function of the input voltage. These two classes, in reality, can be performed by the same circuits. The first class merely clamps the input terminal to a fixed-voltage source, thereby providing a constant output current. In all the circuits to follow, the voltage source can be replaced by an input-voltage signal to change the circuit to a voltage-to-current converter (see Chap. 4).

Current regulators are called upon to handle floating loads (both leads off ground) and grounded loads (one lead grounded). We will present a circuit for each type. Only those types capable of handling load currents under several amps will be discussed.

## 24.7 FLOATING-LOAD CURRENT REGULATOR

**ALTERNATE NAMES** Current source, voltage-to-current converter, voltage-to-current amplifier, transconductance amplifier, current sink, constant-current regulator, controlled current source, unipolar current source.

**EXPLANATION OF OPERATION** Current sources are always made with a simple application of Ohm's law: $I_o = V_o/R_s$. Thus we merely need a voltage amplifier with a fixed output $V_o$, and a sampling resistor $R_s$ for the output current $I_o$. The feedback voltage is generated across $R_s$. Any tendency for $I_o$ to change will be reproduced by a change in $V_o$. This change is fed back to the input through the feedback resistor $R_f$, resulting in a correction which restores both $V_o$ and $I_o$ to their original values. Figure 24.8 shows a typical way this is performed if the load resistor $R_L$ must have both terminals off ground. The input voltage $V_R$ is generated by a zener diode which is chosen according to the stability requirements of $I_o$.

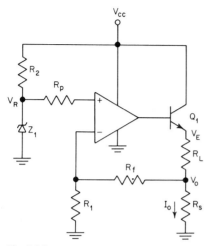

**Fig. 24.8** Current regulator for a floating load (i.e., where both leads of the load are off ground).

The output voltage at $V_o$ is computed from the basic formula for a noninverting amplifier:

$$V_o = \frac{V_R(R_f + R_1)}{R_1}$$

If $R_f$ is at least 100 times larger than $R_s$, the output current is

$$I_o = \frac{V_o}{R_s}$$

Combining these two equations yields

$$I_o = \frac{V_R(R_f + R_1)}{R_1 R_s}$$

Note that $I_o$ is totally independent of $R_L$. However, if $R_L$ gets too large, $Q_1$ will saturate as it attempts to maintain $I_o$ at a constant value. The current will then fall off as $R_L$ is further increased. That is, when $R_L + R_s = V_{cc}/I_o$ is reached, the current $I_o$ will begin to vary according to

$$I_o = \frac{V_{cc}}{R_L + R_s}$$

## DESIGN PARAMETERS

| Parameter | Description |
|---|---|
| $A_{vo}$ | Open-loop gain of op amp at dc |
| $I_o$ | Output current of circuit |
| $I_{Z0}$ | Optimum zener-diode current |
| $P_1(\max)$ | Maximum allowable dissipation for $Q_1$ |
| $Q_1$ | Buffer transistor to increase output-current capability of circuit |
| $R_1$ | Part of feedback network |
| $R_2$ | Resistor used to optimize zener current |
| $R_f$ | Feedback resistor |
| $R_L$ | Load resistor |
| $R_{\text{out}}$ | Output resistance of current regulator |
| $R_s$ | Current-sensing resistor used for feedback control of $I_o$ |
| $V_{cc}$ | Input voltage (positive power supply) |
| $V_o$ | Voltage across sense resistor |
| $V_R$ | Zener voltage |

## DESIGN EQUATIONS

| Description | Equation |
|---|---|
| Output current into load resistor $R_L$ | $I_o = \dfrac{V_R(R_f + R_1)}{R_1 R_s}$ |
| Output resistance of circuit driving $R_L$ (but not including $R_L$) | $R_{\text{out}} = \dfrac{R_f + R_1 A_{vo}}{R_f + R_1 + R_s}$ |

| Description | Equation |
|---|---|
| Maximum allowable $R_L$ after which $I_o$ will not remain constant | $R_L(\text{max}) = \dfrac{V_{cc}}{I_o} - R_s$ |
| Voltage at transistor emitter | $V_E = \dfrac{V_R(R_f R_s + R_f R_L + R_s R_L)}{R_1 R_s}$ |
| Voltage at $V_o$ | $V_o = \dfrac{V_E R_s}{R_s + R_L}$ |
| Resistor values | $R_s \geq \dfrac{I_o V_{cc} - P_1(\text{max})}{I_o^2}$ |
| | $R_f \geq 100\, R_s$ |
| | $R_1 = \dfrac{V_R R_f}{I_o R_s - V_R}$ |
| | $R_p = \dfrac{R_1 R_f}{R_1 + R_f}$ |
| | $R_2 = \dfrac{V_{cc} - V_R}{I_{Z0}}$ |

## REFERENCE

1. Analog Devices, Inc., *Analog Dialogue*, vol. 2, no. 1, p. 4, March 1968.

## 24.8 GROUNDED-LOAD CURRENT REGULATOR

**ALTERNATE NAMES** These are the same as those for the last circuit plus single-ended current source.

**EXPLANATION OF OPERATION** Operation of this regulator is similar to that of the floating-load current regulator. In this case, however, the current-sampling resistor $R_s$ is floating and the load resistor $R_L$ has one end grounded. Since $R_s$ is floating, two feedback resistors $R_f$ and $R_4$ are required. The conventional feedback resistor $R_f$ samples the current and $R_4$ provides a reference voltage from the other side of $R_s$. If $R_4$ was deleted, both $R_s$ and $R_L$ would become sampling resistors. The current through $R_L$ would then be determined by the resistance of $R_L$. A true current regulator puts out a current totally independent of $R_L$.

With $R_4$ connected, the output current is

$$I_o = \dfrac{R_f V_R}{R_1 R_s}$$

if the following is satisfied: $R_1 = R_3$ and $R_f = R_4 + R_s$. If $R_1 + R_f \gg R_s$, the output resistance of this current source is

$$R_{\text{out}} = R_s \dfrac{R}{\Delta R}$$

where $R$ is the least accurate resistor of $R_1$, $R_f$, $R_3$, and $R_4$, and $\Delta R$ is the number of ohms by which that least accurate resistor deviates from the ideal. The ideal resistances are determined from $R_1 = R_3$ and $R_f = R_4 + R_s$. If $R_1$ and $R_3$ have small trimming potentiometers in series with them, $R_{\text{out}}$ can be

## 24-20 REGULATORS

trimmed to almost infinity. However, resistor changes with temperature must be recognized, since a very small change will quickly reduce the output resistance. The voltage source driving $R_3$ has a source resistance which must be added to $R_3$ to determine the true $R_3$. If the voltage follower is used as shown in Fig. 24.9, the source resistance will be much smaller than 1 Ω. If $R_3$ is very large, say more than 100 kΩ, the source resistance will be negligible.

If a negative current source is required, the following changes to Fig. 24.9 must be done: 1. $V_R$ must drive the grounded side of $R_1$ and the left side of $R_3$ must be grounded, 2. $Q_1$ must be replaced with a PNP transistor with a $-V$ applied to its collector.

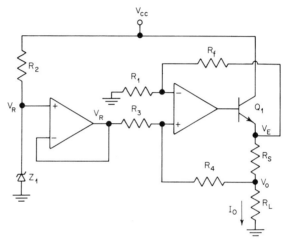

**Fig. 24.9** Grounded-load current regulator.

### DESIGN PARAMETERS

| Parameter | Description |
|---|---|
| $I_o$ | Output current |
| $I_{Z0}$ | Optimum zener-diode current |
| $P_1(\max)$ | Maximum allowable power dissipation of $Q_1$ |
| $Q_1$ | Buffer transistor to provide high output currents |
| $R$ | Any of $R_1$, $R_3$, $R_4$, or $R_f$ |
| $R_1$ | Part of feedback network |
| $R_2$ | Establishes optimum current in zener diode |
| $R_3$ | Part of feedback network |
| $R_4$ | Part of feedback network |
| $R_f$ | Part of feedback network |
| $R_L$ | Load resistance |
| $R_s$ | Current-sensing resistor |
| $\Delta R$ | A small change in any of $R_1$, $R_3$, $R_4$, or $R_f$ |
| $V_{cc}$ | Positive input voltage to circuit |
| $V_o$ | Output voltage at $R_L$ |
| $V_R$ | Reference voltage of zener diode |
| $Z_1$ | Zener diode |

## DESIGN EQUATIONS

| Description | Equation |
|---|---|
| Output current into load resistor $R_L$ | $I_o = \dfrac{V_R R_f}{R_1 R_s}$ |
| Output resistance of current source | $R_{out} = R_s \dfrac{R}{\Delta R}$ |
| Maximum allowable load resistor | $R_L(\max) = \dfrac{V_{cc}}{I_o} - R_s$ |
| Output voltage across $R_L$ | $V_o = I_o R_L$ |
| Resistor values | $R_s \geq \dfrac{V_{cc} I_o - P_1(\max)}{I_o^2}$ $R_f \geq 100 R_s$ $R_4 = R_f - R_s$ $R_1 = R_3 = \dfrac{R_f V_R}{V_{cc} - V_R}$ $R_2 = \dfrac{V_{cc} - V_R}{I_{Z0}}$ |

## REFERENCES

1. Graeme, J. G.: "Application of Operational Amplifiers — Third-Generation Techniques," p. 78, McGraw-Hill Book Company, New York, 1973.
2. Dobkin, R. C.: Bilateral Current Source, Op Amp Circuit Collection, National Semiconductor Corp., February 1970, p. 6.
3. Jung, W.: Low Temperature Coefficient Current Source Becomes Ultra-Compliant with Two Resistors, *Electron. Des.*, June 7, 1973, p. 108.

# Chapter 25

# Sampling Circuits

## INTRODUCTION

Two classes of sampling circuits will be presented in this chapter: (1) real-time sampling circuits, simply called multiplexers, and (2) delayed sampling circuits, called sample-and-hold (S/H) multiplexers. The multiplexer term implies that two or more inputs converge to a common output. The number of inputs is determined by system requirements and could conceivably be 100 or more if certain design rules are followed. However, one must pay a price for many inputs. Sampling time per input channel must be lower. Crosstalk and loading from other channels increase with large numbers of channels. These problems are present in both real-time multiplexers and S/H multiplexers.

This chapter contains design data on three types of sampling circuits. An FET real-time multiplexer using CMOS switches will be presented first. The errors expected in an FET multiplexer will be detailed. The second circuit is a real-time multiplexer which uses precision op amp gates. This circuit has extremely good accuracy but suffers from limited input range.

A sample/hold circuit will also be described in detail. The discussion will explore the conflicting requirements of speed and accuracy. This particular type of S/H circuit was chosen for presentation since it is available from several manufacturers on a single chip (except for the holding capacitor).

## 25.1 FET MULTIPLEXER

**ALTERNATE NAMES** Analog multiplexer, multichannel sampling circuit, analog commutator, multichannel analog switch, CMOS multiplexer, MUX.

**EXPLANATION OF OPERATION** Either junction field-effect transistors (JFETs) or metal-oxide semiconductor FETs (MOSFETs) can be utilized as the switching devices in a multiplexer. If MOSFETs are used, the complementary type, CMOS, is recommended, since it provides the best overall performance. These devices are available in large arrays which are optimized for this application.

Figure 25.1A shows a representative four-channel CMOS multiplexer. The detail of each switch is shown in Fig. 25.1B. By paralleling an N-channel with a P-channel FET, the interaction between the gate voltage and the source voltage is minimized. This interaction is explained as follows: The

## 25-2 SAMPLING CIRCUITS

drain-to-source on resistance $R_{DS}$ in an ideal FET is determined only by the voltage on the gate. In a real FET, however, $R_{DS}$ depends on the voltage between gate and source or between the gate and drain, whichever is smaller. This is not usually a problem in the FET OFF state, since the gate is pulled to a large enough ± voltage that the source or drain voltage cannot bring the FET into the conductive state. When the FET is ON, however, the drain-to-source resistance is modulated by the drain (or source) voltage. This undesirable feature is called resistance modulation. The parallel CMOS switch reduces resistance modulation to a second-order error. This is possible since the resistance vs. $V_{GS}$ curves of the two devices have opposite slopes. The ON resistance remains almost constant as $V_S$ or $V_D$ vary from zero to maximum.

The errors due to a nonzero $R_{DS}(on)$ are further minimized by using a noninverting op amp circuit at the output node of the four switches. The input resistance of a noninverting op amp circuit is approximately

$$R_{in} \approx A_{vo} R_{id}$$

The voltage gain of the circuit from a typical input, say $v_1$, is therefore

$$A_{vc} = \frac{v_o}{v_1} = \frac{R_{in}}{R_{DS1}(on) + R_{in}}$$

If a 1 percent accurate multiplexer is required, $R_{in}(min)$ must be at least 100 times larger than $R_{DS}(on,max)$. Likewise, a 0.1 percent circuit requires an $R_{in}(min)/R_{DS}(on,max)$ ratio greater than 1,000.

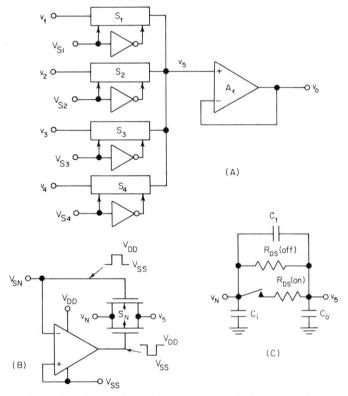

**Fig. 25.1** A four-channel CMOS multiplexer (A), a detailed diagram of one switch and its driver (B), and the equivalent circuit of a switch (C).

Another dc error source which should be considered in a high-accuracy MUX is the leakage current through the OFF switches. These currents can be converted to equivalent leakage resistances to simplify error calculations. Assume that $S_1$ is on and all other switches are off. If $v_2$, $v_3$, and $v_4$ all equal zero (and all have small source resistances), we assume that the three leakage resistances of $S_2$, $S_3$, and $S_4$ are connected in parallel to ground. In this case it appears that a resistance of

$$R_L = \frac{1}{1/R_{DS2}(\text{off}) + 1/R_{DS3}(\text{off}) + 1/R_{DS4}(\text{off})}$$

is shunted from $v_5$ to ground. The voltage gain of the entire circuit, assuming an ideal op amp and nonideal switches, is then

$$A_{vc} = \frac{v_o}{v_1} = \frac{R_L}{R_{DS1}(\text{on}) + R_L}$$

The ratio $R_{DS}(\text{off})/R_{DS}(\text{on})$ must be much larger than 1,000 if a 0.1 percent MUX is required.

Attenuation of high frequencies through the multiplexer can be caused by either op amp limitations or FET output capacitance. The $-3$-dB frequency of the op amp is simply its unity-gain crossover frequency if the noninverting configuration shown in Fig. 25.1A is utilized. The $-3$-dB frequency of each FET switch is $1/2\pi R_{DS}(\text{on})C_o$. Since the output terminals of all four switches are in parallel, $C_o$ must be multiplied by four to determine the actual $-3$-dB frequency due to this cause.

If high-speed commutating operation is required, several other factors must be investigated. The FET switch driving circuit is often slower than the switch itself. The op amp unity-gain bandwidth must also be adequate. Manufacturer's data should be consulted.

## DESIGN PARAMETERS

| Parameter | Description |
| --- | --- |
| $A_1$ | Wide-bandwidth operational amplifier which is stable with 100% feedback |
| $A_{vo}$ | DC open-loop voltage gain of op amp |
| $A_{vc}$ | Closed-loop voltage gain of circuit from any particular input to the output |
| $C_i$ | Input capacitance of an FET switch |
| $C_t$ | Transfer capacitance of an FET switch |
| $C_o$ | Output capacitance of an FET switch |
| $f_{cp}$ | The pole frequency of the circuit (where voltage gain has been reduced 3 dB) |
| MUX | Multiplexer |
| $R_{DSN}(\text{on})$ | On resistance of FET switch No. $N$ |
| $R_{DSN}(\text{off})$ | Off resistance of FET switch No. $N$ |
| $R_{id}$ | Differential input resistance of op amp |
| $R_{in}$ | Closed-loop input resistance of op amp |
| $R_L$ | Parallel resistance seen at op amp input due to OFF switches |
| $S_1$ to $S_4$, $S_N$ | FET switches, $N$th FET switch |
| $v_o$ to $v_5$ | Voltages as shown in Fig. 25.1 |
| $V_{DD}$ | Maximum positive voltage applied to CMOS gates (i.e., drain voltage) |
| $V_{S1}$ to $V_{S4}$, $V_{SN}$ | Drive signals for FET switches, $N$th drive signal |
| $V_{SS}$ | Largest negative voltage applied to CMOS gates (i.e., source voltage) |

## SAMPLING CIRCUITS

### DESIGN EQUATIONS

| Eq. No. | Description | Equation |
|---|---|---|
| 1 | Voltage gain from any input $v_1$ to $v_4$ to $v_o$ assuming ideal op amp and switches | $A_{vc} = \dfrac{v_o}{v_1} = \dfrac{v_o}{v_2} = \dfrac{v_o}{v_3} = \dfrac{v_o}{v_4} = 1$ |
| 2 | Voltage gain from a typical input (say $v_1$) assuming nonideal op amp and switches | $A_{vc} = \dfrac{v_o}{v_1} = \dfrac{R_L \parallel R_{in}}{R_{DS1}(\text{on}) + R_L \parallel R_{in}}$ |
| 3 | Approximate input resistance of op amp | $R_{in} \approx A_{vo} R_{id}$ |
| 4 | Shunting resistance of off switches assuming $S_1$ on and $S_2$ to $S_4$ off | $R_L = \dfrac{1}{1/R_{DS2}(\text{off}) + 1/R_{DS3}(\text{off}) + 1/R_{DS4}(\text{off})}$ |
| 5 | Bandwidth of circuit assuming an ideal op amp ($S_1$ on and $S_2$ to $S_4$ off) | $f_{cp} = \dfrac{1}{2\pi R_{DS1}(\text{on})(C_{o1} + C_{o2} + C_{o3} + C_{o4})}$ |
| 6 | Maximum signal levels allowed for $v_1$ ($v_2$ to $v_4$ are similar) | $V_{SS} \leqq v_1 \leqq V_{DD}$ |

### REFERENCES

1. Bergersen, T. B.: Field Effect Transistors in Analog Switching Circuits, Motorola Application Note AN-220, 1966.
2. Givins, S.: Field Effect Transistors as Analog Switches, *Comput. Des.*, June 1974, p. 106.
3. Fullager, D.: Analog Switches Replace Reed Relays, *Electron. Des.*, June 21, 1973, p. 98.
4. Schmid, H.: Electronic Analog Switches, *Electro-Technology*, June 1968, p. 35.

### 25.2 PRECISION GATE MULTIPLEXER

**ALTERNATE NAMES** Analog multiplexer, analog commutator, multichannel analog switch, multichannel sampling circuit, MUX.

**EXPLANATION OF OPERATION** Although this circuit is useful only for negative input voltages, it is popular because of its simplicity and high accuracy. If only a three-channel input is required, such as that shown in Fig. 25.2, a quad op amp along with several easily available resistors and diodes provides a low-cost multiplexer. This circuit also allows us to adjust the gain of each channel independently (Ref. 1).

The three input circuits are essentially precision rectifiers. For example, if $v_{S1} \leqq 0$, the $A_1$ circuit has a gain of

$$\dfrac{v_4}{v_1} = -\dfrac{R_2}{R_1} \quad \text{if } v_1 \leqq 0$$

or

$$\dfrac{v_4}{v_1} = 0 \quad \text{if } v_1 > 0$$

The diode $D_1$ is reverse-biased in this case, so $v_{S1}$ has no effect on the precision rectifier action. However, if $v_{S1}$ is >1 V, $D_1$ is forward-biased. This

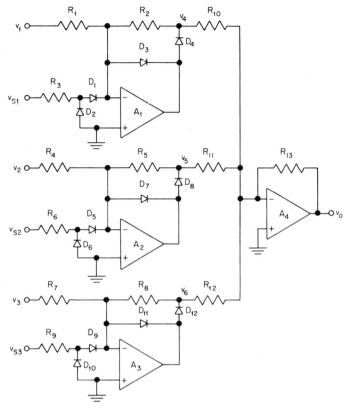

**Fig. 25.2** A three-channel multiplexer utilizing precision gates (useful only for negative input voltages).

clamps the output of $A_1$ to negative saturation, since $D_4$ is reverse-biased and the feedback loop is opened. In this case $v_4 = 0$ for all values of $v_1$ as long as

$$\frac{|v_1|}{R_1} < \frac{v_{S1}}{R_3}$$

If $v_1$ gets more negative than this inequality allows, $v_4$ will become positive. $R_3$ is usually made much smaller than $R_1$ so that the inequality will be easily satisfied.

The output signals from the precision rectifiers are summed into the $A_4$ adder circuit. This provides another phase inversion. Thus $v_o$ will always be negative, since $v_1$, $v_2$, and $v_3$ are required to be negative. Resistors $R_{10}$, $R_{11}$, and $R_{12}$ provide another opportunity for independent adjustment of the gain of each channel.

### DESIGN PARAMETERS

| Parameter | Description |
| --- | --- |
| $A_1$ to $A_3$ | Op amps used in precision rectifiers |
| $A_4$ | Op amp used in adder circuit |
| $D_3, D_4, D_7, D_8, D_{11}, D_{12}$ | Diodes used in $A_1$ to $A_3$ feedback circuits to provide precision rectification |

## 25-6 SAMPLING CIRCUITS

| Parameter | Description |
|---|---|
| $D_1, D_2, D_5, D_6, D_9, D_{10}$ | Diodes used to isolate $v_{S1}$ to $v_{S3}$ from $v_1$ to $v_3$ when $v_{S1}$ to $v_{S3}$ are $\leqq$ zero |
| $R_1, R_2, R_4, R_5, R_7, R_8$ | Resistors which determine gain of $A_1$ to $A_3$ circuits |
| $R_3, R_6, R_9$ | Resistors used to limit currents in $D_1, D_2, D_5, D_6, D_9, D_{10}$ and also to provide a reasonable load resistance for $v_{S1}, v_{S2},$ and $v_{S3}$ |
| $R_{10}, R_{11}, R_{12}, R_{13}$ | Resistors which determine voltage gain of adder circuit |
| $v_o$ to $v_6$ | Voltages as shown in Fig. 25.2 |
| $v_{S1}$ to $v_{S3}$ | Rectangular input waveforms used to control on-off status of $A_1$ to $A_3$ |

### DESIGN EQUATIONS

| Eq. No. | Description | Equation |
|---|---|---|
| 1 | Voltage gain from $v_1$ to $v_o$ assuming $v_{S1} > 1$ V | $\dfrac{v_o}{v_1} = \dfrac{R_2 R_{13}}{R_1 R_{10}}$ |
| 2 | Voltage gain from $v_2$ to $v_o$ assuming $v_{S2} > 1$ V | $\dfrac{v_o}{v_2} = \dfrac{R_5 R_{13}}{R_4 R_{11}}$ |
| 3 | Voltage gain from $v_3$ to $v_o$ assuming $v_{S3} > 1$ V | $\dfrac{v_o}{v_3} = \dfrac{R_8 R_{13}}{R_7 R_{12}}$ |
| 4 | Maximum allowed (negative) $v_1$ | $|v_1|_{\max} < \dfrac{R_1 v_{S1}}{R_3}$ |
| 5 | Maximum allowed (negative) $v_2$ | $|v_2|_{\max} < \dfrac{R_4 v_{S2}}{R_6}$ |
| 6 | Maximum allowed (negative) $v_2$ | $|v_3|_{\max} < \dfrac{R_7 v_{S3}}{R_9}$ |

### REFERENCE

1. "Precision Gate," G. A. Philbrick Researches, Inc., Applications Manual for Computing Amplifiers, Nimrod Press, Inc., Boston, Mass., June 1966, p. 58.

### 25.3 SAMPLE-AND-HOLD CIRCUIT

**ALTERNATE NAMES** Sample/hold circuit, S/H, sampling circuit.

**EXPLANATION OF OPERATION** As shown in Fig. 25.3, an S/H circuit requires a high-output-current op amp ($A_1$), a high-quality switch ($S$), a low-leakage capacitor ($C$), and an output op amp ($A_2$) which has a low input bias current. The input op amp must be capable of driving a capacitive load without any hint of instability. The switch must have a high $R_{DS}(\text{off})/R_{DS}(\text{on})$ ratio so that $C$ can quickly charge to its peak value and maintain that value with minimal droop between sampling times. This switch must also have very small coupling between its digital input and analog output. This coupling would allow switching transients (which occur at $S$ turn-off) to change the final voltage stored on $C$. Lastly, the output op amp bias and offset currents (and their change with temperature) must be small.

Circuit operation is fairly straightforward. The input op amp ($A_1$) maintains $v_2$ at the same potential as $v_1$. During the sampling interval $T_1$, the voltage $v_2$ is deposited on $C$ by the closure of $S$. After $T_1$ terminates (when $S$ opens), the voltage $v_2$ is maintained on $C$ for a duration $T_2$ until the next sampling interval. Op amp $A_2$ holds the transferred value of $v_2$ on its output $v_o$ for the entire $T_2$ duration.

We will now consider the effect of nonideal parameters on the S/H circuit performance. The holding capacitor must be carefully sized. The following constraints limit the maximum size of $C$:

1. Op amp $A_1$ must have a current drive capability of at least

$$I_{max} = \frac{C[v_o(max) - v_o(min)]}{T_1}$$

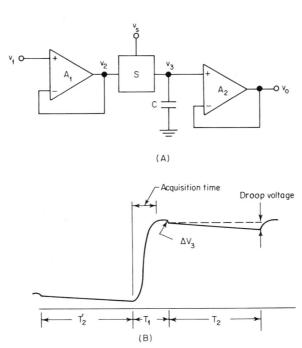

**Fig. 25.3** A basic sample-and-hold circuit (A). Exaggerated output waveform (B).

2. The maximum ON resistance of $S$ and the sampling time $T_1$ controls the upper limit for $C$. If a 1 percent S/H is required, we need $T_1 > 5 R_{DS}(\text{on,max})\, C$. If a 0.1 percent system is desired, make $T_1 > 7 R_{DS}(\text{on,max})\, C$.

Both the above factors tell us that a large $C$ requires a large sampling time $T_1$. A high current-drive capability from $A_1$ and a low ON resistance for $S$ are also mandatory if $C$ is large. At the other end of the scale, however, if $C$ is too small, other problems show up. These problems are listed:

1. During the hold period, $T_2$ leakage currents through $C$, $S$, and the $A_2$ input will easily discharge a small capacitor.

2. The gate-to-source (or gate-to-drain) capacitance transfers charge to/from $C$ when the gate waveform turns $S$ off. This error adds or subtracts from the sample voltage stored on $C$. If $C$ is small, the error is more pronounced.

## 25-8 SAMPLING CIRCUITS

Monolithic circuits are available which contain $A_1$, $A_2$, and $S$ on a single chip. Each of these devices is optimized according to the trade-offs itemized above.

### DESIGN PARAMETERS

| Parameter | Description |
|---|---|
| $A_1$ | High-current-driver op amp |
| $A_2$ | Op amp with high input resistance |
| $A_{vc}$ | Closed-loop voltage gain of entire circuit |
| $C$ | Holding capacitor |
| $C_{GD}$ | Gate-to-drain capacitance of switch |
| $I_{max}$ | Maximum output current available from $A_1$ |
| $N$ | Variable used in accuracy calculations |
| $R_c$ | Leakage resistance of $C$ |
| $R_{DS}$ | Drain-to-source resistance of $S$ |
| $R_{ic}$ | Common-mode input resistance of $A_2$ |
| $S$ | Electronic switch |
| S/H | Sample and hold |
| $T_1$ | Sampling time |
| $T_2$ | Time between samples |
| $v_o$ to $v_3$ | Voltages as indicated in Fig. 25.3 |
| $v_3$(droop) | Voltage droop on $C$ during hold time $T_2$ |
| $\Delta v_3$ | Error voltage added to (or subtracted from) $v_3$ owing to $C_{GD}$ |
| $\Delta v_G$ | Change in gate voltage of $S$ |
| $V^{(\pm)}$ | Power-supply voltages |

### DESIGN EQUATIONS

| Eq. No. | Description | Equation |
|---|---|---|
| 1 | Voltage gain of circuit during sample time | $A_{vc} = \dfrac{v_o}{v_1} = 1$ |
| 2 | Required output-current capability of $A_1$ | $I_{max} = \dfrac{C[v_o(\max) - v_o(\min)]}{T_1(\min)}$ |
| 3 | Required sampling interval $T_1$ | $T_1(\min) = 5R_{DS}(\text{on,max})C$ for a 1% S/H or $T_1(\min) = 7R_{DS}(\text{on,max})C$ for a 0.1% S/H |
| 4 | Magnitude of error voltage deposited on $C$ when $v_G$ changes by $\Delta v_G$ | $\Delta v_3 = \Delta v_o = \dfrac{\Delta v_G C_{GD}}{C + C_{GD}}$ |
| 5 | Maximum droop of voltage $v_3$ during $T_2$ | $v_3(\text{droop}) = v_3(\max)\left[1 - \exp\left(-\dfrac{T_2}{RC}\right)\right]$ where $R = R_{ic} \| R_c \| R_{DS}(\text{off})$ |

### DESIGN PROCEDURE

We begin this procedure by assuming that $T_2$, $A_{vc}$, $v_o(\max)$, $v_o(\min)$, and $\Delta v_G$ are fixed by the system into which this S/H circuit is to be installed. The S/H circuit accuracy and droop are also specified. We also assume $A_1$, $A_2$, $S$, and the type of holding capacitor have been selected. Our job is to compute the capacitor size, the voltage error of $v_3$ (and $v_o$) caused by $\Delta v_G$, and the droop of $v_3$ (and $v_o$) during $T_2$.

## DESIGN STEPS

*Step 1.* Equation 5 is rearranged to determine $C$:

$$C \geq \frac{T_2}{R \ln\{v_3(\max)/[v_3(\max) - v_3(\text{droop})]\}}$$

where $R = R_{ic}\|R_c\|R_{DS}(\text{off})$

A value of $R_c$ can be obtained from the specification sheet for the type of capacitor used even though the exact capacitance of the capacitor is not known until the calculation above is performed.

*Step 2.* Choose a value $N$ from the following table:

| Required Sampling Accuracy, % | N |
|---|---|
| 10 | 3 |
| 1 | 5 |
| 0.1 | 7 |
| 0.01 | 9 |

Calculate a first-cut sampling time $T_1$ using a modified form of Eq. 3:

$$T_1(\min) = NR_{DS}(\text{on,max})C$$

Determine a second-cut sampling time using a rearranged Eq. 2:

$$T_1(\min) = \frac{C[v_o(\max) - v_o(\min)]}{I_{\max}}$$

Use the higher of the two values calculated above for the actual $T_1$.

*Step 3.* Use Eq. 4 to find the approximate holding error due to switch capacitance.

**S/H CIRCUIT DESIGN EXAMPLE** Suppose we want to sample and hold a speech signal before it is sent into an A/D converter. We will assume the highest audio frequency of interest is 5 kHz. According to the rules of sampling, we must sample at a frequency of at least 10 kHz. Thus $T_2 \approx 1/10^4 = 100\ \mu\text{s}$. The maximum input- and output-voltage levels are specified. The gate drive of $S$ is also given.

*Design Requirements*
$T_2 = 100\ \mu\text{s}$
$v_o(\max) = +10\ \text{V}$
$v_o(\min) = 1\ \text{V}$
$\Delta v_G = 15\ \text{V}$
$v_3(\text{droop}) = 0.01\ \text{V}$
$V^{(+)} = +15\ \text{V}$
$V^{(-)} = \text{ground}$
$N = 5$ (1 percent accuracy)

*Device Data*
$I_{\max} = 10\ \text{mA}\ (A_1 = \frac{1}{4}\ \text{LM324})$
$R_{DS}(\text{on,max}) = 1{,}000\ \Omega$
$C_{GD} = 4\ \text{pF}$ $\quad\quad\quad\quad\quad\quad (S = \frac{1}{4}\ \text{CD4016})$
$R_{DS}(\text{off,min}) = 1.5 \times 10^{11}\ \Omega$
$R_{ic} = 1.5 \times 10^{12}\ \Omega$ $\quad\quad\quad (A_2 = \text{CA3130})$
$R_L(\min) = 2{,}000\ \Omega$
$R_c \approx 10^{13}\ \Omega$ $\quad\quad\quad\quad$ (polycarbonate capacitor)

*Step 1.* The holding capacitor is

$$C \geq \frac{T_2}{R \ln\{V_3(\max)/[V_3(\max) - V_3(\text{droop})]\}}$$

$$\geq \frac{10^{-4}}{(1.5 \times 10^{12}) \| 10^{13} \| (1.5 \times 10^{11}) \ln[10/(10 - 0.01)]} \geq 0.74 \text{ pF}$$

This capacitance is unreasonably small because the leakage resistances are so high. Assume a reasonable capacitance of 100 pF is used for $C$.

*Step 2.* We let $N = 5$, since a 1 percent circuit is required. The first-cut $T_1(\min)$ is

$$T_1(\min) = N\, R_{DS}(\text{on},\max)C$$
$$= 5(1{,}000)\, 10^{-10} = 0.5\ \mu\text{s}$$

The second-cut $T_1(\min)$ is

$$T_1(\min) = \frac{C[v_o(\max) - v_o(\min)]}{I_{\max}}$$

$$= \frac{10^{-10}(10 - 1)}{2 \times 10^{-2}} = 0.045\ \mu\text{s}$$

It appears that the risetime of the voltage on $C$ is constrained more by $R_{DS}(\text{on})$ than it is by $I_{\max}$. We therefore let $T_1 = 0.5\ \mu\text{s}$.

*Step 3.* The approximate holding error due to $C_{GD}$ is

$$\Delta v_o \approx \frac{\Delta v_G C_{GD}}{C + C_{GD}}$$

$$\approx \frac{15(4 \times 10^{-12})}{10^{-10} + 4 \times 10^{-12}} = 0.58 \text{ V}$$

This error is 5.8 percent of $v_o(\max)$ and 58 percent of $v_o(\min)$. If we raise $C$ to 1,000 pF, these errors are reduced to 0.58 and 5.8 percent, respectively. Step 3 must also be redone. The new $T_1$ is 5 $\mu$s.

## REFERENCES

1. Jones, D.: Applications of a Monolithic Sample-and-Hold/Gated Operational Amplifier, Harris Semiconductor Application Note 517, March 1974.
2. Buchanan, J. E.: C-MOS Switch Speeds Up Sample-and-Hold Circuit, *Electronics*, Sept. 27, 1973, p. 127.
3. Patstone, W., and C. Dunbar: Choosing a Sample-and-Hold Amplifier Is Not as Simple as It Used to Be, *Electronics*, Aug. 2, 1973, p. 101.

Chapter **26**

# Time and Phase Circuits

## INTRODUCTION

This chapter contains design information on two types of circuits. The first two sections describe circuits which control the phase lead or phase lag of a system without affecting gain. The relationship between frequency and phase will be thoroughly presented.

The last circuit in the chapter presents a method of delaying a bilevel signal by a specified time. This can also be done with a timer IC, but the circuit shown here can be implemented with one-fourth of a quad op amp and several other parts. The required board space is less than required with a timer IC. The use of an FET input op amp allows time delays of days or weeks to be implemented.

### 26.1 PHASE LEAD/LAG CIRCUIT

**ALTERNATE NAMES** Lead/lag circuit, all-pass circuit, constant-amplitude phase shifter.

**EXPLANATION OF OPERATION** This circuit provides a phase adjustment of $v_o$ with respect to $v_1$ from $-180$ to $+180°$ (Ref. 1). The voltage gain remains

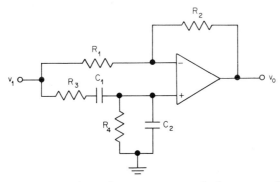

**Fig. 26.1** A single op amp phase-adjustment circuit which can provide either a lead or a lag.

## 26-2 TIME AND PHASE CIRCUITS

constant at 1/5 for changes in either frequency or phase adjustment. The phase adjustment is accomplished by simultaneously changing $R_3$ and $R_4$ or simultaneously changing $C_1$ and $C_2$. If an easier method of adjustment is required, the circuits shown in Sec. 26.2 are recommended.

If $R_1 = 5R_2$, the voltage transfer function of the lead/lag circuit is

$$A_{vc} = \frac{v_o}{v_i} = \frac{1}{5} \frac{1 - jX/3}{1 + jX/3}$$

where $X = 2\pi fRC - 1/2\pi fRC$
$R = R_3 = R_4$
$C = C_1 = C_2$

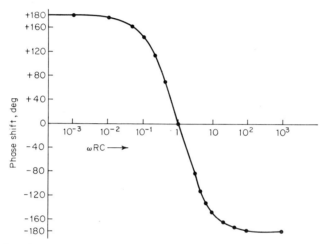

**Fig. 26.2** The output phase of Fig. 26.1 relative to its input phase.

The voltage gain is always 1/5, since the absolute magnitude of $1 - jX/3$ is always equal to that of $1 + jX/3$.

The phase shift of the circuit can be expressed as

$$\phi = -2 \tan^{-1}\left(\frac{X}{3}\right)$$

If $2\pi fRC$ approaches dc, $\phi$ approaches +180°. This provides the normal inverting gain of $-1/5$, since $C$ is open-circuited at dc. As $2\pi fRC$ increases to unity, the phase shift becomes zero. Lastly, when $2\pi fRC$ approaches infinity, the phase shift approaches $-180°$.

### DESIGN PARAMETERS

| Parameter | Description |
|---|---|
| $C_1, C_2$ | Part of phase-shifting network |
| $C$ | Common value for $C_1$ and $C_2$ |
| $f$ | Frequency in hertz |
| $\phi$ | Phase shift in degrees |
| $R_1, R_2$ | Sets voltage gain of circuit at 1/5 |
| $R_3, R_4$ | Part of phase-shifting network |

| Parameter | Description |
|---|---|
| $R$ | Common value for $R_3$ and $R_4$ |
| $X$ | Variable used in Eq. 1 |
| $v_1$ | Input voltage (reference phase) |
| $v_2$ | Output voltage (phase-shifted relative to $v_1$) |

## DESIGN EQUATIONS

| Eq. No. | Description | Equation |
|---|---|---|
| 1 | Voltage transfer function of circuit | $A_{vc} = \dfrac{v_o}{v_1} = \dfrac{1}{5}\dfrac{1 - jX/3}{1 + jX/3}$ <br> where $X = 2\pi fRC - 1/2\pi fRC$ <br> $R = R_3 = R_4$ <br> $C = C_1 = C_2$ |
| 2 | Phase shift of $v_o$ relative to $v_1$ | $\phi = -2\tan^{-1}\left(\dfrac{X}{3}\right)$ |
| 3 | Required relationship between $R_1$ and $R_2$ | $R_1 = 5R_2$ |

## REFERENCES

1. Genin, R.: Realization of an All-Pass Transfer Function Using Operational Amplifiers, *IEEE Proc.*, October 1968, p. 1746.

## 26.2 ADJUSTABLE LEAD/LAG CIRCUITS

**ALTERNATE NAMES** All-pass circuit, constant-amplitude phase shifter, lead network, lag network.

**EXPLANATION OF OPERATION** Figure 26.3A is an adjustable phase-lead circuit with a range of 0 to +180°. Likewise, Fig. 26.3B is a phase-lag circuit which can be adjusted from 0 to −180°. Both these circuits have the advantage of requiring the use of only one resistor to adjust phase. As with the circuit shown in Fig. 26.1, the phase difference between $v_o$ and $v_1$ depends on both the component values and the frequency. Figures 26.4A and B show the phase shift of these circuits as a function of $2\pi fR_3C$ (Ref. 1).

### DESIGN PARAMETERS

| Parameter | Description |
|---|---|
| $C$ | Part of phase-shifting network |
| $f$ | Frequency in hertz |
| $R_1, R_2$ | Establishes gain of circuit |
| $R_3$ | Part of phase-shifting network |
| $v_1$ | Input voltage (reference phase) |
| $v_o$ | Output voltage (phase-shifted relative to $v_1$) |

## 26-4 TIME AND PHASE CIRCUITS

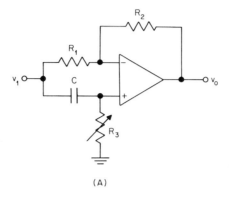

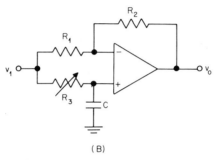

**Fig. 26.3** An adjustable phase-lead circuit (A) and an adjustable phase-lag circuit (B).

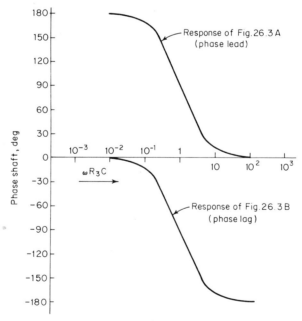

**Fig. 26.4** Phase as a function of $2\pi f R_3 C$ for the lead circuit (A) and the lag circuit (B).

## DESIGN EQUATIONS

| Eq. No. | Description | Equation |
|---|---|---|
| | LEAD CIRCUIT (Fig. 26.3A) | |
| 1 | Voltage transfer function of circuit | $A_{vc} = \dfrac{v_o}{v_i} = \dfrac{j2\pi f R_3 C - 1}{j2\pi f R_3 C + 1}$ |
| 2 | Phase shift of $v_o$ relative to $v_i$ | $\phi = 2\tan^{-1}(1/2\pi f R_3 C)$ |
| 3 | Required relationship between $R_1$ and $R_2$ | $R_1 = R_2$ |
| | LAG CIRCUIT (Fig. 26.3B) | |
| 4 | Voltage transfer function of circuit | $A_{vc} = \dfrac{v_o}{v_i} = \dfrac{1 - j2\pi f R_3 C}{1 + j2\pi f R_3 C}$ |
| 5 | Phase shift of $v_o$ relative to $v_i$ | $\phi = -2\tan^{-1}(2\pi f R_3 C)$ |
| 6 | Required relationship between $R_1$ and $R_2$ | $R_1 = R_2$ |

## REFERENCE

1. Francis, J. R.: Constant-Amplitude Phase Shifter, *EEE*, January 1971, p. 63.

## 26.3 ANALOG TIMER

**ALTERNATE NAMES** Timer, interval timer, RC timer, long-interval timer.

**EXPLANATION OF OPERATION** An op amp can perform some of the functions of the monolithic IC timer. One such application is the generation of long delays lasting hours or days. The FET input op amp is superior to the timer IC in this application because of its high input resistance. The circuit shown in Fig. 26.5 can generate delays of days or weeks if a low-leakage capacitor and an FET input op amp are utilized.

The op amp is used as a comparator with ac positive feedback through $R_5$, $R_6$, and $C$. The switch is in the clear-standby position prior to timer operation. In this state the voltage at the op amp inverting input is

$$V' = V_1 = \dfrac{R_2 V_R}{R_1 + R_2}$$

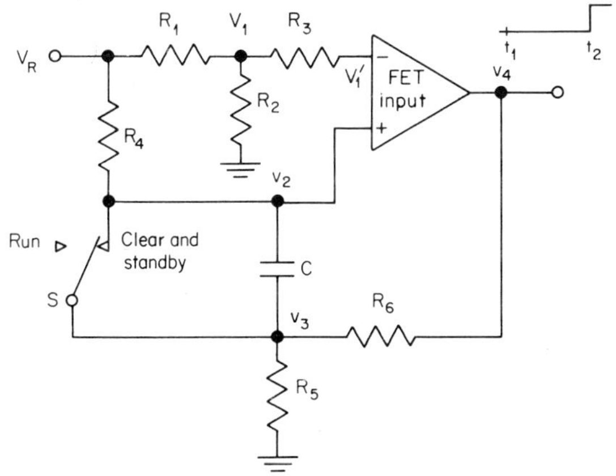

**Fig. 26.5** An analog timer which can provide a time delay up to many days or weeks.

## 26-6 TIME AND PHASE CIRCUITS

and the voltage at the noninverting input is

$$v_2(\text{standby}) = \frac{(R_5 \| R_6) V_R}{R_4 + R_5 \| R_6}$$

Resistor values are chosen such that $v_2(\text{standby}) \ll V_1$ (by at least several volts). This guarantees that the output $v_4$ will be at the LOW state in the standby mode.

When $S$ is placed in the RUN position, the current through $R_4$ begins to charge $C$ toward a final value of $V_R$. Voltages $v_3$ and $v_4$ are near zero during the RUN mode, since the op amp inverting input is more positive than its

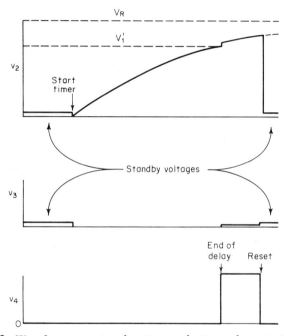

**Fig. 26.6** Waveforms at various locations in the timer shown in Fig. 26.5.

noninverting input. When $v_2$ charges up to approximately $V_1$, the op amp output switches to the HIGH state. A fraction $R_5/(R_5 + R_6)$ of $v_4$ is ac-coupled back to the op amp noninverting input to assure a clean, quick change of state. After the timer period has ended, the voltage $v_2$ continues to rise toward $V_R$ until $S$ is placed in the clear position. Voltage $v_4$ remains high until this clear operation takes place.

### DESIGN PARAMETERS

| Parameter | Description |
| --- | --- |
| $C$ | Provides long-duration time constant along with $R_4$ |
| $R_1, R_2$ | Establishes trip voltage $V_1$ for op amp inputs |
| $R_3$ | Makes driving resistance for op amp inputs equal so that input bias currents do not affect timing accuracy |
| $R_4$ | Main timing resistor used to provide a long-duration time constant |
| $R_5, R_6$ | Provides positive feedback (hysteresis) so that op amp will make a quick, clean change of states at end of timing period |

| Parameter | Description |
|---|---|
| $R_c$ | Capacitor leakage resistance |
| $R_{ic}$ | Common-mode input resistance of op amp |
| $R_L$ | Manufacturer's minimum recommended load resistance for op amp |
| $S$ | Mechanical switch or relay (must have low leakage between contacts when in OFF state) |
| $T$ | Period of timer in seconds |
| $V_1, V_1'$ | Trip voltage at op amp inverting input |
| $v_2$ | Voltage at op amp noninverting input |
| $v_2$(standby) | Voltage at op amp noninverting input while $S$ is in standby position |
| $v_3$ | Positive feedback voltage |
| $v_4$ | Output voltage |
| $V_R$ | Reference input voltage (timer can be no more stable than this voltage) |
| $V^{(\pm)}$ | Power-supply voltages |

## DESIGN EQUATIONS

| Eq. No. | Description | Equation |
|---|---|---|
| 1 | Time delay in seconds | $T = R_4 C \ln\left(\dfrac{V_R}{V_R - V_1}\right)$ |
| 2 | Time delay in seconds if $R_1 = R_2$ (this means that $V_R = 2V_1$) | $T = 0.693 R_4 C$ |
| 3 | Fixed voltage at op amp inverting input (assuming $R_{ic} \gg R_3$) | $V_1' = V_1 = \dfrac{R_2 V_R}{R_1 + R_2}$ |
| 4 | Voltages $v_2$ and $v_3$ during standby mode | $v_2(\text{standby}) = v_3(\text{standby}) = \dfrac{(R_5 \| R_6) V_R}{R_4 + R_5 \| R_6}$ |
| 5 | Voltage $v_2$ as a function of time during run mode | $v_2(t) = V_R \left[1 - \exp\left(-\dfrac{t}{R_4 C}\right)\right]$ |
|   | Recommended resistor values: | |
| 6 | $R_1, R_2$ | $R_1 = R_2$ chosen so they do not load down $V_R$ source |
| 7 | $R_3, R_4$ | $R_3 = R_4 = \dfrac{T}{0.693 C}$ |
| 8 | $R_5$ | $R_5 = 0.1 R_6$ |
| 9 | $R_6$ | $R_6 \gg R_L$ |
| 10 | Recommended capacitor size | Use a stable, low-leakage, high-capacitance device such as a polycarbonate, polystyrene, or polyester capacitor |

## DESIGN PROCEDURE

Since timer ICs are more practical for short-duration timing, we will assume this procedure is for a very long timing interval. The op amp is assumed to be a high-quality FET input device. A capacitor is picked which has the highest possible capacitance-leakage resistance product.

## DESIGN STEPS

*Step 1.* Since circuit size is usually important, these design steps begin with several trade-offs. One approach is to study available capacitors to find which type has the highest value for

$$\frac{(\text{Capacitance})(\text{leakage resistance})(\text{stability})}{\text{volume}}$$

The effects of temperature must be considered in this trade-off. The leakage resistance does not need to be larger than the op amp common-mode input resistance. Good candidates for $C$ are polycarbonate, polystyrene, or polyester capacitors. The polystyrene device has the best capacitance-leakage resistance product, but it turns sour above 80°C.

*Step 2.* Determine values for $R_3$ and $R_4$ according to Eq. 7. These resistors should be no larger than $1/100$ of $R_c$ or $R_{ic}$ if a 1 percent timer is required.

*Step 3.* Let $R_6$ be 10 times the load resistance expected for the circuit. Compute an approximate value for $R_5$ using Eq. 8.

*Step 4.* Choose values for $R_1 = R_2$ which will not excessively load $V_R$.

**EXAMPLE OF A TIMER DESIGN** We will design a 1,000-s timer to illustrate the design steps numerically. Polycarbonate capacitors have been chosen for $C$, and a high-quality FET input op amp is assumed.

*Design Requirements*
  $T = 1{,}000$ s $\pm 1$ percent
  $V_R = 10$ V
  $V^{(+)} = 10$ V
  $R_L = 2{,}000 \, \Omega$

*Device Data* (+25°C)
  $R_{ic} \approx 10^{11} \, \Omega$
  $R_c \approx 10^{10} \, \Omega$
  $C = 10 \, \mu\text{F}$

*Step 1.* We assume a 10-$\mu$F polycarbonate capacitor is available. Its leakage resistance is specified to be approximately $10^{10} \, \Omega$.

*Step 2.* Resistors $R_3$ and $R_4$ are found from

$$R_3 = R_4 = \frac{T}{0.693C} = \frac{1{,}000}{0.693(10^{-5})} = 144 \text{ M}\Omega$$

This is only 69 times smaller than $R_c$, so our 1 percent accuracy requirement will not quite be achieved. Since the $\Omega$F product of a capacitor type is constant at one temperature, a larger polycarbonate device would not help the situation. A polystyrene capacitor has a large enough leakage resistance to give 1 percent accuracy, but it is no good above 80°C. If a timer with much better than 1 percent accuracy is required, a crystal clock with a large number of counters is recommended.

*Step 3.* Since $R_L = 2{,}000 \, \Omega$, we let $R_6 = 20$ k$\Omega$. The value for $R_5$ can be in the vicinity of

$$R_5 = \frac{R_6}{10} = \frac{20{,}000}{10} = 2{,}000 \, \Omega$$

*Step 4.* The $V_R$ source has an output resistance of $<10 \, \Omega$, so we let $R_1 = R_2 = 1{,}000 \, \Omega$.

Chapter 27

# Waveform Generators

## INTRODUCTION

Ramp, triangle, and staircase generators are discussed in this chapter. Square-wave and pulse generators have already been covered in Chap. 20. Sine-wave generators were presented in Chap. 21.

Ramp (or sawtooth) generators, often used in cathode-ray-tube sweep systems, will be presented in Sec. 27.1. Section 27.2 contains detailed design information on a highly adjustable triangle waveform generator. The last section will cover design details of a staircase generator.

## 27.1 VOLTAGE RAMP GENERATOR

**ALTERNATE NAMES** Voltage-sweep generator, sawtooth generator, free-running sweep oscillator.

**EXPLANATION OF OPERATION** The two CMOS transistors $Q_1$ and $Q_2$ operate in the current-mirror configuration. Resistor $R_1$ establishes a current $I_2$ in $Q_2$ of

$$I_2 = \frac{V^{(+)} - V_{T2}}{R_1} + \frac{1 - 2\{KR_1[V^{(+)} - V_{T2}]\}^{1/2}}{2KR_1^2}$$

Transistors $Q_1$ and $Q_2$ should be on the same chip so that their drain characteristics will be approximately equal. Since $V_{T1} = V_{T2}$, we must have $I_1 = I_2$. We therefore have established a constant current with a magnitude which depends on $R_1$. This current-mirror concept allows us to simulate a high-impedance current source using a resistor $R_1$ in an isolated branch.

During the ramp formation $(T_1)$, $Q_3$ is held OFF by the $G_1$, $G_2$ latch. Constant current $I_1$ develops a precisely linear ramp by charging up $C$. Op amp $A_1$ has an extremely high input impedance, so it will not affect the current-mirror-charging circuit. The same ramp voltage developed on $C$ appears at $v_o$. When $v_o$ rises to $V_U$, the output of comparator $A_3$ goes LOW. This flips the latch, which causes $C$ to discharge quickly through $Q_3$. When $v_o$ has discharged down to $V_L$, the output of comparator $A_2$ goes LOW. This again flips the latch, which opens $Q_3$. The ramp is then allowed to begin over again.

## 27-2 WAVEFORM GENERATORS

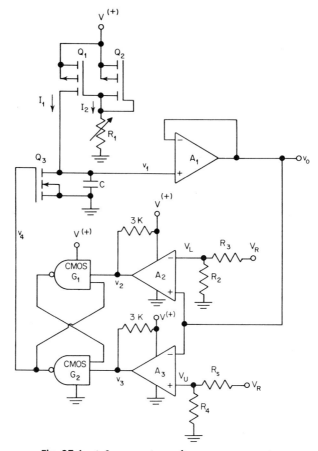

**Fig. 27.1** A free-running voltage-ramp generator.

The upper and lower trip levels ($V_U$ and $V_L$) are determined by the sizes of $R_2$ through $R_5$. If we want $V_L \approx 0$, the inverting input of $A_2$ can be grounded. In this case $A_2$ must be capable of operating with input voltages down to ground potential. If this is not the case, the negative supply terminal of $A_2$ must be returned to a voltage below zero. We must then be careful that the $G_1$ input is driven correctly. A clamp diode will be required to keep $v_2 \geqq$ zero volts.

### DESIGN PARAMETERS

| Parameter | Description |
|---|---|
| $A_1$ | High-input-resistance buffer used to sense voltage ramp across $C$ |
| $A_2, A_3$ | Comparators used to detect $V_U$ and $V_L$ |
| $C$ | Integrating capacitor used to generate ramp |
| $G_1, G_2$ | NAND gates wired as a latch |
| $I_1$ | Reflected current equal to $I_2$ |
| $I_2$ | Current generated by $V^{(+)}$, $R_1$, and $Q_2$ |
| $K$ | CMOS constant ($\approx 0.004$) |
| $Q_1, Q_2$ | Mirror-type current source |

## VOLTAGE RAMP GENERATOR

| Parameter | Description |
|---|---|
| $Q_3$ | Switch used to discharge $C$ |
| $R_1$ | Used to adjust magnitude of $I_1$ and $I_2$ |
| $R_2, R_3$ | Sets level of $V_L$ |
| $R_4, R_5$ | Sets level of $V_U$ |
| $T_1$ | Ramp duration |
| $T_2$ | Ramp retrace time |
| $v_o, v_1$ | Ramp voltage waveform (use $v_o$ for the output, since any load on $v_1$ will degrade the ramp waveshape) |
| $v_2$ | Trigger voltage for latch used to terminate retrace |
| $v_3$ | Trigger voltage for latch used to terminate ramp |
| $v_4$ | Voltage used to hold $Q_3$ on during retrace |
| $V_L$ | Lower limit of ramp waveform |
| $V_R$ | Reference voltage used to create $V_L$ and $V_U$ |
| $V_{T2}$ | Threshold voltage of $Q_2$ FET (usually 2 to 3 V) |
| $V_U$ | Upper limit of ramp waveform |
| $V^{(+)}$ | Positive power-supply voltage |

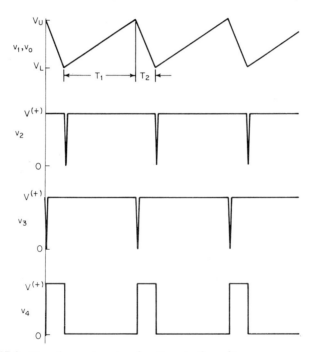

**Fig. 27.2** Waveforms at various locations in the voltage-ramp generator.

## DESIGN EQUATIONS

| Eq. No. | Description | Equation |
|---|---|---|
| 1 | Charging current into $C$ during ramp | $I_1 = \dfrac{V^{(+)} - V_{T2}}{R_1} + \dfrac{1 - 2\{KR_1[V^{(+)} - V_{T2}]\}^{1/2}}{2KR_1^2}$ |
| 2 | Rate of rise of ramp | $\dfrac{\Delta v_1}{\Delta t} = \dfrac{I_1}{C}$ |

| Eq. No. | Description | Equation |
|---|---|---|
| 3 | Upper trip voltage | $V_U = \dfrac{R_4 V_R}{R_4 + R_5}$ |
| 4 | Lower trip voltage | $V_L = \dfrac{R_2 V_R}{R_2 + R_3}$ |
| 5 | Duration of ramp | $T_1 = \dfrac{R_1 C (V_U - V_L)}{V^{(+)} - V_T}$ |
| 6 | Ramp retrace time | $T_2 = R_{DS}(\text{on}) C (\ln V_U - \ln V_L)$ |
| 7 | Frequency of oscillation | $F_o = \dfrac{1}{T_1 + T_2}$ |

## REFERENCES

1. McKinley, R. J.: Dual Op Amp Comparator Controls Ramp Reference, *Electronics*, Oct. 11, 1971, p. 76.
2. Hart, B. L., and R. W. J. Barker: A Precision Timed Ramp Generator with Zero Idling Power Consumption, *IEEE Proc.*, July 1973, p. 1047.

## 27.2 VOLTAGE TRIANGLE GENERATOR

**ALTERNATE NAME** Free-running triangle generator.

**EXPLANATION OF OPERATION** An ideal triangle generator has the following properties:
1. Independent adjustment of positive and negative slopes
2. Peak-to-peak amplitude adjustment which has no effect on positive and negative slopes
3. A dc offset adjustment which can place the triangle waveform at any location within the limits of the two power-supply voltages

The circuit shown in Fig. 27.3 closely approaches this ideal. Resistors $R_2$ and $R_3$ provide independent positive and negative slope adjustment. Resistor $R_4$ determines the peak-to-peak limits of the triangle, and $R_5$ is the offset adjustment.

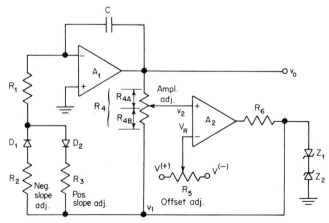

**Fig. 27.3** A voltage-triangle generator which utilizes an integrator and a comparator.

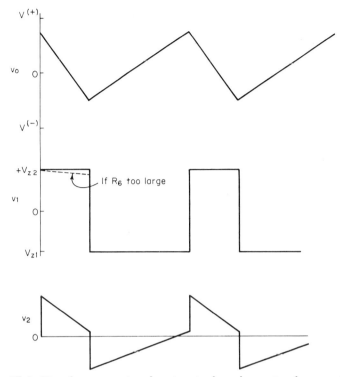

**Fig. 27.4** Waveforms at various locations in the voltage-triangle generator.

Circuit operation is briefly explained as follows: Op amp $A_1$ is an integrator. Its output slews either positive or negative depending on the instantaneous polarity of $v_1$. Figure 27.4 shows the triangle which appears at the output terminal of $A_1$. Voltage $v_1$ is a rectangular waveform with $\pm$ limits established by the zener diodes. When $v_1$ is $V_{Z2}$ (a positive voltage), $v_o$ integrates downward. Likewise, when $v_1$ is negative, $v_o$ integrates upward. The $v_1$ waveform is created by the $A_2$ comparator. This circuit changes states whenever $v_2$ passes through $V_R$. As shown in Fig. 27.4, $v_2$ is a composite waveform made from $v_o$ and $v_1$. If $R_4$ is adjusted upward from center such that $v_o$ dominates the shape of $v_2$, $A_2$ trips sooner than nominal and the peak-to-peak voltage at $v_o$ will decrease. Conversely, if $R_4$ is adjusted such that $v_1$ dominates the shape of $v_2$, $A_1$ has to integrate further before $A_2$ trips. This increases the peak-to-peak amplitude of $v_o$. Potentiometer $R_5$ also controls the trip point of $A_2$. However, the shape of $v_2$ is not affected, since $R_5$ merely controls the dc level of $v_o$ without influencing its shape.

### DESIGN PARAMETERS

| Parameter | Description |
|---|---|
| $A_1$ | Op amp used in integrator stage |
| $A_2$ | Op amp (or comparator) used to control triangle offset and peak-to-peak amplitude |
| $D_1, D_2$ | Diodes which allow separate control of positive and negative triangle slopes |

## 27-6 WAVEFORM GENERATORS

| Parameter | Description |
|---|---|
| $f_{min}$ | Minimum triangle generator output frequency |
| $I_{b2}$ | Input bias current of $A_2$ |
| $I_{Z1}, I_{Z2}$ | Nominal bias current for $Z_1$ and $Z_2$ |
| $R_1$ | Resistor which controls maximum $\pm$ slopes of triangle |
| $R_2, R_3$ | Resistors (variable, if required) which control minimum $\pm$ slopes of triangle |
| $R_4$ | Potentiometer which determines peak-to-peak amplitude of triangle |
| $R_5$ | Potentiometer which controls the triangle dc offset |
| $R_6$ | Resistor used to establish a nominal bias current in $Z_1$ and $Z_2$ |
| $v_1$ | Rectangular waveform from $A_2$ |
| $v_2$ | Composite waveform made from $v_o$ and $v_1$ |
| $v_o$ | Output triangle waveform |
| $v_o$(peak) | The $\pm$ peak value of $v_o$ relative to its average (dc) value (i.e., $\frac{1}{2}$ of its peak-to-peak value) |
| $V_R$ | Reference voltage set by $R_5$ which controls triangle offset |
| $V_{sat}$ | The voltage difference between $V^{(\pm)}$ and the respective $\pm$ maximum output voltages from $A_1$ |
| $V_{Z1}$ | The zener breakdown voltage of $Z_1$ plus the forward breakdown voltage of $Z_2$ |
| $V_{Z2}$ | The zener breakdown voltage of $Z_2$ plus the forward breakdown voltage of $Z_1$ |
| $V^{(\pm)}$ | Power-supply voltages |
| $Z_1, Z_2$ | Zener diodes used to control amplitude of $v_1$ |

## DESIGN EQUATIONS

| Eq. No. | Description | Equation |
|---|---|---|
| 1 | Positive slope of output ramp | $\dfrac{\Delta v_o}{\Delta t} = \dfrac{V_{Z1}}{(R_1 + R_3)C}$ |
| 2 | Negative slope of output ramp | $\dfrac{\Delta v_o}{\Delta t} = \dfrac{-V_{Z2}}{(R_1 + R_2)C}$ |
| 3 | Maximum peak-to-peak amplitude of ramp when $R_{4B} = 0$ | $v_o(\text{max, peak-to-peak}) = V^{(+)} - V^{(-)} - 2V_{sat}$ |
| 4 | Maximum positive offset | $v_o(+ \text{ off}) \leq V^{(+)} - v_o(\text{peak})$ |
| 5 | Maximum negative offset | $v_o(- \text{ off}) \geq V^{(-)} + v_o(\text{peak})$ |
| | Resistor values: | |
| 6 | $R_1$ | $R_1 \approx \dfrac{V_Z - V_D}{C|\Delta v_o/\Delta t|_{max}}$ |
| 7 | $R_2$ | $R_2 = \dfrac{V_{Z2} - V_D}{C|\Delta v_o/\Delta t|_{min}} - R_1$ where $\Delta v_o/\Delta t$ applies to negative slopes only |
| 8 | $R_3$ | $R_3 = \dfrac{|V_{Z1}| - V_D}{C(\Delta v_o/\Delta t)_{min}} - R_1$ where $\Delta v_o/\Delta t$ applies to positive slopes only |
| 9 | $R_4, R_5$ | $10R_6 \leq R_4 = R_5 \leq \frac{1}{100} I_{b2}$ |

| Eq. No. | Description | Equation |
|---|---|---|
| 10 | $R_6$ | $R_6 \approx \dfrac{V^{(+)} - V_{Z2}}{I_{Z2}} \approx \dfrac{|V^{(-)}| - V_{Z1}}{I_{Z1}}$ |
| 11 | Integrating capacitor | $C \approx \dfrac{10^{-7}}{f_{min}}$ |
| 12 | Zener-diode voltages | $2\text{ V} \leq |V_{Z1}| \leq |V_{Z2}| \leq |V^{(\pm)}| - 2$ |

## DESIGN PROCEDURE

A wide range of component-value combinations will provide satisfactory circuit operation. The designer, therefore, has more latitude for component values than the design equations suggest. The approach used here is to choose the integrator capacitor according to a rule of thumb (Eq. 11). The other equations then provide first-cut values for resistor values. As in all circuit designs, some adjustment of component values is always required after the first breadboard is tested.

*Step 1.* Calculate a value for $C$ using Eq. 11. In most cases the actual value for $C$ can be anything within an order of magnitude on either side of the calculated value.

*Step 2.* Choose two identical zener diodes ($Z_1$ and $Z_2$) at the approximate center of the range suggested in Eq. 12.

*Step 3.* Calculate a value for $R_1$ using Eq. 6.

*Step 4.* Compute values for $R_2$ and $R_3$ with Eqs. 7 and 8.

*Step 5.* Find an approximate value for $R_6$ using Eq. 10.

*Step 6.* Resistors $R_4$ and $R_5$ are chosen to lie somewhere within the range suggested in Eq. 9.

**EXAMPLE OF A TRIANGLE GENERATOR DESIGN** We will design a subaudio triangle generator using the six design steps.

*Design Requirements*

$f_{min} = 10^{-2}$ Hz

$\left| \pm \dfrac{\Delta v_o}{\Delta t} \right|_{max} = 10$ V/s

$\left| \pm \dfrac{\Delta v_o}{\Delta t} \right|_{min} = 0.01$ V/s

$V^{(\pm)} = \pm 15$ V

*Device Data*

$I_{b2} = 3 \times 10^{-8}$ A
$V_D = 0.7$ V

*Step 1.* A nominal value for $C$ is

$$C \approx \dfrac{10^{-7}}{f_{min}} = \dfrac{10^{-7}}{10^{-2}} = 10 \ \mu\text{F}$$

*Step 2.* Equation 12 suggests we use zener diodes with breakdown voltages between 2 V and $15 - 2 = 13$ V. We will use 1N707 diodes, which have a breakdown voltage of 8 V at 5 mA. The actual breakdown voltages to be used in the following calculations must have 0.7 V added to the 8 V, since one zener is forward-biased while the other is in the zener breakdown region. We conclude that $|V_{Z1}| = V_{Z2} = 8.7$ V.

*Step 3.* Equation 6 provides us with

$$R_1 \approx \frac{V_Z - V_D}{C|\Delta v_o/\Delta t|_{max}}$$

$$\approx \frac{8.7 - 0.7}{10^{-5}\,(10)} = 80 \text{ k}\Omega$$

*Step 4.* The slope-adjustment resistors are found from Eqs. 7 and 8 (these equations are identical since $V_{Z2} = |V_{Z1}|$):

$$R_2 = R_3 = \frac{V_{Z2} - V_D}{C|\Delta v_o/\Delta t|_{min}} - R_1$$

$$= \frac{8.7 - 0.7}{10^{-5}\,(0.01)} - 80{,}000 = 80 \text{ M}\Omega$$

With such a large input resistor, we must use an FET input op amp for $A_1$. Otherwise the input resistance and bias current of $A_1$ would affect the triangle accuracy at low sweep rates.

*Step 5.* Equation 10 is calculated to provide us with $R_6$:

$$R_6 = \frac{V^{(+)} - V_{Z2}}{I_{Z2}} = \frac{15 - 8.7}{5 \times 10^{-3}} = 1{,}260 \text{ }\Omega$$

Any resistor between 1,000 and 1,500 $\Omega$ should be satisfactory.

*Step 6.* Resistors $R_4$ and $R_5$ must first satisfy

$$R_4 = R_5 \geq 10\,R_6 = 10(1{,}260) = 12{,}600 \text{ }\Omega$$

They must also satisfy

$$R_4 = R_5 \leq \frac{1}{100\,I_{b2}} = \frac{1}{100(3 \times 10^{-8})} = 0.33 \text{ M}\Omega$$

We will let $R_4 = R_5 = 100$ k$\Omega$.

## REFERENCE

1. Larsen, D. G.: Triangle Wave Generator Keeps Slopes Constant as Amplitude Changes, *Electron. Des.*, Sept. 28, 1972, p. 80.

## 27.3 VOLTAGE STAIRCASE GENERATOR

**ALTERNATE NAMES**  Step generator, staircase generator.

**EXPLANATION OF OPERATION**  Each negative transition of $v_1$ causes the timer IC to produce a positive output pulse for a duration $T_2$. This pulse turns on the CMOS switch $S_1$, causing the $A_1$ circuit to integrate upward for the duration of the timer pulse. The voltage added to $v_o$ during $T_2$ is

$$\Delta v_o = \frac{V_R\,T_2}{[R_{DS}(\text{on}) + R_3]\,C_3}$$

Each $\Delta v_o$ step of the staircase is held by the integrator until the next negative transition of $v_1$. Any droop in the step during this hold time $T_1$ is caused by leakage current through $S_2$ or $C_3$.

The staircase generator is reset when the binary counter is full and $G_1$ activates $S_2$. The size of the AND gate must be compatible with the counter size.

This circuit offers several advantages over the conventional D/A approach for staircase generation:

1. All step amplitudes and risetimes are guaranteed to be identical, since they are formed by the same components.

2. The circuit is glitch-free, since the circuit is in the integrate mode during the rising portion of each step. (A glitch is a noise transient caused by timing errors in the counter of a conventional D/A converter.)

3. Step amplitudes are easily adjusted with only one part ($R_3$).

The integrate duration $T_2$ is (assuming timer is the 555 type)

$$T_2 = 1.1\ R_1 C_2$$

Substituting this into the equation for $\Delta v_o$, we get

$$\Delta v_o = \frac{1.1\ R_1 C_2 V_R}{[R_{DS}(\text{on}) + R_3]\ C_3}$$

If $C_2$ and $C_3$ are identical capacitor types with reasonable tracking over temperature, $\Delta v_o$ will be quite stable. Also, if $R_{DS}(\text{on}) \ll R_3$, errors caused by switch-resistance variations will be minimized.

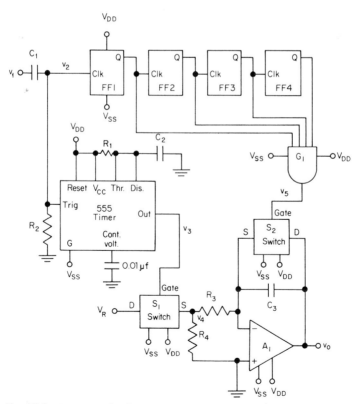

**Fig. 27.5** A triggered-voltage staircase generator with automatic reset.

## 27-10 WAVEFORM GENERATORS

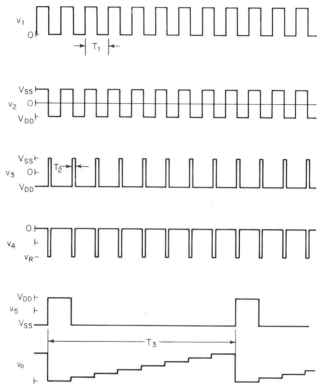

**Fig. 27.6** Waveforms in the staircase generator circuit.

## DESIGN PARAMETERS

| Parameter | Description |
|---|---|
| $A_1$ | Op amp used as integrate-and-hold circuit |
| $C_1$ | Used to make $v_2(\text{av}) = 0$ |
| $C_2$ | Determines time $T_2$ |
| $C_3$ | Determines step size |
| $FF_1$ to $FF_N$ | N-stage counter |
| $G_1$ | N-input AND gate |
| $R_1$ | Determines time $T_2$ |
| $R_2$ | Used to make $v_2(\text{av}) = 0$ |
| $R_3$ | Determines step size |
| $R_4$ | Clamps $v_4$ to ground when $S_1$ is off |
| $R_{DS}(\text{on})$ | On resistance of $S_1$ and/or $S_2$ |
| $S_1$ | Sampling switch which causes integrator to integrate only during rising portion of each step |
| $S_2$ | Switch which clears integrator at end of staircase |
| $T_1$ | Period of $v_1$ waveform |
| $T_2$ | Pulse width of timer output |
| $T_3$ | Duration of staircase |
| $v_1$ to $v_5$ | Voltages as shown in Figs. 27.5 and 27.6 |
| $\Delta v_o$ | Step size |
| $V_{DD}$ | Drain supply voltage for CMOS devices |
| $V_R$ | Reference voltage which helps determine step amplitude |
| $V_{SS}$ | Source supply voltage for CMOS devices |

## DESIGN EQUATIONS

| Eq. No. | Description | Equation |
|---|---|---|
| 1 | Magnitude of each step voltage | $\Delta v_o = \dfrac{V_R T_2}{[R_{DS}(\text{on}) + R_3]C_3}$ |
| 2 | Integrate duration $T_2$ (assuming 555-type timer is utilized) | $T_2 = 1.1\, R_1 C_2$ |
| 3 | Input capacitor $C_1$ size | $C_1 \gg \dfrac{1}{2\pi f_{in} R_2}$ |
| 4 | Timer capacitor $C_2$ | $C_2$ is chosen according to timer design specification |
| 5 | Recommended integrator capacitor size $C_3$ | $C_3 \ll \dfrac{T_1}{R_{DS}(\text{on})}$ |
|  | Resistor values: |  |
| 6 | $R_1$ | $R_1$ is chosen according to timer design specification (see Eq. 2) |
| 7 | $R_2$ | $R_2 = 2$ to $10\ \text{k}\Omega$ |
| 8 | $R_3$ | $R_3 = \dfrac{V_R T_2}{\Delta v_o\, C_3} - R_{DS}(\text{on})$ <br> NOTE: Make $R_3 \gg R_{DS}(\text{on})$ |
| 9 | $R_4$ | $R_4 \gg R_{DS}(\text{on})$ |

## REFERENCE

1. Strange, M.: Staircase Generator Resists Output Drift, *Electronics*, Dec. 4, 1972, p. 90.

Appendix I

# Operational Amplifier Parameters

The following tabulation contains the common parameters found on operational amplifier data sheets. The list is not complete, since some parameters found in particular data sheets are not widely used or their application is for only a very specialized type of op amp. The minimum, typical, and maximum parameter values are for op amps at 25°C and in existence in 1976. Most of these parameters are discussed in Chaps. 1 and 2.

| Parameter | Description | Units | Min | Typical | Max |
|---|---|---|---|---|---|
| $A_{cm}$ | Common-mode voltage gain as a function of frequency | dB | | | |
| $A_{cmo}$ | Common-mode voltage gain at dc | dB | −40 | 0 | 60 |
| $A_v$ | Differential voltage gain as a function of frequency | dB | | | |
| $A_{vo}$ | Differential voltage gain at dc | dB | 53 | 90 | 170 |
| $C_{ic}$ | Common-mode input capacitance | pF | 1 | 2 | 3 |
| $C_{id}$ | Differential input capacitance | pF | 1 | 2 | 30 |
| CMRR | Common-mode rejection ratio | dB | −160 | −90 | −64 |
| $f_f$ | Full-power bandwidth (−3 dB) | kHz | 0.05 | 50 | 50,000 |
| $f_{op1}$ | First pole of op amp | Hz | 1 | 10 | 4,000 |
| $f_{op2}$ | Second pole of op amp | MHz | 0.5 | 1 | 100 |
| $f_u$ | Frequency of unity voltage gain | MHz | 0.25 | 1 | 200 |
| $I_b$ | Input bias current | nA | $10^{-4}$ | 100 | 45,000 |
| $I_{io}$ | Input offset current | nA | $10^{-4}$ | 20 | 6,000 |
| $\Delta I_{io}/\Delta T$ | Temperature coefficient of input offset current | pA/°C | $10^{-3}$ | 100 | 3,000 |

## I-2 OPERATIONAL AMPLIFIER PARAMETERS

| Parameter | Description | Units | Min | Typical | Max |
|---|---|---|---|---|---|
| $I_n$ | Equivalent rms input noise current at 10 Hz | pA/$\sqrt{\text{Hz}}$ | 0.5 | 1 | 1.8 |
| $I_{snk}$ | Maximum sink output current | mA | 0.5 | 1 | $10^3$ |
| $I_{src}$ | Maximum source output current | mA | 5 | 10 | $10^3$ |
| $I^{(\pm)}$ | Nominal power-supply current | mA | 0.03 | 3 | 35 |
| $P_D$ | Power consumption if $V^{(\pm)} = \pm 15$ V and $R_L = \infty$ | mW | 0.5 | 30 | $10^3$ |
| $P_o$ | Percent overshoot for large-signal voltage follower | % | 5 | 10 | 50 |
| PSRR | Power-supply rejection ratio | dB | $-100$ | $-90$ | $-55$ |
| $\phi_m$ | Phase margin | Degrees | 5 | 10 | 50 |
| $R_{ic}$ | Common-mode input resistance | M$\Omega$ | 1 | 10 | $10^8$ |
| $R_{id}$ | Differential input resistance | M$\Omega$ | 0.09 | 1 | $10^7$ |
| $R_o$ | Output resistance at dc | $\Omega$ | 4 | 50 | $10^3$ |
| $S$ | Slew rate | V/$\mu$s | 0.005 | 0.5 | 6,000 |
| $t_r$ | Risetime (10 to 90%) | $\mu$s | 0.002 | 0.35 | 1.4 |
| $t_s$ | Voltage-follower large-signal settling time (to $<0.1\%$) | $\mu$s | 0.05 | 1.2 | 2 |
| $V_{io}$ | Input offset voltage | mV | 0.075 | 3 | 50 |
| $\Delta V_{io}/\Delta T$ | Temperature coefficient of input offset voltage | $\mu$V/°C | 0.6 | 5 | 50 |
| $V_n$ | Equivalent rms input noise voltage at 10 Hz | nV/$\sqrt{\text{Hz}}$ | 8 | 10 | 35 |
| $V_o^{(+)}$ | Maximum positive output swing | V | $V^{(+)} - 4$ | $V^{(+)} - 2$ | $V^{(+)}$ |
| $V_o^{(-)}$ | Maximum negative output swing | V | $V^{(-)}$ | $V^{(-)} + 2$ | $V^{(-)} + 3$ |
| $V_{\min}^{(\pm)}$ | Minimum supply voltages | V | $\pm 1$ | $\pm 3$ | $\pm 12$ |

Appendix II

# Operational Amplifier Maximum Ratings

If the following parameters are exceeded, the operational amplifier may be damaged. Not all these parameters will be found on every op amp data sheet. Only those the manufacturer feels are important for his particular type of op amp are usually given. All ratings assume an ambient temperature of 25°C.

| Maximum rating | Description | Units | Min | Typical | Max |
|---|---|---|---|---|---|
| $I_{in}$ | Maximum (±) input terminal current | mA | 1 | 5 | 10 |
| $I_o$ | Maximum output current (source or sink) | A | 0.005 | 0.01 | 1 |
| $P_D(\max)$ | Maximum internal dissipation | W | 0.2 | 0.5 | 2 |
| $T_s$ | Maximum output short-circuit duration | s | 0 | $\infty$ | $\infty$ |
| $V_{ic+}$ | Maximum positive common-mode input voltage | V | $V^{(+)} - 3$ | $V^{(+)}$ | $V^{(+)} + 10$ |
| $V_{ic-}$ | Maximum negative common-mode input voltage | V | $V^{(-)} - 10$ | $V^{(-)}$ | $V^{(-)} + 1$ |
| $V_{id}$ | Maximum differential input voltage | V | ±1 | ±30 | ±300 |
| $V^{(\pm)}$ | Maximum power-supply voltages | V | ±8 | ±22 | ±45 |

II-1

# Appendix III
# Circuit Fabrication Techniques

## INTRODUCTION

Although op amps are being made more foolproof each year, the designer must still be familiar with protection circuitry and noise-reduction methods. High-performance devices usually need special consideration in these two areas. Also, the passive devices in each circuit must not be forgotten. The performance of most feedback circuits utilizing high-gain op amps depends almost entirely on the quality of the passive components.

## III.1 PROTECTION CIRCUITS

Methods to protect each type of op amp terminal will be considered.

**INPUT TERMINALS** The $\pm$ input terminals of op amps must be protected against excessive currents and voltages. In most cases this protection is implemented by not allowing the input voltages to exceed the manufacturer's maximum ratings. If input voltages are within prescribed limits, maximum input currents will not be exceeded. In those cases where the range of input voltages cannot be guaranteed, one or two of the protective circuits shown in Fig. III.1 may be required. Diodes $D_3$ and $D_4$ in the figure will not allow the op amp differential input voltage to exceed $\pm 0.7$ V. Some super beta op amps (such as the 108) also have internal diodes across the inputs. Since these diodes are on the op amp chip, their current must be limited by the external circuit to prevent excessive chip dissipation. Diodes $D_3$ and $D_4$, in conjunction with $R_3$ and $R_4$, will keep most of the excess current away from the op amp in an overvoltage condition. But since most op amp circuits contain resistors in the op amp input circuits ($R_1$, $R_2$, and $R_5$), these can be used to further limit input currents.

Transient overloads on either the + or − inputs relative to ground can be handled in several ways. Many op amps specify that the common-mode input voltages cannot exceed the supply voltages. Diodes $D_1$, $D_2$, $D_5$, and $D_6$ will constrain op amp input voltages to stay within 0.7 V of $V^{(\pm)}$. These diodes could also be returned to voltages other than $V^{(\pm)}$ if required.

Zener diodes $Z_1$ to $Z_4$ can also be utilized to protect each input from $\pm$ over-

III-1

## III-2 CIRCUIT FABRICATION TECHNIQUES

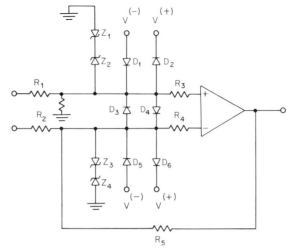

**Fig. III.1** Several ways to limit op amp differential and common-mode input voltages.

voltages. However, zener diodes are more expensive than general-purpose diodes. Either approach provides the same degree of protection.

**OUTPUT TERMINAL** At 25°C most op amps can tolerate a short circuit from the output terminal to ground for an indefinite period. At elevated temperatures the designer should consider the power-dissipation derating curve. For example, suppose we are using an op amp with the following maximum ratings and characteristics:

Maximum internal dissipation, 400 mW
Derating above 25°C ambient, 6 mW/°C
Output current limit, 20 mA
Power-supply voltages, ±15 V

If the output is shorted to ground, the current limiter inside the op amp can inject 20 mA into the short. The op amp dissipation will be 20 mA × 15 V =

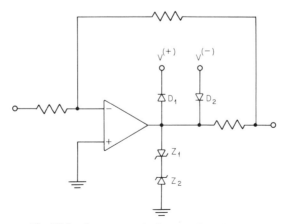

**Fig. III.2** Output terminal protection circuits.

300 mW. This is only 100 mW below the maximum. We divide 100 mW by the derating factor (6 mW/°C) to determine the maximum ambient temperature where this shorted condition is safe: 100 mW/6 mW/°C = 16.7°C. Therefore this op amp should not be operated in ambient temperatures above 25°C + 16.7°C = 41.6°C if the output is shorted to ground.

Many op amps can also allow an output short circuit to either power-supply terminal with no damage. If the output terminal has any possibility of accidental connection to voltages larger than $V^{(\pm)}$, the ideas recommended in Fig. III.2 may be used. Diodes $D_1$ and $D_2$ will clamp the output to a maximum of $V^{(\pm)} \pm 0.7$ V. Zener diodes $Z_1$ and $Z_2$ could be used to keep the output voltage within any arbitrary output range.

**POWER-SUPPLY TERMINALS** Many op amps are destroyed by power-supply transients or accidental reversal of power-supply voltages. Several circuits which protect the op amp from such problems are shown in Fig. III.3 (Ref. 1). Protection against supply reversal is easily implemented with two diodes as shown in Fig. III.3A. Power-supply transient protection can be provided with the circuit of Fig. III.3B. The zener-diode breakdown voltages are chosen to be greater than $V^{(\pm)}$ but less than $V^{(\pm)}$ (max). Thus $Z_1$ and $Z_2$ will dissipate power only in a transient situation. The FET devices $Q_1$ and $Q_2$ are chosen with an $I_{DSS}$ larger than the expected op amp supply current but less than the recommended zener currents for $Z_1$ and $Z_2$. Capacitors $C_1$ and $C_2$ bypass all high frequencies to ground. Some op amps will oscillate if the capacitors are not included.

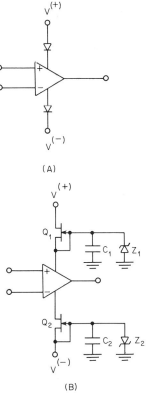

**Fig. III.3** Protection of the op amp power-supply terminals. (A) Protection against supply reversal. (B) Protection against supply transients.

## III.2 NOISE PREVENTION TECHNIQUES

Every designer is familiar with the saying "It worked beautifully as a breadboard but everything went wrong on the prototype." Many of these prototype failures are due to (1) grounding/bypassing problems, (2) shielding/guarding problems, (3) unexpected interface problems, and (4) unwanted coupling problems. Sometimes these potential problems are overlooked during the breadboard phase of a project because "by accident" they were not present. A good designer develops the prototype configuration in the back of his mind all during the breadboard phase. He experiments with the relative positions of parts and wiring in low-level and high-gain circuits to prevent unwanted parasitics from destroying circuit performance. In this section we will consider several ways to anticipate and prevent such problems.

## III-4 CIRCUIT FABRICATION TECHNIQUES

**GROUNDING/BYPASSING PROBLEMS** Good systems are designed with a single ground point which is characterized as having the smallest quantity of noise, ripple, and transients in the system (Ref. 3). By definition both positive and negative power buses have tight ac coupling to this single ground point for a wide range of frequencies. Ideally each part in all circuits which must return to ground is brought over to this single ground point. In practical systems containing tens or hundreds of active devices, this ideal is not easily achieved. Instead, each op amp, or small group of op amps, has a local single ground point. The local ± power buses have good-quality bypass capacitors which return to the local single ground point. Each local single ground point services op amps which will not interfere with each other. In other words, a 10-W op amp should not be operated from the same power-supply potentials and ground point as a low-level-instrumentation op amp.

Circuits which pull large currents from the power supply should have independent ± buses and ground wires which return all the way back to the main power-supply terminals and the system single ground point. Low-level circuits can be "daisy-chained" as shown in Fig. III.4.

Reference 2 provides us with a set of five rules which will circumvent most grounding problems. These rules assume the system is made up of amplifiers, but system performance will also be improved for other types of circuits.

1. The signal ground for all amplifiers should be a flat plane such as a large copper area of a printed circuit board. The signal ground of amplifiers

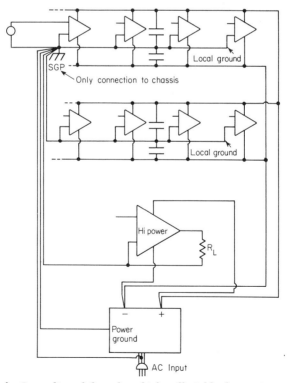

**Fig. III.4** Grounding philosophy which will yield a low-noise system.

on other boards should be connected to the low-level amplifier signal ground with grounding straps or large conductors.

2. Connect all system chassis grounds together with heavy wire or braid.

3. Connect the signal ground of the lowest level amplifier in the system to chassis ground. Make this connection as close as possible to the actual op amp input signal ground.

4. Connect the ground return of the source voltage (external transducer, etc.) for the lowest level amplifier to the same chassis ground mentioned in item 3.

5. Power ground and ± power leads may be "daisy-chained" between amplifiers. Make only *one* connection between power ground and signal ground. This connection should be as close as possible to the cluster of grounds mentioned in items 3 and 4 above (see Fig. III.4).

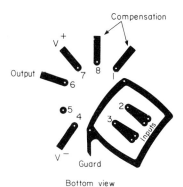

**Fig. III.5** Typical guard ring for a TO-99 op amp where both ± inputs must be protected. (*National Semiconductor Corp.*)

**SHIELDING/GUARDING** The performance of op amp circuits with low-input current requirements can be improved through the use of guarding techniques. Guarding is the preferred method to control leakage currents into the op amp ± input terminals. For example, in most TO-99 op amps, pin 3 is the noninverting input and pin 4 is $V^{(-)}$. If copper traces going to these pins are separated by 0.05 in for a distance of 1 in, then a leakage resistance of $10^{11}$ Ω is possible at a temperature of 125°C (Ref. 4). This assumes a high-quality epoxy glass board which has been properly cleaned. If the circuit is a high-input resistance voltage follower, the error caused by $V^{(-)}$ leaking through $10^{11}$ Ω may be significant. Suppose the voltage follower input is at +10 V and $V^{(-)}$ is −15 V. The leakage current will be

$$I_L = \frac{10 - (-15) \text{ V}}{10^{11} \text{ Ω}} = 250 \text{ pA}$$

This is more leakage than the worst-case offset current of the 208 op amp.

Since most leakage currents on printed circuit boards occur along surfaces, this is where guarding procedures must be implemented. As shown in Fig. III.5, a guard ring is placed around the sensitive terminal (or terminals). This must be done on both sides of the board since the ± inputs usually touch both sides. The guard rings are

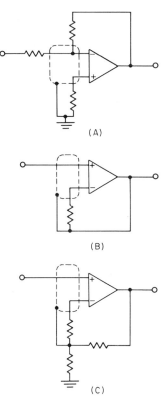

**Fig. III.6** Guard-ring connections for: (*A*) an inverting amplifier, (*B*) a voltage follower, and (*C*) a noninverting amplifier. (*National Semiconductor Corp.*)

electrically connected to a low-impedance source having the same potential as the ± inputs. Thus, for an inverting amplifier, as shown in Fig. III.6A, attach the guard rings to ground. For a voltage follower the guard rings should be connected to the op amp output. Figure III.6C shows how guard rings are biased in a noninverting amplifier.

High-impedance circuits also require shielding to prevent pickup of stray ac electric and magnetic fields. This must be done in a way which will not compromise the dc guarding circuit. In most cases it is best to completely enclose a low-level amplifier (or the input circuitry) with a grounded shield (Ref. 6). If low-frequency magnetic fields (under 100 Hz) are present, a high-permeability magnetic shield should be used. Aluminum or copper shielding may be utilized for electromagnetic interference above 100 Hz.

## III.3 PASSIVE DEVICES IN OP AMP CIRCUITS

The ultimate accuracy and stability of many op amp circuits have a one-to-one correspondence with the accuracy and stability of the passive components in that circuit. A good circuit designer should therefore be familiar with the range of characteristics available for each type of passive device. This section will briefly summarize the range of major parameters for several passive devices.

**Typical Resistor Parameters (25°C)**

| Type | Resistance range available, $\Omega$ | | Purchase tolerance, % | | Temperature coefficient, %/°C | | Drift in 1,000 h, % | |
|---|---|---|---|---|---|---|---|---|
| | Min | Max | Min | Max | Min | Max | Min | Max |
| Carbon composition | 0.1 | $10^{11}$ | 2 | 30 | ±0.01 | ±0.25 | 3 | 10 |
| Metal film | 1 | $2 \times 10^8$ | 0.01 | 10 | ±0.0002 | ±0.01 | $10^{-3}$ | 0.5 |
| Wire-wound | 0.1 | $2 \times 10^5$ | 0.01 | 10 | 0.002 | 0.01 | | |
| Metal glaze | 1 | $2 \times 10^6$ | 1 | 5 | 0.005 | 0.025 | 0.6 | 1 |
| Cermet film | 10 | $10^7$ | 1 | 5 | 0.01 | 0.05 | 0.1 | 0.2 |
| Carbon film | 10 | $10^9$ | 0.25 | 10 | −0.02 | −0.2 | −0.02 | −0.3 |

**Typical Capacitor Parameters (25°C)**

| Type | Capacitance range available, pF $\mu$F | | Purchase tolerance, % | | Temperature coefficient, %/°C | | Drift in 1,000 h, % | |
|---|---|---|---|---|---|---|---|---|
| | Min | Max | Min | Max | Min | Max | Min | Max |
| Ceramic | 5 | 3 | 5 | +80 / −30 | ±0.003 | ±2 | 3 | 20 |
| Mylar | 1,000 | 2 | 1 | 20 | 0.02 | 0.2 | 0.1 | 0.5 |
| Polystyrene | 10 | 3 | 0.05 | 20 | −0.01 | −0.02 | 0.01 | 0.1 |
| Polycarbonate | 1,000 | 20 | 1 | 20 | ±0.005 | ±0.03 | 0.25 | 5 |
| Glass | 0.5 | 0.2 | 1 | 20 | 0.01 | 0.6 | 0.5 | 10 |
| Porcelain | 1 | 0.01 | 1 | 10 | 0.003 | 0.01 | 0.5 | 1 |

# REFERENCES

1. Graeme, J.: Protect Op Amps from Overloads, *Electron. Des.*, vol. 10, p. 96, May 10, 1973.
2. Sheingold, D. H., and EID staff: Grounding Low Level Instrumentation Systems, *Electron. Instr. Dig.*, Jan.–Feb. 1966, p. 16.
3. Thornwall, J. C.: Design Noise Out of Spacecraft Loads, *Electron. Des.*, vol. 8, p. 64, Apr. 11, 1968.
4. National Semiconductor Corp.: Applications Note AN-29, Dec. 1969.
5. Morrison, R.: Protecting Signal Circuits by Grounding and Shielding, *Instr. Technol.*, Oct. 1973, p. 33.
6. Philbrick, G. A.: Researches, Inc., Applications Manual – Computing Amplifiers, 1966, p. 26.

# Appendix IV
# Notation Used in Handbook

| | |
|---|---|
| $A_1$ | Typical op amp symbol |
| $A_{cm}$ | Common mode voltage gain of op amp |
| $A_{cc}$ | Common mode voltage gain of circuit due to resistor unbalance |
| $A_{cmo}$ | Common mode voltage gain of op amp at dc |
| $A_{co}$ | Common mode gain of circuit due to op amp CMRR $\neq \infty$ |
| $A_d$ | Differential voltage gain of circuit |
| ADC | Analog-to-digital converter |
| $A_f$ | Voltage transfer function of feedback network ($\beta$) |
| $A_{gc}$ | Transconductance of voltage-to-current converter |
| $A_{ic}$ | Current gain of current amplifier circuit |
| $A_m$ | Gain margin |
| $A_\pi$ | Gain of op amp when its phase lag $= -\pi$ |
| $A_{rc}$ | Transresistance of current-to-voltage converter |
| $A_v$ | Open-loop voltage gain of op amp |
| $A_{vc}$ | Voltage gain of voltage amplifier circuit |
| $A_{vco}$ | Voltage gain of voltage amplifier circuit at dc |
| $A_{vn}$ | Voltage gain of a passive network |
| $A_{vo}$ | Open-loop voltage gain of op amp at dc |
| $\alpha$ | Common-base current gain of transistor |
| $\beta$ | Feedback circuit voltage gain ($A_f$) |
| $\beta$ | Transistor current gain |
| $C_1$ | Typical capacitor symbol |
| $C_c$ | Compensation capacitor |
| $C_f$ | Feedback capacitor |
| $C_{GD}$ | FET gate-to-drain capacitance |
| $C_{ic}$ | Op amp common-mode input capacitance |
| $C_{id}$ | Op amp differential input capacitance |
| $C_L$ | Load capacitor |
| CMR | Op amp common-mode rejection |
| CMRR | Op amp common-mode rejection ratio |
| $CMRR_c$ | Circuit common-mode rejection ratio |
| DAC | Digital-to-analog converter |
| dB | Decibel |
| $D_1$ | Typical diode symbol |
| $f_o, f_n$ | Resonant or natural frequency |
| $f_\beta$ | Small signal bandwidth of transistor $\beta$ |
| $f_{\beta p1}$ | First pole frequency of feedback network |
| $f_{\beta z1}$ | First zero frequency of feedback network |
| $f_c$ | Carrier or clock frequency |

## IV-2  NOTATION USED IN HANDBOOK

| | |
|---|---|
| $f_{cm}$ | Frequency at which CMRR becomes too low to satisfy circuit requirements |
| $f_{cp1}$ | First pole frequency of closed-loop circuit |
| $f_{cz1}$ | First zero frequency of closed-loop circuit |
| $f_f$ | Maximum frequency of full output voltage swing |
| $f_g$ | Frequency at which loop gain is unity |
| $f_{lp1}$ | First pole of loop gain |
| $f_{lz1}$ | First zero of loop gain |
| $f_m$ | Modulation frequency |
| $f_{max}$ | Maximum frequency for a given circuit performance |
| $f_{op1}$ | First pole frequency of op amp |
| $f_p$ | Frequency of a pole |
| $f_\pi$ | Frequency at which loop phase lag is $-180°$ $(-\pi°)$ |
| $f_u$ | Unity gain crossover frequency of op amp |
| $f_z$ | Frequency of a zero |
| $h_{fe}$ | Transistor common-emitter current gain ($\beta$) |
| $h_{ib}$ | Transistor common-base input resistance |
| $h_{ie}$ | Transistor common-emitter input resistance |
| $h_{oe}$ | Transistor common-emitter output conductance |
| $h_{re}$ | Transistor common-emitter reverse voltage gain |
| $I_1$ | Typical symbol for complex or dc current |
| $i_1$ | Typical symbol for instantaneous current |
| $I_b$ | Op amp input bias current |
| $I_c$ | Capacitor leakage current |
| $I_{CBO}$ | Collector-to-base leakage current |
| $I_{DSS}$ | FET drain-to-source saturation current |
| $I_i, i_i$ | Input current |
| $I_{io}$ | Op amp input offset current |
| $I_n$ | Equivalent rms input noise current |
| $I_o, i_o$ | Output current of circuit |
| $I_z$ | Zener current |
| $k$ | Boltzmann's constant ($1.38 \times 10^{-23}$ joule/K) |
| $L_1$ | Typical inductance symbol |
| LSB | Least significant bit |
| MSB | Most significant bit |
| MUX | Multiplexer |
| NF | Noise figure |
| $\phi$ | Phase angle |
| $\phi_m$ | Phase margin |
| PSR | Power supply rejection |
| PSRR | Power supply rejection ratio |
| PWM | Pulse width modulator |
| $q$ | Electronic charge ($1.6 \times 10^{-19}$ C) |
| $Q$ | Quality factor at resonance |
| $Q_1$ | Typical transistor symbol |
| $R_1$ | Typical resistor symbol |
| $R_c$ | Compensation resistor |
| $R_{DS}$ | FET drain-to-source resistance |
| $R_f$ | Feedback resistor |
| $R_g$ | Generator resistance |
| $R_{ic}$ | Op amp common-mode input resistance |
| $R_{id}$ | Op amp differential input resistance |
| $R_{in}$ | Input resistance of circuit |
| $R_{inc}$ | Common-mode input resistance of circuit |
| $R_{ind}$ | Differential input resistance of circuit |
| $R_L$ | Load resistance |
| $R_o$ | Op amp output resistance |
| $R_{out}$ | Circuit output resistance |
| $R_p$ | Resistor used to cancel effect of input bias current |
| $R_s$ | Source resistance |
| $r_z$ | Dynamic zener resistance |

| | |
|---|---|
| $R_z$ | Static zener resistance |
| $S_1$ | Typical switch symbol (mechanical or electronic) |
| $S$ | Slew rate |
| $s$ | Complex radian frequency ($s = \sigma + j\omega$) |
| $S_A^B$ | Sensitivity of parameter $B$ to a change in parameter $A$ |
| S/H | Sample-and-hold |
| $T$ | Temperature |
| $t_r$ | Op amp risetime (unity gain follower) |
| $T_r$ | Single-shot reset time |
| $T_p$ | Pulse width |
| $V_1$ | Typical complex or dc voltage |
| $v_1$ | Typical instantaneous voltage |
| $V_{cc}$ | Collector supply voltage |
| $V_D$ | Diode forward voltage drop |
| $V_{DD}$ | FET drain supply voltage |
| $V_{EE}$ | Emitter supply voltage |
| $V_i, v_i$ | Circuit input voltage |
| $V_{ic}$ | Common-mode input voltage |
| $V_{io}$ | Op amp input offset voltage |
| $v_n$ | Voltage applied to op amp inverting input |
| $V_n$ | Equivalent rms input noise voltage |
| $V_o, v_o$ | Output voltage of circuit |
| $V_{on}$ | Output noise voltage |
| $V_{oo}$ | Output offset voltage |
| $v_p$ | Voltage applied to noninverting input of op amp |
| $V_p$ | FET pinch-off voltage |
| $V_R$ | Reference dc voltage; resistor noise (Johnson noise) |
| $V_{SS}$ | FET source-supply voltage |
| $V_z$ | Zener breakdown voltage |
| $\omega$ | Radian frequency |
| $\omega_n, \omega_o$ | Natural or resonant radian frequency |
| $X_c$ | Capacitive reactance |
| $y_{11}, Y_{11}$ | Input admittance of a circuit with the output shorted (incremental, lower-case; complex, uppercase) |
| $y_{12}, Y_{12}$ | Reverse transfer admittance, i.e., the input short-circuit current with 1 V applied to output terminals |
| $y_{21}, Y_{21}$ | Forward transfer admittance, i.e., the output short-circuit current with 1 V applied to input terminals |
| $y_{22}, Y_{22}$ | Output admittance of circuit when input terminals are shorted |
| $Y_{ti}$ | Transfer admittance of input circuit |
| $Y_{tf}$ | Transfer admittance of feedback circuit |
| $Z_1$ | Typical zener-diode symbol |
| $Z_f$ | Short-circuit feedback impedance |
| $Z_{id}$ | Op amp input impedance |
| $Z_{in}$ | Circuit input impedance |

## Appendix V
# Decibel Calculations

Calculations involving op amp gain and loop gain are conveniently performed using decibels. Since most op amp data sheets tabulate many of their parameters in decibels, a good designer should be able to convert quickly between ratios and decibels. Table V.1 can be used for these conversions in the absence of a calculator having log functions. This table is merely a numerical listing of the function

$$dB = 20 \log \left(\frac{V_2}{V_1}\right)$$

Use of Table V.1 is best described by two examples.

**TABLE V.1  Numerical Listing of**
$dB = 20 \log (V_2/V_1)$

| $V_2/V_1$ | dB |
|---|---|
| 1.001152 | 0.01 |
| 1.002305 | 0.02 |
| 1.003460 | 0.03 |
| 1.004616 | 0.04 |
| 1.005773 | 0.05 |
| 1.006932 | 0.06 |
| 1.008092 | 0.07 |
| 1.009253 | 0.08 |
| 1.010416 | 0.09 |
| 1.011579 | 0.1 |
| 1.023293 | 0.2 |
| 1.035142 | 0.3 |
| 1.047129 | 0.4 |
| 1.059254 | 0.5 |
| 1.071519 | 0.6 |
| 1.083927 | 0.7 |
| 1.096478 | 0.8 |
| 1.109175 | 0.9 |
| 1.122018 | 1.0 |
| 1.258925 | 2.0 |

## V-2 DECIBEL CALCULATIONS

| $V_2/V_1$ | dB |
|---|---|
| 1.412538 | 3.0 |
| 1.584893 | 4.0 |
| 1.778279 | 5.0 |
| 1.995262 | 6.0 |
| 2.238721 | 7.0 |
| 2.511886 | 8.0 |
| 2.818383 | 9.0 |
| 3.162278 | 10.0 |
| 10.00000 | 20.0 |
| 31.62278 | 30.0 |
| 100.0000 | 40.0 |
| 316.2278 | 50.0 |
| 1,000.000 | 60.0 |
| 3,162.278 | 70.0 |
| 10,000.00 | 80.0 |
| 31,622.78 | 90.0 |
| $10^5$ | 100 |
| $3.162278 \times 10^5$ | 110 |
| $10^6$ | 120 |
| $3.162278 \times 10^6$ | 130 |
| $10^7$ | 140 |
| $3.162278 \times 10^7$ | 150 |
| $10^8$ | 160 |

### EXAMPLE 1: CONVERTING A RATIO TO dB

Suppose a voltage gain (ratio) of $V_2/V_1 = 875,000$ must be converted to dB. In Table V.1 find the largest ratio less than 875,000. This is obviously 316,228, which corresponds to 110 dB. The following division is performed:

$$\frac{875,000}{316,228} = 2.767$$

We note in Table V.1 that 2.5119, corresponding to 8 dB, is the next number below 2.767. The numerical flow chart below shows how the process continues.

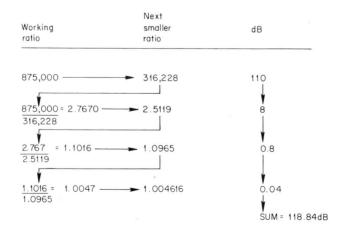

## EXAMPLE 2: CONVERTING dB TO A RATIO

An op amp is specified to have an open-loop dc voltage gain of 98 dB. What is the corresponding voltage ratio? Table V.1 indicates that 31,623 corresponds to 90 dB. This leaves 8 dB, which is equivalent to a ratio of 2.5119. We multiply these two ratios to find the ratio corresponding to 98 dB.

$$31,623 \times 2.5119 = 79,434$$

Thus 79,434 is equivalent to 98 dB.

# Appendix VI
# RC Circuit Characteristics

This appendix is useful for developing circuits having a wide variety of gain-vs.-frequency characteristics. This procedure was described in detail in Sec. 2.3. In order to save space, only gain-vs.-frequency plots are shown. Plots of phase vs. frequency can be constructed using the method described in Sec. 1.9. In the following pages

$$s = \sigma + j\omega \quad \text{and} \quad \omega = 2\pi f$$

Numbers next to each slope on the gain-vs.-frequency plots indicate multiples of $\pm 20$ dB/decade (i.e., $1 = +20$ dB/decade, $-3 = -60$ dB/decade, etc.). Whenever possible, the magnitude of each plateau is also given.

### REFERENCES

1. Scott, N. R.: "Analog and Digital Computer Technology," pp. 36–39, McGraw-Hill, Inc., 1960.
2. Moschytz, G. S.: FEN Filter Design Using Tantalum and Silicon Integrated Circuits, *Proc. IEEE*, vol. 58, pp. 550–566, Apr. 1970.
3. Truxal, J. G.: "Control Engineers Handbook," pp. 6-32 to 6-42, McGraw-Hill, Inc., 1958.
4. Bradley, F. R., and R. McCoy: Driftless DC Amplifier, *Electronics*, Apr. 1952, pp. 144–148.

## VI-2 RC CIRCUIT CHARACTERISTICS

| | ONE-ELEMENT NETWORKS | |
|---|---|---|
| Network | Open-circuit voltage gain $A_{vn} = V_2/V_1$ | Short-circuit transfer admittance $A_t = I_2/V_1$ |
| $V_1 \circ\!\!-\!\!\!\overset{R \;\;\; \overset{I_2}{\leftarrow}}{\wedge\!\wedge\!\wedge}\!\!-\!\!\circ V_2$ $\circ\!\!-\!\!\!\!-\!\!\!\!-\!\!\circ$ | 1  $A_{vn} \uparrow \; 1 \text{———}$  $\omega \rightarrow$ | $-\dfrac{1}{R}$  $\|A_t\| \uparrow \; \dfrac{1}{R} \text{———}$  $\omega \rightarrow$ |
| $V_1 \circ\!\!-\!\!\!\overset{C \;\;\; \overset{I_2}{\leftarrow}}{\|\;\|}\!\!-\!\!\circ V_2$ $\circ\!\!-\!\!\!\!-\!\!\!\!-\!\!\circ$ | 1  $A_{vn} \uparrow \; 1 \text{———}$  $\omega \rightarrow$ | $-sC$  $\|A_t\| \uparrow \; 1 \text{- - -}\; \dfrac{1}{C}$  $\omega \rightarrow$ |

# RC CIRCUIT CHARACTERISTICS VI-3

| | TWO-ELEMENT NETWORKS | |
|---|---|---|
| Network | Open circuit voltage gain $A_{vn} = V_2/V_1$ | Short circuit transfer admittance $A_t = I_2/V_1$ |
| (R series, C shunt) T=RC | $\dfrac{1}{1+sT}$ | $-\dfrac{1}{R}$ |
| | $\|A_{vn}\|$ vs $\omega$: flat at 1 until $\frac{1}{T}$, then −1 slope | $\|A_t\|$ vs $\omega$: flat at $\frac{1}{R}$ |
| (C series, R shunt) T=RC | $\dfrac{sT}{1+sT}$ | $-sC$ |
| | $A_{vn}$ vs $\omega$: rising slope up to $\frac{1}{T}$, then flat at 1 | $\|A_t\|$ vs $\omega$: rising, passing through 1 at $\frac{1}{C}$ |

## VI-4  RC CIRCUIT CHARACTERISTICS

| Network | Two-Element Networks | |
|---|---|---|
| | Open circuit voltage gain $A_{vn} = V_2/V_1$ | Short circuit transfer admittance $A_t = I_2/V_1$ |
| R, C series-shunt network (R series, C shunt to ground at output side); $T = RC$ | $1$ | $-\dfrac{1+sT}{R}$ |
| | (Bode: $A_{vn} = 1$, flat) | (Bode: $|A_t|$ flat at $1/R$ until $\omega = 1/T$, then rising with slope 1) |
| R, C series network (R then C in series from $V_1$ to $V_2$); $T = RC$ | $1$ | $-\dfrac{sC}{1+sT}$ |
| | (Bode: $A_{vn} = 1$, flat) | (Bode: $|A_t|$ rising with slope 1 until $\omega = 1/T$, then flat at $1/R$) |

# RC CIRCUIT CHARACTERISTICS   VI-5

| Network | Open-circuit voltage gain $A_{vn} = V_2/V_1$ | Short-circuit transfer admittance $A_t = I_2/V_1$ |
|---|---|---|
| THREE-ELEMENT NETWORKS | | |

| Network | Open-circuit voltage gain $A_{vn} = V_2/V_1$ | Short-circuit transfer admittance $A_t = I_2/V_1$ |
|---|---|---|
| Circuit with $R_1$ series, $R_2$ and $C$ in shunt branch; $V_1$ input, $V_2$ output, $I_2$ output current. $T_1 = (R_1+R_2)C$, $T_2 = R_2 C$ | 1 | $-\dfrac{1+sT_1}{R_1(1+sT_2)}$ |
| | Plot: $A_{vn}$ vs $\omega$, constant at 1 | Plot: $\lvert A_t \rvert$ vs $\omega$, low-frequency asymptote $\dfrac{1}{R_1}$, corner at $\dfrac{1}{T_1}$, rising to corner at $\dfrac{1}{T_2}$, high-frequency asymptote $\dfrac{1}{R_1}+\dfrac{1}{R_2}$ |
| Circuit with $C_1$ series, $R$ and $C_2$ in shunt branch; $V$ input, $V_2$ output, $I_2$ output current. $T_1 = RC_2$, $T_2 = \dfrac{RC_1 C_2}{C_1+C_2}$ | 1 | $-\dfrac{s(C_1+C_2)(1+sT_2)}{1+sT_1}$ |
| | Plot: $A_{vn}$ vs $\omega$, constant at 1 | Plot: $\lvert A_t \rvert$ vs $\omega$, rising to 1 at corner $\dfrac{1}{T_1}$, flat to corner $\dfrac{1}{T_2}$, then rising to 1 |

## VI-6  RC CIRCUIT CHARACTERISTICS

| THREE-ELEMENT NETWORKS | | |
|---|---|---|
| Network | Open circuit voltage gain $A_{vn}=V_2/V_1$ | Short circuit transfer admittance $A_t=I_2/V_1$ |
| $R_1$—$R_2$ series, $C$ shunt after $R_2$ (to ground); $V_1$ in, $V_2$ out; $I_2$ at $V_2$ <br><br> $T_1 = \dfrac{R_1 R_2 C}{R_1+R_2}$ <br><br> $T_2 = R_2 C$ | 1 <br><br> (plot: $A_{vn}$ vs $\omega$, flat at 1) | $\dfrac{-(1+sT_2)}{(R_1+R_2)(1+sT_1)}$ <br><br> (plot: $|A_t|$ vs $\omega$; level $\dfrac{1}{R_1+R_2}$ up to $\dfrac{1}{T_1}$, slope $-1$ down to $\dfrac{1}{T_2}$, then level $\dfrac{1}{R_1}$) |
| $C_1$—$C_2$ series, $R$ shunt after $C_2$ (to ground); $V_1$ in, $V_2$ out; $I_2$ at $V_2$ <br><br> $T_1 = R(C_1+C_2)$ <br><br> $T_2 = RC_2$ | 1 <br><br> (plot: $A_{vn}$ vs $\omega$, flat at 1) | $-\dfrac{sC_1(1+sT_2)}{1+sT_1}$ <br><br> (plot: $|A_t|$ vs $\omega$; rising slope to $\dfrac{1}{T_1}$, level between $\dfrac{1}{T_1}$ and $\dfrac{1}{T_2}$, then rising slope) |

# RC CIRCUIT CHARACTERISTICS  VI-7

| THREE-ELEMENT NETWORKS ||||
|---|---|---|---|
| Network || Open-circuit voltage gain $A_{vn} = V_2/V_1$ | Short-circuit transfer admittance $A_t = I_2/V_1$ |
| $R_1, R_2$ network with $C$ to ground<br><br>$T_1 = R_1 C$<br>$T_2 = \dfrac{R_1 R_2 C}{R_1 + R_2}$ || $\dfrac{1}{1 + sT_1}$ | $\dfrac{-1}{(R_1+R_2)(1+sT_2)}$ |
| ^ || (Bode plot: $A_{vn}$ magnitude 1, corner at $1/T_1$, slope $-1$) | (Bode plot: $|A_t|$ magnitude $\dfrac{1}{R_1+R_2}$, corner at $1/T_2$, slope $-1$) |
| $C_1, C_2$ in series with $R$ to ground<br><br>$T_1 = R(C_1 + C_2)$<br>$T_2 = RC_1$ || $\dfrac{sT_2}{1 + sT_2}$ | $\dfrac{-s^2 R C_1 C_2}{1 + sT_1}$ |
| ^ || (Bode plot: $A_{vn}$ slope $+1$ up to $1/T_2$, then flat at 1) | (Bode plot: $|A_t|$ slope $+2$ up to $1/T_1$, then slope $+1$ to 1) |

## VI-8  RC CIRCUIT CHARACTERISTICS

| | THREE–ELEMENT NETWORKS | |
|---|---|---|
| Network | Open circuit voltage gain $A_{vn} = V_2/V_1$ | Short circuit transfer admittance $A_t = I_2/V_1$ |
| (Circuit: $V_1$ — $R_1$ — $C$ — $V_2$, with $R_2$ to ground) $T_1 = (R_1+R_2)C$ $T_2 = R_2 C$ $T_3 = R_1 C$ | $\dfrac{sT_2}{1+sT_1}$ | $\dfrac{-sC}{1+sT_3}$ |
| | Bode plot: $A_{vn}$ vs $\omega$, breakpoint at $\dfrac{1}{T_1}$, asymptote $\dfrac{R_2}{R_1+R_2}$ | Bode plot: $|A_t|$ vs $\omega$, breakpoint at $\dfrac{1}{T_3}$, asymptote $\dfrac{1}{R_1}$ |
| (Circuit: $V_1$ — $R_1$ — $C_1$ — $V_2$, with $C_2$ to ground) $T_1 = \dfrac{RC_1 C_2}{C_1+C_2}$ $T_2 = RC_1$ | $\dfrac{C_1/(C_1+C_2)}{1+sT_1}$ | $\dfrac{-sC_1}{1+sT_2}$ |
| | Bode plot: $A_{vn}$ vs $\omega$, low-freq asymptote $\dfrac{C_1}{C_1+C_2}$, slope $-1$, breakpoint at $\dfrac{1}{T_1}$ | Bode plot: $|A_t|$ vs $\omega$, breakpoint at $\dfrac{1}{T_2}$, asymptote $\dfrac{1}{R}$ |

## RC CIRCUIT CHARACTERISTICS

### THREE-ELEMENT NETWORK

| Network | Open circuit voltage gain $A_{vn} = V_2/V_1$ | Short circuit transfer admittance $A_t = I_2/V_1$ |
|---|---|---|
| $R_1$, $C$, $R_2$ network with $I_2$, $V_1$, $V_2$<br><br>$T_1 = (R_1 + R_2)C$<br>$T_2 = R_2 C$ | $\dfrac{1+sT_2}{1+sT_1}$<br><br>(Bode plot: $A_{vn}$ vs $\omega$, break at $1/T_1$ slope $-1$, break at $1/T_2$, level $R_2/(R_1+R_2)$) | $-\dfrac{1}{R_1}$<br><br>(Bode plot: $|A_t|$ vs $\omega$, level at $1/R_1$) |
| $C_1$, $C_2$, $R$ network with $I_2$, $V_1$, $V_2$<br><br>$T_1 = RC_2$<br>$T_2 = \dfrac{RC_1 C_2}{C_1 + C_2}$ | $\dfrac{C(1+sT_1)}{(C_1+C_2)(1+sT_2)}$<br><br>(Bode plot: $A_{vn}$ vs $\omega$, level $C_1/(C_1+C_2)$, break at $1/T_1$ slope $+1$, break at $1/T_2$, level $1$) | $-sC$<br><br>(Bode plot: $|A_t|$ vs $\omega$, slope $+1$ crossing $1$ at $1/C_1$) |

## VI-10  RC CIRCUIT CHARACTERISTICS

### THREE-ELEMENT NETWORKS

| Network | Open circuit voltage gain $A_{vn} = V_2/V_1$ | Short circuit transfer admittance $A_t = I_2/V_1$ |
|---|---|---|
| Network 1: $R_1$ series, $R_2$ and $C$ in shunt branch; $T = \dfrac{R_1 R_2 C}{R_1 + R_2}$ | $\dfrac{T}{R_1 C(1+sT)}$  <br> Asymptotic plot: $A_{vn}$ vs $\omega$, low-frequency level $\dfrac{R_2}{R_1+R_2}$, slope $-1$ at corner $\dfrac{1}{T}$ | $-\dfrac{1}{R_1}$ <br> Asymptotic plot: $|A_t|$ vs $\omega$, flat at $\dfrac{1}{R_1}$ |
| Network 2: $C_1$ series, $R$ and $C_2$ in shunt branch; $T_1 = R(C_1+C_2)$, $T_2 = RC_1$ | $\dfrac{sT_2}{1+sT_1}$ <br> Asymptotic plot: $A_{vn}$ vs $\omega$, slope $+1$ below corner $\dfrac{1}{T_1}$, high-frequency level $\dfrac{C_1}{C_1+C_2}$ | $-sC_1$ <br> Asymptotic plot: $|A_t|$ vs $\omega$, slope $+1$ passing through $1$ at $\dfrac{1}{C_1}$ |

## RC CIRCUIT CHARACTERISTICS    VI-11

| | THREE-ELEMENT NETWORKS | |
|---|---|---|
| Network | Open circuit voltage gain $A_{vn} = V_2/V_1$ | Short circuit transfer admittance $A_t = I_2/V_1$ |
| Network 1: $R_1$ series, $C$ shunt at input, $R_2$ shunt at output. $T_1 = R_1 C$, $T_2 = \dfrac{R_1 R_2 C}{R_1 + R_2}$ | $\dfrac{R_2(1+sT_1)}{(R_1+R_2)(1+sT_2)}$ | $-\dfrac{1+sT_1}{R_1}$ |
| | Bode plot: $A_{vn}$ starts at $\dfrac{R_2}{R_1+R_2}$, breaks at $\dfrac{1}{T_1}$ and $\dfrac{1}{T_2}$, asymptote to 1 | Bode plot: $|A_t|$ starts at $\dfrac{1}{R_1}$, breaks at $\dfrac{1}{T_1}$, slope +1 |
| Network 2: $R$ series, $C_1$ and $C_2$ shunts. $T_1 = R(C_1+C_2)$, $T_2 = RC_1$ | $\dfrac{1+sT_2}{1+sT_1}$ | $-\dfrac{1+sT_2}{R}$ |
| | Bode plot: $A_{vn}$ starts at 1, slope $-1$ between $\dfrac{1}{T_1}$ and $\dfrac{1}{T_2}$, levels at $\dfrac{C_1}{C_1+C_2}$ | Bode plot: $|A_t|$ starts at $\dfrac{1}{R}$, breaks at $\dfrac{1}{T_2}$, slope +1 |

## VI-12  RC CIRCUIT CHARACTERISTICS

| FOUR-ELEMENT NETWORKS | | |
|---|---|---|
| Network | Open circuit voltage gain $A_{vn} = V_2/V_1$ | Short circuit transfer admittance $A_t = I_2/V_1$ |
| Circuit: $V_1$ — $R_1$ — $C_1$ — $V_2$ (with $I_2$), $R_2$ — $C_2$ branch to ground<br><br>$T_1 = R_1 C_1$<br>$T_2 = \dfrac{(R_1+R_2)C_1 C_2}{C_1+C_2}$<br>$T_3 = R_2 C_2$<br>$T_3 < T_2 < T_1$ | 1<br><br>Plot: $A_{vn}$ vs $\omega$, flat at 1 | $\dfrac{-s(C_1+C_2)(1+sT_2)}{(1+sT_1)(1+sT_3)}$<br><br>Plot: $|A_t|$ vs $\omega$, breakpoints at $\dfrac{1}{T_1}, \dfrac{1}{T_2}, \dfrac{1}{T_3}$, asymptote $\dfrac{R_1+R_2}{R_1 R_2}$ |
| Circuit: $V_1$ — $R_1$ — $R_2$ — $V_2$ (with $I_2$), $C_1, C_2$ to ground<br><br>$T_1 = R_1 C_1$<br>$T_2 = \dfrac{R_1 R_2(C_1+C_2)}{R_1+R_2}$<br>$T_3 = R_2 C_2$<br>$T_3 < T_2 < T_1$ | 1<br><br>Plot: $A_{vn}$ vs $\omega$, flat at 1 | $\dfrac{-(1+sT_1)(1+sT_3)}{(R_1+R_2)(1+sT_2)}$<br><br>Plot: $|A_t|$ vs $\omega$, levels at $\dfrac{1}{R_1+R_2}$, breakpoints $\dfrac{1}{T_1}, \dfrac{1}{T_2}, \dfrac{1}{T_3}$, asymptote 1 |

## FOUR-ELEMENT NETWORKS

| Network | Open circuit voltage gain $A_{vn} = V_2/V_1$ | Short circuit transfer admittance $A_t = I_2/V_1$ |
|---|---|---|
| Circuit: $R_1$, $C_1$ in series (with $I_2$) from $V_1$ to $V_2$; $R_2$ and $C_2$ in parallel to ground. <br><br> $T_2 = R_1 C_1$ <br> $T_1 T_3 = R_1 R_2 C_1 C_2$ <br> $T_1 + T_3 = R_1 C_1 + R_2 C_2 + R_2 C_1$ <br> $T_3 < T_2 < T_1$ | $1$ <br><br> Plot: $A_{vn}$ vs $\omega$, flat at $1$ | $\dfrac{-(1+sT_1)(1+sT_3)}{R_2(1+sT_2)}$ <br><br> Plot: $|A_t|$ vs $\omega$; levels at $\dfrac{1}{R_2}$, breakpoints at $\dfrac{1}{T_1}$, $\dfrac{1}{T_2}$, $\dfrac{1}{T_3}$, final slope to $1$ |
| Circuit: $R_1$, $C_1$, $R_2$ in series (with $I_2$) from $V_1$ to $V_2$; $C_2$ to ground. <br><br> $T_2 = R_2 C_2$ <br> $T_1 T_3 = R_1 R_2 C_1 C_2$ <br> $T_1 + T_3 = R_1 C_1 + R_2 C_2 + R_2 C_1$ <br> $T_3 < T_2 < T_1$ | $1$ <br><br> Plot: $A_{vn}$ vs $\omega$, flat at $1$ | $\dfrac{-sC_1(1+sT_2)}{(1+sT_1)(1+sT_3)}$ <br><br> Plot: $|A_t|$ vs $\omega$; breakpoints at $\dfrac{1}{T_1}$, $\dfrac{1}{T_2}$, $\dfrac{1}{T_3}$, levels at $\dfrac{1}{R_1}$ |

## VI-14    RC CIRCUIT CHARACTERISTICS

| FOUR-ELEMENT NETWORKS ||| 
|---|---|---|
| Network | Open circuit voltage gain $A_{vn} = V_2/V_1$ | Short circuit transfer admittance $A_t = I_2/V_1$ |
| (Circuit: $V_1$ — $C_1$ — $V_2$ with $R_1$, $R_2$ branch and $C_2$ to ground)<br><br>$T_2 = \dfrac{R_1 R_2 C_2}{R_1 + R_2}$<br>$T_1 T_3 = R_1 R_2 C_1 C_2$<br>$T_1 + T_3 = R_1 C_1 + R_2 C_2 + R_2 C_1$<br>$T_3 < T_2 < T_1$ | 1<br><br>(plot: $A_{vn}$ vs $\omega$, flat at 1) | $\dfrac{-(1+sT_1)(1+sT_3)}{(R_1+R_2)(1+sT_2)}$<br><br>(plot: $\|A_t\|$ vs $\omega$, starts at $\dfrac{1}{R_1+R_2}$, breakpoints at $\dfrac{1}{T_1}$, $\dfrac{1}{T_2}$, $\dfrac{1}{T_3}$, rises to 1) |
| (Circuit: $V_1$ — $R_1$ — $V_2$ with $C_1$, $R_2$ branch and $C_2$ to ground)<br><br>$T_2 = R_2(C_1 + C_2)$<br>$T_1 T_3 = R_1 R_2 C_1 C_2$<br>$T_1 + T_3 = R_1 C_1 + R_2 C_2 + R_2 C_1$<br>$T_3 < T_2 < T_1$ | 1<br><br>(plot: $A_{vn}$ vs $\omega$, flat at 1) | $\dfrac{-(1+sT_1)(1+sT_3)}{R_1(1+sT_2)}$<br><br>(plot: $\|A_t\|$ vs $\omega$, starts at $\dfrac{1}{R_1}$, breakpoints at $\dfrac{1}{T_1}$, $\dfrac{1}{T_2}$, $\dfrac{1}{T_3}$, rises to 1) |

## RC CIRCUIT CHARACTERISTICS   VI-15

| FOUR-ELEMENT NETWORKS | | |
|---|---|---|
| Network | Open circuit voltage gain $A_{vn} = V_2/V_1$ | Short circuit transfer admittance $A_t = I_2/V_1$ |
| (circuit: $V_1$ — $C_1$ — $R_1$ — $I_2$ — $V_2$, with $R_2$, $C_2$ branch) | 1 | $\dfrac{-sC_1(1+sT_2)}{(1-sT_1)(1+sT_3)}$ |
| $T_2 = (R_1+R_2)C_2$ <br> $T_1 T_3 = R_1 R_2 C_1 C_2$ <br> $T_1+T_3 = R_1 C_1 + R_2 C_2 + R_1 C_2$ <br> $T_3 < T_2 < T_1$ | $A_{vn}$: flat at 1 vs $\omega$ | $\|A_t\|$: rising plot with corners at $\frac{1}{T_1}, \frac{1}{T_2}, \frac{1}{T_3}$, asymptote $\frac{R_1+R_2}{R_1 R_2}$ |
| (circuit: $V_1$ — $R_1$ — $C_1$ — $I_2$ — $V_2$, with $R_2$, $C_2$ branch) | 1 | $\dfrac{-s(C_1+C_2)(1+sT_2)}{(1+sT_1)(1+sT_3)}$ |
| $T_2 = \dfrac{R_2 C_1 C_2}{C_1+C_2}$ <br> $T_1 T_3 = R_1 R_2 C_1 C_2$ <br> $T_1+T_3 = R_1 C_1 + R_2 C_2 + R_1 C_2$ <br> $T_3 < T_2 < T_1$ | $A_{vn}$: flat at 1 vs $\omega$ | $\|A_t\|$: rising plot with corners at $\frac{1}{T_1}, \frac{1}{T_2}, \frac{1}{T_3}$, asymptote $\frac{1}{R_1}$ |

## VI-16  RC CIRCUIT CHARACTERISTICS

| FOUR-ELEMENT NETWORKS |||
|---|---|---|
| Network | Open circuit voltage gain $A_{vn} = V_2/V_1$ | Short circuit transfer admittance $A_t = I_2/V_1$ |
| Circuit: $V_1$ — $R_1$ — • — $R_1$ — $V_2$ (with $I_2$), capacitor $C$ from node to $R_2$ to ground <br><br> $T_1 = (R_1+R_2)C$ <br> $T_2 = (R_2+R_1/2)C$ <br> $T_3 = R_2 C$ <br> $T_3 < T_2 < T_1$ | $\dfrac{1+sT_3}{1+sT_1}$ <br><br> Bode plot of $A_{vn}$: starts at 1, breaks down at $1/T_1$ with slope $-1$, levels at $R_2/(R_1+R_2)$ after $1/T_3$ | $\dfrac{-(1+sT_3)}{2R_1(1+sT_2)}$ <br><br> Bode plot of $|A_t|$: starts at $1/(2R_1)$, breaks at $1/T_2$ slope $-1$, levels at $R_2/(R_1(2R_2+R_1))$ after $1/T_3$ |
| Circuit: $V_1$ — $C_1$ — • — $C_1$ — $V_2$, capacitor $C_2$ from node, then $R$ to ground <br><br> $T_1 = RC_2$ <br> $T_2 = \dfrac{2RC_1C_2}{2C_1+C_2}$ <br> $T_3 = \dfrac{RC_1C_2}{C_1+C_2}$ <br> $T_3 < T_2 < T_1$ | $\dfrac{C_1(1+sT_1)}{(C_1+C_2)(1+sT_3)}$ <br><br> Bode plot of $A_{vn}$: starts at $C_1/(C_1+C_2)$, breaks up at $1/T_1$ slope $+1$, levels at 1 after $1/T_3$ | $\dfrac{-sC_1^2(1+sT_1)}{(2C_1+C_2)(1+sT_2)}$ <br><br> Bode plot of $|A_t|$: slope $+1$, breaks at $1/T_1$, then $1/T_2$ |

## RC CIRCUIT CHARACTERISTICS   VI-17

### FOUR ELEMENT NETWORKS

| Network | Open circuit voltage gain $A_{vn} = V_2/V_1$ | Short circuit transfer admittance $A_t = I_2/V_1$ |
|---|---|---|
| Circuit 1: $R_1$, $C_1$ in series with $V_1$ to $V_2$; $R_2$, $C_2$ shunt to ground.<br><br>$T_1 = (R_1+R_2)C_2$<br>$T_2 = R_2 C_2$<br>$T_4 = R_2 C_2$<br>$T_3 T_5 = R_1 R_2 C_1 C_2$<br>$T_3 + T_5 = R_1 C_1 + R_2 C_2 + R_1 C_2$<br>$T_2 < T_1$<br>$T_5 < T_4 < T_3$ | $\dfrac{1+sT_2}{1+sT_1}$<br><br>Asymptotic Bode plot: level at 1, breaks down at $1/T_1$ with slope $-1$, levels off at $R_2/(R_1+R_2)$ at $1/T_2$. | $\dfrac{-sC_1(1+sT_4)}{(1+sT_3)(1+sT_5)}$<br><br>Asymptotic Bode plot: rises with slope 1, breaks at $1/T_3$, $1/T_4$, $1/T_5$, asymptote to $1/R_1$. |
| Circuit 2: $R_1$ in series with $V_1$ to $V_2$, with $C_1$ shunt; $R_2$, $C_2$ shunt to ground.<br><br>$T_1 = R_1 C_1$<br>$T_2 = R_2 C_2$<br>$T_3 = R_1 C_2$ | $\dfrac{(1+sT_1)(1+sT_2)}{s^2 T_1 T_2 + s(T_1+T_2+T_3) + 1}$<br><br>Bode: level 1, break down at $1/T_1$, $1/T_2$ to $(T_1+T_2)/(T_1+T_2+T_3)$, back up to 1. | $\dfrac{1+sT_1}{R}$<br><br>Bode: level at $1/R_1$, breaks up at $1/T_1$ with slope 1. |

## VI-18  RC CIRCUIT CHARACTERISTICS

| Network | FOER—ELEMENT NETWORKS | |
|---|---|---|
| | Open circuit voltage gain $A_{vn} = V_2/V_1$ | Short circuit transfer admittance $A_t = I_2/V_1$ |
| (Network 1: $C_1$, $R_1$, $C_2$ in series path; $R_2$ shunt) $\omega_1^2 = \dfrac{1}{R_1 R_2 C_1 C_2}$ $A = \dfrac{1/\omega_1}{(R_1+R_2)C_1 + R_2 C_2}$ $T_1 = (R_1+R_2)C_1$ $T_2, T_3 = \dfrac{2A}{\omega_1 \pm \omega_1 (1-4A^2)^{1/2}}$ | $\dfrac{sR_2 C_1}{1+sT_1}$ (plot: $A_{vn}$ vs $\omega$, slope +1 breaking at $1/T_1$ to level $R_2/(R_1+R_2)$) | $\dfrac{-s^2/R_1}{s^2 + s(\omega_1/A) + \omega_1^2}$ (plot: $|A_t|$ vs $\omega$, +2 slope, peak, breaks at $1/T_2$ and $1/T_3$ to level $1/R_1$) |
| (Network 2: $C_1$, $R_1$, $R_2$ in series path; $C_2$ shunt) $\omega_1^2 = \dfrac{1}{R_1 R_2 C_1 C_2}$ $A = \dfrac{1/\omega_1}{(R_1+R_2)C_1 + R_2 C_2}$ $T_1 = \dfrac{R_1 C_1 C_2}{C_1 + C_2}$ $T_2, T_3 = \dfrac{2A}{\omega_1 \pm \omega_1 (1-4A^2)^{1/2}}$ | $\dfrac{C_1/(C_1+C_2)}{1+sT_1}$ (plot: $A_{vn}$ vs $\omega$, flat at $C_1/(C_1+C_2)$ then −1 slope breaking at $1/T_1$) | $\dfrac{-s/R_1 R_2 C_2}{s^2 + s(\omega_1/A) + \omega_1^2}$ (plot: $|A_t|$ vs $\omega$, +1 then peak then −1, breaks at $1/T_2$ and $1/T_3$) |

RC CIRCUIT CHARACTERISTICS   VI-19

| FOUR-ELEMENT NETWORKS | | |
|---|---|---|
| Network | Open circuit voltage gain $A_{vn} = V_2/V_1$ | Short circuit transfer admittance $A_t = I_2/V_1$ |
| Network 1: $V_1$ — $R_1$ — $R_2$ — $C_2$ — $V_2$, with $C_1$ to ground; $I_2$ out<br><br>$T_1 T_2 = R_1 R_2 C_1 C_2$<br>$T_1 + T_2 = R_1 C_1 + R_2 C_2 + R_1 C_2$<br>$T_3 = R_1 C_1$<br>$T_1 \neq T_2$ | $\dfrac{1}{1+sT_3}$ | $\dfrac{-sC_2}{(1+sT_1)(1+sT_2)}$ |
| | [Bode plot: $A_{vn}$ vs $\omega$, flat at 1, breaks down at $1/T_3$ with slope $-1$] | [Bode plot: $|A_t|$ vs $\omega$, rises with slope $1$ to $1/T_1$, flat, then slope $-1$ after $1/T_2$] |
| Network 2: $V_1$ — $C_1$ — $C_2$ — $R_2$ — $V_2$, with $R_1$ to ground; $I_2$ out<br><br>$T_1 T_2 = R_1 R_2 C_1 C_2$<br>$T_1 + T_2 = R_1 C_1 + R_2 C_2 + R_1 C_2$<br>$T_3 = R_1 C_1$<br>$T_1 \neq T_2$ | $\dfrac{sT_3}{1+sT_3}$ | $\dfrac{-s^2 C_2 (T_1 T_2)^{1/2}}{(1+sT_1)(1+sT_2)}$ |
| | [Bode plot: $A_{vn}$ vs $\omega$, slope $1$ up to $1/T_3$, then flat at 1] | [Bode plot: $|A_t|$ vs $\omega$, slope $2$ up to $1/T_1$, slope $1$ to $1/T_2$, then flat at $C_2/(T_1 T_2)^{1/2}$] |

## VI-20  RC CIRCUIT CHARACTERISTICS

| | FOUR-ELEMENT NETWORKS | |
|---|---|---|
| Network | Open circuit voltage gain $A_{vn} = V_2/V_1$ | Short circuit transfer admittance $A_t = I_2/V_1$ |
| Network 1: $R_1$, $C_1$ in series from $V_1$ to $V_2$; $C_2$ to ground, $R_2$ to ground<br><br>$\omega_1, \omega_2 = \dfrac{B \pm (B^2 - 4A)^{1/2}}{2A}$<br>$T_3 = R_1 C_1$<br>$A = R_1 R_2 C_1 C_2$<br>$B = R_1 C_1 + R_2 C_2 + R_2 C_1$ | $\dfrac{sR_2C_1}{(1+sT_1)(1+sT_2)}$<br><br>Bode plot: $A_{vn}$ vs $\omega$, slopes +1, -1 with corners at $\omega_1, \omega_2$ | $\dfrac{sC_1}{1+sT_3}$<br><br>Bode plot: $|A_t|$ vs $\omega$, slope +1 up to $\dfrac{1}{T_3}$, then flat at $\dfrac{1}{R_1}$ |
| Network 2: $C_1$ from $V_1$ to $V_2$; $R_1$ and $R_2$ in series to ground, $C_2$ to ground<br><br>$T_1$ or $T_2 = R_1 C_1$<br>$T_1$ or $T_2 = \dfrac{R_1 R_2 (C_1 + C_2)}{R_1 + R_2}$<br>$T_3 = R_1 C_1$ | $\dfrac{R_2(1+sT_1)}{(R_1+R_2)(1+sT_2)}$<br><br>$\dfrac{C_1}{C_1+C_2} > \dfrac{R_2}{R_1+R_2}$<br><br>Bode: $A_{vn}$ starts at $\dfrac{R_2}{R_1+R_2}$, slope +1 at $\dfrac{1}{T_1}$, corner at $\dfrac{1}{T_2}$, with $R_1C_1 = R_2C_2$<br><br>$\dfrac{C_1}{C_1+C_2} < \dfrac{R_2}{R_1+R_2}$ | $-\dfrac{1+sT_3}{R_1}$<br><br>Bode: $|A_t|$ flat at $\dfrac{1}{R_1}$, slope +1 above $\dfrac{1}{T_3}$ |

# RC CIRCUIT CHARACTERISTICS  VI-21

| Network | Open circuit voltage gain $A_{vn} = V_2/V_1$ | Short circuit transfer admittance $A_t = I_2/V_1$ |
|---|---|---|
| Circuit: $V_1$ — C — C — $V_2$, with $R_2$ from midpoint to $R_1$ to ground. $I_2$ out at $V_2$. <br><br> $T_1$ or $T_2 = R_2 C$ <br> $T_1$ or $T_2 = \dfrac{2 R_1 R_2 C}{R_1 \, R_2}$ <br> $T_3 = R_1 C$ | $\dfrac{s T_3}{1 + s T_3}$ <br><br> (plot of $A_{vn}$ vs $\omega$, breakpoint at $1/T_3$, slope +1) | $\dfrac{-s R_1 C (1 + s T_1)}{(R_1 + R_2)(1 + s T_2)}$ <br><br> (plot of $|A_t|$ vs $\omega$; $R_1 < R_2$, $R_1 = R_2$, $R_2 < R_1$; breakpoints $1/T_1$, $1/T_2$) |
| Circuit: $V_1$ — R — R — $V_2$, with C from midpoint to ground. $I_2$ out at $V_2$. <br><br> $T_1 = R C_1$ <br> $T_2$ or $T_3 = R C_2$ <br> $T_2$ or $T_3 = \dfrac{R(C_1 + C_2)}{2}$ | $\dfrac{1}{1 + s T_1}$ <br><br> (plot of $A_{vn}$ vs $\omega$, breakpoint at $1/T_1$, slope $-1$) | $\dfrac{-(1 + s T_2)}{2R(1 + s T_3)}$ <br><br> (plot of $|A_t|$ vs $\omega$, level $\dfrac{1}{2R}$; $C_1 < C_2$, $C_1 = C_2$, $C_2 < C_1$; breakpoints $1/T_1$, $1/T_2$) |

FOUR ELEMENT NETWORKS

## VI-22  RC CIRCUIT CHARACTERISTICS

### FOUR-ELEMENT NETWORKS

| Network | Open-circuit voltage gain $A_{vn} = V_2/V_1$ | Short-circuit transfer admittance $A_t = I_2/V_1$ |
|---|---|---|
| Circuit: $R_1$ and $R_2$ in series, with $C$ and $R_3$ in series forming a shunt branch from the node between $R_1,R_2$ and output; $V_1$ input, $V_2$ output, $I_2$ output current. <br><br> $T_1 = R_2 C$ <br> $T_2 = \dfrac{R_2(R_1+R_3)C}{R_1+R_2+R_3}$ <br> $T_3 = \dfrac{R_1 R_2 C}{R_1+R_2}$ | $\dfrac{R_3(1+sT_1)}{(R_1+R_2+R_3)(1+sT_2)}$ <br><br> Bode plot: $A_{vn}$ vs $\omega$, low-frequency asymptote $\dfrac{R_3}{R_1+R_2+R_3}$, break at $\dfrac{1}{T_1}$, rising segment, break at $\dfrac{1}{T_2}$, high-frequency asymptote $\dfrac{R_3}{R_1+R_3}$ | $-\dfrac{1+sT_1}{(R_1+R_2)(1+sT_3)}$ <br><br> Bode plot: $|A_t|$ vs $\omega$, low-frequency asymptote $\dfrac{1}{R_1+R_2}$, break at $\dfrac{1}{T_3}$, slope $-1$, break at $\dfrac{1}{T_1}$, high-frequency asymptote $\dfrac{1}{R_1}$ |
| Circuit: $R_1$ in series, with $R_2$ and $C$ in series shunting to $R_3$ to ground; $V_1$ input, $V_2$ output, $I_2$ output current. <br><br> $T_1 = \dfrac{R_2(R_1+R_3)C}{R_1+R_2+R_3}$ <br> $T_2 = \dfrac{R_2 R_3 C}{R_2+R_3}$ | $\dfrac{(R_2+R_3)(1+sT_2)}{(R_1+R_2+R_3)(1+sT_1)}$ <br><br> Bode plot: $A_{vn}$ vs $\omega$, low-frequency asymptote $\dfrac{R_2+R_3}{R_1+R_2+R_3}$, break at $\dfrac{1}{T_1}$, slope $-1$, break at $\dfrac{1}{T_2}$, high-frequency asymptote $\dfrac{R_3}{R_1+R_3}$ | $-\dfrac{1}{R_1}$ <br><br> Bode plot: $|A_t|$ vs $\omega$, flat at $\dfrac{1}{R_1}$ |

# RC CIRCUIT CHARACTERISTICS

## FOUR-ELEMENT NETWORKS

| Network | Open circuit voltage gain $A_{vn} = V_2/V_1$ | Short circuit transfer admittance $A_t = I_2/V_1$ |
|---|---|---|
| Circuit 1: $R_1$ in series from $V_1$ to $V_2$; $R_2$ shunt, $C$ shunt, $R_3$ shunt at output. $T_1 = \dfrac{(R_1R_2+R_1R_3+R_2R_3)C}{R_1+R_3}$ $T_2 = R_2 C$ | $\dfrac{R_3(1+sT_2)}{(R_1+R_3)(1+sT_1)}$ | $-\dfrac{1}{R_1}$ |
| (asymptotic Bode plots) | $A_{vn}$: low-frequency level $\dfrac{R_3}{R_1+R_3}$, slope $-1$ between $\dfrac{1}{T_1}$ and $\dfrac{1}{T_2}$, high-frequency level $\dfrac{R_2\|R_3}{R_1+R_2\|R_3}$ | $\|A_t\|$: flat at $\dfrac{1}{R_1}$ |
| Circuit 2: $R_1$ in series from $V_1$ to $V_2$, with $C$ in parallel with $R_1$; $R_2$ shunt at input side of $R_1$; $R_3$ shunt at output. $T_1 = (R_1+R_2)C$ $T_2 = \dfrac{(R_1R_2+R_1R_3+R_2R_3)C}{R_1+R_3}$ $T_3 = R_2 C$ | $\dfrac{R_3(1+sT_1)}{(R_1+R_3)(1+sT_2)}$ | $-\dfrac{1+sT_1}{R_1(1+sT_3)}$ |
| (asymptotic Bode plots) | $A_{vn}$: low-frequency level $\dfrac{R_3}{R_1+R_3}$, slope $+1$ between $\dfrac{1}{T_1}$ and $\dfrac{1}{T_2}$, high-frequency level $\dfrac{R_3}{R_1\|R_2+R_3}$ | $\|A_t\|$: level $\dfrac{1}{R_1}$, slope $+1$ between $\dfrac{1}{T_1}$ and $\dfrac{1}{T_3}$, high-frequency level $\dfrac{R_1+R_2}{R_1R_2}$ |

## VI-24   RC CIRCUIT CHARACTERISTICS

| Network | Four-Element Networks Open circuit voltage gain $A_{vn} = V_2/V_1$ | Short circuit transfer admittance $A_t = I_2/V_1$ |
|---|---|---|
| (Network: $R_1$, $R_2$ series with $C_1$, $C_2$ to ground; $V_1$ in, $V_2$ out, $I_2$) <br><br> $R = R_1 = R_2$ <br> $C = C_1 = C_2$ <br> $T_1 = \dfrac{R_1 R_2 C_1}{R_1 + R_2}$ | $\dfrac{\omega_0^2}{(s+2.618\omega_0)(s+0.382\omega_0)}$ <br><br> (Bode plot: $A_{vn}$ vs $\omega$, breaks at $0.382\omega_0$ slope $-1$, and $2.618\omega_0$ slope $-2$) | $\dfrac{-1}{(R_1+R_2)(1+sT)}$ <br><br> (Bode plot: $|A_t|$ vs $\omega$, level at $\dfrac{1}{R_1+R_2}$, break at $\dfrac{1}{T_1}$ slope $-1$) |
| (Network: $C_1$, $C_2$ series with $R_1$, $R_2$ to ground; $V_1$ in, $V_2$ out, $I_2$) <br><br> $\omega_0 = \dfrac{1}{RC}$ <br> $R = R_1 = R_2$ <br> $C = C_1 = C_2$ <br> $T_1 = R_1(C_1+C_2)$ | $\dfrac{s^2}{(s+0.117\omega_0)(s+2.82\omega_0)}$ <br><br> (Bode plot: $A_{vn}$ vs $\omega$, slope $+2$ then break at $0.177\omega_0$ slope $+1$, then $2.82\omega_0$ level at $1$) | $\dfrac{-s^2 R_1 C_1 C_2}{1+sT_1}$ <br><br> (Bode plot: $|A_t|$ vs $\omega$, slope $+2$ then break at $\dfrac{1}{T_1}$ slope $+1$) |

# Index

Ac-dc converter, **16**-5
Active inductorless filters:
  bandpass, **12**-1 to **12**-12
  bandstop, **13**-1 to **13**-8
  high-pass, **11**-1 to **11**-10
  low-pass, **10**-1 to **10**-12
Active *RC* filters:
  bandpass, **12**-1 to **12**-12
  bandstop, **13**-1 to **13**-8
  high-pass, **11**-1 to **11**-10
  low-pass, **10**-1 to **10**-12
Active resonator, **12**-1, **12**-7
A/D (analog-to-digital) converter, **6**-1 to **6**-8
ADC (analog-to-digital converter), **6**-1 to **6**-8
Adder (*see* Amplifiers, summing)
Adding/substracting amplifier (*see* Amplifiers, summing)
Adjustable lead/lag circuits, **26**-3 to **26**-5
Admittance:
  input, **2**-23
  output, **2**-23
  transfer, **2**-23
All-pass circuit, **26**-1, **26**-3
AM demodulator, synchronous, **7**-1 to **7**-4
AM detector, precision, **16**-5
AM modulator, **18**-1 to **18**-7
Amplifiers, **4**-1 to **4**-21
  ac-coupled, **4**-17 to **4**-18
  active bandpass, **12**-1 to **12**-12
  antilog, **17**-8 to **17**-12
  capacitive transducer (*see* charge-sensitive *below*)
  capacitor-coupled (*see* ac-coupled *above*)
  charge-sensitive, **4**-18 to **4**-20
  current, **1**-4, **4**-12 to **4**-14
  data, **9**-1
  dc-isolated (*see* ac-coupled *above*)
  differential, **1**-2, **9**-1 to **9**-10
    inside op amp, **1**-26 to **1**-27

Amplifiers (*Cont.*):
  differentiating, **15**-1
  error, **9**-1
  gated, **18**-7
  instrumentation, **9**-8 to **9**-10
  integrating, **15**-9
  inverting, **1**-2, **1**-12 to **1**-13, **2**-1 to **2**-2, **4**-1 to **4**-10
  level-shifting, **1**-27
  limited, **16**-1
  linear, **1**-5
  log, **17**-1
  logarithmic, **17**-1
  noninverting, **1**-2, **1**-14, **2**-2, **2**-3, **4**-10 to **4**-12
  nonlinear, **1**-5
  output-power, inside op amp, **1**-27
  photodiode (*see* transresistance *below*)
  summing, **1**-3, **4**-20 to **4**-21
  switching-mode, **18**-9
  transadmittance (*see* transconductance *below*)
  transconductance, **1**-3, **4**-16 to **4**-17, **24**-17, **24**-19
  transducer, **9**-1
  transimpedance (*see* transresistance *below*)
  transresistance, **4**-14 to **4**-15
  two-state, **18**-9
  voltage-controlled, **19**-1
  voltage-to-current, **24**-17, **24**-19
Amplitude leveler, **16**-1
Amplitude limiter, **16**-1 to **16**-4
Amplitude modulator, **18**-1 to **18**-7
Analog commutator, **25**-1, **25**-4
Analog compressor, **17**-1
Analog divider, **19**-6
Analog gate, **18**-7, **25**-1, **25**-4
Analog integrator, **15**-9
Analog multiplexer, **25**-1, **25**-4
Analog multiplier, **19**-6
Analog switch, **25**-1, **25**-4
Analog timer, **26**-5 to **26**-8

1

# INDEX

Analog-to-digital converter, 6-1 to 6-8
Antilog converter, 17-8
Antilogarithmic amplifier, 17-8 to 17-12
Astable multivibrator, 20-1 to 20-4

Bandpass filter:
  multiple-feedback, 12-1 to 12-7
  state-variable, 12-7 to 12-12
Bandstop filter:
  active inductor, 13-1 to 13-4
  twin-tee, 13-4 to 13-8
Bandwidth, 1-10, 2-16, 3-31, I-1
Bandwidth/power booster, op amp, 23-1 to 23-6
Basic rules of op amp circuit design, 2-1
Bessel filter:
  high-pass, 11-1 to 11-11
  low-pass, 10-1 to 10-12
Binary counter, 20-4
Bipolar flip-flop, 20-4
Biquadratic filter, 12-7
Bistable multivibrator, 20-4
Bode approximation, 1-22 to 1-23
Buck regulator, 24-15
Buffer, op amp, 23-1, 23-7
Bufferred zener source, 24-1
Butterworth filter:
  high-pass, 11-1 to 11-11
  low-pass, 10-1 to 10-12
Bypassing/grounding problems, 3-26, III-4 to III-5

Capacitance:
  input, 3-7, 3-24 to 3-25, I-1
  stray, effect on feedback stability, 3-26
Capacitive transducer amplifier, 4-18
Capacitor-coupled amplifier, 4-17
Capacitor multiplier, 22-1 to 22-3
Charge-sensitive amplifier, 4-18 to 4-20
Charge-to-voltage converter, 4-18
Chebyshev filter:
  high-pass, 11-1 to 11-11
  low-pass, 10-1 to 10-12
Circuit fabrication techniques, III-1 to III-7
Closed-loop gain, general method to compute, 2-21 to 2-25
CMR (common-mode rejection), 9-2 to 9-3
CMRR (*see* Common-mode rejection ratio)
Common-mode gain, 2-19 to 2-20, 9-1 to 9-10
  circuit, 9-2 to 9-10
  op amp, 9-2 to 9-10
  and circuit, 9-1 to 9-4
Common-mode rejection, 9-2 to 9-3

Common-mode rejection ratio; op amp, 2-18 to 2-21, 9-2 to 9-10, I-1
  and circuit, 9-3 to 9-7
Commutator, analog, 25-1, 25-4
Comparators, 1-5, 5-1
  frequency, 14-6
  latching, 5-5, 5-11
  regenerative, 5-5, 5-11
  zero-crossing, 5-1, 5-5
Compensation:
  data sheet recommendation, 3-19
  effect on slew rate, 2-17 to 2-18
  lag, 3-11 to 3-13
  lead, 3-13 to 3-15, 3-30 to 3-31
  lead-lag, 3-15 to 3-18, 3-32 to 3-35
Complementary transistor op amp booster, 23-1, 23-7
Converters:
  ac-dc, 16-5
  analog-to-digital, 6-1 to 6-8
  antilog, 17-8
  capacitance-to-inductance, 22-3
  charge-to-voltage, 4-18
  current-to-current, 4-12
  current-to-voltage, 4-14
  digital-to-analog, 6-8 to 6-10
  exponential, 17-8
  logarithmic, 17-1
  phase-to-dc, 8-4
  voltage-to-current, 24-17, 24-19
  voltage-to-frequency, 21-5
  voltage-to-pulse width, 18-9
Counter, binary, 20-4
Current:
  input bias, 1-11, 2-5, I-1
  input offset, 1-11, 2-8, I-1
Current regulator:
  floating-load, 24-17 to 24-19
  grounded-load, 24-19 to 24-21
Current sink, 24-17
Current source:
  controlled, 24-17
  (*See also* Amplifiers, transconductance)
Current-to-current converter (*see* Amplifiers, current)
Current-to-voltage converter (*see* Amplifiers, transresistance)

D/A (digital-to-analog) converter, 6-8 to 6-10
DAC (digital-to-analog converter), 6-8 to 6-10
Damping factor, 10-2 to 10-7, 11-3 to 11-6
Data amplifier, 9-1
Data compressor, 17-1
Data expander, 17-8
Dc-isolated amplifier (*see* Amplifiers, ac-coupled)

# INDEX 3

Decibels, calculations using, 1-21, V-1 to V-3
Demodulators:
  AM, 7-1, 16-5
  FM, 7-4 to 7-8
  phase-sensitive, 7-1
  suppressed-carrier, 7-1
  synchronous AM, 7-1 to 7-4
Design equations:
  adjustable lead/lag circuits, 26-5
  amplifier: ac-coupled, 4-18
    charge-sensitive, 4-20
    current, 4-13
    inverting, 4-3 to 4-4
    noninverting, 4-11 to 4-12
    summing, 4-21
    transconductance, 4-17
    transresistance, 4-15
  amplitude limiter, 16-4
  amplitude modulator, 18-4
  analog timer, 26-7
  antilog amplifier, 17-10 to 17-11
  astable multivibrator, 20-3 to 20-4
  bandpass filter: multiple-feedback, 12-3
    state-variable, 12-9
  bandstop filter: active inductor, 13-3
    twin-tee, 13-6 to 13-7
  bistable multivibrator, 20-6
  capacitance multipler, 22-3
  converter: analog-to-digital, 6-5
    digital-to-analog, 6-10
  current-limited voltage regulator, 24-4 to 24-5
  demodulator: FM, 7-7
    synchronous AM, 7-4
  detector: level, 5-11
    with hysteresis, 5-15
    peak, 8-3
    phase, 8-6 to 8-7
    zero-crossing, 5-5
    with hysteresis, 5-8
  differential amplifier, 9-4
  differentiator, 15-4 to 15-5
  discriminator, pulse-width, 7-10 to 7-11
  dual voltage regulator, 24-14
  FET-controlled multiplier, 19-3 to 19-4
  FET multiplexer, 25-4
  floating-load current regulator, 24-18 to 24-19
  frequency-difference detector, 14-8 to 14-9
  frequency doubler, 14-3 to 14-4
  grounded-load current regulator, 24-21
  half-wave rectifier, 16-7 to 16-8
  high-pass filter: second-order, 11-3 to 11-4
    third-order, 11-8
  high-voltage regulator, 24-10 to 24-11
  inductance simulator, 22-6

Design equations (*Cont.*):
  integrator, 15-11 to 15-12
  log-antilog multiplier/divider, 19-9 to 19-10
  logarithmic amplifier, 17-4 to 17-6
  low-pass filter: second-order, 10-4 to 10-5
    third-order, 10-10
  monostable multivibrator, 20-8 to 20-9
  op amp bandwidth/power booster, 23-5
  op amp output-voltage booster, 23-10
  phase lead/lag circuit, 26-3
  precision gate multiplexer, 25-6
  precision voltage reference, 24-13
  pulse-amplitude modulator 18-8 to 18-9
  pulse-width modulator, 18-12 to 18-13
  sample-and-hold circuit, 25-8
  shunt voltage regulator, 24-12
  sine-wave oscillator, 21-3
  switching voltage regulator, 24-16 to 24-17
  voltage-controlled oscillator, 21-8
  voltage ramp generator, 27-3 to 27-4
  voltage staircase generator, 27-11
  voltage triangle generator, 27-6 to 27-7
Design examples:
  amplifier, inverting, 4-5 to 4-10
    gain of 10, 3-27 to 3-31
    wide-bandwidth, 3-31 to 3-35
  amplitude modulator, 18-5 to 18-7
  analog timer, 26-8
  antilog amplifier, 17-11 to 17-12
  bandpass filter: multiple-feedback, 12-5 to 12-7
    state-variable, 12-10 to 12-12
  bandstop filter: active inductor, 13-4
    twin-tee, 13-7 to 13-8
  converter, analog-to-digital, 6-6 to 6-7
  current-limited voltage regulator, 24-6 to 24-9
  detector: level, with hysteresis, 5-16 to 5-18
    phase, 8-9 to 8-11
  differential amplifier, 9-6 to 9-8
  differentiator, 15-7 to 15-8
  discriminator, pulse-width, 7-14 to 7-16
  FET-controlled multiplier, 19-4 to 19-6
  frequency-difference detector, 14-10 to 14-11
  frequency doubler, 14-5
  half-wave rectifier, 16-10 to 16-12
  high-pass filter: second-order, 11-5 to 11-6
    third-order, 11-9 to 11-10
  inductance simulator, 22-7 to 22-8
  log-antilog multiplier/divider, 19-10 to 19-11
  logarithmic amplifier, 17-6 to 17-8

## 4  INDEX

Design examples (*Cont.*):
  low-pass filter:
    second-order, 10-6 to 10-7
    third-order, 10-11
  monostable multivibrator, 20-12 to 20-13
  op amp bandwidth/power booster, 23-5 to 23-6
  pulse-width modulator, 18-13 to 18-14
  sample-and-hold circuit, 25-9 to 25-10
  sine-wave oscillator, 21-4 to 21-5
  voltage-controlled oscillator, 21-9 to 21-10
  voltage triangle generator, 27-7 to 27-8
Design parameters:
  adjustable lead/lag circuits, 26-3
  amplifier: current, 4-14
    inverting, 4-2 to 4-3
    transconductance, 4-16
    transresistance, 4-15
  amplitude limiter, 16-3
  amplitude modulator, 18-3
  analog timer, 26-6 to 26-7
  antilog amplifier, 17-10
  astable multivibrator, 20-2
  bandpass filter: multiple-feedback, 12-2 to 12-3
    state-variable, 12-8 to 12-9
  bandstop filter: active inductor, 13-3
    twin-tee, 13-6
  bistable multivibrator, 20-6
  capacitance multiplier, 22-2
  converter: analog-to-digital, 6-4
    digital-to-analog, 6-9 to 6-10
  current-limited voltage regulator, 24-3 to 24-4
  demodulator, FM, 7-7
  detector: level, 5-10 to 5-11
    with hysteresis, 5-13 to 5-14
    peak, 8-3
    phase, 8-5 to 8-6
    zero-crossing, 5-4
    with hysteresis, 5-7 to 5-8
  differential amplifier, 9-3
  differentiator, 15-3
  discriminator, pulse-width, 7-10
  FET-controlled multiplier, 19-3
  FET multiplexer, 25-3
  floating-load current regulator, 24-18
  frequency-difference detector, 14-8
  frequency doubler, 14-3
  grounded-load current regulator, 24-20
  half-wave rectifier, 16-7
  high-pass filter: second-order, 11-3
    third-order, 11-8
  inductance simulator, 22-6
  integrator, 15-11
  log-antilog multiplier/divider, 19-8 to 19-9
  logarithmic amplifier, 17-4

Design parameters (*Cont.*):
  low-pass filter: second-order, 10-4
    third-order, 10-10
  monostable multivibrator, 20-7 to 20-8
  op amp bandwidth/power booster, 23-3 to 23-4
  op amp output-voltage booster, 23-9 to 23-10
  phase lead/lag circuit, 26-2 to 26-3
  precision gate multiplexer, 25-5 to 25-6
  pulse-amplitude modulator, 18-7 to 18-8
  pulse-width modulator, 18-12
  sample-and-hold circuit, 25-8
  sine-wave oscillator, 21-2 to 21-3
  voltage-controlled oscillator, 21-7
  voltage ramp generator, 27-2 to 27-3
  voltage staircase generator, 27-10
  voltage triangle generator, 27-5 to 27-6
Design procedure and steps:
  amplifier, inverting, 4-4 to 4-5
  amplitude modulator, 18-4 to 18-5
  analog timer, 26-7 to 26-8
  antilog amplifier, 17-11
  bandpass filter: multiple-feedback, 12-4 to 12-5
    state-variable, 12-10
  bandstop filter: active inductor, 13-3 to 13-4
    twin-tee, 13-7
  converter, analog-to-digital, 6-5 to 6-6
  current-limited voltage regulator, 24-5 to 24-6
  detector: level, with hysteresis, 5-15 to 5-16
    phase, 8-7 to 8-9
  differential amplifier, 9-4 to 9-6
  differentiator, 15-6 to 15-7
  discriminator, pulse-width, 7-11 to 7-14
  FET-controlled multiplier, 19-4
  frequency-difference detector, 14-9 to 14-10
  frequency doubler, 14-4 to 14-5
  half-wave rectifier, 16-8 to 16-10
  high-pass filter: second-order, 11-4 to 11-5
    third-order, 11-8 to 11-9
  inductance simulator, 22-6 to 22-7
  log-antilog multiplier/divider, 19-10
  logarithmic amplifier, 17-6
  low-pass filter: second-order, 10-5 to 10-6
    third-order, 10-10 to 10-11
  monostable multivibrator, 20-9 to 20-12
  op amp bandwidth/power booster, 23-5
  pulse-width modulator, 18-13
  sample-and-hold circuit, 25-8 to 25-9

# INDEX 5

Design procedure and steps (*Cont.*):
  sine-wave oscillator, 21-3 to 21-4
  voltage-controlled oscillator, 21-8 to 21-9
  voltage triangle generator, 27-7
Detectors:
  FM, 7-4
  frequency-difference, 14-6 to 14-11
  level: inverting, 5-9 to 5-11
    with hysteresis, 5-11 to 5-18
    noninverting, 5-10 to 5-11
    with hysteresis, 5-13 to 5-15
  peak, 8-1 to 8-3
  phase, 8-4 to 8-11
  phase-error, 8-4
  phase-sensitive, 7-1
  phase-shift, 8-4
  positive-peak, 8-1 to 8-3
  pulse-width, 7-8
  zero-crossing: inverting, 5-1 to 5-5
    with hysteresis, 5-5 to 5-8
    noninverting, 5-3 to 5-5
    with hysteresis, 5-6 to 5-8
Design steps (*see* Design procedure and steps)
Difference amplifier, 9-1
Differential amplifier:
  inside op amp, 1-25 to 1-27
  linear, 9-1 to 9-10
  logarithmic, 17-1
Differential gain, 9-1 to 9-10
Differentiating amplifier, 15-1
Differentiator, 1-3, 15-1 to 15-8
Digital-to-analog converter, 6-8 to 6-10
Discriminators:
  AM, 7-1, 16-5
  FM, 7-4, 14-6
  pulse-width, 7-8 to 7-16
Dividers:
  analog, 19-6
  (*See also* Multiplier/divider)
Doubler, frequency, 14-1
Dual power supply, 24-13 to 24-14
Dual-slope A/D converter, 6-1

Error amplifier, 9-1
Errors, operational amplifier, 2-3 to 2-21
Exponential amplifier, 17-8
Exponential converter, 17-8

Feedback:
  design examples, 3-27 to 3-35
  positive and negative, 3-2, 3-3
Feedback equation, 1-17 to 1-19, 3-5
Feedback instability, seven causes of, 3-1, 3-18 to 3-27
Feedback limiter, 16-1
Feedback network, transfer admittance, 2-22 to 2-25

Feedback network calculations, 3-6 to 3-9, 3-24 to 3-25
Feedback stability:
  first-cut analysis, 3-3
  graphical analysis, 3-4, 3-20
Feedback-stability-design examples, 3-27 to 3-35
Feedback-stability measurements (*see* Measurements, feedback-stability)
Feedback theory, 3-1 to 3-18
FET-controlled multiplier, 19-1 to 19-6
FET multiplexer, 25-1 to 25-4
Field-effect transistor (*see* FET-controlled multiplier; FET multiplexer)
Filters:
  bandpass: multiple-feedback, 12-1 to 12-7
    state-variable, 12-7 to 12-12
  bandstop: active inductor, 13-1 to 13-4
    twin-tee, 13-4 to 13-8
  high-pass: second-order, 11-1 to 11-6
    third-order, 11-7 to 11-10
  low-pass: second-order, 10-1 to 10-7
    third-order, 10-8 to 10-11
First-derivative circuit, 15-1
Flip-flop, 20-4
Floating-load current regulator, 24-17 to 24-19
FM demodulator, 7-4 to 7-8
FM detector, 7-4
Follower, voltage (*see* Amplifiers, noninverting)
Frequency comparator, 14-6
Frequency-difference detector, 14-6 to 14-11
Frequency discriminator, 7-4, 14-6
Frequency doubler, 14-1 to 14-5
Frequency meter, 7-4
Frequency multiplier, 14-1

Gain:
  closed-loop, how to compute, 2-21
  common-mode, 2-19 to 2-20, 9-1 to 9-10
  effect of input resistance on, 1-16
  effect of open-loop gain on, 1-14 to 1-15
  effect of output resistance on, 1-17
  loop (*see* Loop gain and phase)
  minimum stable closed-loop, 3-21 to 3-22
Gain margin, 3-10 to 3-38
Gated amplifier, 18-7
Generators:
  pulse, 20-7, 21-5
  ramp, 27-1 to 27-4
  rectangular, 20-1
  sawtooth, 27-1

Generators (*Cont.*):
  sine-wave, 21-1
  square-wave, 20-1
  staircase, 27-8
  step, 27-8
  triangle, 27-4
  triggered pulse, 20-7
  voltage-controlled, 21-5
Grounding/bypassing problems, 3-26, III-4 to III-5

Half-wave rectifier, 16-5 to 16-12
Harmonic generator, 14-1
Heterodyne circuit, 14-6
High-pass filter:
  differentiator, 15-1
  second-order, 11-1 to 11-6
  third-order, 11-7 to 11-10
High-voltage op amp booster, 23-7
High-voltage regulator, 24-9 to 24-11
Hum-reduction circuit, 13-1, 13-4
Hysteresis:
  in level detector, 5-12
  in zero-crossing detector, 5-5

Ideal diode (half-wave rectifier), 16-5
Impedance transformer (*see* Amplifiers, current)
Inductance simulator, 22-3 to 22-8
Inductorless bandpass filter, 12-1, 12-7
Inductorless bandstop filter, 13-1, 13-4
Inductorless high-pass filter, 11-1, 11-7
Inductorless low-pass filter, 10-1, 10-8
Instability, feedback, seven causes of, 3-1, 3-18 to 3-27
Instrumentation amplifier:
  high-quality, 9-8 to 9-10
  low-cost, 9-1
Integral amplifier, 15-9
Integrating amplifier, 15-9
Integrator, 1-3, 15-9 to 15-12
Interval timer, 26-5
Inverse log amplifier, 17-8
Inverter (*see* Amplifiers, inverting)

Lag circuit, 26-1, 26-3
Large-signal behavior of op amps, 1-19 to 1-20
Latch, bilevel, 20-4
Lead circuit, 26-1, 26-3
Lead/lag circuit, 26-1, 26-3
Level detector (*see* Detectors, level)
Limited amplifier, 16-1
Limiter, amplitude, 16-1 to 16-4
Linear amplitude modulator, 18-1
Linear modulator, 18-1, 18-7
Linear multiplier, 19-1
Load, capacitive, effect on stability, 3-22 to 3-24

Lock-in amplifier, 7-1
Logarithmic amplifier, 1-6, 17-1 to 17-8
Logarithms, 1-21, 17-1
Log ratio circuit, 17-1
Log subtracting circuit, 17-1
Loop gain and phase, 1-24 to 1-26, 3-3 to 3-11
  with lag compensation, 3-11 to 3-13
  with lead compensation, 3-13 to 3-15
  with lead/lag compensation, 3-15 to 3-18
Low-pass filter:
  integrator, 15-9
  second-order, 10-1
  third-order, 10-8

Margin:
  gain, 3-10 to 3-38
  phase, 1-25, 3-10 to 3-38
Maximum ratings of op amps, II-1
Measurements, feedback-stability:
  closed-loop ac method, 3-37 to 3-38
  loop-gain method, 3-35 to 3-36
  transient-response method, 3-38
Memory, one-bit, 20-4
Modulators:
  AM, 18-1 to 18-7
  FM, 21-5
  linear amplitude, 18-1
  pulse-amplitude, 18-7 to 18-9
  pulse-duration, 18-9
  pulse-width, 18-9 to 18-14
Monostable multivibrator, 20-7 to 20-13
Multiple-feedback bandpass filter, 12-1 to 12-7
Multiplexer:
  analog, 18-7, 25-1
  FET (CMOS), 25-1 to 25-4
  precision gate, 25-4 to 25-6
  single-channel, 18-7
Multiplier/divider:
  log-antilog, 19-6 to 19-11
  single-quadrant, 19-6
Multipliers:
  analog, 19-1, 19-6
  capacitor, 22-1 to 22-3
  FET-controlled, 19-1 to 19-6
  linear, 19-1
Multivibrators, 1-6
  astable, 20-1 to 20-4
  bistable, 20-4 to 20-6
  monostable, 20-7 to 20-13
MUX (*see* Multiplexer)

Noise:
  input current, 2-9, I-2
  input voltage, 2-9, I-2
Noise prevention techniques, III-3 to III-6

# INDEX 7

Notation, **IV**-1 to **IV**-3
Notch filter, **13**-1, **13**-4

One-shot, **20**-7
Operational amplifiers:
 applications, **1**-5
 bandwidth, **1**-10, **1**-12, **1**-15, **2**-16, **I**-1
  closed-loop, **2**-16
  full-power, **2**-16 to **2**-17, **I**-1
  unity-gain, **2**-16, **I**-1
 bandwidth/power booster, **23**-1 to **23**-6
 circuits inside, **1**-25 to **1**-28
 closed-loop characteristics, **1**-23 to **1**-24
 common-mode gain, **2**-19 to **2**-20, **9**-1 to **9**-10
 common-mode rejection ratio, **2**-18 to **2**-21, **9**-2 to **9**-10, **I**-1
 differential gain, **2**-19, **I**-1
 equivalent input noise, **2**-9 to **2**-12, **I**-2
 errors, minimizing, **2**-3 to **2**-21
 full-power output, **2**-16
 input bias current, **1**-11, **2**-5 to **2**-7, **I**-1
 input capacitance, **2**-12 to **2**-13, **I**-1
 input offset current, **1**-11, **2**-8 to **2**-9, **I**-1
 input offset voltage, **1**-11, **2**-3 to **2**-6, **I**-1
 input resistance, **1**-11, **1**-12, **1**-16, **2**-12, **I**-2
 large-signal behavior, **1**-19 to **1**-20, **5**-3
 maximum ratings, **II**-1
 narrow-band spot-noise, **2**-11, **I**-2
 noise-figure, **2**-11
 open-loop characteristics, **1**-21
 output resistance, **1**-12, **1**-17, **2**-13, **2**-14, **I**-2
 ouptut voltage booster, **23**-7 to **23**-10
 packaging, **1**-8 to **1**-10
 parameters, **1**-4, **I**-1 to **I**-2
 phase-shift, open-loop, **1**-20 to **1**-22
 power-supply rejection ratio, **2**-21, **I**-2
 protection circuits, **III**-1 to **III**-3
 real versus ideal, **1**-14
 rise time, **2**-16, **I**-2
 slew rate, **1**-11, **2**-17 to **2**-18, **I**-2
 symbol, illustrated, **1**-1, **1**-4
 voltage gain, **1**-9, **1**-12, **1**-14, **1**-21, **2**-14 to **2**-16, **I**-1
Oscillators, **1**-7
 sine-wave, **21**-1
 staircase, **27**-8
 sweep, free-running, **27**-1
 voltage-controlled, **21**-5 to **21**-10

Packaging, op amp, **1**-8
 discrete parts, **1**-10
 dual in-line, illustrated, **1**-10
 flat package, illustrated, **1**-10
 TO-99 package, illustrated, **1**-8

PAM (*see* Pulse-amplitude modulator)
Parallel regulator, **24**-11
Parameters, op amp, **1**-9 to **1**-11, **I**-1 to **I**-2
Parasitic suppressor, **13**-1, **13**-4
Passive devices, **III**-6
Peak detector, **8**-1 to **8**-3
Peak-signal tracker, **8**-1
Phase detector, **8**-4 to **8**-11
Phase-difference detector, **8**-4
Phase-error detector, **8**-4
Phase inverter (*see* Amplifiers, inverting)
Phase lag circuit, **26**-1
Phase lead circuit, **26**-1
Phase lead/lag circuit, **26**-1 to **26**-3
Phase margin, **1**-25, **3**-10 to **3**-38
Phase-sensitive demodulator, **7**-1
Phase-shift detector, **8**-4
Phase-shift oscillator, **21**-1
Phase-to-dc converter, **8**-4
Photodiode amplifier (*see* Amplifiers, transresistance)
Polarity selector, **16**-5
Pole, gain and phase of, **1**-22
Pole locations in $s$ plane:
 high-pass filter, second-order, **11**-2 to **11**-4
 low-pass filter: second-order, **10**-2 to **10**-4
 third-order, **10**-9
Positive peak detector, **8**-1
Power booster, op amp, **23**-1, **23**-7
Power supply:
 bypass caps for, **3**-26 to **3**-27
 voltage-regulated, **24**-1
Power-supply rejection ratio, **2**-21, **I**-2
Precision gate multiplexer, **25**-4
Precision half-wave rectifier, **16**-5 to **16**-12
Precision limiter, **16**-1
Precision voltage supply, **24**-1, **24**-13
PSRR (*see* Power-supply rejection ratio)
Pulse-amplitude modulator, **18**-7 to **18**-9
Pulse catcher, **7**-8
Pulse-counting FM demodulator, **7**-4
Pulse-duration modulator, **18**-9
Pulse generator:
 triggered, **20**-7
 voltage-controlled, **21**-5
Pulse-height modulator, **18**-7
Pulse-width demodulator, **7**-8
Pulse-width detector, **7**-8
Pulse-width discriminator, **7**-8 to **7**-16
Pulse-width modulator, **18**-9 to **18**-14
Push-pull buffer, op amp, **23**-1, **23**-7
PWM (*see* Pulse-width modulator)

$Q$ of filter:
 high-pass, **11**-3
 low-pass, **10**-2 to **10**-4

Ramp generator, voltage, 27-1 to 27-4
*RC* circuit characteristics, VI-1 to VI-24
*RC* timer, 26-5
Rectangular-waveform generator, 20-1
Rectifier, precision half-wave, 16-5
Reference, precision voltage, 24-1, 24-13
Regulators, 1-7
   current: floating-load, 24-17 to 24-19
      grounded-load, 24-19 to 24-21
   dual voltage, 24-13
   high-voltage, 24-9 to 24-11
   shunt, voltage, 24-11 to 24-12
   switching, voltage, 24-15
   tracking, dual, 24-13
   voltage, current-limited, 24-1
Resistance, op amp input, 1-11, 2-12, I-2
Resistor-capacitor timer, 26-5
Resonator, active, 12-1
Rise time, 1-11, I-2
Rules, basic for op amp circuit design, 2-1

Sample-and-hold circuit, 25-6 to 25-10
Sampling circuit, 1-8, 18-7, 25-1, 25-4
Sampling gate, 18-7, 25-1, 25-4
Sawtooth generator, 27-1
Schmitt trigger (*see* Detectors, level: zero-crossing)
Second-order filter:
   high-pass, 11-1 to 11-6
   low-pass, 10-1 to 10-7
S/H (sample-and-hold) circuit, 25-6 to 25-10
Shielding/guarding, III-5 to III-6
Short-circuit-proof voltage regulator, 24-1
Shunt regulator, 24-11
Simulator:
   capacitor, 22-1
   inductor, 22-3 to 22-8
Sine-wave oscillator, 21-1
Single-shot, 20-7
Slew rate, 1-11, 2-17, I-2
Slew-rate booster, 23-1
Source, current, 4-16 to 4-17, 24-17 to 24-21
Square-wave generator, 20-1
Stability analysis, first-cut, 3-3
Staircase generator, 27-8
State-variable bandpass filter, 12-7 to 12-12
Step generator, 27-8
Suppressed-carrier AM demodulator, 7-1
Sweep generator, 27-1
Switching-mode amplifier, 18-9
Switching voltage regulator, 24-15 to 24-17
Synchronous detector, 7-1
Synchronous switching demodulator, 7-1

Third-order filter:
   high-pass, 11-7 to 11-10
   low-pass, 10-8 to 10-11
Time-averaging FM demodulator, 7-4
Timer, 26-5 to 26-8
Tracking regulator, 24-13
Transadmittance amplifier (*see* Amplifiers, transconductance)
Transfer admittance, 2-23
Transimpedance amplifier (*see* Transresistance amplifier)
Transresistance amplifier, 4-14 to 4-15
Triangle generator, 27-4
Trigger, Schmitt (*see* Detectors, level: zero-crossing)
Triggered pulse generator, 20-7
Twin-tee bandstop filter, 13-4
Two-state modulator, 18-9

Unity-gain filter:
   high-pass: second-order, 11-1 to 11-6
      third-order, 11-7 to 11-10
   low-pass: second-order, 10-1 to 10-7
      third-order, 10-8 to 10-11

VCO (voltage-controlled oscillator), 21-5 to 21-10
V/F (voltage-to-frequency) converter, 21-5
VFC (voltage-to-frequency converter), 21-5
Voltage, input offset, 1-11, 2-3, I-2
Voltage-controlled amplifier, 19-1
Voltage-controlled oscillator, 21-5 to 21-10
Voltage follower, 4-10
Voltage ramp generator, 27-1 to 27-4
Voltage reference, precision, 24-13
Voltage regulators:
   current-limited, 24-1 to 24-9
   dual, 24-13 to 24-14
   high-voltage, 24-9 to 24-11
   shunt, 24-11 to 24-12
   switching, 24-15 to 24-17
Voltage source, dual, 24-13
Voltage staircase generator, 27-8 to 27-11
Voltage step generator, 27-8
Voltage-sweep generator, 27-1 to 27-4
Voltage-to-current converter, 4-16, 24-17, 24-19
Voltage-to-frequency converter, 21-5
Voltage-to-pulse width converter, 18-9
Voltage triangle generator, 27-4 to 27-8
Volume compressor, 16-1

Waveform generator:
   rectangular, 20-1
   sawtooth, 27-1, 27-4

Waveform generator (*Cont.*):
   square-wave, 20-1
   staircase, 27-8
Wien-bridge sine-wave oscillator, 21-1 to 21-5

Y parameters:
   in closed-loop gain calculation, 2-24

Y parameters (*Cont.*):
   how to compute, 2-23
   relation to Z parameters, 2-21 to 2-22

Zero, gain and phase of, 1-22 to 1-23
Zero-bound circuit, 16-5
Zero-crossing comparator, 5-1, 5-5
Zero-crossing detector (*see* Detectors, zero-crossing)